Fish Karyotypes

A Check List

Ryoichi Arai
Doctor of Science
Honorary permanent curator
Department of Zoology, University Museum, University of Tokyo
7-3-1 Hongo, Bunkyo-ku, Tokyo 113-0033, Japan
araryo@um.u-tokyo.ac.jp

Library of Congress Control Number: 2010936627

ISBN: 978-4-431-53876-9 Springer Tokyo Berlin Heidelberg New York

Printed in Japan

Printed on acid-free paper

Springer is a part of Springer Science+Business Media
springer.com

Contents

Introduction

In many vertebrate groups, the study of karyotypes and genome size has contributed along with analyses of mitochondrial and nuclear gene sequences to the resolution of challenges in biology systematics and evolution. However, in fishes—the most diverse of all vertebrate groups—higher taxa traditionally have been classified largely by morphology and paleontology, with a much smaller input of cytogenetic information. In part, this is because karyotypes can be obtained only from living specimens, tissues, or cells, which makes it challenging to study the karyotypes of fishes that are difficult to collect alive (e.g., deep-sea fishes). Of course, even fresh material provides no guarantee that reliable chromosome figures can be obtained easily.

DNA sequence data are exerting an increasingly strong influence in modern fish systematics, for example, by leading to proposed challenges in the classification of numerous higher taxa ranging from genera to orders. However, the most fruitful approach is certain to be one that involves synthetic analyses of morphology, molecular phylogenetics, comparative karyology, and genome size, rather than focusing on only one or a few of these sources of data. For example, although it may be very difficult to establish homology of karyotypes analyzed by Giemsa staining and several banding methods among taxa, the polarity of karyotype states nonetheless can still be inferred by analysis of Robertsonian fusion/fission, tandem fusion, pericentric inversion, paracentric inversion, aneuploidy, or polyploidy in any monophyletic taxon, even when the polarity of DNA sequences is unclear. The role of cytogenetic data is likely to increase further because powerful new methods such as fluorescence in situ hybridization (FISH) (Phillips 2007: P-53) are implemented in fishes beyond model species (e.g., zebrafish, medaka, sticklebacks, and pufferfish). Given their limited scope, FISH data are not included here, but their future importance is clearly acknowledged.

The purpose of this book is to facilitate the implementation of an integrative approach to fish systematics by providing karyotype information for 3,425 species/subspecies of extant jawless, cartilaginous, actinopterygian, and lobe-finned fishes. This presentation fills an important need, as fish karyotypes tend to be published not only in well-known and easily accessible journals, but also in museum journals of more regional significance or in other venues that are difficult to obtain. Several books on fish chromosomes have been published in the past (Denton 1973: D-7; Chiarelli and Capanna 1973: C-1; Ojima 1983: O-73; Vasiliev 1985: V-72; Klinkhardt et al. 1995: K-114), but this volume represents the first in nearly 15 years and is the most comprehensive. Such an update is clearly warranted, given the historical growth in the numbers of karyotyped species/subspecies listed in Table 1.

Table 1 Historical transition of the numbers of karyotyped species/subspecies

Author(s)	Denton	Park	Ojima et al.	Sola et al.	Ojima	Vasiliev	Klinkhardt et al.	Arai
Year	1973	1974	1976	1981	1983	1985	1995	This
Reference no.	D-7	P-6	O-22	S-88	O-73	V-72	K-114	book
Myxini	4	2	3	0	3	5	0	8
Petromyzontida	7	6	8	0	7	13	14	14
Chondrichthyes								
Holocephali	1	2	0	0	2	2	2	2
Elasmobranchii	15	8	4	0	13	17	52	68
Actinopterygii								
Cladistia	2	2	2	4	7	6	7	7
Chondrostei	1	4	0	5	7	12	12	21
Neopterygii								
Holostei	2	2	2	2	3	1	3	3
Teleostei	446	505	417	798	1,035*	1,258	2,182*	3,296
Sarcopterygii	3	1	0	1	2	5	5	6
Total	481	532	436	810	1,079*	1,318	2,277*	3,425

*Synonymous species were counted as different species.

Substantial revisions to the species names and higher taxonomy of many fishes in the intervening years also make an updated compendium necessary. In this book, synonymies are circumvented by providing both the name reported in the original karyotype papers and currently recognized names. In particular, the classification of taxa above the species level follows Nelson (2006: N-68), and species names conform to the up-to-date list in the *Catalogue of Fishes* by Eschmeyer (http://research.calacademy.org/redirect?url=http://researcharchive.calacademy.org/research/ichthyology/catalog/fishcatmain.asp, E-13). The classification of chromosomes followed Levan et al. (1964: L-25): chromosomes were classified to metacentrics (M), submetacentrics (SM), subtelocentrics (ST), and acrocentrics (A). As shown in Table 1 (Levan et al. 1964), M, m, sm, st, t, and T do not denote chromosomes but rather centromeric position; e.g., acrocentrics is characterized by t. Levan et al. (1964) do not propose abbreviations for chromosomes. Two kinds of abbreviations for chromosomes, M-SM-ST-A and m-sm-st-t(-T), have been adopted in karyotypes. I adopt M-SM-ST-A because chromosome classification can be differentiated from centromeric position by these abbreviations and thus the problematic treatment of T in the other system can be avoided. In addition to providing data for extant fishes, ancestral chromosome numbers (ACN = NAN sensu Arai and Nagaiwa 1976: A-64) are proposed in this book.

To date, karyotype data have been made available for 53 orders (84% of the known total) and 269 families (52% of total). In addition, genome size has been reported in 52 orders (83%) and 264 families (51%), which are available in standardized form as part of the Animal Genome Size Database (http://www.genomesize.com, G-85). As a result, it is possible to compare these two characters, at least at the family level (Tables 2, 3).

In some cases, a given karyotype consists of chromosomes that cannot easily be distinguished as either SM or ST, such that different authors may report different information. For example, the arm number (NF_1) of *Oncorhynchus keta* ($2n = 74$) is reported as 100 in Sasaki et al. (1968: S-14) versus 102 in Ueda (1985: U-72), and the NF_2 of *Alburnus alburnus* ($2n = 50$) is given as 92 in Cataudella et al. (1977: C-34) versus 86 in Hafez et al. (1978: H-3). These different reports on karyotypes between conspecific populations were included in the database (Tables 4–7) without comment, although it should be noted that these differences may derive from artifacts of preparation technique or taxonomic problems, rather than representing real polymorphisms. If so, then it will be important to resolve these discrepancies before meaningful comparisons can be made. Similarly, there may be issues relating to differences in reported karyotypes resulting from different degrees of chromosome condensation, to the lack of a uniform terminology among authors, or even to some miscalculations (number of arms, NOR position, etc.). Some examples of these sources of error were detected when examining the literature; e.g., the same researcher(s) have at times defined M/SM as two-arm chromosomes in some taxa, but M/SM/ST as two-arm chromosomes in other taxa. To prevent such confusion, these two-arm definitions were differentiated in the present book; i.e., NF_1 means M/SM as two-arm chromosomes and NF_2 means M/SM/ST as two-arm chromosomes.

There are several fundamental questions that remain to be answered with regard to phylogenetic karyology in fishes. For example: (1) Why is polyploidy in teleost fishes only observed in freshwater species and not in marine species? (2) What is the phylogenetic significance of differences in the number and location of NORs as shown by different banding methods (especially by the FISH method with 18S rDNA and 5S rDNA probes)? (3) In cases in which both large and small B chromosomes have been reported, are both, neither, or only the large B chromosomes to be counted in the diploid chromosome number? Or should this be considered on a case-by-case basis? (4) What is the biological significance of microchromosomes, and what is their relationship (if any) with B chromosomes? (5) Can sex chromosomes be differentiated in more species by using FISH methods with probes of genes specific to sex chromosomes?

Many individuals helped me in various ways with the preparation of this book. For papers on karyotypes, I would like to thank Ana Lucia Dias, Eliana Feldberg, Pedro Manoel Galetti, Jr., John R. Gold, Anisur Rahman Khuda-Bukhsh, Naresh Sahebrao Nagpure, Claudio de Oliveira, Gento Shinohara, Akihiko Shinomiya, Akinori Takai, Toru Taniuchi, Takayoshi Ueda, Han-lin Wu, and numerous researchers who sent kindly me copies of their paper(s). I also wish to express my sincere gratitude to Petr Ráb, who plowed through the whole manuscript and gave me valuable comments, and to T. Ryan Gregory for improving the English text.

Availability of Fish Karyotypes

Karyotypes have been reported for 3,425 species/subspecies of fishes (including jawless, cartilaginous, ray-finned, and lobe-finned fishes) (see Table 1). This total represents a coverage of about 12.2% of extant fish diversity (62 orders, 515 families, and 27,977 species); however, these are not sampled evenly from among fish groups and instead are biased toward freshwater taxa such as the Cypriniformes, Characiformes, Siluriformes, Cyprinodontiformes, and Cichlidae in the Perciformes. Specifically, numbers of karyotyped species/subspecies are 747 (21.8%) in Cypriniformes, 341 (10.0%) in Characiformes, 362 (10.6%) in Siluriformes, 345 (10.1%) in Cyprinodontiformes, and 130 (3.8%) in Cichlidae (Perciformes). On the other hand, 4 orders (Orectolobiformes, Echinorhiniformes, Pristiophoriformes, and Pristiformes) and 30 families in cartilaginous fishes and 5 orders (Albuliformes, Saccopharyngiformes, Ateleopodiformes, Lampriformes, and Polymixiiformes) and 216 families in actinopterygian fishes currently lack any karyotype data. Of course, relative coverage depends on the number of recognized taxa, which has been updated significantly in recent times consequent to molecular analyses that may or may not reflect undue inflation (Nelson 2006: N-68; Kottelat and Freyhof 2007: K-130).

Fish Vouchering and Identification

To verify the identification of fishes included in karyotype studies, voucher specimens should be deposited in a museum, curated university collection, or another appropriate institute. Unfortunately, vouchered specimens are not available for most species for which karyotypes have been reported, meaning that their initial identification cannot be confirmed. Moreover, there are many papers in which the localities of material fishes purchased from fish dealers were not described, which further increases the prospect of taxonomic errors. This problem is especially relevant when the classification of a taxon in question has been revised, making it difficult to link old and new names. For example, two different karyotypes were reported in a labrid, *Pseudolabrus japonicus* (Arai and Koike 1980: A-75; Ojima and Kashiwagi 1979: O-27). Thereafter, *Pseudolabrus japonicus* was separated to two different species, *P. eosthinus* and *P. sieboldi* by revisional study of *P. japonicus* (Mabuchi and Nakabo 1997: M-1). By examination of *P. japonicus* material deposited at a museum, it was known that the two karyotypes corresponded to these two different species (Mabuchi et al. 2002: M-2).

Classification of Extant Fishes

Fish systematics has developed greatly during the past 10 years, mainly based on molecular phylogenetic studies. Although results in molecular phylogenetics are not always agreed upon by all researchers, it is necessary that a consistent taxonomic system be used in a database such as this. In this regard, the higher taxonomic classification proposed by Nelson (2006: N-68) has been used with the following updates:

Orders Batrachoidiformes and Lophiiformes were changed from the superorder Paracanthopterygii to the superorder Acanthopterygii (Miya et al. 2003, 2005: M-138, 139).

Although polyphyly of the order Gasterosteiformes and the suborder Labroidei (order Perciformes) has been reported (Kawahara et al. 2008: K-105; Mabuchi et al. 2007: M-17) and

Stylephoriformes (Miya et al. 2007: M-140), close relationship between Alepocephaliformes and Otocephala (Lavoué et al. 2008: L-24), and Gobiiformes (Thacker 2009: T-63) were proposed, these hypotheses were treated as pending problems in this book.

The higher-level classification used here is as follows:

Genus- and species-level designations given here are those presented by Eschmeyer (2009: E-13). Again, there have been many suggested changes to fish taxonomy at this level as well. For example, in European cyprinid fishes, many species previously placed in *Leuciscus* are now in the genera *Squalius*, *Telestes*, and *Petroleuciscus*, and many species previously placed in *Chondrostoma* are now in the genera *Protochondrostoma*, *Pseudochondrostoma*, *Parachondrostoma*, *Achondrostoma*, and *Iberochondrostoma* (Kottelat and Freyhof 2007: K-130). Changes such as these are noted in the karyotype database (Tables 4–7).

Historical Transition of Numbers of Karyotyped Species/Subspecies

The number of karyotyped species/subspecies has increased rapidly since the early 1970s. For example, in 1973 karyotypes were available for 481 species/subspecies; in 1985, 1,318 species/subspecies had been karyotyped; and at the time of this writing, data exist for 3,425 species/subspecies (see Table 1). Since the last compendium in 1995 (K-114), the number of karyotyped taxa in Elasmobranchii, Chondrostei, and Teleostei increased but that in Petromyzontida, Holocephali, Cladistia, and Holostei did not increase. In some cases, an increase in the number of karyotyped taxa in Elasmobranchii and Teleostei has been caused by the erection of numerous additional taxa and through revised techniques for chromosome preparation (Klinkhardt 1991: K-79).

Relationship Between Karyotype and Genome Size

Relationships between karyotypes and genome size in families were also explored (see Tables 2, 3). Sources of karyotypes and genome size were usually different. Data of karyotypes and genome size were extracted from Tables 4–7 and the up-to-date list of Gregory (www.genomesize.com, G-85).

As with karyotype data, there are significant gaps in the fish genome size dataset.

Two orders (Echinorhiniformes and Pristiophoriformes) and 21 families in cartilaginous fishes and eight orders (Alubuliformes, Saccopharyngiformes, Gonorynciformes, Ateleopodiformes, Lampriformes, Polymixiiformes, Percopsiformes, and Stephanoberyciformes) and 230 families in actinopterygian fishes have no information on genome size. Two orders (Echinorhiniformes and Pristiophoriformes) and 19 families in cartilaginous fishes and five orders (Albuliformes, Saccopharyngiformes, Ateleopodiformes, Lampriformes, Polymixiiformes) and 181 families in actinopterygian fishes have neither information on karyotypes nor information on genome size.

As for jawless fishes, karyotypes and genome size in the Myxiniformes differ from those in the Petromyzontiformes, i.e., $2n \leq 36$ and genome size >5.0 pg/cell in Myxiniformes versus $2n = 76$ or >140 and genome size <4.3 pg/cell in Petromyzontiformes. Myxiniformes exhibits different chromosome numbers and DNA amounts between somatic cells and spermatogonia, i.e., $2n =$ 14–36 in somatic cells versus $2n =$ 16–96 in spermatogonia. B chromosomes have been reported in spermatogonia in Myxiniformes (Nakai et al. 1995: N-62; Kojima et al. 2010: K-140).

As for jawed fishes, diploid chromosome numbers and genome size in cartilaginous fishes except Chimaeriformes are larger than those in actinopterygian fishes. The Polypteriformes and the Coelacanthiformes possess about 7.0–9.0 pg/cell. The Ceratodontiformes has extraordinary large genome sizes (more than 80.0 pg/cell).

Acipenseriformes is of special interest among Actinopterygii in their possession of 105 to 372 chromosomes and 2.4 to 13.8 pg/cell. Ploidy levels including diploidy (2X), tetraploidy (4X), and hexaploidy (6X) have been inferred in this order. However, the high number of chromosomes suggests another possible series, i.e., tetraploidy, octaploidy, and dodecaploidy (Fontana et al. 2007: F-61).

Diploid chromosome numbers and genome size in Neopterygii except the Lepisosteiformes, Osteoglossiformes, Characiformes, Siluriformes, Osmeriformes, and Salmoniformes are usually 48 or 50 and lower than 3.0 pg/cell, respectively. However, more than 3.0 pg/cell in Neopterygii has been reported in Anguilliformes, Cypriniformes, Characiformes, Siluriformes, Argentiniformes, Esociformes, Myctophiformes, Gasterosteiformes, Batrachoidiformes, Synbranchiformes, and Perciformes, i.e., Muraenidae, Cyprinidae (Cyprininae, part of Gobioninae, part of Leuciscinae, part of Rasborinae, Schizothoracinae, and polyploid species in Barbinae), polyploid species of Cobitidae, Catostomidae, part of Characidae, Curimatidae, Prochilodontidae, Ariidae, part of Callichthyidae, Doradidae, part of Loricariidae, Plotosidae, Microstomatidae, part of Umbridae, Myctophidae, part of Syngnathidae, Batrachoididae, part of Synbranchidae, Scaridae, Artedidraconidae, and Channichthyidae.

Among taxa described above, freshwater fishes such as part of Cypriniformes (Catostomidae, part of Cyprinidae, part of Cobitidae), and part of Siluriformes (part of Callichthyidae) may be of polyploid origin (Allendorf and Thorgaard 1984: A-16; Ferris 1984: F-23).

Complex relationships between ploidy and genome size have been reported in the Cobitoidea. In one example, the Leptobotiinae-Balitoridae exhibit $2n = 48–50$ and genome size about 1.0 pg/cell, whereas the Botiinae possess $2n = 98–100$ and genome size about 2.0 pg/cell. A second example is provided by comparing *Cobitis* species with $2n = 48–50$ and about 4.0 pg/cell versus those with $2n = 96–100$ and about 7.0 pg/cell. A third example comes from non-*Cobitis* cobitids with $2n = 48–50$ with about 2.0 pg/cell versus those with $2n = 96–100$ and about 4.5 pg/cell (Suzuki 1996: S-143).

The diploid chromosome number in marine neopterygian fishes, which have various genome sizes (0.8–4.4 pg/cell), was generally 46, 48, or 50 with the exception of $2n = 56$ in Osmeriformes and did not show polyploidy, although triploidy was exceptionally reported in the Zoarcidae (Perciformes) (Morescalchi et al. 1996: M-89). In other words, it is difficult to infer polyploidy in marine fishes from genome size.

In Teleostei, genome size is relatively conserved in families, with the important exception of those including recent or ancient polyploids. Diploid chromosome numbers have been thought to vary across taxa above the family level, but when they were analyzed in terms of Robertsonian translocation and tandem fusion, they were found to be more conservative in each family. On the other hand, there are several problems in karyological analysis. The finding of B chromosomes might cause change of $2n$, NF, and ancestral chromosome number (ACN) in taxa that have been reported to possess no B chromosomes. It is likely that many more species, when analyzed with sufficient intensity, will be found to possess B chromosomes (Camacho et al. 2000: C-2). B chromosomes have been studied actively in Characiformes and neotropical Siluriformes (Carvalho et al. 2008: C-98), but there have been very few data on B chromosomes in the other taxa.

Table 2 **Numbers of karyotyped genera and species/subspecies (sp/ssp), diploid chromosome number (2n) and genome size in families of jawless fishes**

FCM = flow cytometry, FD = Feulgen densitometry, BFA = bulk fluorometric assay, SCF = static cell fluorometry.

Order/family/subfamily	No. of genera		No. of sp/ssp		2n	Genome size (pg/cell)			
	total	studied	total	studied		FCM	FD	BFA	SCF
PHYLUM CHORDATA									
Subphylum Craniata									
Superclass Myxinomorphi									
Class MYXINI									
MYXINIFORMES				8					
Myxinidae				8					
Myxininae	4	2	25	4	14, 28, 34, 36				6.9–9.2
Eptatretinae	3	1	45	4	34, 36	5.4	5.5	5.6	4.6–6.9
Superclass Petromyzontomorphi									
Class PETROMYZONTIDA									
PETROMYZONTIFORMES				14					
Petromyzontidae	8	5	34	11	142–168	2.6, 4.2	2.6–2.9, 4.2	3.2	
Geotriidae	1	1	1	1	ca. 180		3.1		
Mordaciidae	1	1	3	2	76		2.8		

Table 3 Numbers of karyotyped genera and species/subspecies (sp/ssp), diploid chromosome number (2n) and genome size in families of jawed fishes

FCM = flow cytometry, FD = Feulgen densitometry, FIA = Feulgen image analysis densitometry, BFA = bulk fluorometric assay.

Table 3.1 Class CHONDRICHTHYES

Order/family/subfamily/genus	No. of genera		No. of sp/ssp		2n	Genome size (pg/cell)			
	total	studied	total	studied		FCM	FD	FIA	BFA
Superclass Gnathostomata									
Grade Chondrichthiomorphi									
Class CHONDRICHTHYES									
Subclass Holocephali									
CHIMAERIFORMES				2					
Callorhinchidae	1	0	3	0				3.9	
Rhinochimaeridae	3	0	8	0				3.2	
Chimaeridae	2	2	22	2	58, 86		3.0	4.0	3.2
Subclass Elasmobranchii									
Subdivision Selachii									
HETERODONTIFORMES				2					
Heterodontidae	1	1	8	2	102	17.5	14.5, 29.6		13.6
ORECTOLOBIFORMES				0					
Brachaeluridae	2	0	2	0					
Ginglymostomatidae	3	0	3	0		7.6, 11.4	10.9	8.5	8.0
Hemiscylliidae	2	0	12	0				9.1–11.0	
Orectolobidae	3	0	6	0				10.1	
Parascylliidae	2	0	7	0					
Rhincodontidae	1	0	1	0					
Stegostomatidae	1	0	1	0					
LAMNIFORMES				2					
Alopiidae	1	0	3	0					
Cetorhinidae	1	0	1	0					
Lamnidae	1	1	5	1	82	12.9	13.4	10.0	
Megachasmidae	1	0	1	0					
Mitsukurinidae	1	0	1	0					
Odontaspididae	2	1	3	1	ca. 84	10.9			
Pseudocarchariidae	1	0	1	0					
CARCHARHINIFORMES				17					
Carcharhinidae	12	4	50	7	74–90	6.7–8.3	5.5–13.1	5.7–9.9	6.8–8.6
Hemigaleidae	4	0	7	0					
Leptochariidae	1	0	1	0					
Proscylliidae	3	0	5	0					
Pseudotriakidae	2	0	2	0					
Scyliorhinidae	16	2	113	5	62, 64, 72	18.1	11.3–14.7		15.4
Sphyrnidae	2	1	8	1	78–86	6.6	8.9	6.1	7.0, 7.8
Triakidae				4					
Galeorhininae	6	0	10	0			17.3		
Triakinae	3	2	28	4	68, 72, 80		8.6–9.8		9.0–12.8
HEXANCHIFORMES				3					
Chlamydoselachidae	1	1	1	1	ca. 100		9.2		
Heptranchiidae	1	1	1	1	72				
Hexanchidae	1	0	2	0			10.7		
Notorynchidae	1	1	1	1	104	8.8			
ECHINORHINIFORMES				0					
Echinorhinidae	1	0	2	0					

Table 3.1 Class CHONDRICHTHYES (continued)

Order/family/subfamily/genus	No. of genera		No. of sp/ssp		2n	Genome size (pg/cell)			
	total	studied	total	studied		FCM	FD	FIA	BFA
SQUALIFORMES				4					
Centrophoridae	2	0	14	0			13.1–14.2	14.5	
Dalatiidae	7	0	10	0			18.2		
Etmopteridae	5	1	41	2	86	16.2	32.3	23.8–25.4	
Oxynotidae	1	1	5	1	62		34.1	25.0	
Somniosidae	7	0	17	0			16.4–26.0	19.5–23.2	
Squalidae	2	1	10	1	58–60, 78	14.0	13.7–14.4	11.6	12.0
SQUATINIFORMES				1					
Squatinidae	1	1	15	1	88			32.8	18.6–19.6
PRISTIOPHORIFORMES				0					
Pristiophoridae	2	0	5	0					
Subdivision Batoidea									
TORPEDINIFORMES				6					
Narcinidae	9	2	37	2		14.7, 24.1	21.0–24.0		8.4
Narcine					28				8.4
Narke					54		21.0–24.0		
Torpedinidae	2	1	22	4	66, 82, 86	14.1	14.0–15.0		14.6
PRISTIFORMES				0					
Pristidae	2	0	7	0					5.6
RAJIFORMES				11					
Rajidae	26	3	238	7	58, 96–98	5.8–7.2	5.4–8.1	5.5–6.9	5.6–7.2
Rhinidae	1	0	1	0					
Rhinobatidae	4	1	42	4	59–64, 84, 92	4.9–8.0	5.9–8.3	6.0	8.0
Rhynchobatidae	1	0	4	0					
MYLIOBATIFORMES				22					
Suborder Platyrhinoidei				1					
Platyrhinidae	2	1	3	1	64	9.0, 15.5			15.4
Suborder Zanobatoidei				0					
Zanobatidae	1	0	2	0			11.7		
Suborder Myliobatoidei				21					
Dasyatidae	6	2	68	8	58, 64–78	7.8–10.1	8.5–13.7	6.8–13.4	9.4
Gymnuridae	2	1	11	2	56	10.0, 11.4			16.2
Hexatrygonidae	1	0	1	0		10.0			
Myliobatidae				6					
Mobulinae	2	1	10	1	66	9.6–10.2	9.4		
Myliobatinae	4	1	20	4	52–54	10.4–11.9	8.7–10.8		9.8
Rhinopterinae	1	1	7	1	64	10.0–10.2			10.4
Plesiobatidae	1	0	1	0					
Potamotrygonidae	3	2	20	3	66, 90				
Urolophidae	2	1	24	2	52, 72	15.5	13.1		13.0
Urotrygonidae	2	0	16	0					

Table 3.2 Class ACTINOPTERYGII. Part 1 Cladistia and Chondrostei

Order/family	No. of genera		No. of sp/ssp		2n	Genome size (pg/cell)			
	total	studied	total	studied		FCM	FD	FIA	BFA
Grade Teleostomi (OSTEICHTHYES)									
Class ACTINOPTERYGII									
Subclass Cladistia									
POLYPTERIFORMES				7					
Polypteridae	2	2	16	7	36, 38		9.1–14.5	7.4	9.4–9.8
Subclass Chondrostei									
ACIPENSERIFORMES				21					
Acipenseridae	4	4	25	20	105–372				
(2X)					105–120	2.4–4.7	3.2–3.6	4.4	
(4X)					240–260	8.8–9.5	6.1–9.1		
(6X)					372	13.1		13.8	
Polyodontidae	2	1	2	1	120	3.2–4.9			

Table 3.3 Class ACTINOPTERYGII. Part 2 Neopterygii

Order/family/subfamily/genus	No. of genera		No. of sp/ssp		2n	Genome size (pg/cell)			
	total	studied	total	studied		FCM	FD	FIA	BFA
Subclass Neopterygii									
Division Holostei									
LEPISOSTEIFORMES				2					
Lepisosteidae	2	1	7	2	56, ca. 68	2.8	2.9	2.8	2.4
AMIIFORMES				1					
Amiidae	1	1	1	1	46		2.3–2.5	2.3	
Division Teleostei									
Subdivision Osteoglossomorpha									
HIODONTIFORMES				2					
Hiodontidae	1	1	2	2	50		1.2		
OSTEOGLOSSIFORMES				14					
Osteoglossidae	4	4	7	7	40–56			1.6–1.8	2.0
Pantodontidae	1	1	1	1	48				1.5
Mormyridae	18	2	201	2	48				2.0–2.4
Notopteridae	4	4	8	4	34, 42	2.2			2.6
Gymnarchidae	1	0	1	0					
Subdivision Elopomorpha									
ELOPIFORMES				3					
Elopidae	1	1	6	1	48				2.4
Megalopidae	1	1	2	2	46, 50–52			2.0	
ALBULIFORMES				0					
Suborder Albuloidei				0					
Albulidae	1	0	3	0					
Suborder Notacanthoidei				0					
Halosauridae	3	0	15	0					
Notacanthidae	3	0	10	0					
ANGUILLIFORMES				33					
Suborder Anguilloidei				6					
Anguillidae	1	1	15	5	38	2.2		2.0	2.8–3.3
Heterenchelyidae	2	0	4	0					
Moringuidae	2	1	6	1	50				

Table 3.3 Class ACTINOPTERYGII. Part 2 Neopterygii (continued)

Order/family/subfamily/genus	No. of genera		No. of sp/ssp		2n	Genome size (pg/cell)			
	total	studied	total	studied		FCM	FD	FIA	BFA
Suborder Muraenoidei				12					
Chlopsidae	8	0	18	0				3.9	
Muraenidae	15	3	185	12	36, 42	4.6		3.8	4.4–5.1
Myrocongridae	1	0	4	0					
Suborder Congroidei				15					
Colocongridae	1	0	5	0					
Congridae	22	2	160	5	34, 38	2.4–3.6			
Derichthyidae	2	0	3	0					
Muraenesocidae	4	1	8	1	38			2.3	
Nemichthyidae	3	0	9	0					
Nettastomatidae	6	0	38	0					
Ophichthidae	52	7	290	8	38–48				
Serrivomeridae	2	0	10	0					
Synaphobranchidae	10	1	32	1	26			3.2	
SACCOPHARYNGIFORMES				0					
Suborder Cyematoidei				0					
Cyematidae	2	0	2	0					
Suborder Saccopharyngoidei				0					
Eurypharyngidae	1	0	1	0					
Monognathidae	1	0	15	0					
Saccopharyngidae	1	0	10	0					
Subdivision Otocephala (= Ostarioclupeomorpha)									
CLUPEIFORMES				30					
Suborder Denticipitoidei				0					
Denticipitidae	1	0	1	0					
Suborder Clupeoidei				30					
Pristigasteridae	9	0	34	0					
Chirocentridae	1	0	2	0				1.6	
Clupeidae				23					
Alosinae	7	3	31	9	46, 48		2.0		2.2–2.8
Clupeinae	16	6	72	9	28, 44–48, 50–54		2.0–2.7	1.8–2.1	1.5
Dorosomatinae	6	4	22	5	48	2.0		1.8	
Pellonulinae	23	0	44	0				2.2	
Engraulidae	16	4	139	7	42, 44, 48	2.9	3.0		3.8
GONORYNCHIFORMES				2					
Suborder Chanoidei				1					
Chanidae	1	1	1	1	32				
Suborder Gonorynchoidei				0					
Gonorynchidae	1	0	5	0					
Suborder Knerioidei				1					
Kneriidae	4	0	30	0					
Phractolaemidae	1	1	1	1	28				
CYPRINIFORMES				747					
Superfamily Cyprinoidea				630					
Cyprinidae	220	180	2420	628					
Acheilognathinae	3	3	50	34	42–48	2.1–2.3	1.8–2.1		
Barbinae	?	30	?	154	48–150				
(2X)		16		75	48–50	1.5–2.2	1.4–2.5		1.9–2.4
(4X)		9		57	96–100	2.7–3.7	3.4–3.5, 4.6		
(6X)		5		22	150				
Cultrinae	?	15	?	30	48	2.6	1.8–2.4		
Cyprininae	6	4	25	22	100–162				
(4X)		4		22	100	3.4–4.5	3.1–3.8, 4.8	3.4	
(6X)		1		3	150–162	5.4	4.6–4.8, 6.1		

Table 3.3 Class ACTINOPTERYGII. Part 2 Neopterygii (continued)

Order/family/subfamily/genus	No. of genera		No. of sp/ssp		2n	Genome size (pg/cell)			
	total	studied	total	studied		FCM	FD	FIA	BFA
Gobiobotinae	?	2	?	7	50				
Gobioninae	?	21	?	65	50	2.4–3.7	2.8–3.3		
Hypophthalmichthyinae	2	2	3	2	48		1.9–2.1		
Labeoninae	?	12	?	50	44, 48, 50		2.8	2.2–2.5	2.2–2.6
Leuciscinae	?	68	?	200	48, 50	2.0–3.7	1.8–3.2	2.5–2.8	
Rasborinae (= Danioninae)	?	18	?	46	48–78				
(2X)		16		41	48–52	2.2, 3.4–4.6	1.8–2.9, 4.4	2.8	3.2, 3.6–4.4
(2X)		3		5	70–78	2.3	3.3		
Schizothoracinae	?	12	?	29	90–148, >400				
(4X)		11		22	90–92, 98–100		3.0–3.1, 4.9		
(6X)		1		6	148		6.5–7.0		
(18X)		1		1	417–470		17.2		
Squaliobarbinae	?	3	?	3	48	2.0	2.0–2.2		
Tincinae	1	1	1	1	48	2.3	1.7–2.1		
Xenocyprinae	4	4	?	7	48		2.1–2.8		
Psilorhynchidae	2	1	6	2	50				
Superfamily Cobitoidea				117					
Gyrinocheilidae	1	1	3	1	48	1.2	1.0	1.3	1.3
Catostomidae	13	9	72	15	96–100	3.7–5.5	4.0	4.2–5.1	
Cobitidae	26	19	177	76					
Botiinae	5	5	?	20	98–100	1.6–2.1	1.8–1.9		1.7
Leptobotiinae	2	2	?	11	50–52		1.1		
Cobitinae	19	12	130	45	48–100				
Cobitis (2X)				16	48, 50	2.8–4.0	4.3		
Cobitis (3X)				1	74–75	6.0			
Cobitis (4X)				3	96–98, 100	6.8–7.6	7.2		
non-*Cobitis* (2X)				27	48, 50		2.0–2.2		2.4–2.8
non-*Cobitis* (4X)				3	100		4.5–4.6		
Balitoridae	59	13	590	24					
Nemacheilinae	29	8	420	19	44–50, 75		1.0–1.1, 2.2		
Balitorinae	29	5	170	5	48, 50		0.9		
Vaillantellidae	1	1	1	1	50				
CHARACIFORMES				341					
Suborder Citharinoidei				0					
Citharinidae	3	0	8	0					
Distichodontidae	17	0	90	0					
Suborder Characoidei				341					
Acestrorhynchidae	1	1	15	3	50		1.7, 3.1		
Alestiidae (Alestidae)	18	1	110	1	56				2.4
Anostomidae	12	7	137	39	54		2.6–3.5		2.8–3.4
Characidae				194					
Aphyocharacinae	2	2	10	4	50		2.5–2.7		3.4
Bryconinae	3	2	43	12	50		2.4		
Chalceinae	1	1	2	1	52, 54		2.2		2.2
Characinae	12	6	70	13	46, 50, 52		2.2–3.1		3.4
Cheirodontinae	15	9	46	16	32, 42, 50, 52		3.7		
Glandulocaudinae	19	2	50	5	52		2.5, 3.1		
Iguanodectinae	2	1	2	1	50		2.4		
Paragoniatinae	7	1	8	1	52				
Serrasalminae	15	11	80	35	54, 58, 60–64		3.2–3.6	3.0	3.2–3.4
Stethaprioninae	4	3	12	4			3.5		
Tetragonopterinae	?	20	?	91	36, 38, 46–54	2.5–4.0	2.1–4.2		3.2–4.2
Triporthinae	?	1	?	11	50, 52		2.7, 3.5		
Chilodontidae	2	2	7	2	54				
Crenuchidae	12	1	74	9	50				
Ctenoluciidae	2	2	7	2	36				
Curimatidae	8	7	95	37	46, 54, 56, 102		2.8–3.8		
Cynodontidae	4	0	14	0			2.0–2.1		

Table 3.3 Class ACTINOPTERYGII. Part 2 Neopterygii (continued)

Order/family/subfamily/genus	No. of genera		No. of sp/ssp		2n	Genome size (pg/cell)			
	total	studied	total	studied		FCM	FD	FIA	BFA
Erythrinidae	3	3	14	11	39–42, 48–54		2.3		2.8
Gasteropelecidae	3	3	9	5	48, 52, 54		2.2–2.6		2.8
Hemiodontidae	6	3	28	7	54				
Hepsetidae	1	0	1	0					
Lebiasinidae	7	4	61	11	22–46		2.0		2.4
Parodontidae	3	2	21	8	54		2.0–2.5		
Prochilodontidae	3	2	21	12	54		3.1–3.7		
SILURIFORMES				362					
Akysidae	4	0	42	0					
Amblycipitidae	3	2	26	9	20–42				
Amphiliidae	2	0	26	0					
Ariidae	21	9	150	15	52–56	4.5		4.9	4.8–5.0
Aspredinidae	12	1	36	1	50				
Astroblepidae	1	0	54	0					
Auchenipteridae	20	4	94	5	56, 58				
Auchenoglanididae	6	0	28	0					
Austroglanidae	1	0	3	0					
Bagridae	18	10	170	33	44, 48–60	1.8	1.9–2.1		1.8–2.2
Callichthyidae				48					
Callichthyinae	5	4	13	7	56–64				
Dianema, Hoplosternum				5	60–62		1.2–1.4		
Callichthys				1	56–58		1.9		3.4
Megalechis				1	64	3.2			
Corydoradinae	4	4	164	41	40–134				
Corydoras (2X)				12	56–62, 74		1.3–3.6	3.2	
Corydoras (2X)				8	40–50, 56		4.5–4.9		4.6–6.0
Corydoras (4X)				5	92, 120–134	8.4	6.3–8.8	6.4	8.4–8.8
Cetopsidae	7	0	23	0					
Chacidae	1	0	3	0					
Clariidae	14	2	90	8	50–56, 100		1.8		2.4
Claroteidae	7	0	59	0					
Cranoglanididae	1	1	3	1	74				
Diplomystidae	2	2	6	3	56		2.6		
Doradidae	30	11	72	13	56, 58, 66		3.5		3.2
Erethistidae	6	1	14	1	50				
Heptapteridae	25	6	175	20	46, 52–58, 87		1.8–2.2		
Heteropneustidae	1	1	3	1	56				
Ictaluridae	7	5	46	33	40–62	2.0	1.9–2.1		2.4
Loricariidae				74					
Ancistrinae	27	5	217	10	38–40, 48–52				3.6
Hypoptopomatinae	16	8	79	12	54, 58, 72		1.8, 2.7		4.2
Hypostominae	?	6	?	23	52–54, 64–80		3.2–4.5	3.2–3.4	4.2
Lithogeneinae	1	0	2	0					
Loricariinae	31	7	209	18	36–74			3.0	3.2
Neoplecostominae	?	5	?	10	54		2.3		
Upsilodinae	?	1	?	1	96				
Malapteruridae	2	0	19	0					2.0
Mochokidae	11	2	179	9	54, 56			1.9, 2.8	2.2–2.4
Nematogenyidae	1	0	1	0					
Pangasiidae	3	2	28	5	58, 60				
Pimelodidae	31	14	85	30	50, 54, 56		1.1, 2.0–2.8		2.4
Plotosidae	10	1	35	2	36–38, 48			3.5	
Pseudopimelodidae	5	5	26	6	54		2.2–2.5		
Schilbeidae	15	4	56	4	58, 66				2.0
Scoloplacidae	1	1	4	1	50				
Siluridae	11	4	97	13	42, 54–60, 86	2.3	1.7–2.9		1.8

Table 3.3 Class ACTINOPTERYGII. Part 2 Neopterygii (continued)

Order/family/subfamily/genus	No. of genera		No. of sp/ssp		2n	Genome size (pg/cell)			
	total	studied	total	studied		FCM	FD	FIA	BFA
Sisoridae				9					
Sisorinae	6	2	23	2	42, 46–48				
Glyptosterninae	11	4	89	7	36, 42, 50–56				
Trichomycteridae	41	6	201	18	32, 54, 81		2.3–2.6		
GYMNOTIFORMES				17					
Suborder Gymnotoidei				6					
Gymnotidae	2	2	33	6	40–48, 52–54				2.0
Suborder Sternopygoidei				11					
Apteronotidae	13	2	45	3	22, 24, 52				1.4
Hypopomidae	7	3	16	4	36–38, 42, 50				
Rhamphichthyidae	3	1	12	1	52			1.9	
Sternopygidae	5	2	28	3	28–40, 46–48				2.0
Subdivision Euteleostei									
ARGENTINIFORMES				5					
Suborder Argentinoidei				5					
Argentinidae	2	1	23	1	44		1.7		
Microstomatidae	11	4	38	4	36, 54, 60, 62		3.4, 6.3		
Opisthoproctidae	6	0	11	0					
Suborder Alepocephaloidei				0					
Alepocephalidae	23	0	90	0					
Bathylaconidae	2	0	3	0					
Platytroctidae	13	0	37	0					
OSMERIFORMES				19					
Galaxiidae	8	2	52	10	22, 30–44		1.8–2.1		
Osmeridae				9					
Hypomesinae	1	1	6	3	56				
Osmerinae	4	3	8	3	54, 56	1.2	1.5–1.7	1.4	
Plecoglossinae	1	1	1	1	56				
Salanginae	5	2	16	2	56				
Retropinnidae	3	0	5	0					
SALMONIFORMES				85					
Salmonidae				85					
Coregoninae	3	3	32	27	60–64, 72–86	6.0–7.1	5.1–6.9	4.0–4.9	
Thymallinae	1	1	5	3	98–110	4.3		4.0	
Salmoninae	7	7	73	55	52–92				
Brachymystax				1	90–92				
Hucho				2	82–84				
Parahucho				1	62				
Oncorhynchus				21	52, 58–70, 74	4.9–6.4	4.1–6.6	4.7–5.0	6.0–6.6
Salmo				12	54–58, 78–84	5.2–6.4	5.5–5.8	6.5	
Salvelinus				18	78–84	5.7	7.4–7.5	5.8–6.7	7.0
ESOCIFORMES				11					
Esocidae	1	1	6	6	50	2.3	2.2–2.7	1.8–2.3	
Umbridae	3	3	7	5	22, 44, 48, 78				
Dallia				1	78		2.5		
Novumbra				1	48		2.1		
Umbra				3	44, 22		4.8–5.0	5.1	5.4
STOMIIFORMES				7					
Diplophidae	3	0	8	0					
Suborder Gonostomatoidei				7					
Gonostomatidae	5	3	23	3	12, 48				
Sternoptychidae	10	2	67	4	35, 48–52				

Table 3.3 Class ACTINOPTERYGII. Part 2 Neopterygii (continued)

Order/family/subfamily/genus	No. of genera		No. of sp/ssp		2n	Genome size (pg/cell)			
	total	studied	total	studied		FCM	FD	FIA	BFA
Suborder Phosichthyoidei				0					
Phosichthyidae	7	0	20	0					
Stomiidae	28	0	273	0				2.6	
ATELEOPODIFORMES				0					
Ateleopodidae	4	0	12	0					
AULOPIFORMES				9					
Suborder Synodontoidei				7					
Aulopidae	2	0	10	0					
Paraulopidae	1	0	10	0					
Pseudotrichonotidae	1	0	1	0					
Synodontidae	4	3	57	7	26–27, 48	2.2–2.9	1.6–2.5	2.7, 3.1	2.4
Suborder Chlorophthalmoidei				2					
Bathysauroididae	1	0	1	0					
Bathysauropsidae	1	0	3	0					
Chlorophthalmidae	2	1	19	2	24, 36				
Ipnopidae	5	0	29	0					
Notosudidae	3	0	19	0					
Suborder Alepisauroidei				0					
Alepisauridae	2	0	3	0					
Evermannellidae	3	0	7	0					
Paralepididae	13	0	56	0					
Scopelarchidae	4	0	17	0					
Suborder Giganturoidei				0					
Bathysauridae	1	0	2	0					
Giganturidae	1	0	2	0					
MYCTOPHIFORMES				28					
Neoscopelidae	3	1	6	1	48		2.6	5.0	
Myctophidae	32	16	240	27	44, 48		3.8–4.0		
LAMPRIFORMES				0					
Lampridae (= Lamprididae)	1	0	2	0					
Lophotidae	2	0	3	0					
Radiicephalidae	1	0	1	0					
Regalecidae	2	0	2	0					
Stylephoridae	1	0	1	0					
Trachipteridae	3	0	10	0					
Veliferidae	2	0	2	0					
POLYMIXIIFORMES				0					
Polymixiidae	1	0	10	0					
Superorder Paracanthopterygii									
PERCOPSIFORMES				1					
Amblyopsidae	5	0	6	0					
Aphredoderidae	1	1	1	1	48				
Percopsidae	1	0	2	0					
GADIFORMES				15					
Bregmacerotidae	1	0	15	0					
Euclichthyidae	1	0	1	0					
Gadidae				12					
Gadinae	12	6	25	10	26, 38–48			1.7–1.9	1.8
Lotinae	3	1	5	1	48				
Ranicipitinae	1	1	1	1	48				
Macrouridae	27	0	350	0				1.5–1.9	
Melanonidae	1	0	2	0					

Table 3.3 Class ACTINOPTERYGII. Part 2 Neopterygii (continued)

Order/family/subfamily/genus	No. of genera		No. of sp/ssp		2n	Genome size (pg/cell)			
	total	studied	total	studied		FCM	FD	FIA	BFA
Merlucciidae	1	0	13	0				1.7–1.9	1.9–2.0
Moridae	18	0	105	0				1.5, 1.9	
Muraenolepididae	1	1	4	1	48				
Phycidae	5	2	25	2	28, 48			1.7	
OPHIDIIFORMES				3					
Suborder Ophidioidei				3					
Carapidae	7	1	31	1	36				
Ophidiidae	48	2	222	2	43, 44			1.2–1.3	1.4, 1.7
Suborder Bythitoidei				0					
Aphyonidae	6	0	22	0					
Bythitidae	16	0	62	0					
Parabrotulidae	2	0	3	0					
Superorder Acanthopterygii									
MUGILIFORMES				18					
Mugilidae	17	8	72	18	24, 28, 48	1.6	1.4	1.6–2.0	2.0
ATHERINIFORMES				21					
Suborder Atherinopsoidei				12					
Atherinopsidae	11	6	108	12	44, 46, 48			1.3	2.2
Suborder Atherinoidei				9					
Atherinidae	12	2	60	2	48			2.1	
Atherionidae	1	1	3	1	48				
Melanoteaniidae	17	3	113	6	46, 48				2.6
Notocheiridae	2	0	6	0					
Phallostethidae	5	0	22	0					
BELONIFORMES				26					
Suborder Adrianichthyoidei				13					
Adrianichthyidae				13					
Adrianichthyinae	2	0	5	0					
Horaichthyinae	1	0	1	0					
Oryziinae	1	1	22	13	28–42, 46, 48	2.1–2.2	1.5–1.9		2.2
Suborder Belonoidei				13					
Belonidae	10	4	34	7	48, 50, 54	2.3	2.2	2.0	2.2–2.4
Exocoetidae	8	1	52	1	48	1.9			
Hemiramphidae	12	4	109	4	40, 46, 48, 52	2.0	1.5	1.9–2.4	1.5–2.2
Scomberesocidae	2	1	4	1	42				
CYPRINODONTIFORMES				346					
Suborder Aplocheiloidei				205					
Aplocheilidae	2	2	7	6	38, 48		1.5	1.5	1.4
Nothobranchiidae	?	10	250	155	16–42, 46–50				2.4
Rivulidae	28	12	236	44	20, 34–48, 54				3.0
Suborder Cyprinodontoidei				141					
Anablepidae	3	1	15	1	46				
Cyprinodontidae	9	6	104	35	48, 50, 52				3.2
Fundulidae	4	2	50	24	32–34, 40–48	2.7–3.0		2.6	2.8–3.0
Goodeidae	16	16	36	36	24–30, 42, 48				
Poeciliidae				44					
Aplocheilichthyinae	1	1	?	2	48				
Procatopodinae	9	3	78	4	48				
Poeciliinae	27	8	225	38	42–48 (69, 72)	1.3–2.0	1.3–2.1	2.8	1.7–1.9
Profundulidae	1	0	5	0					
Valenciidae	1	1	2	1	48				

Table 3.3 Class ACTINOPTERYGII. Part 2 Neopterygii (continued)

Order/family/subfamily/genus	No. of genera		No. of sp/ssp		2n	Genome size (pg/cell)			
	total	studied	total	studied		FCM	FD	FIA	BFA
STEPHANOBERYCIFORMES				5					
Barbourisiidae	1	0	1	0					
Cetomimidae	9	0	20	0					
Gibberichthyidae	1	0	2	0					
Hispidoberycidae	1	0	1	0					
Megalomycteridae	4	0	5	0					
Melamphaidae	5	4	36	5	42, 46–50, 58				
Mirapinnidae	3	0	5	0					
Rondeletiidae	1	0	2	0					
Stephanoberycidae	3	0	3	0					
BERYCIFORMES				9					
Suborder Trachichthyoidei				5					
Anomalopidae	6	0	8	0					
Anoplogastridae	1	1	2	1	48				
Diretmidae	3	1	4	2	44–46, 70			2.9	
Monocentridae	2	1	4	1	48				
Trachichthyidae	7	1	39	1	48			1.3	
Suborder Berycoidei				1					
Berycidae	2	1	9	1	48	1.7		2.0	
Suborder Holocentroidei				3					
Holocentridae	8	3	78	3	48, 50	2.0	1.3	1.5–1.7	1.8
ZEIFORMES				2					
Suborder Cyttoidei				0					
Cyttidae	1	0	3	0				1.5	
Suborder Zeioidei				2					
Grammicolepididae	3	0	3	0					
Oreosomatidae	4	1	10	1	42			2.5–2.6	
Parazenidae	3	0	4	0					
Zeidae	2	1	5	1	42–44			2.5	
Zeniontidae	3	0	7	0					
GASTEROSTEIFORMES				19					
Suborder Gasterosteoidei				10					
Aulorhynchidae	2	0	2	0					
Gasterosteidae	5	4	10	10	42, 46	1.2		1.3	1.2–1.4
Hypoptychidae	1	0	1	0					
Indostomidae	1	0	3	0					
Suborder Syngnathoidei				9					
Aulostomidae	1	0	3	0			1.4		
Centriscidae	2	0	4	0				0.9–1.1	
Fistulariidae	1	0	4	0		1.5–1.8		1.4	
Macroramphosidae	3	1	11	2	48			1.1	
Pegasidae	2	0	5	0					
Solenostomidae	1	0	5	0					
Syngnathidae	52	4	232	7	36, 44, 48, 58			1.1–2.7	
Hippocampus, Syngnathus				4	44, 48	0.9–1.1		0.9–2.1	1.3
Nerophis				1	58	3.6–3.9			
BATRACHOIDIFORMES				9					
Batrachoididae	22	5	78	9	44, 46, 48				3.4–6.0
SYNBRANCHIFORMES				8					
Suborder Mastacembeloidei				4					
Chaudhuriidae	6	0	9	0					
Mastacembelidae	5	3	73	4	48		1.6		1.5

Table 3.3 Class ACTINOPTERYGII. Part 2 Neopterygii (continued)

Order/family/subfamily/genus	No. of genera		No. of sp/ssp		2n	Genome size (pg/cell)			
	total	studied	total	studied		FCM	FD	FIA	BFA
Suborder Synbranchoidei				4					
Synbranchidae	4	3	17	4					
Monopterus				2	24, 42		1.2–1.6		
Synbranchus				1	42–46		5.6–8.5		
SCORPAENIFORMES				101					
Suborder Dactylopteroidei				1					
Dactylopteridae	2	1	7	1	48				
Suborder Scorpaenoidei				36					
Scorpaenidae	56	12	418	36					
Sebastinae	7	4	133	18	46, 48	1.8–1.9		1.9–2.0	1.9–2.2
Scorpaeninae	20	5	185	16	34–48				
Scorpaena				2	34–36, 40–48	2.8	1.8	2.9	2.8
Pterois				1	48			2.0	
Apistinae	3	0	3	0					
Tetraroginae	11	2	38	2	47–48, 50	2.1			
Synanceiinae	9	1	35	1	48	1.2, 1.8			
Caracanthidae	1	0	4	0					
Aploactinidae	17	0	38	0					
Pataecidae	3	0	3	0					
Gnathanacanthidae	1	0	1	0					
Congiopodidae	4	0	9	0					
Suborder Platycephaloidei				9					
Triglidae	10	3	105	3	47–48			1.6	1.6, 2.0
Peristediidae	4	0	36	0					
Bembridae	5	0	10	0					
Platycephalidae	18	4	65	6	48			1.4–1.8	
Hoplichthyidae	1	0	10	0					
Suborder Anoplopomatoidei				1					
Anoplopomatidae	2	1	2	1	30			1.4	1.7
Suorder Hexagrammoidei				5					
Hexagrammidae	5	2	12	5	48	1.7	1.5	1.4–1.9	1.8–1.9
Suborder Normanichthyoidei				0					
Normanichthyidae	1	0	1	0					
Suborder Cottoidei				49					
Rhamphocottidae	1	0	1	0					2.2
Ereuniidae	2	0	3	0					
Cottidae	70	16	275	34	32, 37–48, 52		1.5	1.4–1.9	1.8–1.9
Comephoridae	1	1	2	2	48				
Abyssocottidae	7	6	23	10	48				
Hemitripteridae	3	1	8	1	46			1.8–2.0	1.9
Agonidae	22	1	47	1	48			1.5	
Psychrolutidae	8	0	35	0				1.8–2.0	
Bathylutichthyidae	1	0	1	0					
Cyclopteridae	6	1	28	1	50			1.7–1.9	
Liparidae	29	0	334	0				1.5–1.8	
PERCIFORMES				867					
Suborder Percoidei				314					
Acropomatidae	8	0	31	0					
Ambassidae	9	3	46	4	40, 44, 48			1.1	
Aplodactylidae	1	0	5	0					
Apogonidae	23	4	273	13	34–38, 46	2.9		1.6–2.6	
Arripidae	1	0	4	0				1.4	
Banjosidae	1	0	1	0					
Bathyclupeidae	1	0	5	0					
Bramidae	7	1	22	1	54				
Caesionidae	4	0	20	0				2.2–2.3	
Callanthiidae	2	0	12	0					

Table 3.3 Class ACTINOPTERYGII. Part 2 Neopterygii (continued)

Order/family/subfamily/genus	No. of genera		No. of sp/ssp		2n	Genome size (pg/cell)			
	total	studied	total	studied		FCM	FD	FIA	BFA
Carangidae	32	13	140	26	46–50, 56	1.7	1.2–1.4	1.2–1.8	1.4–1.7
Caristiidae	2	0	5	0					
Centracanthidae	2	1	8	1	44–48				
Centrarchidae	8	8	31	23	40, 46, 48	1.9–2.1	1.9–2.2	1.5–2.3	
Centrogeniidae	1	0	1	0					
Centropomidae	1	1	12	1	48				
Cepolidae	4	0	19	0					
Chaetodontidae	11	2	122	11	48	1.5–1.6	1.1	1.4–1.7	1.7
Cheilodactylidae	5	0	22	0				1.3–1.5	
Chironemidae	2	0	5	0					
Cirrhitidae	12	0	33	0		1.5			
Coryphaenidae	1	0	2	0				1.2	
Dichistiidae	1	0	2	0					
Dinolestidae	1	0	1	0					
Dinopercidae	2	0	2	0					
Drepaneidae	1	0	3	0				1.4	
Echeneidae	4	1	8	1	42			1.4	1.4
Emmelichthyidae	3	0	15	0		2.1		1.5	
Enoplosidae	1	0	1	0					
Epigonidae	6	0	25	0					
Gerreidae	8	4	44	8	48	1.4	0.9	1.2	1.6
Glaucosomatidae	1	0	4	0					
Grammatidae	2	0	12	0					
Haemulidae	17	6	145	17	48	1.6	1.2–1.5	1.7	1.7–2.0
Inermiidae	2	0	2	0					
Kuhliidae	1	1	10	2	48				
Kyphosidae	16	4	45	7	48	1.8		1.8, 2.1	1.6, 2.2
Lactariidae	1	0	1	0					
Latidae	2	2	9	2	48			1.4	
Latridae	3	0	8	0					
Leiognathidae	4	3	30	3	48			1.0–1.4	
Leptobramidae	1	0	1	0					
Lethrinidae	5	1	39	2	48	2.8–3.2		2.2–2.5	
Lobotidae	2	1	5	1	48				
Lutjanidae	7	3	105	18	47–48	2.9	2.0–2.2	1.4–2.7	1.9, 2.6
Malacanthidae	5	0	40	0					2.0
Menidae	1	0	1	0					
Monodactylidae	2	1	5	2	48				1.8
Moronidae	3	3	8	6	48	1.6		1.9	1.8
Mullidae	6	4	62	6	44, 48	1.2–1.3	1.0	1.1–1.3	
Nandidae	4	3	21	4	46, 48				
Nematistiidae	1	0	1	0					
Nemipteridae	5	0	64	0		2.2		1.5–1.7	
Notograptidae	1	0	3	0					
Opistognathidae	3	0	78	0				2.1	
Oplegnathidae	1	1	7	2	48	1.9			
Ostracoberycidae	1	0	3	0					
Pempheridae	2	1	26	1	48	1.4			
Pentacerotidae	7	0	12	0		1.6		1.4–1.5	
Percichthyidae	11	2	34	8	48	1.9		1.7	
Percidae	10	8	201	25	48	2.3–2.4		1.8–2.1	2.4
Perciliidae	1	0	2	0					
Plesiopidae	11	1	46	1	48				
Polycentridae	4	1	4	1	46				
Polynemidae	8	1	41	1	48				
Polyprionidae	2	0	5	0				1.5–1.8	
Pomacanthidae	8	3	82	14	48, 52			1.4	
Pomatomidae	1	1	1	1	48			1.6	1.9
Priacanthidae	4	1	18	1	52	1.5		1.7–1.8	2.2

Table 3.3 Class ACTINOPTERYGII. Part 2 Neopterygii (continued)

Order/family/subfamily/genus	No. of genera		No. of sp/ssp		2n	Genome size (pg/cell)			
	total	studied	total	studied		FCM	FD	FIA	BFA
Pseudochromidae	20	0	119	0				1.4–1.9	
Rachycentridae	1	0	1	0					1.5
Sciaenidae	70	25	270	38	46, 48	1.3, 1.9	1.2–1.6	1.3–1.5	1.5–2.0
Scombropidae	1	0	3	0					
Serranidae	64	8	475	29	48	2.1–2.7	1.8–2.2	1.3–2.5	2.4–2.6
Sillaginidae	3	1	31	1	48			1.3	
Sparidae	33	14	115	28	48	1.9	1.0–1.7	1.3–1.5	1.9–2.0
Symphysanodontidae	1	0	6	0					
Terapontidae	16	2	48	4	48	1.7		1.3–1.6	
Toxotidae	1	0	6	0		1.5			
Suborder Elassomatoidei				1					
Elassomatidae	1	1	6	1	48				
Suborder Labroidei				241					
Cichlidae	112	54	?	130					
American cichlids	?	32	?	82	38–52, 60	2.0–2.5	2.4		2.0–2.4
Asian cichlids	1	1	3	2	46, 48				
African cichlids	?	21	?	46	38–48	1.9	1.6–2.4		2.0–2.4
Embiotocidae	13	3	23	3	48				1.5–2.0
Pomacentridae	28	12	348	46					
Amphiprioninae	1	1	27	3	48	2.3–2.4		1.6–2.1	
Chrominae	5	2	?	10	28–48	2.1	2.6	1.7–2.1	2.2
Lepidozyginae	1	0	1	0					
Pomacentrinae	21	10	?	33	36, 42, 48	2.3–3.4	1.5–1.7	1.4–2.1	
Labridae	68	21	453	58	22, 32–48	1.5–3.5	1.6–2.9	1.3–2.8	1.8–2.0
Odacidae	4	0	12	0					
Scaridae	10	4	88	4	46, 48	4.2	2.5–2.8	2.9–3.2	3.8–4.6
Suborder Zoarcoidei				13					
Bathymasteridae	3	1	7	1	26				1.9
Zoarcidae	46	3	230	4	48, 72		2.9–3.2	1.6–2.2	
Stichaeidae	37	6	76	6	28, 46–48, 56				1.6
Cryptacanthodidae	1	0	4	0					
Pholidae	3	1	15	2	26, 46				
Anarhichadidae	2	0	5	0					
Ptilichthyidae	1	0	1	0					
Zaproridae	1	0	1	0					
Scytalinidae	1	0	1	0					
Suborder Notothenioidei				63					
Artedidraconidae	4	3	25	9	46			4.1	
Bathydraconidae	11	7	16	7	20, 36–38, 44–48			2.8	
Bovichtidae	3	2	11	4	48				
Channichthyidae	11	10	15	13	47–48		3.7–4.4		
Eleginopidae	1	1	1	1	48				
Harpagiferidae	1	1	6	1	48				
Nototheniidae	14	11	50	27	22–32, 46–50, 58			2.0–3.6	
Pseudaphritidae	1	1	1	1	48				
Suborder Trachinoidei				8					
Ammodytidae	8	1	23	1	46				
Champsodontidae	1	0	13	0					
Cheimarrhichthyidae	1	0	1	0					
Chiasmodontidae	4	0	15	0					
Creediidae	7	0	16	0					
Leptoscopidae	3	0	5	0					
Percophidae	11	0	44	0					
Pinguipedidae	5	1	54	3	26, 42, 48			1.1–1.2	
Trachinidae	2	2	6	2	48				
Trichodontidae	2	1	2	1	48				
Trichonotidae	1	0	8	0					
Uranoscopidae	8	1	50	1	26–32	1.5		1.4	

Table 3.3 Class ACTINOPTERYGII. Part 2 Neopterygii (continued)

Order/family/subfamily/genus	No. of genera		No. of sp/ssp		2n	Genome size (pg/cell)			
	total	studied	total	studied		FCM	FD	FIA	BFA
Suborder Pholidichthyoidei				0					
Pholidichthyidae	1	0	2	0					
Suborder Blennioidei				27					
Blenniidae	56	13	360	25	40–48	1.6	1.2–2.4	1.0–1.7	1.7–1.9
Chaenopsidae	13	0	86	0					
Clinidae	4	1	12	1	48				
Dactyloscopidae	9	0	43	0					
Labrisomidae	15	1	105	1	48				2.0
Tripterygiidae	23	0	150	0		1.7–2.6		1.7	
Suborder Icosteoidei				0					
Icosteidae	1	0	1	0					
Suborder Gobiesocoidei				5					
Gobiesocidae	36	4	140	5	42, 46, 48				
Suborder Callionymoidei				5					
Callionymidae	13	2	182	5	32, 36–38, 42	1.2, 1.6		1.4, 2.0	
Draconettidae	2	0	12	0					
Suborder Gobioidei				125					
Rhyacichthyidae	1	0	2	0					
Odontobutidae	5	3	15	4	44	2.2–2.4	2.5		
Eleotridae	35	10	155	14					
Butinae	13	3	?	5	46, 48			2.5	
Eleotrinae	22	5	?	9	46, 48				
Ptereleotridae	5	1	36	1	46			1.2	
Xenisthmidae	6	0	12	0					
Kraemeriidae	2	0	8	0					
Gobiidae	210	49	1950	106					
Gobiinae	130	21	?	52	30, 38–50, 52		0.8–2.0	1.7–3.4	
Gobionellinae	56	15	?	37	34, 40–48, 52	2.4–3.0		2.5	2.4–2.8
Sicydiinae	7	1	?	1	44				
Oxudercinae	10	8	?	14	38, 42–48			1.9	
Amblyopinae	10	2	?	2	38, 44, 46				
Microdesmidae (= Cerdalidae)	5	0	30	0					
Schindleriidae	1	0	3	0					
Suborder Kurtoidei				1					
Kurtidae	1	1	2	1	44				
Suborder Acanthuroidei				12					
Acanthuridae	6	3	80	6	34, 36, 48	1.6–1.7	1.4	1.3–2.0	
Ephippidae	8	1	16	1	48			1.5–1.6	1.9
Luvaridae	1	0	1	0					
Scatophagidae	2	2	4	2	48			1.4	1.5
Siganidae	1	1	27	3	42, 48	1.2–1.4		1.2–1.4	
Zanclidae	1	0	1	0					
Suborder Scombrolabracoidei				0					
Scombrolabracidae	1	0	1	0					
Suborder Scombroidei				10					
Sphyraenidae	1	1	21	1	48	1.2	1.4	1.1–1.3	1.7, 2.4
Gempylidae	16	0	24	0				1.6	
Trichiuridae	10	0	39	0				1.8	
Scombridae	15	4	51	9	48	1.7		1.6–1.8	1.8–2.2
Xiphiidae	1	0	1	0				1.3	1.8
Istiophoridae	3	0	11	0				1.3	
Suborder Stromateoidei				0					
Amarsipidae	1	0	1	0					
Centrolophidae	7	0	28	0				1.4–1.6	
Nomeidae	3	0	16	0					
Ariommatidae	1	0	7	0					
Tetragonuridae	1	0	3	0					
Stromateidae	3	0	15	0					1.6
Suborder Anabantoidei (Labyrinthici, in part)				30					

Table 3.3 Class ACTINOPTERYGII. Part 2 Neopterygii (continued)

Order/family/subfamily/genus	No. of genera		No. of sp/ssp		2n	Genome size (pg/cell)			
	total	studied	total	studied		FCM	FD	FIA	BFA
Anabantidae	4	3	33	9	46, 48				
Helostomatidae	1	1	1	1	48				1.8
Osphronemidae	14	8	86	20	16, 42, 46, 48	1.1–1.9		1.3	1.2–1.6
Suborder Channoidei				11					
Channidae	2	2	29	11	32–48, 66, 104		1.3–1.9		2
Suborder Caproidei				1					
Caproidae	2	1	11	1	42–46				
PLEURONECTIFORMES				62					
Suborder Psettodoidei				0					
Psettodidae	1	0	3	0				1.4	
Suborder Pleuronectoidei				62					
Citharidae	5	0	6	0					
Scophthalmidae	4	2	8	3	40, 44			1.3, 1.7	
Paralichthyidae	16	6	105	14	28, 38, 46, 48	1.4	1.6	1.0–1.1	1.5–2.0
Pleuronectidae	23	12	60	21	44, 46, 48	1.2–1.3	1.3–1.5	1.1–1.7	1.3–1.9
Bothidae	20	2	140	3	38, 44	1.2			
Paralichthodidae	1	0	1	0					
Poecilopsettidae	3	0	20	0					
Rhombosoleidae	9	0	19	0					
Achiropsettidae	4	0	6	0					
Samaridae	3	0	20	0					
Achiridae	7	5	33	7	34–42				1.3
Soleidae	35	5	130	7	30, 42, 46–48	1.5		1.5–2.1	
Cynoglossidae	3	3	127	7	34, 38–42, 46	1.5		1.2	2.2
LOPHIIFORMES				4					
Antennariidae	12	2	42	3	46, 48				1.6
Lophiidae	4	1	25	1	46			2.1	2.0
Tetrabrachiidae	1	0	1	0				1.5	
Lophichthyidae	1	0	1	0					
Brachionichthyidae	1	0	4	0					
Chaunacidae	2	0	14	0					
Ogcocephalidae	10	0	68	0					1.5
Caulophrynidae	2	0	5	0					
Neoceratiidae	1	0	1	0					
Melanocetidae	1	0	5	0					
Himantolophidae	1	0	18	0					
Diceratiidae	2	0	6	0					
Oneirodidae	16	0	62	0			2.0		
Thaumatichthyidae	2	0	7	0					
Centrophrynidae	1	0	1	0					
Ceratiidae	2	0	4	0					
Gigantactinidae	2	0	22	0					
Linophrynidae	5	0	27	0					
TETRAODONTIFORMES				65					
Suborder Triacanthodoidei				0					
Triacanthodidae	11	0	21	0					
Suborder Balistoidei				31					
Triacanthidae	4	2	7	2	48			1.0	
Balistidae	11	8	40	14	40–46	1.4–1.5	1.1	1.2–1.3	1.4
Monacanthidae	32	7	102	10	33–40	1.1–1.2		0.9–1.3, 1.6	1.3–1.4
Ostraciidae	14	2	33	5	34–36, 48–50	2.0–2.2	1.9		1.7, 2.2
Suborder Tetraodontoidei				34					
Triodontidae	1	0	1	0					
Tetraodontidae	19	7	130	29	28, 34–46	0.8–0.9		0.8–1.0	0.8–1.0
Diodontidae	6	2	19	4	46, 52	1.6–1.7			1.8
Molidae	3	1	4	1	46	1.7–1.9			

Table 3.4 Class SARCOPTERYGII

Order/family/species	No. of genera		No. of sp/ssp		2n	Genome size (pg/cell)			
	total	studied	total	studied		FCM	FD	FIA	BFA
Class SARCOPTERYGII									
Subclass Coelacanthimorpha									
COELACANTHIFORMES				1					
Latimeriidae	1	1	2	1	48=32+16MC		7.2		
Subclass Dipnotetrapodomorpha									
CERATODONTIFORMES				5					
Suborder Ceratodontoidei				1					
Ceratodontidae	1	1	1	1	54=34+20MC	106.0	150.0	109.0	
Suborder Lepidosirenoidei				4					
Lepidosirenidae	1	1	1	1	38	161.0	226.0–248.0		
Protopteridae	1	1	4	3	34, 68	125.0–131.0	80.0–266.0		
Protopterus annectens					34		81.0		
Protopterus dolloi					68		163.0		

Cytogenetic Approach to Fish Systematics

To clarify interrelationships of fishes cytogenetically, genome size has been studied in fishes (Hinegardner and Rosen 1972: H-13; Ojima and Yamamoto 1990: O-48; Hardie and Hebert 2004: H-40; Pie et al. 2007: P-61; Smith and Gregory 2009: S-191). However, the question of the C-value enigma has been a puzzle for almost half a century, which suggests that a simple comparison of taxa and their genome size may be insufficient for the study of genome evolution in fishes. Concerning genome size, the transposable elements, the spectrum of size and frequency of small spontaneous nucleotide insertions and deletions, and genome duplication are the important parameters in the long-term evolution of genome size (Petrov 2001: P-5; Gregory 2005: G-22).

As inferred by gene mapping analysis, inter- and intra-chromosomal rearrangements by Robertsonian translocation, tandem fusion, pericentric- and paracentric-inversion have occurred in fishes and a higher rate of chromosomal rearrangements in teleosts compared to other vertebrates has been hypothesized based on a comparison of the medaka genome with the zebrafish, pufferfish, and human genomes (Ravi and Venkatesh 2008: R-117). Therefore, synthetic analyses of karyotypes, genome sizes, and DNA sequences, and stepwise study inferring the karyotype and genome size of the latest common ancestor in monophyly from lower to higher taxa, may be necessary to clarify fish systematics.

Recently, the early fish proto-karyotype has been studied (Jaillon et al. 2004: J-21; Naruse et al. 2004: N-77; Woods et al. 2005: W-37; Kohn et al. 2006: K-141; Nakatani et al. 2007: N-75). According to Sato and Nishida (2010: S-205), whole-genome duplication (WGD), which generates many thousands of duplicate genes, is believed to be one of the major evolutionary events that shaped the genomes of vertebrates including fishes and tetrapods. Interestingly, the analysis of teleost fish genomes has revealed that teleosts experienced an additional WGD (3R-WGD), whereas tetrapods experienced only 1R- and 2R-WGD; exceptionally, some lineages of amphibians and reptiles have experienced an additional WGD. The chromosomal distribution of the homologous genes can be compared between tetrapods and teleosts by whole-genome sequence analysis, and this information can then be used to infer the karyotype of the osteichthyan ancestor. Nakatani et al. (2007: N-75) hypothesized the following karyotype evolution model in fishes based on reconstruction of the vertebrate ancestral genome. Before the first round of WGD, the vertebrate ancestor karyotype was $2n = 20$–26, and the subsequent 2R-WGD and some genome rearrangements yielded the jawed vertebrate ancestor of $2n = 80$. After the divergence of Osteichthyes and Chondrichthyes, genome rearrangements reduced the number of chromosomes in the osteichthyan ancestor to $2n = 62$. After the divergence of ray-finned and lobe-finned fishes, in the lineage of ray-finned fishes (Actinopterygii), chromosome fusions reduced the number of chromosomes and produced the teleost ancestor with $2n = 26$. Subsequently, the whole-genome duplication (3R-WGD) in the teleost ancestor doubled the number of chromosomes to $2n = 52$. The number of chromosomes in the teleost lineage has remained nearly unchanged during evolution, and the chromosome numbers of extant teleost species peak at $2n = 48$ or 50.

Database of Karyotypes: How to Use the Database

The database of fish karyotypes (Tables 4–7) is organized in the form of tables subdivided into 12 columns (A to L) as follows.

1. Column A contains current scientific names of karyotyped taxon. Classification of species, as a rule, followed Eschmeyer's *Catalogue of Fishes* (E-13). Classification of higher taxa than species, as a rule, followed Nelson (N-68). Hybrids were not included. Synonymy of species/subspecies followed, as a rule, Eschmeyer (E-13).
2. Column B includes the names used in the original karyotype papers in cases in which these differ from currently accepted classification.
3. Column C shows the sex of fishes studied. The majority of fishes reproduce bisexually. However, sex chromosome systems unequivocally identified by karyotypes are known only in a limited number of species. Datasets for such heterosomes were given separately for both sexes and were marked with 'F' for females and 'M' for males. For possible further items, see also column J.
4. Column D contains diploid chromosome number ($2n$), marked with an asterisk when inferred from a haploid number. B chromosomes, as a rule, were excluded from diploid chromosome number.
5. Column E includes the karyotype. Classification of chromosomes followed Levan et al. (L-25): M, metacentrics, SM, submetacentrics, ST, subtelocentrics, A, acrocentrics. When these could not be clearly derived from source publications, classification was as follows: meta- and/or submetacentric (M/SM), submeta- and/or subtelocentric (SM/ST), and subtelo- and/or acrocentric (ST/A). Difference in the karyotype could be attributed to different degrees of chromosome condensation, leading to differences in chromosome classification among authors. In karyotypes of cartilaginous fishes and ancient fishes such as lobe-finned, acipenseriform, and lepisosteiform fishes, small dot-like microchromosomes (MC) have been observed. They are so small that they could not be identified to any type of chromosomes defined by Levan et al. (1964: L-25) at present. It is unknown whether MC is different from M, SM, ST, and A. In this book, MC was added as an additional type to M, SM, ST, and A. As for papers in which the description of karyotype disagrees with the figures, karyotypes based on the figures, as a rule, were adopted.
6. Column F (NF_1) shows fundamental arm number, when M and SM are counted as two-armed.
7. Column G (NF_2) contains fundamental arm number, when M, SM, and ST are counted as two-armed. The arm number by Scheel (1972: S-24) differs from NF_2. As Scheel counted all chromosomes with a short arm as two-arms, acrocentrics with short arms were counted as two-arms, i.e., NF sensu Scheel $\geq NF_2$. Therefore, Scheel's arm number is shown in parentheses.
8. Column H includes the number of Ag-NORs. The number and position of NORs can differ by different methods such as chromomycin A_3 and silver staining. Silver staining is the method specific to NORs and studied widely. The number of Ag-NORs in the embryo tends to be larger than that of adults. Recently, 18S rDNA and 5S rDNA, which are components of NORs, have been examined by fluorescence in situ hybridization (FISH). However, the number of species for which NORs have been studied by FISH is limited.
9. Column I shows genome size (pg/cell). Following Gregory (G-85), the methods used to estimate genome size were listed in five categories: flow cytometry (FCM), Feulgen densitometry (FD), Feulgen image analysis densitometry (FIA), bulk fluorometric assay (BFA), and static cell fluorometry (SCF). To facilitate comparison of genome size (pg/cell), the genome sizes of standard species were updated according to Gregory (G-85): *Acipenser ruthenus* (3.8 pg), *Carassius auratus* (3.5 pg), *Cyprinus carpio* (3.4 pg), *Gallus domesticus*

(2.5 pg), *Homo sapiens* (7.0 pg), *Myxine garmani* (9.2 pg), *Mus musculus* (6.5 pg), *Oncorhynchus mykiss* (5.2 pg), *Salmo salar* (6.0 pg), *Scyliorhinus canicula* (11.4 pg), *Thymallus thymallus* (4.3 pg), and *Tinca tinca* (2.0 pg). Therefore, genome size by Ojima and Yamamoto (O-48) was revised to be 76% of their genome size in this book. The revised genome size was marked with an asterisk. Genome size, which is difficult to specify to one of different karyotypes in a given species, was shown in parentheses.

10. Column J contains cytogenetic information on sex chromosomes, ploidy, B chromosomes, and diploid chromosome number of the latest common ancestor (ACN, ancestral chromosome number) and others: B, B chromosomes; ploidy, 2X, 3X, 6X. ACN was inferred based on $2n$ and the number of large chromosomes (LC), which were inferred to be formed by Robertsonian fusion, being $2n$ + number of LC. The definition of ACN is the same as NAN sensu Arai and Nagaiwa (1976: A-64).
11. Column K includes the locality of fish analyzed in the karyotype papers. If the locality was not provided in the original source, known distributions for the species appear in parentheses.
12. Column L shows references numbered as provided in the References.

Table 4 Jawless fishes

Table 4.1 Class MYXINI. Order MYXINIFORMES

A	B	C	D	E	F	G	H	I	J	K	L
Current scientific name of taxon Family/subfamily/species	Reported in karyotype paper	Sex	2n	Karyotype	NF_1	NF_2	Ag-NORs	Genome size (pg/cell)	Comments	Locality	Reference
Myxinidae											
Eptatretinae											
Eptatretus burgeri		F, M	36	(spermatogonia 2n=52)				6.0 SCF	B in spermatogonia	Japan (Kanagawa, Ibaraki)	N-6, N-62
Eptatretus burgeri	*burgeriare*	F	36	36A	36	36				Japan (Kanagawa)	K-66, K-67
Eptatretus cirrhatus	Type A	F, M	34	(spermatogonia 2n=72)				4.6 SCF	B in spermatogonia	New Zealand	N-62
Eptatretus cirrhatus	Type B	M	34	(spermatogonia 2n=80)					B in spermatogonia	New Zealand	N-62
Eptatretus okinoseanus		F, M	34	(spermatogonia 2n=54)				5.4 SCF	B in spermatogonia	Japan (Kanagawa, Ibaraki)	N-6, N-62
Eptatretus stoutii		M	34	(spermatogonia 2n=48-54)				5.6 SCF, (5.5 FD)	B in spermatogonia	Canada (off Bamfield)	N-62, A-106
Eptatretus stoutii								5.4 FCM, (5.6 BFA)		N. Pacific	T-28, T-73, H-37
Myxininae											
Myxine garmani		F, M	14	(spermatogonia 2n=16)				9.2 SCF		Japan (Kanagawa, Ibaraki)	N-6, N-62
Myxine glutinosa		F, M	28	(germ cells 2n=42-44)	28	28				Sweden (Gullmaren Fjord)	N-47
Myxine glutinosa		F, M	28	(spermatogonia 2n=44)				8.6 SCF		Sweden (Baltic)	N-62
Paramyxine atami		F, M	34	(spermatogonia 2n=48)				6.9 SCF	B in spermatogonia	Japan (Kanagawa, Ibaraki)	N-6, N-62
Paramyxine atami		F	36	36A	36	36				Japan (Kanagawa)	K-66, K-67
Paramyxine atami		F	34	34A	34	34				Japan (Kanagawa)	K-66, K-67
Paramyxine sheni		M	34	(spermatogonia 2n=66-96)					B in spermatogonia	Taiwan (Taitung)	S-68, K-140

Table 4.2 Class PETROMYZONTIDA. Order PETROMYZONTIFORMES

A		B	C	D	E	F	G	H	I	J	K	L
Current scientific name of taxon Family/species		Reported in karyotype paper	Sex	2n	Karyotype	NF_1	NF_2	Ag-NORs	Genome size (pg/cell)	Comments	Locality	Reference
Petromyzontidae												
Eudontomyzon	*mariae*			168							Europe	H-44
Ichthyomyzon	*fossor*			166 (mode)					2.7* FD		USA (Great lakes)	R-85, R-116
Ichthyomyzon	*gagei*	Petromyzonidae		164 (mode)	164 ST/A	164			2.6* FD		USA (AL)	H-35, R-116
Lampetra	*aepyptera*			161–168	all ST/A	161–168				ammocoetes	USA (AL)	H-25
Lampetra	*fluviatilis*			164 (mode)					2.6 FCM, 2.9* FD		UK	R-85, R-116, V-86
Lampetra	*lamottei*	*lamottenii*		166 (mode)					2.8* FD		USA (Great lakes)	R-85, R-116
Lampetra	*planeri*			164 (mode)					2.7, 2.8* FD		UK	R-85, R-116, A-106
Lampetra	*zanandreai*			142							Europe	Z-40
Lethenteron	*camtschatica*	*Entosphenus japonicus*		144–162							Japan (Hokkaido)	K-67
Lethenteron	*reissneri*	*Entosphenus*	F, M	165–174							Japan (Hokkaido)	S-13
Petromyzon	*marinus*			168	168 M/SM/ST/A				4.2 FCM, 4.2* FD		UK (England)	P-60, T-73, R-116
Petromyzon	*marinus*								3.2 BFA			H-37
Geotriidae												
Geotria	*australis*			180					3.1* FD		S. W. Australia	R-86, R-116
Mordaciidae												
Mordacia	*mordax*	Petromyzonidae		76	numerous M/SM				2.8* FD		S. E. Australia	P-59, R-116
Mordacia	*praecox*		M	76							Australia (N.S.W.)	R-84

Table 5 Class CHONDRICHTHYES

Table 5.1 Order CHIMAERIFORMES

A	B	C	D	E	F	G	H	I	J	K	L
Current scientific name of taxon Family/species	Reported in karyotype paper	Sex	2n	Karyotype	NF_1	NF_2	Ag-NORs	Genome size (pg/cell)	Comments	Locality	Reference
Chimaeridae											
Chimaera monstrosa			86	86 ST/A/MC	86					Sweden	N-46
Hydrolagus colliei		F	ca. 58	all A/MC	ca. 58			3.0 FD, 3.2 BFA		USA (Los Angeles)	O-7, H-37

Table 5.2 Order HETERODONTIFORMES

A	B	C	D	E	F	G	H	I	J	K	L
Current scientific name of taxon Family/species	Reported in karyotype paper	Sex	2n	Karyotype	NF_1	NF_2	Ag-NORs	Genome size (pg/cell)	Comments	Locality	Reference
Heterodontidae											
Heterodontus francisci		F	102	26 non-A + 76A		128		17.5 FCM, 14.5 FD, 13.6 BFA		(E. Pacific)	S-40, S-185, H-37
Heterodontus japonicus		M	102	10 M/SM + 92 ST/A	112			27.2, 31.9 FD		Japan (Sagami Bay)	I-6

Table 5.3 Order LAMNIFORMES

A	B	C	D	E	F	G	H	I	J	K	L
Current scientific name of taxon Family/species	Reported in karyotype paper	Sex	2n	Karyotype	NF_1	NF_2	Ag-NORs	Genome size (pg/cell)	Comments	Locality	Reference
Lamnidae											
Carcharodon carcharias		M	82	48 non-A + 34A		130		12.9 FCM	XY?	Atlantic	S-40, M-114
Odontaspididae											
Carcharias taurus			ca. 84					10.9 FCM		(global)	S-41

Table 5.4 Order CARCHARHINIFORMES

A		B	C	D	E	F	G	H	I	J	K	L
Current scientific name of taxon Family/species		Reported in karyotype paper	Sex	2n	Karyotype	NF_1	NF_2	Ag-NORs	Genome size (pg/cell)	Comments	Locality	Reference
Carcharhinidae												
Carcharhinus	*acronotus*		M	84	32 non-A + 52A		116		6.7 FCM, 6.8 BFA		USA (NC)	S-41, H-37
Carcharhinus	*limbatus*		M	ca. 86	ca. 50 A and others		ca. 120		7.8 FCM, 7.4 BFA		USA (NC)	S-41, H-37
Carcharhinus	*limbatus*			86	33 M/SM + 53A	119			8.2 FD		Mediterranean	S-115
Carcharhinus	*obscurus*			ca. 78	ca. 20 M/SM + others	ca. 98			5.5, 6.0 FD		Japan (Chiba)	A-105, M-133
Carcharhinus	*plumbeus*			ca. 74	ca. 18 M/SM + others	ca. 92			6.0 FD		Japan (Ogasawara)	A-105
Galeocerdo	*cuvieri*		F, M	86	38 non-A + 48A		124		8.3 FCM		USA (NC)	S-40, M-114
Galeocerdo	*cuvieri*			86	40 M/SM + 46A	126	126		(7.6* FCM)		Mediterranean	S-115, O-48
Prionace	*glauca*		F, M	86	30 M/SM + 56 ST/A	116			8.6 FD		Japan (off Sanriku)	A-105
Prionace	*glauca*			78	28 M/SM + 50 ST/A	106			8.6 FD		(global)	A-102
Prionace	*glauca*		M	78	4M + 6SM + 20 ST/A and others				(8.6 BFA)		N. Pacific	Y-4, H-37
Rhizoprionodon	*terraenovae*		M	80	44 non-A + 36A		124		7.2 FCM	sex chrom?	USA (NC)	S-40, M-114
Rhizoprionodon	*terraenovae*			90	32 M/SM + 58A	122	122					S-115
Scyliorhinidae												
Cephaloscyllium	*umbratile*		M	64	34 M/SM + 30 ST/A	98			14.7 FD		Japan (Suruga Bay)	A-101
Cephaloscyllium	*ventriosum*			64	46 non-A + 18A		110		18.1 FCM, 15.4 BFA		(E. Pacific)	S-40, H-37
Scyliorhinus	*canicula*			62	18M + 24SM + 20 A/MC	104		2	11.3 FD		Italy	S-111, R-87
Scyliorhinus	*stellaris*			72	30M + 20SM + 22 MC	122		2	12.3 FD		Italy	S-111, R-87
Scyliorhinus	*torazame*		M	64	26 M/SM + 38 ST/A	90			13.2 FD		Japan (Shimokita)	A-101
Sphyrnidae												
Sphyrna	*lewini*		F	86	20 M/SM + 66 ST/A	106			(6.1 FIA)		Japan (Ogasawara)	A-105, H-40
Sphyrna	*lewini*			78	18 non-A + 60A		96		6.6 FCM, (7.0 BFA)		(global)	S-40, H-37
Triakidae												
Mustelus	*canis*			80	44 M/SM + 36A	124	124		9.6 FD, 9.2 BFA		(W. Atlantic)	S-115, H-37
Mustelus	*manazo*		F	68	44 M/SM + 24 ST/A	112			9.3 FD		Japan (Chiba)	A-102
Triakis	*scyllia*		M	72	36 M/SM + 36 ST/A	108			9.8 FD		Japan, Pacific	A-102
Triakis	*semifasciata*			72	52 non-A + 20 A		124		9.6 BFA	XY?	(E. Pacific)	S-40, M-114, H-37

Table 5.5 Order HEXANCHIFORMES

A		B	C	D	E	F	G	H	I	J	K	L
Current scientific name of taxon Family/species		Reported in karyotype paper	Sex	2n	Karyotype	NF_1	NF_2	Ag-NORs	Genome size (pg/cell)	Comments	Locality	Reference
Chlamydoselachidae												
Chlamydoselachus anguineus			M	100	12 M/SM + 88 ST/A	112			9.2 FD		Japan (Suruga Bay)	I-6
Heptranchiidae												
Heptranchias	*perlo*			ca. 72	6M + 66 ST/A	78					(global)	I-3
Notorynchidae												
Notorynchus	*cepedianus*	*maculatus*	F	104	4 non-A + 100A		108		8.8 FCM		(global)	S-40, M-114

Table 5.6 Order SQUALIFORMES

A		B	C	D	E	F	G	H	I	J	K	L
Current scientific name of taxon Family/species		Reported in karyotype paper	Sex	2n	Karyotype	NF_1	NF_2	Ag-NORs	Genome size (pg/cell)	Comments	Locality	Reference
Etmopteridae												
Etmopterus	*pusillus*			86					16.1* FCM		(global)	O-48
Etmopterus	*spinax*	Squalidae	M	86	86 A/MC	86			32.3 FD		Atlantic	N-46, S-112
Oxynotidae												
Oxynotus	*centrina*			62	56 M/SM + 6A	118	118		34.0 FD		(Mediterranean)	S-113, S-185, O-74
Squalidae												
Squalus	*acanthias*		M	78	38 M/SM + 40 ST/A	116					Sweden	N-46
Squalus	*acanthias*		M	60	60 non-A		120		12.0 BFA		USA (off NC)	S-40, H-37
Squalus	*acanthias*		F	64	32 M/SM + 32 ST/A	96			14.0 FCM	ACN=66	USA (off NC)	M-114
Squalus	*acanthias*			58	56 M/SM + 2A	114	114		(11.6 FIA)			P-50, H-40

Table 5.7 Order SQUATINIFORMES

A		B	C	D	E	F	G	H	I	J	K	L
Current scientific name of taxon Family/species		Reported in karyotype paper	Sex	2n	Karyotype	NF_1	NF_2	Ag-NORs	Genome size (pg/cell)	Comments	Locality	Reference
Squatinidae												
Squatina	*californica*			88	26 non-A + 62A		114		18.6 BFA	XY/XX	(E. Pacific)	S-40, M-114, H-37

Table 5.8 Order TORPEDINIFORMES

A	B	C	D	E	F	G	H	I	J	K	L
Current scientific name of taxon Family/species	Reported in karyotype paper	Sex	2n	Karyotype	NF_1	NF_2	Ag–NORs	Genome size (pg/cell)	Comments	Locality	Reference
Narcinidae											
Narcine brasiliensis	Rajiformes	F, M	28	18M + 4SM + 6ST	50	56		8.4 BFA		Gulf of Mexico	D-17, H-37
Narke japonica		F	54	28 M/SM + 26 ST/A	82			24.0 FD		Japan (Izu)	I-5
Torpedinidae											
Torpedo californica		F	82	4 M/SM + 78 ST/A	86			14.6 BFA		Japan (Iwate)	I-5, H-37
Torpedo marmorata			86	86 ST/A	86			14.0 FD		Italy	O-74, S-111
Torpedo marmorata		F, M	86	66A + 20 MC	86		3			Italy (Tyrrhenian Sea)	S-114
Torpedo tokionis		M	86	86A	86	86				Japan (Suruga Bay)	A-103
Torpedo torpedo	*ocellata*	F, M	66	12 M/SM + 14A + 40 MC	78		3	15.0 FD		Italy (Tyrrhenian Sea)	O-74, S-111, S-114

Table 5.9 Order RAJIFORMES

A	B	C	D	E	F	G	H	I	J	K	L
Current scientific name of taxon Family/species	Reported in karyotype paper	Sex	2n	Karyotype	NF_1	NF_2	Ag–NORs	Genome size (pg/cell)	Comments	Locality	Reference
Rajidae											
Amblyraja radiata	*Raja*		98							Sweden	N-43
Dipturus batis	*Raja*	M	98	6 M/SM + 92A	104			6.2 FD		Sweden	N-43, N-46, S-111
Raja asterias			98	6 M/SM + 92A	104		8	7.0 FD		Italy	O-74, R-118, S-113
Raja clavata		M	98	4 M/SM + 94 ST/A/MC	102			6.3 FD		Sweden	N-43, N-46, S-112
Raja eglanteria		F, M	58	30 non-A + 28A		88		6.5 FCM	ACN=60	USA (NC)	S-40, M-114
Raja montagui		M	96	18 M/SM + 16ST + 62A	114	130	12	6.9 FD		Italy (near Naples)	R-88, R-118, S-111
Raja polystigma		F, M	96	18 M/SM + 16ST + 62A	114	130				Italy (near Naples)	R-112
Rhinobatidae											
Rhinobatos hynnicephalus		F	60	26M + 25SM + 6ST + 3A	111	117	4	7.6* FCM		Japan (Osaka)	K-49, O-48
Rhinobatos hynnicephalus		M	59	26M + 24SM + 6ST + 3A	109	115	4			Japan (Osaka)	K-49
Rhinobatos lentiginosus		F	84							Gulf of Mexico	D-17
Rhinobatos productus		F, M	92	44 non-A + 48A		136		8.0 FCM, 8.0 BFA	XY/XX	(E. Pacific)	S-40, M-114, H-37
Rhinobatos schlegelii		F	64	55 M/SM + 9 ST/A	119			5.9 FD	ACN=70	Japan	A-110
Rhinobatos schlegelii		M	63	54 M/SM + 9 ST/A	117			4.9 FCM, 5.9 FD	ACN=69	Japan	A-110, C-100

Table 5.10 Order MYLIOBATIFORMES

A	B	C	D	E	F	G	H	I	J	K	L
Current scientific name of taxon Suborder/family/subfamily/specie	Reported in karyotype paper	Sex	2n	Karyotype	NF_1	NF_2	Ag-NORs	Genome size (pg/cell)	Comments	Locality	Reference
Suborder Platyrhinoidei											
Platyrhinidae											
Platyrhinoidis triseriata		F, M	64	32 non-A + 32A		96		15.5 FCM, 15.4 BFA	XY/XX	(E. Pacific)	S-40, M-114, H-37
Suborder Myliobatoidei											
Dasyatidae											
Dasyatis akajei		F	72	34 M/SM + 38 ST/A	106			8.0, 8.4* FCM, 8.3 FD		Japan (Chiba)	A-100, C-100, O-48
Dasyatis americana		F	78	9 M/SM + 69A	87	87		9.3 FCM, 11.1 FD		USA (NC)	M-114, S-115, S-185
Dasyatis kuhlii			64					9.9* FCM		(Indo-West Pacific)	O-48, S-115
Dasyatis matsubarai		M	64	40 M/SM + 24 ST/A	104			9.6 FD		Japan (Shimokita)	A-103
Dasyatis sabina	Rajiformes	F, M	68	28 M/SM + 40 ST/A	96			10.1 FCM		Gulf of Mexico	D-17, T-73
Dasyatis sayi	Rajiformes	F	68	34 M/SM + 34 ST/A	102			9.4 BFA		Gulf of Mexico	D-17, H-37
Dasyatis violacea			58	46 M/SM + 12A	104	104		13.7 FD		Italy	O-74, S-113, S-115
Dasyatis violacea		M						9.6 FD		Pacific	A-103
Taeniura lymma		M	64	38 M/SM/ST + 26A		102	4	13.4 FIA		Bali	R-87, R-118, H-40
Gymnuridae											
Gymnura japonica			56	32 M/SM + 24 ST/A	88					Japan	A-104
Gymnura micrura			56	44 non-A + 12A		100		11.4 FCM, 16.2 BFA		(W. Atlantic)	S-40, H-37

Table 5.10 Order MYLIOBATIFORMES (continued)

A		B	C	D	E	F	G	H	I	J	K	L
Current scientific name of taxon Suborder/family/subfamily/specie		Reported in karyotype paper	Sex	2n	Karyotype	NF_1	NF_2	Ag–NORs	Genome size (pg/cell)	Comments	Locality	Reference
Myliobatidae												
Mobulinae												
Mobula	*japonica*		F	66	26M + 12SM + 28 ST/A	104			9.3 FD, 9.6 FCM		Japan (Sanriku)	A-104, C-100
Mobula	*japonica*		M	66	26M + 11SM + 29 ST/A	103			9.4 FD		Japan (Sanriku)	A-104
Myliobatinae												
Myliobatis	*aquila*			52	32 M/SM + 20A	84	84		10.8 FD		(Atlantic to Indian)	S-113
Myliobatis	*californica*		F	52	50 non-A + 2A		102		10.4 FCM, 9.8 BFA		(E. Pacific)	S-40, M-114, H-37
Myliobatis	*freminvillii*		M	52	50 non-A + 2A		102		10.6 FCM, 9.8 BFA	XY?	USA (NC)	S-40, M-114, H-37
Myliobatis	*tobijei*		M	54	40 M/SM + 14 ST/A	94			8.7 FD, 10.7 FCM		Japan (Misaki)	A-100, C-100
Rhinopterinae												
Rhinoptera	*bonasus*			64	42 non-A + 22A		106		10.0 FCM		(W. Atlantic)	S-40
Potamotrygonidae												
Paratrygon	*aiereba*		F, M	90	4M + 2SM + 10ST + 74A	96	106	3		ACN=92	Brazil (AM)	V-77
Potamotrygon	*motoro*		F, M	66	18M + 12SM + 10ST + 26A	96	106	7		ACN=92	Brazil (AM)	V-77
Potamotrygon	*orbignyi*		F	66	22M + 10SM + 8ST + 26A	98	106	8		ACN=92	Brazil (AM)	V-77
Urolophidae												
Urolophus	*aurantiacus*		F	52	44 M/SM + 8 ST/A	96			15.5 FCM, 13.1 FD		Japan	A-100, C-100
Urolophus	*halleri*			72	20 M/SM + 52A	92			13.0 BFA		(E. Pacific)	S-40, H-37

Table 6 Class ACTINOPTERYGII (OSTEICHTHYES)

Table 6.1 Order POLYPTERIFORMES

A		B	C	D	E	F	G	H	I	J	K	L
Current scientific name of taxon Family/species		Reported in karyotype paper	Sex	2n	Karyotype	NF_1	NF_2	Ag–NORs	Genome size (pg/cell)	Comments	Locality	Reference
Polypteridae												
Erpetoichthys	*calabaricus*	*Calamoichthys*		36	28M + 6SM + 2ST	70	72	2			(Africa)	S-131, S-132
Erpetoichthys	*calabaricus*	*Calamoichthys*	M	36	32M + 4ST	68	72			ACN=46	(Africa)	C-13
Erpetoichthys	*calabaricus*	*Calamoichthys*	F, M	36	30M + 6SM	72	72				(Africa)	D-8
Erpetoichthys	*calabaricus*	*Calamoichthys*	F, M	36	36 M/SM	72	72		(9.1 FD, 9.8 BFA)	ACN=46	Cameroon	V-37, G-85, H-37
Polypterus	*delhezi*		F	36	26M + 10SM	72	72				(Africa)	U-52
Polypterus	*delhezi*		F, M	36	36 M/SM	72	72		9.7 FD	ACN=46	Zaire	V-37
Polypterus	*delhezi*		F	36	32M + 4SM	72	72			ACN=46	Africa (Zaire R.)	C-92
Polypterus	*endlicheri*	*endlicheri congicus*	F, M	36	32M + 4SM	72	72			ACN=46	Africa (Zaire R.)	C-92
Polypterus	*ornatipinnis*			36	26M + 10SM	72	72				(Africa)	U-52
Polypterus	*ornatipinnis*		F, M	36	36 M/SM	72	72		9.6 FD	ACN=46	(Africa)	V-37
Polypterus	*palmas*		F, M	36	24M + 12SM	72	72				(W. Africa)	D-8
Polypterus	*palmas*			36	26M + 10SM	72	72			ACN=46	(W. Africa)	U-52
Polypterus	*palmas*		F	36	32M + 4SM	72	72			ACN=46	Africa (Zaire R.)	C-92
Polypterus	*palmas*		F, M	36	36 M/SM	72	72		8.9 FD, (7.4 FIA)	ACN=46	(Africa)	V-37, H-41
Polypterus	*senegalus*			36	28M + 8SM	72	72	2			(Africa)	S-131, S-132
Polypterus	*senegalus*		F, M	36	26M + 10SM	72	72	2		ACN=46	Nigeria	M-127, U-52
Polypterus	*weeksii*		F, M	38	34 M/SM + 4A	72	72		9.8 FD	ACN=46	(Africa)	V-37

Table 6.2 Order ACIPENSERIFORMES

A		B	C	D	E	F	G	H	I	J	K	L
Current scientific name of taxon Family/species		Reported in karyotype paper	Sex	2n	Karyotype	NF_1	NF_2	Ag–NORs	Genome size (pg/cell)	Comments	Locality	Reference
Acipenseridae												
Acipenser	*baeri*			ca. 248				8	8.3 FCM	4X	Russia (Siberia)	V–9, V–10, B–68, F–70
Acipenser	*brevirostrum*			372	178 M/SM + 194 ST/A	550		10	13.1 FCM, 13.8 FIA	6X	USA (FL)	K–54, B–69, H–41
Acipenser	*brevirostrum*								18.6 FIA		N. America	H–41
Acipenser	*fulvescens*		F	264	134 M/SM + 70 ST/A + 60 MC	398			8.9 FCM	4X	USA (WI)	B–69, F–62
Acipenser	*gueldenstaedtii*			250	92 M/SM + 158 A/MC	342		6–8	7.9 FCM	4X	Russia (Volga R.)	B–37, B–68, V–72
Acipenser	*medirostris*			249					8.8 FCM	4X	N. America	B–69, V–110
Acipenser	*mikadoi*	*medirostris*							14.3 FCM	6X	Far East	B–68, L–78
Acipenser	*naccarii*			ca. 239		390		8	6.3 FD	4X	Italy	F–58, F–61, F–70
Acipenser	*nudiventris*			118	54M + 5A + 59 A/MC	172			3.9 FCM	2X	Russia (Black Sea)	A–88, B–68
Acipenser	*oxyrinchus*			122	78 M/SM + 44 A/MC	200			(4.4 FIA)	2X	(Atlantic)	F–63, H–41
Acipenser	*oxyrinchus*	*oxyrhynchus desotoi*							4.6 FCM		N. America	B–69
Acipenser	*persicus*			262	134 M/SM + 128 A/MC	396				4X	Iran	N–38
Acipenser	*ruthenus*			118	82 M/SM + 36 A/MC	200		2–3	3.7 FCM	2X	Russia (Volga R.)	V–72, B–37, B–68
Acipenser	*ruthenus*			118	58M + 4A + 56 MC	176				2X	Danube R.	R–10
Acipenser	*schrenckii*			240					6.1 FD	4X	Amur R.	F–61
Acipenser	*sinensis*			264	78M + 20SM + 26 ST/A + 140 MC	362				4X	China (Hubei)	Y–15
Acipenser	*sinensis*		F, M	240	76M + 80SM + 20 ST/A + 64 MC	396			9.1 FD	4X	China (Hubei)	X–2
Acipenser	*stellatus*			118	70 M/SM + 48 A/MC	188		2–3	3.7, 4.7 FCM	2X	Caspian Sea	V–72, B–37, B–68, L–78
Acipenser	*sturio*			ca. 116		ca. 186			3.2 FD	2X	(Atlantic)	F–58, M–133
Acipenser	*transmontanus*		M	ca. 248				8	9.5 FCM	4X	N. America	B–69, F–61, F–70, V–110
Huso	*dauricus*			120					3.8 FCM	2X	Amur R.	B–68
Huso	*huso*			118	60 M/SM + 58 A/MC	178		2–3	2.4 FCM	2X	Russia (Volga R.)	B–37, B–68, V–72
Huso	*huso*			118	84 M/SM + 34 A/MC	202		4	3.6 FD	2X	(E. Europe)	F–32, F–61
Pseudoscaphirhynchus	*kaufmanni*								3.5 FCM		Amu Darya R.	B–68
Scaphirhynchus	*platorynchus*	*platorhynchus*	M	112	50 M/SM + 14A + 48 MC	162			3.5 FD	2X	USA (Great Lakes)	O–7
Scaphirhynchus	*platorynchus*								4.7 FCM		N. America	B–69
Polyodontidae												
Polyodon	*spathula*		M	120	32M + 8SM + 8A + 72 MC	160		4	3.2, 3.9 FCM	2X	USA (AL)	D–14, B–68, T–73
Polyodon	*spathula*								4.9 FCM		N. America	B–69

Table 6.3 Order LEPISOSTEIFORMES

A	B	C	D	E	F	G	H	I	J	K	L
Current scientific name of taxon Family/species	Reported in karyotype paper	Sex	2n	Karyotype	NF_1	NF_2	Ag-NORs	Genome size (pg/cell)	Comments	Locality	Reference
Lepisosteidae											
Lepisosteus osseus		F, M	56	22M + 12SM + 6ST + 16 MC	90		2			USA (Mississippi R.)	R-120
Lepisosteus osseus			56	12M + 22SM + 22 ST/A	90		4			(N. America)	O-32
Lepisosteus oculatus	*productus*	F	68	28 M/SM + 14A + 26 MC	96			2.8* FCM, 2.9 FD, 2.8 FIA		USA (Great Lakes)	O-7, O-48, H-40

Table 6.4 Order AMIIFORMES

A	B	C	D	E	F	G	H	I	J	K	L
Current scientific name of taxon Family/species	Reported in karyotype paper	Sex	2n	Karyotype	NF_1	NF_2	Ag-NORs	Genome size (pg/cell)	Comments	Locality	Reference
Amiidae											
Amia calva		F	46	20 M/SM + 26 ST/A	66			2.6 FD, 2.3 FIA	ACN=56	USA (Great Lakes)	O-7, H-40
Amia calva		F	46	2M + 24 SM/ST + 20A		72	2	2.0 FD		(USA)	S-135

Table 6.5 Order HIODONTIFORMES

A	B	C	D	E	F	G	H	I	J	K	L
Current scientific name of taxon Family/species	Reported in karyotype paper	Sex	2n	Karyotype	NF_1	NF_2	Ag-NORs	Genome size (pg/cell)	Comments	Locality	Reference
Hiodontidae											
Hiodon alosoides	Osteoglossiformes		50	40 M/SM + 10ST	90	100		1.2 FD	ACN=56?	USA, Canada	U-68, B-14
Hiodon tergisus	Osteoglossiformes		50	42 M/SM + 8 ST/A	92			1.2 FD	ACN=56	Canada (Manitoba)	B-14

Table 6.6 Order OSTEOGLOSSIFORMES

A		B	C	D	E	F	G	H	I	J	K	L
Current scientific name of taxon Family/species		Reported in karyotype paper	Sex	2n	Karyotype	NF_1	NF_2	Ag-NORs	Genome size (pg/cell)	Comments	Locality	Reference
Osteoglossidae												
Arapaima	*gigas*			56	4M + 12 SM/ST + 40A		72		(1.6 FIA, 2.0 BFA)	ACN=56	(S. America)	U-51, H-13, H-40
Arapaima	*gigas*		F, M	56	28 M/SM + 28 ST/A	84		2		ACN=56	Brazil (MT)	M-147
Heterotis	*niloticus*			40	26M + 10SM + 2ST + 2A	76	78	2	2.0 FD	ACN=56	(W. Africa)	H-14
Osteoglossum	*bicirrhosum*		F	56	1SM + 1ST + 54A	57	58		1.8 FIA, 2.0 BFA	ACN=56	(S. America)	U-68, H-13, H-40
Osteoglossum	*bicirrhosum*			56	3ST + 53A	56	59			ACN=56	(S. America)	S-125
Osteoglossum	*ferreirai*			54	2M + 4SM + 14ST + 34A	60	74			ACN=56	(S. America)	S-125
Scleropages	*formosus*			50	4 M/SM + 10ST + 36A	54	64		1.9 FD	ACN=56	(SE Asia)	U-51, H-14
Scleropages	*jardini*			48	16M + 6SM + 26A	70	70	2	2.0 FD	ACN=56	(New Guinea)	H-14
Scleropages	*leichardti*			44	16M + 8SM + 6ST + 14A	68	74	2	2.0 FD	ACN=56	(N. Australia)	H-14
Pantodontidae												
Pantodon	*buchholzi*	*bucholzi*		48	12M + 12 SM/ST + 24A		72		1.5 BFA	ACN=56?	(W. Africa)	U-68, H-13
Mormyridae												
Gnathonema	*petersii*			48	10M + 6SM + 32A	64	64		2.4 BFA	ACN=56	Africa	U-68, H-13
Marcusenius	*brachistius*			48	1M + 4SM + 2ST + 41A	53	55				Africa	U-68
Notopteridae												
Chitala	*chitala*	*Notopterus*		42	42A	42	42	2	2.1* FCM	ACN=50	(Asia)	U-68, T-22, O-48
Notopterus	*notopterus*		F, M	42	42A	42	42	2			India (Haryana, WB)	R-60, R-72, K-42
Papyrocranus	*afer*			34	4SM + 30A	38	38			ACN=46	Africa	U-68
Xenomystus	*nigri*			42	42A	42	42		2.6 BFA	ACN=50	Africa	U-68, H-13

Table 6.7 Order ELOPIFORMES

A		B	C	D	E	F	G	H	I	J	K	L
Current scientific name of taxon Family/species		Reported in karyotype paper	Sex	2n	Karyotype	NF_1	NF_2	Ag-NORs	Genome size (pg/cell)	Comments	Locality	Reference
Elopidae												
Elops	*saurus*		F, M	48	6 M/SM + 42 ST/A	54			2.4 BFA	ACN=54	USA (LA)	D-26, H-13
Megalopidae												
Megalops	*atlanticus*			50	50A	50	50			ACN=54	USA (Atlantic)	D-23
Megalops	*cyprinoides*	*cundinga*		46	46A	46	46				India (WB)	K-46
Megalops	*cyprinoides*		F	52	52A	52	52		(2.0 FIA)		India (Bombay)	R-66, H-40

Table 6.8 Order ANGUILLIFORMES

A		B	C	D	E	F	G	H	I	J	K	L
Current scientific name of taxon Suborder/family/subfamily/species		Reported in karyotype paper	Sex	2n	Karyotype	NF_1	NF_2	Ag-NORs	Genome size (pg/cell)	Comments	Locality	Reference
Suborder Anguilloidei												
Anguillidae												
Anguilla	*anguilla*		F	38	12M + 10SM + 16A	60	60			ZW, ACN=53	Poland (Krakow)	P-13, P-14
Anguilla	*anguilla*		M	38	12M + 10SM + 16A	60	60			ZZ, ACN=54	Poland (Krakow)	P-13, P-14
Anguilla	*anguilla*		F	38	12M + 10SM + 16A	60	60			ZW, ACN=53	Germany	P-10
Anguilla	*anguilla*		M	38	12M + 10SM + 16A	60	60			ZZ, ACN=54	Germany	P-10
Anguilla	*anguilla*		F, M	38	22 M/SM + 16A	60	60	2	(3.2 BFA)		Italy	S-87, H-13
Anguilla	*anguilla*	Type I		38	20 M/SM + 18A	58	58			ACN=52	(Europe)	K-88
Anguilla	*anguilla*	Type II		38	21 M/SM + 17A	59	59			ACN=53	(Europe)	K-88
Anguilla	*australis*		F, M	38	22 M/SM + 16A	60					New Zealand	S-87
Anguilla	*australis*			38	20 M/SM + 18A	58					(New Zealand)	N-28
Anguilla	*japonica*			38	20 M/SM + 18A	58				ACN=54	Japan	N-28, K-88
Anguilla	*japonica*		F, M	38	10M + 10SM + 18A	58	58				China (Wuhan)	Y-15
Anguilla	*japonica*		F	38	12M + 8SM + 18A	58	58			ZW, ACN=53	Korea (Han R.)	P-7, P-9
Anguilla	*japonica*		M	38	12M + 8SM + 18A	58	58			ZZ, ACN=54	Korea (Han R.)	P-9
Anguilla	*marmorata*			38	20 M/SM + 18A	58	58			ACN=54	Japan (Amami-oshima)	K-93
Anguilla	*rostrata*		F	38	12M + 10SM + 16A	60	60			ZW, ACN=53	USA (Philadelphia)	P-10
Anguilla	*rostrata*		M	38	12M + 10SM + 16A	60	60			ZZ, ACN=54	USA (Philadelphia)	P-10
Anguilla	*rostrata*		F, M	38	22 M/SM + 16A	60	60	2	(2.8 BFA)		USA (Boston Bay)	S-87, H-13
Anguilla	*rostrata*			38	14M + 6SM + 18A	58					USA (FL)	O-10
Anguilla	*rostrata*		F	38	10M + 10SM + 18A	58	58		(2.0 FIA)	ZW, ACN=53	USA (ME)	P-14, H-41

Table 6.8 Order ANGUILLIFORMES (continued)

A	B	C	D	E	F	G	H	I	J	K	L
Current scientific name of taxon Suborder/family/subfamily/species	Reported in karyotype paper	Sex	2n	Karyotype	NF_1	NF_2	Ag-NORs	Genome size (pg/cell)	Comments	Locality	Reference
Moringuidae											
Moringua linearis			50							India (Porto Novo)	S-118
Suborder Muraenoidei											
Muraenidae											
Enchelycore nigricans		F, M	42	6M + 8SM + 12ST + 16A	56	68	2		ACN=54	Brazil (Atlantic)	V-106
Enchelycore pardalis	*Muraena*		42	8M + 2SM + 32A	52	52	2	4.6* FCM	ACN=52	Japan (Wakayama)	T-5, T-6, T-79, O-48
Gymnothorax eurostus		F, M	42	12 M/SM + 30A	54	54	2		XX/XY, ACN=52	Japan (Okinawa)	T-5, T-6
Gymnothorax kidako			36	16M + 8SM + 12A	60	60	2		ACN=52	Japan (Wakayama)	T-5, T-6, T-79
Gymnothorax miliaris		F	42	14M + 18SM + 10ST	74	84	2		ACN=54	Brazil (Bahia)	V-106
Gymnothorax ocellatus		F, M	42	16M + 18SM + 8A	76	76	2		ACN=54	Brazil (SP)	P-43
Gymnothorax pictus	*Sideria picta*		42	14M + 14SM + 14A	70	70		3.8 FIA		India (WB)	K-46, H-41
Gymnothorax pictus	*Sideria picta*		42	42 ST/A	42					India (Andaman Is.)	R-45
Gymnothorax reevesi		F, M	42	34 M/SM + 8A	76	76				China (Guangdong)	R-90
Gymnothorax unicolor		F, M	42	12 M/SM + 30A	54	54	2	4.4 BFA	ACN=54	Italy	D-5, S-9, G-85
Gymnothorax vicinus			42	8M + 6SM + 28A	56	56	2		ACN=54	Brazil (RN)	V-106
Muraena helena		F, M	42	18 M/SM + 24A	60	60	2	5.1 BFA	ACN=54	Mediterranean	C-99, S-9, G-85
Muraena pavonina		F	42	6M + 4SM + 32A	52	52	2		ACN=54	Brazil (Atlantic)	V-106
Suborder Congroidei											
Congridae											
Ariosoma anagoides	*Alloconger*	F	34	10M + 8SM + 6ST + 10A	52	58			ACN=54	Japan (Wakayama)	T-5
Ariosoma anago	*Anago*		38	18 M/SM + 20A	56	56	2	3.6* FCM		Japan (Kobe)	T-8, O-48
Conger conger		F, M	38	8M + 4SM + 26 ST/A	50		2	3.3 FCM	ACN=48	Mediterranean	S-47, G-85
Conger conger			38	12 M/SM + 26 ST/A	50		2			Spain (Malaga)	A-46
Conger japonicus		F	38	10M + 10SM + 18A	58	58		2.4* FCM	ACN=52	Japan (Wakayama)	T-5, O-48
Conger myriaster	*Astroconger*	F	38	12M + 8 SM + 18A	58	58			ZW, ACN=53	Korea (Yellow Sea)	P-7, P-9
Conger myriaster	*Astroconger*	M	38	12M + 8 SM + 18A	58	58			ZZ, ACN=54	Korea (Yellow Sea)	P-9
Conger myriaster		F	38	14M + 4SM + 20A	56	56			ACN=54	Japan (Yamaguchi)	N-32
Conger myriaster		F, M	38	8M + 10SM + 20A	56	56		2.4* FCM	ZW/ZZ, ACN=54	Japan (Kobe)	O-39, O-48
Conger myriaster			38	8M + 10SM + 20A	56	56				China (Shandong)	Y-20

Table 6.8 Order ANGUILLIFORMES (continued)

A	B	C	D	E	F	G	H	I	J	K	L
Current scientific name of taxon Suborder/family/subfamily/species	Reported in karyotype paper	Sex	2n	Karyotype	NF_1	NF_2	Ag-NORs	Genome size (pg/cell)	Comments	Locality	Reference
Muraenesocidae											
Muraenesox cinereus			38	12M + 4SM + 6ST + 16A	54	60		(2.3 FIA)		China	Z-23, H-41
Ophichthidae											
Myrophinae											
Muraenichthys gymnotus		F	48	4ST + 44A	48	52			$X_1X_1X_2X_2$, ACN=48	Japan (Suruga Bay)	M-108
Muraenichthys gymnotus		M	47	1M + 4ST + 42A	48	52			X_1X_2Y, ACN=48	Japan (Suruga Bay)	M-108
Ophichthinae											
Dalophis imberbis		F	46	2M + 6SM + 5ST + 33A	54	59	3-4		ZW, ACN=53	Italy (Sardinia)	S-182
Dalophis imberbis		M	46	2M + 6SM + 6ST + 32A	54	60	3-4		ZZ, ACN=54	Italy (Sardinia)	S-182
Echelus myrus			38	20 M/SM + 18 ST/A	58		2			Spain (Malaga)	A-46
Myrichthys ocellatus			38	8M + 14SM + 10ST + 6A	60	70	2		ACN=54	Brazil (RN)	V-106
Ophichthus altipennis	*Pisoödonophis zophistius*		38	10M + 16SM + 4ST + 8A	64	68			ACN=54	Japan (Yamaguchi)	N-32
Ophisurus macrorhynchus		F	38	20M + 14SM + 4ST	72	76			ACN=54	Japan (Yamaguchi)	N-32
Ophisurus serpens	*Ophysurus*		38	16M + 20ST + 2A	54	74	2			Spain (Malaga)	T-35
Pisodonophis boro			38	18M + 4SM + 4ST + 12A	60	64				India (WB)	K-42
Pisodonophis boro		M	40							India (Portonovo)	N-13
Synaphobranchidae											
Synaphobranchus kaupi		F	26	22 M/SM + 2ST + 2A	48	50				Japan (Hokkaido)	I-13

Table 6.9 Order CLUPEIFORMES

A		B	C	D	E	F	G	H	I	J	K	L
Current scientific name of taxon Suborder/family/subfamily/species		Reported in karyotype paper	Sex	2n	Karyotype	NF_1	NF_2	Ag-NORs	Genome size (pg/cell)	Comments	Locality	Reference
Suborder Clupeoidei												
Clupeidae												
Alosinae												
Alosa	*kessleri pontica*			48	2ST + 46A	48	50				former USSR	V-72
Alosa	*pseudoharengus*		F, M	48	48A	48	48		2.8 BFA	ACN=48	USA (ME, MA, NY)	M-53, H-13
Brevoortia	*aurea*		F	46	2M + 2SM + 42A	50	50	2		$X_1X_1X_2X_2$	Brasil (RJ)	B-55
Brevoortia	*aurea*		M	45	3M + 2SM + 40A	50	50	2		X_1X_2Y, ACN=48	Brasil (RJ)	B-55
Brevoortia	*patronus*		F, M	46	2M + 2SM + 42A	50	50			ACN=48	USA (Atlantic)	D-23
Brevoortia	*pectinata*			46	2M + 2SM + 42A	50	50	2			Brazil	B-86
Brevoortia	*smithi*		F, M	46	2M + 2SM + 42A	50	50			ACN=48	USA (Atlantic)	D-23
Brevoortia	*tyrannus*		F, M	46	2M + 2SM + 42A	50	50			ACN=48	USA (Atlantic)	D-23
Gudusia	*chapra*	*Gadusia*	M	46		46				ACN=46	India (WB)	K-31
Gudusia	*variegata*	*Gadusia*		46	46A	46	46			ACN=46	(Myanmer)	M-162
Clupeinae												
Clupea	*harengus*			52							USA (ME)	R-82
Clupea	*harengus*		F, M	52	6M + 2ST + 44A	58	60	2		ACN=52	N. Europe	K-81, K-82
Clupea	*harengus*		F, M	52	6M + 2ST + 44A	58	60	2			Russia (White Sea)	K-81, K-82
Clupea	*harengus*			52	8M + 4ST + 40A	60	64	2			Norwegian Sea	K-81, K-82
Clupea	*harengus*		F, M	50	6M + 2ST + 42A	56	58	2			Germany (Baltic Sea)	K-81, K-82
Clupea	*harengus harengus*			54	12-16 M/SM + 42-38 ST/A	66-70					Russia (Baltic Sea)	S-72
Clupea	*pallasi*		F, M	52	6 M/SM + 46 ST/A	58			1.9 FD	ACN=52	Japan	I-8
Clupea	*harengus pallasi*			52	8 M/SM + 44 ST/A	60			(1.8 FIA, 1.5 BFA)		Russia (White Sea)	S-72, H-13, H-40
Clupea	*pallasi marisalbi*			52	6M + 2SM + 44A	60	60			ACN=52	Russia (White Sea)	L-86
Clupea	*pallasi marisalbi*			51	7M + 2SM + 42A	60	60			ACN=52	Russia (White Sea)	L-86
Clupea	*pallasi marisalbi*			50	8M + 2SM + 40A	60	60			ACN=52	Russia (White Sea)	L-86
Corica	*soborna*			56	16M + 20SM + 20A	92	92				India (WB)	K-46
Harengula	*clupeola*			28	22M + 2SM + 2ST + 2A	52	54			ACN=50	USA (Atlantic)	D-23
Herklotsichthys	*quadrimaculatus*			48	48A	48	48		1.5* FCM		Japan	O-48
Sardinella	*melanura*			44	8 M/SM + 36 ST/A	52					India (Andaman Is.)	R-45
Sardinella	*zunasi*		F, M	48	48 ST/A	48			2.3 FD	ACN=48	Japan (Tokyo Bay)	I-8
Sardinella	*zunasi*			48	48A	48	48			ACN=48	China (Shandong)	W-7
Sardinops	*sagax*	*melanostictus*	F, M	48	48 ST/A	48			2.7 FD	ACN=48	Japan (Iwate)	I-8

Table 6.9 Order CLUPEIFORMES (continued)

A		B	C	D	E	F	G	H	I	J	K	L
Current scientific name of taxon Suborder/family/subfamily/species		Reported in karyotype paper	Sex	2n	Karyotype	NF_1	NF_2	Ag-NORs	Genome size (pg/cell)	Comments	Locality	Reference
Dorosomatinae												
Clupanodon	*punctatus*			48	2M + 46A	50	50			ACN=48	China (Shandong)	W-7
Dorosoma	*cepedianum*		F, M	48	2SM + 4ST + 42A	50	54		2.0 FCM	ACN=50	USA (LA)	F-26, G-85
Dorosoma	*petenense*		F, M	48	2SM + 2ST + 44A	50	52			ACN=50	USA (LA)	F-26
Gonialosa	*manmina*			38	38A	38	38				(India)	M-162
Konosirus	*punctatus*			48	48A	48	48		1.6 FCM	ACN=48	Korea (Busan)	K-125
Engraulidae												
Coilinae												
Coilia	*nasus*	*brachygnathus*	F	47	47A	47	47	2		ZO, ACN=47	China (Hubei)	H-21, Y-15
Coilia	*nasus*	*brachygnathus*	M	48	48A	48	48	2		ZZ, ACN=48	China (Hubei)	H-21, Y-15
Thryssa	*baelama*	*Thrissina*		42	42 ST/A	42					India (Andaman Is.)	R-45
Engraulinae												
Anchoa	*mitchilli*		F, M	48	48A	48	48			ACN=48	Atlantic	D-23
Engraulis	*compressa*	*Anchoa*		48		48					USA (CA)	C-46
Engraulis	*encrasicholus ponticus*			44							Russia	I-20
Engraulis	*japonicus*			48	48A	48	48		2.8* FCM		Japan	O-48
Engraulis	*mordax*			48	48A	48	48		3.0 FD, 3.8 BFA	ACN=48	(E. Pacific)	O-6, O-8, H-13

Table 6.10 Order GONORYNCHIFORMES

A		B	C	D	E	F	G	H	I	J	K	L
Current scientific name of taxon Suborder/family/subfamily/species		Reported in karyotype paper	Sex	2n	Karyotype	NF_1	NF_2	Ag-NORs	Genome size (pg/cell)	Comments	Locality	Reference
Suborder Chanoidei												
Chanidae												
Chanos	*chanos*			32	14M + 4SM + 14 ST/A	50				ACN=50	Taiwan	A-63
Suborder Knerioidei												
Phractolaemidae												
Phractolaemus	*ansorgei spinosus*		F, M	28	24M + 2SM + 2A	54			3.0 FD		Zaire	V-34

Table 6.11 Order CYPRINIFORMES

A		B	C	D	E	F	G	H	I	J	K	L
Current scientific name of taxon Superfamily/family/subfamily/species		Reported in karyotype paper	Sex	2n	Karyotype	NF_1	NF_2	Ag-NORs	Genome size (pg/cell)	Comments	Locality	Reference
Superfamily Cyprinoidea												
Cyprinidae												
Acheilognathinae												
Acheilognathus	*chankaensis*	*Acanthorhodeus*	F, M	44	14M + 14SM + 16 ST/A	72			1.8* FD	ACN=50	China (Hubei)	H-19, Y-15, C-83
Acheilognathus	*chankaensis?*	*Acanthorhodeus gracilis*	F, M	44	24 M/SM + 20 ST/A	68				ACN=50	Korea (Sam-rye)	L-9, L-11
Acheilognathus	*cyanostigma*		F, M	44	28 M/SM + 16 ST/A	72		2		ACN=50	Japan (Lake Biwa)	O-20, T-6, T-11
Acheilognathus	*gracilis*		F, M	42	16M + 12SM + 14ST	70	84			ACN=50	China (Wuhan)	H-23, Y-15
Acheilognathus	*imberbis*	*Paracheilognathus*	F, M	44	14M + 18SM + 12 ST/A	76				ACN=50	China (Wuhan)	H-19, Y-15
Acheilognathus	*longipinnis*		F, M	44	28 M/SM + 16 ST/A	72				ACN=50	Japan (Yodo R.)	O-20
Acheilognathus	*macropterus*	*Acanthorhodeus asmussi*	F, M	44	24 M/SM + 20 ST/A	68				ACN=50	Korea (Eui-ryeong)	L-9, L-11
Acheilognathus	*macropterus*	*Acanthorhodeus*	F, M	44	14M + 18SM + 12 ST/A	76				ACN=50	China (Wuhan)	H-19, Y-15
Acheilognathus	*macropterus*			44	14M + 16SM + 14ST	74	88	8 in embryo		ACN=50	China (Amur R.)	U-24
Acheilognathus	*melanogaster*	*moriokae*	M	44	28 M/SM + 16 ST/A	72				ACN=50	Japan (Ibaraki)	O-20
Acheilognathus	*peihoensis*	*Acanthorhodeus*	F, M	44	14M + 12SM + 8ST + 10A	70	78			ACN=50	China (Guilin)	Y-15
Acheilognathus	*rhombeus*	*rhombea*	F, M	44	28 M/SM + 16 ST/A	72				ACN=50	Korea (Ma-ryeong)	L-9, L-11
Acheilognathus	*rhombeus*		F, M	44	10M + 20 SM/ST + 14A		74			ACN=50	Korea (Geum R.)	U-33
Acheilognathus	*rhombeus*		F, M	44	14M + 16SM + 14ST	74	88			ACN=50	South Korea	U-80, U-81
Acheilognathus	*rhombeus*		F, M	45	29 M/SM + 16ST	74	90			ACN=50	South Korea	U-80, U-81
Acheilognathus	*rhombeus*			46	28 M/SM + 18ST	74	92			ACN=50	South Korea	U-80, U-81
Acheilognathus	*rhombeus*			47	27 M/SM + 20ST	74	94			ACN=50	South Korea	U-80, U-81
Acheilognathus	*rhombeus*	*rhombea*	F, M	44	14M + 14SM + 16 ST/A	72		2	(2.3* FCM, 2.0 FD)	ACN=50	Japan (Lake Biwa)	O-20, O-48, T-6, T-11
Acheilognathus	*rhombeus*	*rhombea*	F, M	44	12M + 20 SM/ST + 12A		76			ACN=50	Japan (Lake Biwa)	O-18
Acheilognathus	*tabira erythropterus*	Akahire-tabira	M	44	28 M/SM + 16 ST/A	72			(1.9 FD)	ACN=50	Japan (Ibaraki)	O-20, S-141
Acheilognathus	*tabira nakamurae*	Seboshi-tabira	F, M	44	28 M/SM + 16 ST/A	72				ACN=50	Japan (Fukuoka)	O-20
Acheilognathus	*tabira tabira*		F, M	44	14M + 14SM + 16 ST/A	72		2		ACN=50	Japan (Yodo R., Yoshii R.)	O-20, O-48, T-6, T-11
Acheilognathus	*tabira tabira*		F, M	44	12M + 20 SM/ST + 12A		76			ACN=50	Japan (Lake Biwa)	O-18
Acheilognathus	*tonkinensis*	*Acanthorhodeus*	F, M	44	14M + 14SM + 16 ST/A	72				ACN=50	China (Guilin)	Y-15
Acheilognathus	*tonkinensis*		M	44	14M + 14SM + 8ST + 8A	72	80			ACN=50	(China)	A-85
Acheilognathus	*tonkinensis*		F	44							China (Wuhan)	H-19
Acheilognathus	*yamatsutae*			44	12M + 16SM + 16 ST/A	72				ACN=50	Korea	L-9
Acheilognathus	*yamatsutae*		M	44	24 M/SM + 20 ST/A	68				ACN=50	Korea (Eui-ryeong)	L-9, L-11
Acheilognathus	*yamatsutae*	*cyanostigma*	F	44	12M + 16SM + 16 ST/A	72				ACN=50	Korea (Go-san)	L-9, L-10
Acheilognathus	*typus*	*Pseudoperilampus*	M	44	28 M/SM + 16 ST/A	72			1.9 FD	ACN=50	Japan (Chiba)	O-20, S-141
Rhodeus	*atremius atremius*	*atremius*	F, M	46	4 SM + 42 ST/A	50				ACN=50	Japan (Fukuoka)	O-20
Rhodeus	*fangi*	*atremius fangi*		46	4SM + 42ST	50	92	15 in embryo		ACN=50	China (Fujian)	U-24
Rhodeus	*fangi*	*atremius*	F, M	46	4 M/SM + 42 ST/A	50		2 (6 in embryo)		ACN=50	China (Zhejiang)	U-18
Rhodeus	*atremius suigensis*	*suigensis*	F, M	46	4SM + 42 ST/A	50				ACN=50	Japan (Yoshii R.)	O-20
Rhodeus	*notatus*	*atremius suigensis*		46	4SM + 42ST	50	92	8 in embryo		ACN=50	China (Liaoning)	U-80, U-81
Rhodeus	*notatus*	*suigensis*	F	46	4 M/SM + 42 ST/A	50				ACN=50	Korea (Go-san)	L-9, L-11

Table 6.11 Order CYPRINIFORMES (continued)

A	B	C	D	E	F	G	H	I	J	K	L
Current scientific name of taxon Superfamily/family/subfamily/species	Reported in karyotype paper	Sex	2n	Karyotype	NF_1	NF_2	Ag-NORs	Genome size (pg/cell)	Comments	Locality	Reference
Rhodeus notatus		F, M	46	4SM + 42 ST/A	50				ACN=50	Korea (Go-san)	L-9, L-10
Rhodeus ocellatus		F, M	48	28 M/SM + 20 ST/A	76				ACN=50	Korea (Jang-heung)	L-9, L-11
Rhodeus ocellatus		F, M	48	10M + 24SM + 14ST	82	96			ACN=50	China (Hubei)	H-19, Y-15
Rhodeus ocellatus kurumeus	*ocellatus smithii*	F, M	48	28 M/SM + 20 ST/A	76				ACN=50	Japan (Fukuoka)	O-20
Rhodeus ocellatus kurumeus			48	18M + 10SM + 20ST	76	96	2		ACN=50	Japan (Osaka)	K-22
Rhodeus ocellatus kurumeus			48	14M + 14SM + 20ST	76	96	2		ACN=50	Japan (Yanagawa)	K-22
Rhodeus ocellatus kurumeus			48	8M + 20SM + 20ST	76	96	1-2		ACN=50	Japan (Saga)	S-93
Rhodeus ocellatus ocellatus		F, M	48	8M + 20SM + 20A	76	76	2	(2.2* FCM)	ACN=50	Japan (Kobe, Lake Biwa)	O-20, O-48, T-4, T-11
Rhodeus ocellatus ocellatus		F, M	48	8M + 20 SM/ST + 20A		76			ACN=50	Japan (Osaka)	O-18
Rhodeus ocellatus ocellatus			48	8M + 20SM + 20A	76	76	2 (2 in embryo)		ACN=50	China, Korea (Gyonggi)	U-24
Rhodeus pseudosericeus			48	8M + 20SM + 20ST	76	96			ACN=50	Korea (Gangwon-do)	A-86
Rhodeus sericeus amarus	*amarus*		48	14M + 24 SM/ST + 10A		86		2.1 FD	ACN=50	France	H-2, H-4
Rhodeus sericeus amarus			48	6M + 26SM + 4ST + 12A	80	84			ACN=50	Hungary	B-54
Rhodeus sericeus amarus			48	6M + 26SM + 16A	80				ACN=50	Hungary	M-64
Rhodeus sericeus amarus		F, M	48	34 M/SM + 14 ST/A	82				ACN=50	Bosnia-Herzegovina	S-77
Rhodeus sericeus amarus		F, M	48	14M + 12SM + 12ST + 10A	74	86				Russia (Don R.)	R-22
Rhodeus sericeus sericeus	*sericeus*	F	48							China (Hubei)	H-19
Rhodeus sinensis			48	8M + 20SM + 20ST	76	96	8 (15 in embryo)		ACN=50	China (Amur R.), Korea	U-24
Rhodeus sinensis	*Pseudoperilampus lighti*	F, M	48	12M + 22SM + 8ST + 6A	82	90			ACN=50	China (Sichuan)	Y-15
Rhodeus sinensis	*uyekii*	F, M	48	28 M/SM + 20 ST/A	76				ACN=50	Korea (Go-san)	L-9, L-10
Rhodeus sp.	*sinensis*		48	14M + 28SM + 6 ST/A	90				ACN=50	China (Kunming)	L-38
Rhodeus sp.	*sinensis*	F, M	48	12M + 28SM + 8 ST/A	88				ACN=50	China (Shandong)	W-34
Rhodeus sp.	*sinensis*	M	48							China (Wuhan)	H-19
Tanakia himantegus chii	*himantegus*		48	8M + 20SM + 18ST + 2A	76	94			ACN=50	China (Shanghai)	U-22
Tanakia himantegus chii			48	8M + 20SM + 18ST + 2A	76	94	2-3		ACN=50	China (Shanghai)	U-71
Tanakia himantegus himantegus			48	8M + 20SM + 18ST + 2A	76	94	2-3		ACN=50	Taiwan	U-71
Tanakia himantegus himantegus	*Paracheilognathus himantegus*	F, M	48	10M + 18SM + 6ST + 14A	76	82			ACN=50	(Taiwan)	A-83
Tanakia intermedia intermedia	*Acheilognathus*	F, M	48	24 M/SM + 24 ST/A	72				ACN=50	Korea (Jang-heung)	L-9, L-11
Tanakia intermedia subsp.	*Acheilognathus*	F, M	48	28 M/SM + 20 ST/A	76				ACN=50	Korea (Gong-ju)	L-9, L-11
Tanakia koreensis			48	8M + 20SM + 20ST	76	96	4		ACN=50	Korea (Gyongsangbuk-do)	U-24
Tanakia koreensis	*Acheilognathus limbata*		48	8M + 20SM + 20ST	76	96			ACN=50	Korea (Go-san)	L-9, L-10
Tanakia lanceolata	*Acheilognathus*	F, M	48	10M + 18SM + 20 ST/A	76		2	(2.1* FCM, 2.1FD)	ACN=50	Japan (Lake Biwa, Fukuoka)	O-20, O-48, T-6, T-11
Tanakia lanceolata	*Acheilognathus*		48	8M + 20 SM/ST + 20A		76			ACN=50	Japan (Lake Biwa)	O-18
Tanakia limbata	*Acheilognathus*	F, M	48	10M + 18SM + 20 ST/A	76		2	(2.1* FCM)	ACN=50	Japan (Fukuoka, Hyogo)	O-20, O-48, T-6, T-11
Tanakia limbata	*Acheilognathus*	F, M	48	28 M/SM + 20 ST/A	76		4		ACN=50	Japan	U-13
Tanakia limbata			48	8M + 20SM + 20ST	76	96	1-3		ACN=50	Japan (Mie)	S-93
Tanakia signifer			48	8M + 20SM + 16ST + 4A	76	92	2		ACN=50	Korea (Gangwon-do)	U-24
Tanakia signifer	*Acheilognathus*		48	8M + 12SM + 28ST	68	96			ACN=50	Korea	L-15
Tanakia somjinensis			48	8M + 20SM + 20ST	76	96	2-4		ACN=50	Korea	U-71

Table 6.11 Order CYPRINIFORMES (continued)

A		B	C	D	E	F	G	H	I	J	K	L
Current scientific name of taxon Superfamily/family/subfamily/species		Reported in karyotype paper	Sex	2n	Karyotype	NF_1	NF_2	Ag-NORs	Genome size (pg/cell)	Comments	Locality	Reference
Tanakia	*tanago*		F, M	48	28 M/SM + 20 ST/A	76				ACN=50	Japan (Chiba)	O-20
Barbinae												
Acrossocheilus	*wenchowensis beijiangensis*		F, M	50	14M + 16SM + 14ST + 6A	80	94			ACN=50	China (Guangdong)	G-69, Y-15
Acrossocheilus	*deauratus*			50	10M + 12SM + 28A	72				ACN=50	N.E. Thailand	A-84
Acrossocheilus	*fasciatus*		F, M	50	14M + 16SM + 10ST + 10A	80	90			ACN=50	China (Guangdong)	G-69, Y-15
Acrossocheilus	*hemispinus hemispinus*		F, M	50	10M + 16SM + 8ST + 16A	76	84			ACN=50	China (Guangxi)	Y-15
Acrossocheilus	*iridescens zhujiangensis*		F, M	50	14M + 16SM + 10ST + 10A	80	90			ACN=50	China (Guangdong)	G-69, Y-15
Acrossocheilus	*labiatus*			50	16M + 12SM + 4ST + 18A	78	82			ACN=50	Taiwan	A-78
Acrossocheilus	*parallens*		F, M	50	14M + 16SM + 14ST + 6A	80	94			ACN=50	China (Guangdong)	G-69, Y-15
Acrossocheilus	*yunnanensis*			50	18M + 16SM + 16 ST/A	84			2.5 FD		China (Yunnan)	Z-5, Z-8
Acrossocheilus	*yunnanensis*		F	50	10M + 18SM + 12ST + 10A	78	90			ACN=50	China (Sichuan)	L-43, Y-15
Aulopyge	*huegeli*			100	48 M/SM + 52 ST/A	148				4X, ACN=100	Bosnia	B-20
Balantiocheilos	*melanopterus*			50	10M + 12SM + 28A	72			2.1* FCM		(S.E. Asia)	O-48
Balantiocheilos	*melanopterus*		F, M	50	6M + 18SM + 16ST + 10A	74	90	2		ACN=50	(S.E. Asia)	K-136
Barbodes	*carnaticus*			100	22M + 40SM + 22ST + 16A	162	184	6		4X, ACN=100	India (W. Ghats)	N-55
Barbonymus	*altus*	*Puntius*		50	10M + 24SM + 4ST + 12A	84	88			ACN=50	Thailand (Ayuthaya)	M-9
Barbonymus	*gonionotus*	*Puntius*		50	2M + 20SM + 4ST + 24A	72	76		(2.2 FD)	ACN=50	Thailand (Ayuthaya)	M-9, S-141
Barbonymus	*gonionotus*	*Puntius*	F, M	50	12M + 12SM + 4ST + 22A	74	78			ACN=50	Thailand	W-31
Barbonymus	*schwanenfeldii*	*Puntius*		50	6M + 28 SM/ST + 16A		84		2.2 BFA	ACN=50	(Asia)	T-60, H-13
Barbus	*ablabes*		F	50	18M + 30SM + 2 ST/A	98		2			Guinea, Africa	R-21
Barbus	*amatolicus*			48	22 M/SM + 26 ST/A	70					S. Africa	S-193
Barbus	*andrewi*			100	34 M/SM + 66 ST/A	134				4X, ACN=100	S. Africa	T-55
Barbus	*anema*			50	42 M/SM + 8A	92	92			ACN=50	Ethiopia (Alvero R.)	G-47
Barbus	*anoplus*			48	30 M/SM + 18 ST/A	78		2			S. Africa	S-193
Barbus	*anoplus*			50							S. Africa	T-59
Barbus	*argenteus*			50	44 M/SM + 6 ST/A	94		2			S. Africa	S-193
Barbus	*barbus*			100	12M + 48 SM/ST + 40A		160		3.5 FD	4X, ACN=100	France	H-2, H-4
Barbus	*bigornei*		F, M	48	18M + 30SM	96	96	2		ACN=50	Guinea	R-21
Barbus	*brevipinnis*			48	40 M/SM + 8 ST/A	88		2			S. Africa	S-193
Barbus	*bynni*		F, M	150	70 M/SM + 80 ST/A	220				6X, ACN=150	Ethiopia (L. Abaya)	G-47
Barbus	*bynni bynni*			150	50 M/SM + 100A	200				6X	Ethiopia (Baro R.)	K-108
Barbus	*bynni bynni*			150	70 M/SM + 80A	220				6X	Ethiopia (L. Abaya)	K-108
Barbus	*bynni occidentalis*		F	148						6X	W. Africa, Guinea	G-66
Barbus	*bynni waldroni*			150						6X	W. Africa	G-84
Barbus	*calidus*			100		126				6X	S. Africa	S-193
Barbus	*callensis*	*setivimensis*		100						4X	Africa	G-84
Barbus	*canis*			150	76M + 24ST + 50A	226	250			6X	Israel (Jordan R.)	G-59
Barbus	*cyclolepis*		F, M	100	26M + 16SM + 36ST + 22A	142	178	4		4X	Macedonia	R-24
Barbus	*erubescens*			100						4X	Africa	D-27

Table 6.11 Order CYPRINIFORMES (continued)

A		B	C	D	E	F	G	H	I	J	K	L
Current scientific name of taxon Superfamily/family/subfamily/species		Reported in karyotype paper	Sex	2n	Karyotype	NF_1	NF_2	Ag-NORs	Genome size (pg/cell)	Comments	Locality	Reference
Barbus	*ethiopicus*		M	150	40 M/SM + 110 ST/A	190				6X	Ethiopia (Meki R.)	G-47
Barbus	*eutaenia*			48		82,84					S. Africa	S-193
Barbus	*fasciolatus*	*bariloides*		48	32M + 16SM	96	96			ACN=48	(Africa)	R-4
Barbus	*goktschaicus*		F, M	100	6M + 18SM + 76 ST/A	124				4X	Armenia (Lake Sevan)	K-109
Barbus	*gurneyi*			50	26 M/SM + 24 ST/A	76		2			S. Africa	S-193
Barbus	*holotaenia*		F, M	50	24M + 26 SM/ST		100			ACN=50	(Africa)	R-4
Barbus	*hospes*			96	46 M/SM + 50 ST/A	142		2		4X	S. Africa	S-193
Barbus	*intermedius*			150	90M + 60A	240				6X	Ethiopia	G-47
Barbus	*intermedius*			150	66M + 84A	216				6X	Ethiopia	G-47
Barbus	*issenensis*			100						4X	N. Africa	G-84
Barbus	*kerstenii*		F	50	34 M/SM + 16A	84				ACN=50	Ethiopia (Lake Abaya)	G-47
Barbus	*ksibi*			100						4X	NW. Africa	G-84
Barbus	*kubanicus*	*tauricus kubanicus*		100						4X	(Russia)	V-72
Barbus	*macrops*			50	14M + 28SM + 8 ST/A	92		2		ACN=50	Guinea	R-21
Barbus	*massaensis*			100						4X	N. Africa	G-84
Barbus	*meridionalis*		M	100	22M + 20SM + 12ST + 46A	142	154			4X	Italy	C-34
Barbus	*motebensis*			50	24 M/SM + 26 ST/A	74		2			S. Africa	S-193
Barbus	*moulouyensis*			100						4X	NW. Africa	G-84
Barbus	*nasus*			100						4X	N. Africa	G-84
Barbus	*paludinosus*			50	46 M/SM + 4A	96	96			ACN=50	Ethiopia (Bulbula R.)	G-47
Barbus	*paludinosus*	developed dorsal spine		50	44 M/SM + 6 ST/A	94				ACN=50	Ethiopia (Omo R.)	G-49
Barbus	*paludinosus*	reduced dorsal spine		50	44 M/SM + 6 ST/A	94				ACN=50	Ethiopia (Omo R.)	G-49
Barbus	*paludinosus*			50	30 M/SM + 20 ST/A	80		2			S. Africa	S-193
Barbus	*parawaldroni*			150						6X	W. Africa	G-84
Barbus	*peloponnesius petenyi*	*meridionalis petenyi*	F, M	100	50 M/SM + 50 ST/A	150				4X	Bosnia-Herzegovina	S-76
Barbus	*petitjeani*		F, M	150	36M + 90 SM/ST + 24A		276			6X	W. Africa, Guinea	G-66
Barbus	*plebejus*	*barbus plebejus*	F, M	100	26M + 18SM + 18ST + 38A	144	162			4X	Italy	C-34
Barbus	*pleurogramma*			50	44 M/SM + 6 ST/A	94				ACN=50	Ethiopia (Lake Tana)	G-49
Barbus	*sacratus*			150						6X	W. Africa	G-84
Barbus	*serra*			100						4X	S. Africa	T-55
Barbus	*trevelyani*			100		128				4X	S. Africa	T-55
Barbus	*trevelyani*			96	32 M/SM + 64 ST/A	128		4		4X	S. Africa	S-193
Barbus	*trimaculatus*			48	30 M/SM + 18 ST/A	78		2			S. Africa	S-193
Barbus	*wurtzi*		F, M	150						6X	W. Africa, Guinea	G-84
Barbus	sp. 1			50	44 M/SM + 6A	94	94			ACN=50	Ethiopia (Alvero R.)	G-47
Barbus	sp. 2		F, M	50	44 M/SM + 6A	94	94			ACN=50	Ethiopia (Lake Abaya)	G-47
Barbus	sp. 3		F, M	50	48 M/SM + 2A	98	98			ACN=50	Ethiopia (Hare R.)	G-47
Caecobarbus	*geertsi*		F, M	50	12M + 28SM + 10ST	90	100			ACN=50	Zaire	V-36
Capoeta	*capoeta*			150	24M + 60SM + 14ST + 52A	234	248			6X	Iran (Roudbar)	S-1

Table 6.11 Order CYPRINIFORMES (continued)

A		B	C	D	E	F	G	H	I	J	K	L
Current scientific name of taxon Superfamily/family/subfamily/species		Reported in karyotype paper	Sex	2n	Karyotype	NF_1	NF_2	Ag-NORs	Genome size (pg/cell)	Comments	Locality	Reference
Capoeta	*capoeta*			150	24M + 56SM + 14ST + 56A	230	244			6X	Iran (Golestan Natn. Park)	S-1
Capoeta	*capoeta sevangi*	*Varicorhinus*	M	150	10M + 30SM + 110A	190	190			6X	Armenia (Lake Sevan)	K-109
Capoeta	*capoeta umbla*		F, M	150	86 M/SM + 64 ST/A	236				6X	Turkey (Tigris R.)	K-51
Capoeta	*damascina*			148	78 M/SM + 32ST + 38A	226	258			6X	Jordan	G-59
Capoeta	*trutta*		F, M	150	70 M/SM + 80 ST/A	220				6X	Turkey (Tigris R.)	K-51
Carasobarbus	*canis*	*Barbus*		150	76 M/SM + 24ST + 50A	226	250			6X	Middle East	G-59
Catla	*catla*		F, M	50	8M + 16SM + 14ST + 12A	74	88			ACN=50	India (WB)	M-27
Catla	*catla*			50	4M + 12SM + 34A	66	66	4		ACN=50	India	L-2
Catla	*catla*			50	10M + 16SM + 8ST + 16A	76	84	4		ACN=50	India (near Lucknow)	N-4
Catlocarpio	*siamensis*			98	18M + 54 SM/ST + 26A		170		3.5 FD	4X, ACN=98	(Asia)	S-127
Chagunius	*chagunio*		F	50	16M + 28SM + 6A	94	94			ACN=50	India (Assam)	K-46, C-108
Cyclocheilichthys	*apogon*			50	12M + 8SM + 6ST + 24A	70	76			ACN=50	Thailand (Ayuthaya)	M-9
Cyclocheilichthys	*enoplos*			50	10M + 30SM + 4ST + 6A	90	94	4		ACN=50	Thailand (Uthai Thani)	M-12
Cyprinion	*macrostomus*		F, M	50	6M + 24SM + 12ST + 8A	80	92	4		ACN=50	Turkey	Y-24
Folifer	*brevifilis brevifilis*	*Tor* (*Folifer*)	F, M	50	14M + 14SM + 16ST + 6A	78	94			ACN=50	China (Guangdong)	G-69, Y-15
Folifer	*brevifilis brevifilis*	*Tor* (*Folifer*)		50	14M + 16SM + 20 ST/A	80				ACN=50	China	Z-8
Hypselobarbus	*curmuca*	*Gonoproktopterus*		100	18M + 38SM + 28ST + 16A	156	184	6		4X, ACN=100	India (W. Ghats)	N-55
Hypsibarbus	*wetmorei*	*Puntius darphani*		50	12M + 8SM + 6ST + 24A	70	76			ACN=50	Thailand (Ayuthaya)	M-9
Labeobarbus	*aeneus*	*Barbus*		148	48 M/SM + 100 ST/A	196				6X	S. Africa (Ciskei)	O-2
Labeobarbus	*capensis*	*Barbus*		150	58 M/SM + 92 ST/A	208				6X	S. Africa (S.W. Cape)	O-2
Labeobarbus	*capensis*		M	150	16M + 58SM + 42ST + 34A	224	266			6X	S. Africa (Rondegat R.)	N-67
Labeobarbus	*intermedius*	*Barbus*	F, M	150	66 M/SM + 84 ST/A	216				6X	Ethiopia (L. Tana, Kulfo R.)	G-47, K-108
Labeobarbus	*intermedius*	*Barbus*	F, M	150	90 M/SM + 60 ST/A	240				6X	Ethiopia (Awash R.)	G-47
Labeobarbus	*kimberleyensis*	*Barbus*		148	56 M/SM + 92 ST/A	204				6X	S. Africa	O-2
Labeobarbus	*marequensis*		F, M	150	26M + 44SM + 42ST + 38A	220	262			6X	S. Africa (Marico R.)	N-67
Labeobarbus	*natalensis*	*Barbus*		150	50 M/SM + 100 ST/A	200				6X	S. Africa (Mgeni R.)	O-2
Labeobarbus	*polylepis*	*Barbus*		150	56 M/SM + 94 ST/A	206				6X	S. Africa	O-2
Labeobarbus	*polylepis*		F	150	18M + 60SM + 42ST + 30A	228	270			6X, ACN=150	S. Africa (Elands R.)	N-67
Labeobarbus	*polylepis*	*Barbus*		148	62 M/SM + 86 ST/A	210		4		6X	S. Africa	S-193
Leptobarbus	*hoevenii*			50	10M + 34 SM/ST + 6A		94			ACN=50	(Asia)	S-126
Leptobarbus	*hoevenii*			50	16M + 30SM + 2ST + 2A	96	98				(India, Thailand)	Z-29
Luciobarbus	*bocagei*	*Barbus*	F, M	100	64 M/SM + 36A	164			3.8 FCM	4X	Portugal	C-72, C-75
Luciobarbus	*brachycephalus*	*Barbus*		100	24M + 46 SM/ST + 30A		170			4X	(Caspian Sea)	V-72
Luciobarbus	*comizo*	*Barbus*	F, M	100	12M + 60SM + 28 ST/A	172			2.7 FCM	4X	Portugal	C-72, C-75
Luciobarbus	*microcephalus*	*Barbus*	F, M	100	18M + 50SM + 32 ST/A	168			3.7 FCM	4X	Portugal	C-72, C-75
Luciobarbus	*sclateri*	*Barbus*	F, M	100	10M + 44SM + 46 ST/A	154			3.7 FCM	4X	Portugal	C-72, C-75
Luciobarbus	*steindachneri*	*Barbus*	F, M	100	10M + 48SM + 42 ST/A	158			3.6 FCM	4X	Portugal	C-72, C-75
Mystacoleucus	*marginatus*			50	16M + 10SM + 24A	76	76			ACN=50	C. Thailand	A-84
Neolissochilus	*dukai*	*Barbus*		100	26M + 50 SM + 6ST + 18A	176	182			4X	India (Assam)	K-46

Table 6.11 Order CYPRINIFORMES (continued)

A		B	C	D	E	F	G	H	I	J	K	L
Current scientific name of taxon Superfamily/family/subfamily/species		Reported in karyotype paper	Sex	2n	Karyotype	NF_1	NF_2	Ag-NORs	Genome size (pg/cell)	Comments	Locality	Reference
Neolissochilus	*hexagonolepis*	*Acrossocheilus*		100	16M + 26SM + 58 ST/A	142				4X	Nepal	M-48
Neolissochilus	*hexagonolepis*			100	32M + 16SM + 6ST + 46A	148	154			4X	India (A.P.)	S-173
Neolissochilus	*hexagonolepis*			100	20M + 18SM + 14ST + 48A	138	152	4		4X	India (Meghalaya)	M-160
Neolissochilus	*sumatranus*	*Acrossocheilus*		98	8M + 36SM + 16ST + 38A	142	158			4X	(Asia)	S-123
Neolissochilus	sp.	*Barbus*		100	44M + 12SM + 44A	156	156			4X	India	K-46
Onychostoma	*elongatus*	*Varicorhinus*	F, M	50	12M + 12SM + 14ST + 12A	74	88			ACN=52	China (Guangdong)	G-69, Y-15
Onychostoma	*gerlachi*	*Varicorhinus*	F, M	50	12M + 12SM + 14ST + 12A	74	88			ACN=52	China (Guangdong)	G-69, Y-15
Onychostoma	*sima*	*Varicorhinus simus*	F, M	50	10M + 16SM + 16ST + 8A	76	92			ACN=52	China (Sichuan)	L-43, Y-15
Percocypris	*pingi pingi*			98	42M + 30SM + 26 ST/A	170			4.6 FD	4X, ACN=98	China (Yunnan)	Z-5, Z-8
Percocypris	*pingi regani*			98	40M + 18SM + 40 ST/A	156				4X, ACN=98	China (Yunnan)	Z-8
Poropuntius	*sinensis*	*Barbodes daliensis*		50	10M + 22SM + 18 ST/A	82				ACN=50	China (Yunnan)	Z-5
Poropuntius	*chonglingchungi*	*Barbodes lacustris*		50	12M + 18SM + 20 ST/A	80			2.3 FD	ACN=50	China (Yunnan)	Z-5, Z-8
Probarbus	*jullieni*			98	18M + 28 SM/ST + 52A		144			4X	(Asia)	S-126
Pseudobarbus	*afer*		F, M	100	12M + 40SM + 38ST + 10A	152	190			4X	S. Africa	N-63, T-55
Pseudobarbus	*afer*		F, M	100	12M + 42SM + 36ST + 10A	154	190	3		4X	S. Africa (Blindekloof)	N-63
Pseudobarbus	*asper*		F, M	100	14M + 46SM + 32ST + 8A	160	192			4X	S. Africa (Groot R.)	N-63
Pseudobarbus	*burchelli*	B type	F, M	100	10M + 42SM + 34ST + 14A	152	186			4X	S. Africa (Bainskloof)	N-63
Pseudobarbus	*burchelli*	S type	F, M	100	10M + 40SM + 36ST + 14A	150	186	3		4X, ACN=100	S. Africa (Wolvekloof)	N-63
Pseudobarbus	*burgi*		F, M	100	14M + 36SM + 38ST + 12A	150	188			4X	S. Africa (Wemmers R.)	N-63
Pseudobarbus	*phlegethon*		F, M	100	14M + 40SM + 38ST + 8A	154	192	4		4X	S. Africa (Thee R.)	N-63
Pseudobarbus	*quathlambae*			100						4X	S. Africa	O-2
Pseudobarbus	*tenuis*		F, M	100	14M + 40SM + 32ST + 14A	154	186	4		4X	S. Africa (Grobbelaars R.)	N-63
Puntioplites	*proctozysron*			50	16M + 10SM + 24A	76	76			ACN=50	Thailand (Uthai Thani)	M-12
Puntius	*arulius*			50	6M + 26 SM/ST + 18A		82			ACN=50	(Asia)	T-61
Puntius	*binotatus*			50	8M + 34 SM/ST + 8A		92			ACN=50	(Asia)	T-60
Puntius	*brevis*	*japonicus*	M, F	50	6M + 14SM + 8ST + 22A	70	78				(Asia)	K-27
Puntius	*chola*			50	2M + 4 SM/ST + 44A		56			ACN=50	(Asia)	T-61
Puntius	*chola*			50	2M + 2SM + 46A	54					India (A.P.)	S-173
Puntius	*chola*	*tetrarupagus*	M	50	2M + 2SM + 4ST + 42A	54	58			ACN=50	India (J & K)	T-52
Puntius	*conchonius*		M	50	16M + 24SM + 2ST + 8A	90	92		(2.0* FCM)		India (Simlipal Hills)	K-41, O-48
Puntius	*conchonius*			50	6M + 38 SM/ST + 6A		94		(1.9 BFA)	ACN=50	(Asia)	T-61, H-13
Puntius	*conchonius*			50	4M + 40SM + 6A	94	94				(Asia)	T-12
Puntius	*conchonius*			50					(1.7 FCM)		(Asia)	V-72, V-86
Puntius	*conchonius*		F, M	48	10M + 20SM + 10ST + 8A	78	88			ACN=50	India (J & K)	S-49
Puntius	*cumingi*			50	18M + 26 SM/ST + 6A		94			ACN=50	(Asia)	T-61
Puntius	*denisonii*			50	4M + 20SM + 18ST + 8A	74	92	8		ACN=50	India (Kerala)	N-57
Puntius	*everetti*			50	6M + 30 SM/ST + 14A		86		1.8 FCM, 2.0 BFA	ACN=50	(Asia)	T-60, V-86, H-13
Puntius	*fasciatus*			50	6M + 26 SM/ST + 18A		82			ACN=50	(Asia)	T-60
Puntius	*fasciatus*	*Barbus*	M	50	30 M/SM + 4ST + 16A	80	84		1.5 FD	2B	(Asia)	O-5

Table 6.11 Order CYPRINIFORMES (continued)

A		B	C	D	E	F	G	H	I	J	K	L
Current scientific name of taxon Superfamily/family/subfamily/species		Reported in karyotype paper	Sex	2n	Karyotype	NF_1	NF_2	Ag-NORs	Genome size (pg/cell)	Comments	Locality	Reference
Puntius	*filamentosus*			50	8M + 26 SM/ST + 16A		84			ACN=50	(Asia)	T-61
Puntius	*filamentosus*			50	12M + 16SM + 12ST + 10A	78	90	8		ACN=50	India (Kerala)	N-56
Puntius	*lateristriga*			50	6M + 32 SM/ST + 12A		88			ACN=50	(Asia)	T-60
Puntius	*manipurensis*		F, M	50	22M + 14SM + 6ST + 8A	86	92			ACN=50	India (Manipur)	S-74
Puntius	*melanampyx*			50	12M + 12SM + 14ST + 12A	74	86			ACN=50	India (Tamil Nadu)	K-41
Puntius	*nigrofasciatus*			50	16M + 34 SM/ST		100			ACN=50	(Asia)	T-61
Puntius	*oligolepis*			50	8M + 30 SM/ST + 12A		88			ACN=50	(Asia)	T-60
Puntius	*orphoides*			50	14M + 16SM + 4ST + 16A	80	84			ACN=50	N. Thailand	A-84
Puntius	*orphoides*			50	6M + 36 SM/ST + 8A		92		1.5 FD	ACN=50	(Asia)	T-60, S-127
Puntius	*partipentazona*			50	6M + 34 SM/ST + 10A		90			ACN=50	(Asia)	T-60
Puntius	*pentazona*			50	22M + 26 SM/ST + 2A		98			ACN=50	(Asia)	T-60
Puntius	*sarana*		M	50	12M + 14SM + 12ST + 12A	76	88				India (Haryana)	R-55
Puntius	*sarana subnasutus*			50	12M + 26SM + 8ST + 4A	88	96	8		ACN=50	India (Kerala)	N-57
Puntius	*semifasciolatus*	*Capoeta semifasciolata*	F, M	50	12M + 14SM + 14ST + 10A	76	90			ACN=50	China (Guangdong)	G-69, Y-15
Puntius	*semifasciolatus*			50	8M + 18 SM/ST + 24A		76		(1.7 FCM)	ACN=50	(Asia)	S-136, V-86
Puntius	*sophore*		F	50	2M + 4SM + 44A	56	56			ACN=50	India (Tamil Nadu)	K-41
Puntius	*sophore*		F, M	48	2M + 4SM + 42A	54	54				India (Haryana)	R-51
Puntius	*sophore*		F, M	48	4M + 2ST + 42A	52	54				India (Haryana)	R-75
Puntius	*sophore*	*chrysopterus*	F	48	4M + 6ST + 38A	52	58			ACN=50	India (J & K)	T-52
Puntius	*sophore*	*stigma*	F, M	48	4M + 2SM + 42A	54	54				India (WB)	K-42
Puntius	*sophore*	*stigma*	F, M	48	2M + 2 SM/ST + 44A		52				India (Haryana)	R-46
Puntius	*sophoroides*			50	2M + 2SM + 46A	54	54			ACN=50	Thailand (Ayuthaya)	M-9
Puntius	*stoliczkanus*			50	22M + 22SM + 4ST + 2A	94	98			ACN=50	Thailand (Mae Hong Son)	M-9
Puntius	*tetrazona*	*Barbus*	M	50	34 M/SM + 6ST + 10A	84	90		1.4 FD		(Asia)	O-5
Puntius	*tetrazona*			50					(1.5 FCM)		India	K-102, V-86
Puntius	*tetrazona*			50	6M + 28 SM/ST + 16A		84		2.0 FD, (1.9 BFA)	ACN=50	(Asia)	T-60, S-141, H-13
Puntius	*tetrazona partipentazona*	*partipentazona*		50	6M + 34 SM/ST + 10A		90			ACN=50	(Asia)	T-60
Puntius	*ticto*		F, M	50	20M + 12SM + 10ST + 8A	82	92		(1.8 FCM)		India (Jammu)	S-58, V-86
Puntius	*ticto*			50	28M + 22 SM/ST		100			ACN=50	(Asia)	T-61
Puntius	*ticto*			50	28M + 16SM + 6ST	94	100				India (A.P.)	S-173
Puntius	*titteya*			50	20M + 28 SM/ST + 2A		98		2.4 BFA	ACN=50	(Asia)	T-61, H-13
Sinocyclocheilus	*grahami grahami*			96	22M + 36SM + 38 ST/A	154				4X	China (Kunming)	L-38
Sinocyclocheilus	*grahami tingi*			96	20M + 32SM + 44 ST/A	148				4X	China (Yunnan)	L-38
Sinocyclocheilus	*grahami tingi*			96	18M + 34SM + 44 ST/A	148			4.6 FD	4X	China (Yunnan)	Z-5, Z-7, Z-8
Sinocyclocheilus	*grahami tingi*			96	14M + 34SM + 48ST/A	144					China (Yunnan)	Z-5
Sinocyclocheilus	*maculatus*			96	18M + 32SM + 46 ST/A	146			4.6 FD	4X	China (Yunnan)	Z-5, Z-7, Z-8
Spinibarbus	*hollandi*	*Barbodes caldwelli*	F, M	100	18M + 32SM + 26ST + 24A	150	176			4X, ACN=100	China (Guangdong)	G-68, Y-15
Spinibarbus	*denticulatus denticulatus*	*Barbodes*	F, M	100	18M + 32SM + 26ST + 24A	150	176			4X, ACN=100	China (Guangdong)	G-68, Y-15
Spinibarbus	*sinensis*	*Barbodes*	F, M	100	18M + 32SM + 26ST + 24A	150	176			4X, ACN=100	China (Wuhan)	G-68, Y-15

Table 6.11 Order CYPRINIFORMES (continued)

A	B	C	D	E	F	G	H	I	J	K	L
Current scientific name of taxon Superfamily/family/subfamily/species	Reported in karyotype paper	Sex	2n	Karyotype	NF_1	NF_2	Ag-NORs	Genome size (pg/cell)	Comments	Locality	Reference
Tor chelynoides			100	20M + 30SM + 24ST + 26A	150	174	4		4X, ACN=100	India (Uttarakhand)	S-189, M-160
Tor douronensis			100	22M + 30SM + 48 ST/A	152				4X, ACN=100	China	Z-8
Tor khudree		F, M	100	16M + 28SM + 6ST + 50A	144	150			4X, ACN=100	India (Jammu)	K-36
Tor khudree			100	16M + 16SM + 8ST + 60A	132	140			4X, ACN=100	India (Maharashtra)	L-4
Tor khudree			100	18M + 16SM + 44ST + 22A	134	178	10		4X, ACN=100	India (Maharastra)	K-111
Tor khudree			100	20M + 14SM + 22ST + 44A	134	156	4		4X, ACN=100	India (Karnataka)	M-160
Tor mussullah			100	22M + 24SM + 24ST + 30A	146	170	4		4X, ACN=100	India (Maharastra)	K-111, M-160
Tor progeneius			100	20M + 20SM + 20ST + 40A	140	160	4		4X, ACN=100	India (Assam)	S-189, M-160
Tor putitora		F, M	100	10M + 24SM + 14ST + 52A	134	148			4X, ACN=100	India (U.P.)	K-33
Tor putitora			100	32M + 16ST + 52A	132	148			4X, ACN=100	India (U.P.)	L-4
Tor putitora			100	12M + 18SM + 70 ST/A	130				4X, ACN=100	India (U.P.)	R-71
Tor putitora			100	12M + 22SM + 14ST + 52A	134	148	4		4X, ACN=100	India (Uttarakhand)	S-189, M-160
Tor sinensis			100	18M + 30SM + 52 ST/A	148				4X, ACN=100	China	Z-8
Tor soro			100	24M + 20SM + 6ST + 50A	144	150			4X, ACN=100	Thailand (Kanchana Buri)	M-12
Tor tor	*mosal mahanadicus*		100	44M + 14SM + 42 ST/A	158				4X, ACN=100	India (Orissa)	K-41
Tor tor			100	10M + 30SM + 60 ST/A	140				4X, ACN=100	India (Haryana)	R-71
Tor tor		F, M	100	24M + 24SM + 6ST + 46A	148	154			4X, ACN=100	India (U.P.)	K-36
Tor tor		M	100	26M + 18ST + 56A	126	144			4X, ACN=100	India (U.P.)	L-4
Tor tor			100	20M + 24SM + 24ST + 32A	144	168	8		4X, ACN=100	India (Madhya Pradesh)	S-189, M-160
Varicorhinus beso		F, M	150	66 M/SM + 84 ST/A	216				6X, ACN=150	Ethiopia (Blue Nile, L. Tana)	G-47, K-108
Varicorhinus nelspruitensis			ca. 150						6X	S. Africa	O-2
Cultrinae											
Anabarilius alburnops			48	12M + 24SM + 12ST	84	96			ACN=50	China (Kunming Lake)	Z-3
Anabarilius alburnops			48	14M + 20SM + 14ST	82	96			ACN=50	China (Qilu Lake)	Z-3
Anabarilius andersoni			48	12M + 24SM + 12ST	84	96			ACN=50	China (Kunming)	Z-3
Anabarilius grahami			48	14M + 20SM + 14ST	82	96			ACN=50	China (Yunnan)	Z-3
Anabarilius macrolepis			48	12M + 24SM + 12ST	84	96			ACN=50	China (Yunnan)	Z-3
Ancherythroculter kurematsui		F, M	48	18M + 24SM + 6ST	90	96			ACN=50	China (Sichuan)	Y-15
Ancherythroculter nigrocauda		F, M	48	20M + 24SM + 4ST	92	96			ACN=50	China (Sichuan)	L-32, Y-15
Ancherythroculter wangi		F, M	48	18M + 26SM + 4ST	92	96			ACN=50	China (Sichuan)	L-32, Y-15
Aphyocypris chinensis		M	48	16M + 26SM + 6ST	90	96		(2.3 FD)		Japan (Kyushu)	M-50, S-141
Aphyocypris chinensis		F, M	48	16M + 26SM + 6 ST/A	90					Korea (Yesan-gun)	K-55, L-12
Chanodichthys dabryi	*Erythroculter*	F, M	48	16M + 28SM + 4ST	92	96	7		ACN=50	China (Hubei)	L-40, R-43
Chanodichthys dabryi	*Erythroculter hypselonotus*	F, M	48	16M + 24SM + 8ST	88	96			ACN=50	China (Guangdong)	L-32
Chanodichthys erythropterus	*Culter*	F, M	48	16M + 26SM + 6ST	90	96	6	2.1* FD	ACN=50	China (Hubei)	L-40, R-42, C-83
Chanodichthys erythropterus	*Erythroculter ilishaeformis*	F, M	48	16M + 26SM + 6ST	90	96	6	1.6* FD	ACN=50	China (Hubei)	L-40, R-43, R-105, C-83
Chanodichthys erythropterus	*Erythroculter ilishaeformis*		48	20M + 28 SM/ST		96			ACN=50	(Hubei, China)	L-53
Chanodichthys mongolicus	*Erythroculter*	F, M	48	14M + 28SM + 6ST	90	96		2.0* FD	ACN=50	China (Hubei)	L-40, Y-15, C-83
Chanodichthys mongolicus	*Erythroculter*		48	20M + 28 SM/ST		96			ACN=50	(Hubei, China)	L-53

Table 6.11 Order CYPRINIFORMES (continued)

A		B	C	D	E	F	G	H	I	J	K	L
Current scientific name of taxon Superfamily/family/subfamily/species		Reported in karyotype paper	Sex	2n	Karyotype	NF_1	NF_2	Ag-NORs	Genome size (pg/cell)	Comments	Locality	Reference
Culter	*oxycephaloides*	*Erythroculter*	F, M	48	20M + 24SM + 4ST	92	96			ACN=50	China (Hubei)	L-40
Hemiculter	*bleekeri bleekeri*		M	48	16M + 26SM + 6ST	90	96			ACN=50	China (Hubei)	L-40, Y-15
Hemiculter	*eigenmanni*			48		90				ACN=50	USSR	V-3
Hemiculter	*leucisculus*		F, M	48	16M + 26SM + 6ST	90	96		2.2* FD	ACN=50	China (Hubei, Guangdong)	L-40, Y-15, C-83
Hemiculter	*nigromarginis*		F, M	48	18M + 26SM + 4ST	92	96			ACN=50	China (Sichuan)	L-32, Y-15
Hemiculterella	*sauvagei*		F, M	48	18M + 24SM + 6ST	90	96			ACN=50	China (Sichuan)	L-32, Y-15
Hemiculterella	sp.		F	48							China (Pearl R.)	C-85
Ischikauia	*steenackeri*		F, M	48	10M + 38 SM/ST		96	6	2.6* FCM	ACN=50	Japan (Osaka, Kobe)	T-4, O-48
Megalobrama	*amblycephala*		F, M	48	18M + 26SM + 4ST	92	96	4		ACN=50	China (Hubei)	L-40, R-43, R-105, Z-1
Megalobrama	*amblycephala*		F, M	48	16M + 24SM + 6ST + 2 special				(2.4, 2.0* FD)	ACN=50	China (Hubei)	L-41, L-73, C-83
Megalobrama	*amblycephala*			48	20M + 24SM + 4ST	92	96			ACN=50	China (Hubei)	Z-1
Megalobrama	*amblycephala*			48	26M + 22SM	96	96	4	(2.9* FCM)	ACN=50	China (Shashi)	Z-22, F-5
Megalobrama	*amblycephala*			48	16M + 20SM + 12ST	84	96			ACN=50	China (Beijing)	Y-11
Megalobrama	*terminalis*			48	18M + 22SM + 8ST	88	96			ACN=50	China (Hunan)	L-63
Megalobrama	*terminalis*		F, M	48	14M + 26SM + 8ST	88	96			ACN=50	China (Hubei)	L-40, Y-15
Osteobrama	*belangeri*			50	6M + 16SM + 12ST + 16A	72	84	2-4		ACN=50	India (Manipur)	K-137
Osteobrama	*cotio*	*Rohtee*	F	48	22M + 14SM + 8ST + 4A	84	92			ACN=50	India (WB)	M-27
Osteobrama	*cotio*		F	48	18M + 24SM + 6A	90	90			ACN=50	India (Jammu)	S-56
Osteobrama	*cotio cotio*			48	24M + 14SM + 8ST + 2A	86	94			ACN=50	India (WB)	K-46
Parabramis	*pekinensis*		F, M	48	14M + 26SM + 8ST	88	96	4	1.7* FD	ACN=50	China (Hubei)	L-40, R-43, C-83
Parapelecus	*argenteus*		F, M	48	18M + 22SM + 8ST	88	96			ACN=50	China (Sichuan)	Y-15
Parapelecus	*engraulis*		F, M	48	18M + 24SM + 6ST	90	96			ACN=50	China (Sichuan)	L-32, Y-15
Pseudohemiculter	*dispar*		F, M	48	18M + 24SM + 6ST	90	96			ACN=50	China (Guangdong)	L-32, Y-15
Pseudolaubuca	*sinensis*			48					2.4* FD		China (Wuhan)	C-83
Sinibrama	*changi*		F, M	48	14M + 26SM + 8ST	88	96			ACN=50	China (Sichuan)	L-32
Sinibrama	*macrops*		F, M	48	22M + 20SM + 6ST	90	96			ACN=50	China (Guangxi)	Y-15
Sinibrama	*melrosei*		F, M	48	20M + 24SM + 4ST	92	96			ACN=50	China (Guangdong)	L-32
Cyprininae												
Carassioides	*cantonensis*		F, M	100	18M + 32SM + 18ST + 32A	150	168			4X, ACN=100	China (Guangdong)	G-68, Y-15
Carassius	*auratus*	goldfish	F, M	100	24M + 30SM + 46 ST/A	154				4X, ACN=100	China (Beijing)	W-1
Carassius	*auratus auratus*	goldfish		100	20M + 40 SM/ST + 40A		160		(4.1* FCM)	4X, ACN=100	Japan	K-115, O-48
Carassius	*auratus auratus*	*carassius auratus*, goldfish		100	20M + 28 SM/ST + 52A		148		3.7 FD	4X, ACN=100	France	H-2, H-4
Carassius	*auratus*			100	22M + 30SM + 48 ST/A	152			3.2 FD	XX/XY	China (Yunnan)	Z-2, Z-8
Carassius	*auratus*			104	20M + 84 SM/ST/A					4X, ACN=100	Italy	C-60
Carassius	*auratus auratus*		F, M	100	12M + 36SM + 52 ST/A	148		2		4X, ACN=100	China	O-29
Carassius	*auratus auratus*		F, M	100	22m + 34SM + 22ST + 22A	156	178		(3.5 FCM)	4X, ACN=100	China	Y-14, R-105, V-86
Carassius	*auratus auratus*			100	30M + 34SM + 36 ST/A	164			(3.4 FIA, 4.0 BFA)	4X, ACN=100	China (Amur)	S-62, S-64, H-13, H-41
Carassius	*auratus auratus*			100	52 M/SM + 48 ST/A	152		1-3		4X, ACN=100	Croatia	A-23, M-54
Carassius	*auratus*	var. Dongtingking		100	30M + 20SM + 26ST + 24A	150	176			4X, ACN=100	China (Lake Dongting)	Y-25

Table 6.11 Order CYPRINIFORMES (continued)

A		B	C	D	E	F	G	H	I	J	K	L
Current scientific name of taxon Superfamily/family/subfamily/species		Reported in karyotype paper	Sex	2n	Karyotype	NF_1	NF_2	Ag-NORs	Genome size (pg/cell)	Comments	Locality	Reference
Carassius	*auratus*	back-low type	F, M	100	22M + 30SM + 48 ST/A	152		4	3.2 FD	XX/XY	China (Kunming Lake)	W-10, Z-4, Z-8
Carassius	*auratus*	back-high type		156	30M + 46SM + 80 ST/A	232		6		6X, ACN=150	China (Yunnan)	W-10
Carassius	*auratus*	back-high type		162	36M + 56SM + 70 ST/A	254		6		6X	China (Yunnan, Er hai Lake)	W-11
Carassius	*auratus*	back-high type	F	162	33M + 53SM + 76 ST/A	248			4.6 FD	6X	China (Kunming Lake)	Z-4, Z-8
Carassius	*auratus*	Xiji colour-crucian carp		100	24M + 30SM + 46 ST/A	154				4X, ACN=100	China (Xiji)	W-2
Carassius	*auratus auratus*	Chinese crucian carp		100	12M + 36SM + 52 ST/A	148		2		4X, ACN=100	China	O-30, T-18
Carassius	*auratus auratus*	Hibuna	F	100	12M + 36SM + 52 ST/A	148		2		4X, ACN=100	Japan (Hokkaido)	O-45
Carassius	*auratus auratus*	Hibuna	F	151						3B, 6X	Japan (Hokkaido)	O-45
Carassius	*auratus auratus*	Wakin	F, M	100	20M + 40 SM/ST + 40A		160	2		4X, ACN=100	Japan (Tokyo)	K-115
Carassius	*auratus auratus*	Wakin	F, M	100	12M + 36SM + 52 ST/A	148		2		4X, ACN=100	Japan	O-12, O-30
Carassius	*auratus auratus*	Ryukin	F, M	100	12M + 36SM + 52 ST/A	148				4X, ACN=100	Japan	O-12
Carassius	*auratus auratus*	Comet	F, M	100	12M + 36SM + 52 ST/A	148				XX/XY	Japan	O-26, O-37
Carassius	*auratus auratus*	Shubunkin	F, M	100	12M + 36SM + 52 ST/A	148				XX/XY	Japan	O-26, O-37
Carassius	*auratus auratus*	Kuro-demekin		100	16M + 84 SM/ST/A					4X, ACN=100	Japan (Tokyo)	A-70
Carassius	*auratus auratus*	Ranchuu		100	16M + 42SM + 42 ST/A	158				4X, ACN=100	Japan (Tokyo)	A-70
Carassius	*auratus auratus*	Sanshiki-demekin		100	16M + 40SM + 44 ST/A	156				4X, ACN=100	Japan (Tokyo)	A-70
Carassius	*auratus auratus*	Tetsugyo	F, M	100						4X, ACN=100	Japan (Niigata)	S-156
Carassius	*auratus auratus*	Tetsugyo	F	153						6X, ACN=150	Japan (Niigata)	S-156
Carassius	*auratus buergeri*	Nagabuna	F	100	12M + 36SM + 52 ST/A	148		2		4X, ACN=100	Japan (Lake Suwa)	O-30
Carassius	*auratus gibelio*		F	162	34M + 58SM + 42ST + 28A	254				6X, ACN=150	China	Y-14, Y-15
Carassius	*auratus gibelio*			150					(5.5* FD)	6X	China (Wuhan)	C-83
Carassius	*auratus gibelio*		F, M	156	42M + 74SM + 40 ST/A	272				6X, ACN=150	China (Amur)	S-62
Carassius	*auratus gibelio*		F, M	156	44M + 64SM + 48 ST/A	264				6X, ACN=150	China (Amur, Mishan)	S-64
Carassius	*auratus gibelio*		F, M	156	42M + 74SM + 40 ST/A	272				6X, ACN=150	China (Amur, Fangzheng)	S-64
Carassius	*auratus gibelio*		F, M	162	48M + 56SM + 18ST + 40A	266	284		(6.6* FCM)	6X	China (Amur, Fangzheng)	S-46, F-5
Carassius	*auratus gibelio*		F	156	50M + 64SM + 42 ST/A	270				6X, ACN=150	China (Amur, Dedou)	S-64
Carassius	*auratus gibelio*			162	32M + 52SM + 78 ST/A	246		4		6X	China (Amur, Fangzheng)	W-10
Carassius	*auratus gibelio*		M	100	20M + 40SM + 40 ST/A	160				4X, ACN=100	Russia (Amur R.)	K-85, K-94
Carassius	*auratus gibelio*		F	156	34M + 62SM + 60 ST/A	252				6X, ACN=150	Russia (Amur R.)	K-85, K-94
Carassius	*auratus gibelio*		F	150						6X, ACN=150	Bosnia	S-80
Carassius	*auratus gibelio*			98	48 M/SM/ST + 50A		146			4X, ACN=100	Rumania (Danube plaine)	R-36
Carassius	*auratus gibelio*		F	160	46M + 82 SM/ST + 32A		288		(5.5 FCM)	6X	Czeck	P-22, G-85
Carassius	*auratus gibelio*		F, M	100	14M + 24SM + 10ST + 52A	138	148	1-4	(3.7 FCM)	4X, ACN=100	Poland	B-50, G-85
Carassius	*auratus gibelio*		F	150	26M + 50SM + 20ST + 54A	226	246	2-8		6X, ACN=150	Poland	B-50
Carassius	*auratus gibelio*	clone A		150	36M + 54SM + 36ST + 24A	240	276			6B, 6X	N. China	Z-36
Carassius	*auratus gibelio*	clone P		150	36M + 54SM + 36ST + 24A	240	276			6B, 6X	China (Jiangxi)	Z-36
Carassius	*auratus gibelio*	clone D		156	42M + 54SM + 36ST + 24A	252	288			6B, 6X	N. China	Z-36
Carassius	*auratus grandoculis*	Nigorobuna	F, M	100	12M + 36SM + 52 ST/A	148		2	2.7-4.1 FD	4X, ACN=100	Japan (Lake Biwa)	M-96, O-30, S-141
Carassius	*auratus grandoculis*	Nigorobuna	F, M	100	20M + 40SM + 40 ST/A	160				4X, ACN=100	Japan (Lake Biwa)	K-86

Table 6.11 Order CYPRINIFORMES (continued)

A		B	C	D	E	F	G	H	I	J	K	L
Current scientific name of taxon Superfamily/family/subfamily/species		Reported in karyotype paper	Sex	2n	Karyotype	NF_1	NF_2	Ag-NORs	Genome size (pg/cell)	Comments	Locality	Reference
Carassius	*auratus langsdorfii*	Ginbuna	F	150	28M + 56SM + 66 ST/A	234		8	(5.4* FCM)	6B, 6X	Japan (Lake Biwa)	O-30, O-48
Carassius	*auratus langsdorfii*	Ginbuna	F	150	18M + 66SM + 66 ST/A	234				4B, 6X	Japan (Lake Biwa)	T-15
Carassius	*auratus langsdorfii*	Ginbuna	F, M	100	20M + 40 SM/ST + 40A		160		(4.1* FCM, 3.4 FD)	4X, ACN=100	Japan, Taiwan, Russia	K-94, K-115, O-48, S-141
Carassius	*auratus langsdorfii*	Ginbuna	F	156	34M + 62 SM/ST + 60A		252			6X, ACN=150	Japan, Taiwan, Russia	K-94, K-115
Carassius	*auratus langsdorfii*	Ginbuna	F	206	44M + 82 SM/ST + 80A		332			8X, ACN=200	Japan (Kanto District)	K-94, K-115
Carassius	*auratus langsdorfii*	Ginbuna	F, M	100	20M + 40SM + 40A	160	160			4X, ACN=100	Japan (Hokkaido, Honshu)	M-96
Carassius	*auratus langsdorfii*	Ginbuna	F, M	ca. 150						6X, ACN=150	Japan (Hokkaido, Honshu)	M-96
Carassius	*auratus buergeri*	Ookinbuna		100	12M + 36SM + 52 ST/A	148				4X, ACN=100	Japan (Okayama)	T-15
Carassius	*auratus* subsp. 2	Kinbuna		100	12M + 36SM + 52 ST/A	148		2	3.8* FCM	XX/XY	Japan (Lake Jonuma)	O-30, O-37, O-48, U-1
Carassius	*auratus* subsp. 2	Kinbuna	F, M	100	20M + 40 SM/ST + 40A		160	2		4X, ACN=100	Japan (Ibaraki)	K-115
Carassius	*auratus* subsp. 1	*buergeri*, Nagabuna	F, M	100	20M + 40 SM/ST + 40A		160			4X, ACN=100	Japan (Lake Suwa)	K-86, K-117
Carassius	*auratus* subsp. 1	*buergeri*, Nagabuna	F	156	34M + 62 MS/ST + 60A		252			6X, ACN=150	Japan (Hokkaido)	K-86
Carassius	*auratus* subsp.		M	100	12M + 36SM + 52 ST/A	148				4X, ACN=100	Japan (Ishigaki Is.)	T-13
Carassius	*auratus*	white crucian carp	F	100	20M + 28SM + 38ST + 14A	148	186			4X, ACN=100	China (Jiangsu)	Z-18
Carassius	*carassius*		F, M	100	20M + 40SM/ST + 40A		160	1-3		4X, ACN=100	Netherlands	K-115, M-54
Carassius	*carassius*			100	48 M/SM + 52 ST/A	148			4.3 FCM	4X, ACN=100	Russia	V-72, V-86
Carassius	*carassius*			50	20M + 12SM + 10ST + 8A	82	92			2X	Rumania (Danube plaine)	R-36
Carassius	*carassius*			100	52 M/SM + 48 ST/A	152				4X, ACN=100	Bosnia	S-80
Carassius	*carassius*			100	20M + 44 SM/ST + 36A		164		3.8 FD	4X, ACN=100	France	H-2, H-4
Carassius	*cuvieri*	Gengorobuna	F, M	100	12M + 36SM + 52 ST/A	148		2	4.4* FCM, 4.8 FD	4X, ACN=100	Japan (Lake Biwa)	O-12, O-30, O-48, S-141
Carassius	*cuvieri*	*auratus cuvieri*	F, M	100	20M + 40SM + 40A	160	160			4X, ACN=100	Japan (Lake Biwa)	M-96
Cyprinus	*carpio*			104					(3.4 FCM), 3.4 BFA	4X, ACN=100		T-73, H-13
Cyprinus	*carpio*		F, M	100	12M + 36SM + 52 ST/A	148		2	4.1* FCM	4X, ACN=100	Japan	T-14, O-13, O-48
Cyprinus	*carpio*			100	20M + 32 SM/ST + 48A		152		3.6 FD	4X, ACN=100	France	H-2
Cyprinus	*carpio*	mirror carp		100	34 M/SM + 30ST + 36A	134	164	2		4X, ACN=100	Germany	A-51
Cyprinus	*carpio*			98	50 M/SM + 48 ST/A	148				4X, ACN=100	former Yugoslavia	A-27, A-28
Cyprinus	*carpio*		F, M	100	24M + 24 SM/ST + 52A		148			4X, ACN=100	Rumania, Hungary, Ukraina	R-33
Cyprinus	*carpio*			98	54 M/SM + 44 ST/A	152				4X, ACN=100	Croatia	A-24
Cyprinus	*carpio*			100					2.9*, 3.5 FD		China (Hubei)	L-41, C-83
Cyprinus	*carpio carpio*		F, M	100	22M + 34SM + 22ST + 22A	156	178			4X, ACN=100	China	Y-14, R-105
Cyprinus	*carpio carpio*	Hebao carp		100	28M + 22SM + 50 ST/A	150		2	4.1* FCM	4X, ACN=100	China (Jiangxi)	W-9, F-5
Cyprinus	*carpio carpio*	Xinguo red carp		100	28M + 22SM + 50 ST/A	150		2	4.1* FCM	4X, ACN=100	China (Jiangxi)	W-9, F-5
Cyprinus	*carpio carpio*	Pili carp		100	28M + 22SM + 50 ST/A	150		2		4X, ACN=100	China (Jiangxi)	W-9
Cyprinus	*carpio carpio*	Blue carp		100	28M + 22SM + 50 ST/A	150		2		4X, ACN=100	China (Jiangxi)	W-9
Cyprinus	*carpio*	Scattered mirror carp	M	100	20M + 26SM + 30ST + 24A	146	176		4.1* FCM	4X, ACN=100	China (Jiangsu)	Z-18, F-5
Cyprinus	*carpio*	Ogon	F	99	12M + 35 SM + 52 ST/A	146			ACN=100	4X, ACN=100	Japan	O-35
Cyprinus	*carpio*	Ogon	M	99	12M + 35 SM + 52 ST/A	146			1B, ACN=100	4X, ACN=100	Japan	O-35
Cyprinus	*carpio*	Kohaku	F, M	99	12M + 35 SM + 52 ST/A	146			1B, ACN=100	4X, ACN=100	Japan	O-35
Cyprinus	*carpio*	Kohaku	M	99	12M + 35 SM + 52 ST/A	146			2B, ACN=100	4X, ACN=100	Japan	O-35

Table 6.11 Order CYPRINIFORMES (continued)

A		B	C	D	E	F	G	H	I	J	K	L
Current scientific name of taxon Superfamily/family/subfamily/species		Reported in karyotype paper	Sex	2n	Karyotype	NF_1	NF_2	Ag-NORs	Genome size (pg/cell)	Comments	Locality	Reference
Cyprinus	*carpio*	Sanshoku	F	99	12M + 35 SM + 52 ST/A	146			1B, ACN=100	4X, ACN=100	Japan	O-35
Cyprinus	*carpio*	Sanshoku	M	99	12M + 35 SM + 52 ST/A	146			2-3 B, ACN=100	4X, ACN=100	Japan	O-35
Cyprinus	*carpio chilia*			100	22M + 30SM + 48 ST/A	152		2	3.3 FD	4X, ACN=100	China (Yunnan)	W-9, Z-3, Z-8
Cyprinus	*carpio chilia*			150	28M + 52SM + 70 ST/A	230				6X, ACN=150	China (Yunnan)	Z-8
Cyprinus	*carpio communis*		M	100	24M + 24SM + 52A	148				4X, ACN=100	India (WB)	K-42
Cyprinus	*carpio haematopterus*		M	100	20M + 30SM + 50 ST/A	150				XY, 4X	China	Z-8
Cyprinus	*carpio haematopterus*		F, M	100	22M + 34SM + 22ST + 22A	156	178			4X, ACN=100	China (Wuhan)	Y-15
Cyprinus	*carpio haematopterus*		F, M	100	28M + 38SM + 22ST + 12A	166	188	2		4X, ACN=100	USSR (Amur R.)	R-14
Cyprinus	*carpio rubrofuscus*			100	22M + 30SM + 48 ST/A	152			3.4 FD	XX/XY, 4X	China	Z-2, Z-8
Cyprinus	*carpio rubrofuscus*			100	26M + 36SM + 38 ST/A	162				4X, ACN=100	China	S-67
Cyprinus	*daliensis*	*yunnanensis daliensis*		100	22M + 30SM + 48 ST/A	152			3.1 FD	XX/XY, 4X	China (Yunnan)	Z-7, Z-8
Cyprinus	*longipectoralis*			100	22M + 30SM + 48 ST/A	152		2	3.7 FD	XX/XY, 4X	China (Yunnan)	W-9, Z-3, Z-8
Cyprinus	*megalophthalmus*			100	22M + 30SM + 48 ST/A	152				XX/XY, 4X	China (Yunnan)	Z-3
Cyprinus	*micristius fuxianensis*			100	22M + 30SM + 48 ST/A	152				XX/XY, 4X	China (Yunnan)	Z-3
Cyprinus	*pellegrini barbatus*			100	22M + 30SM + 48 ST/A	152			3.5 FD	XX/XY, 4X	China (Yunnan)	Z-3, Z-8
Cyprinus	*pellegrini pellegrini*			100	22M + 30SM + 48 ST/A	152				XX/XY, 4X	China (Yunnan)	Z-7, Z-8
Procypris	*rabaudi*		F, M	100	22M + 26SM + 22ST + 30A	148	170			4X, ACN=100	China (Nanchong)	Y-14, Y-15
Gobiobotinae												
Gobiobotia	*abbreviata*		F, M	50	22M + 22SM + 6ST	94	100	2		ACN=50	China (Sichuan)	L-45, Y-15, R-43
Gobiobotia	*boulengeri*		M	50	24M + 14SM + 12 ST/A	88		2		ACN=50	China (Sichuan)	L-38, R-43
Gobiobotia	*boulengeri*		F, M	50	14M + 26SM + 10ST	90	100			ACN=52	China (Sichuan)	L-45, Y-14, Y-15
Gobiobotia	*brevibarba*		M	50	12M + 30 SM/ST + 8A		92			ACN=50	Korea (Han R.)	U-33
Gobiobotia	*ichangensis*			50	32M + 12SM + 6 ST/A	94				ACN=50	China (Sichuan)	L-38
Gobiobotia	*longibarba meridionalis*		F, M	50	22M + 18SM + 10ST	90	100			ACN=50	China (Guangdong)	L-45, Y-15
Microphysogobio	*longidorsalis*		M	50	18M + 32 SM/ST		100			ACN=50	Korea (Han R.)	U-33
Microphysogobio	*longidorsalis*			50	18M + 32SM	100	100			ACN=50	Korea	L-15
Microphysogobio	*yaluensis*		M	50	18M + 32 SM/ST		100			ACN=50	Korea (Geum R.)	U-33
Microphysogobio	*yaluensis*			50	16M + 32SM + 2ST	98	100			ACN=50	Korea	L-15
Gobioninae												
Abbottina	*kiatingensis*		M	50							China (Pearl R.)	C-85
Abbottina	*labeoides*		M	50	24M + 24SM + 2ST	98	100			ACN=52	China (Guangxi, Guangdong)	Y-15
Abbottina	*rivularis*		F	50	24M + 24SM + 2ST	98	100		2.7* FD	ACN=50	China (Hubei)	L-29, Y-15, C-83
Abbottina	*rivularis*		F, M	50	22M + 24SM + 4ST	96	100			ACN=50	China (Amur)	W-14
Abbottina	*rivularis*			50	12M + 32SM + 6ST	94	100			ACN=50	Korea	L-15
Belligobio	*nummifer*		F, M	50	18M + 20SM + 10ST + 2A	88	98			ACN=52	China (Sichuan)	L-43, Y-15
Coreius	*guichenoti*		F, M	50	16M + 22SM + 10ST + 2A	88	98			ACN=50	China (Wuhan)	L-29, Y-15
Coreius	*heterodon*		F, M	50	16M + 22SM + 10ST + 2A	88	98			ACN=50	China (Hubei)	L-29, Y-15
Coreoleuciscus	*splendidus*		F, M	50	12M + 30SM + 8 ST/A	92				ACN=50	Korea (Ko-san)	L-77
Coreoleuciscus	*splendidus*		M	50	14M + 30 SM/ST + 6A		94			ACN=50	Korea (Han R.)	U-33

Table 6.11 Order CYPRINIFORMES (continued)

A		B	C	D	E	F	G	H	I	J	K	L
Current scientific name of taxon Superfamily/family/subfamily/species		Reported in karyotype paper	Sex	2n	Karyotype	NF_1	NF_2	Ag-NORs	Genome size (pg/cell)	Comments	Locality	Reference
Coreoleuciscus	*splendidus*		F, M	100	36M + 64 SM/ST		200			4X, ACN=100	Korea (Geum R.)	K-10
Gnathopogon	*caerulescens*			50	12M + 24SM + 14 ST/A	86		4		ACN=50	Japan (Lake Biwa)	T-17
Gnathopogon	*caerulescens*		F, M	50	14M + 36 SM/ST		100	4		ACN=50	Japan (Lake Biwa)	U-40
Gnathopogon	*caerulescens*	*elongatus caerulescens*		50	12M + 26SM + 12 ST?A	88		2		ACN=50	Japan (Kobe, Lake Biwa)	T-4
Gnathopogon	*elongatus*			50	12M + 24SM + 14 ST/A	86		4	2.4* FCM	ACN=50	Japan (Kobe)	T-17, O-48
Gnathopogon	*elongatus*		F, M	50	14M + 36 SM/ST		100	4		ACN=50	Japan (Shiga)	U-40
Gnathopogon	*elongatus elongatus*		F, M	50	12M + 32 SM/ST + 6A		94			ACN=50	Japan (Gifu)	O-18
Gnathopogon	*imberbis*		F, M	50	22M + 24SM + 4ST	96	100			ACN=52	China (Sichuan)	L-43, Y-15
Gnathopogon	*strigatus*		F, M	50	14M + 30SM + 6 ST/A	94				ACN=50	Korea (Ko-san)	L-77
Gobio	*delyamurei*	*tauricus*		50	22M + 22SM + 6 ST/A	94				ACN=50	Ukraine (Crimea)	V-82
Gobio	*gobio*			50	22M + 26 SM/ST + 2A		98		2.8 FD	ACN=50	France	H-2, H-4
Gobio	*gobio*			50	22M + 24SM + 2ST + 2A	96	98		(3.3 FCM)	ACN=50	Rumania (Bucharest)	R-34, C-75
Gobio	*gobio*		F, M	50	38 M/SM + 12 ST/A	88				ACN=50	former Yugoslavia	S-78
Gobio	*gobio*			50	20M + 26SM + 2ST + 2A	96	98			ACN=50	former Yugoslavia	V-83
Gobio	*gobio*			50	24M + 24SM + 2 ST/A	98				ACN=50	Russia (Volga basin)	V-83
Gobio	*kubanicus*			50	20M + 28SM + 2 ST/A	98				ACN=52	Russia (Kuban R.)	V-83
Gobio	*soldatovi*		F, M	50	18M + 26SM + 4ST + 2A	94	98			ACN=50	China (Amur)	W-14
Gobio	*tenuicorpus*		F, M	50	18M + 26SM + 4ST + 2A	94	98			ACN=50	China (Amur)	W-14
Hemibarbus	*barbus*			50	16M + 26 SM + 8 ST/A	92		4	(2.5* FCM)	ACN=50	Japan (Kobe)	T-17, O-48
Hemibarbus	*barbus*		F, M	50	14M + 28 SM/ST + 8A		92			ACN=50	Japan (Lake Biwa)	O-18
Hemibarbus	*labeo*		F, M	50	18M + 22 SM/ST + 10A		90		2.1 FD	ACN=50	Korea (Han R.)	K-4
Hemibarbus	*labeo*		F, M	50	16M + 16SM + 14ST + 4A	82	96		2.1* FD	ACN=50	China (Guangdong)	L-29, Y-15, C-83
Hemibarbus	*labeo*			50							China (Amur)	W-14
Hemibarbus	*longirostris*		F, M	50	18M + 18SM + 10ST + 4A	86	96			ACN=50	China (Guangdong)	L-29, Y-15
Hemibarbus	*longirostris*		M	50	14M + 28SM + 8 ST/A	92				ACN=50	Korea (Ko-san)	L-77
Hemibarbus	*longirostris*		F, M	50	14M + 28 SM/ST + 8A		92		2.2 FD	ACN=50	Korea (Han R.)	K-4
Hemibarbus	*longirostris*		F, M	50	16M + 28 SM/ST + 6A		94			ACN=50	Korea (Geum R.)	U-33
Hemibarbus	*maculatus*		F, M	50	16M + 14SM + 16ST + 4A	80	96	2	2.1* FD	ACN=50	China (Wuhan)	L-29, R-43, C-83
Hemibarbus	*maculatus*			50							China (Amur)	W-14
Hemibarbus	*mylodon*	*Gonoproktopterus*		50	12M + 28 SM/ST + 10A		90			ACN=50	Korea (Han R.)	U-33
Huigobio	*chenhsienensis*		F, M	50	24M + 24SM + 2ST	98	100			ACN=50	China (Guangdong)	H-22, Y-15
Paracanthobrama	*guichenoti*		F, M	50	18M + 20SM + 10ST + 2A	88	98	4	2.6 FD	ACN=50	China (Hubei)	L-29, L-41, R-43, Z-25
Paracanthobrama	*umbrifer*		F, M	50	20M + 12SM + 4ST + 14A	82	86			ACN=50	China (Guilin)	Y-15
Paraleucogobio	*strigatus*		F, M	50	14M + 16SM + 20ST	80	100			ACN=52	China (Amur)	W-14
Paraleucogobio	*strigatus*			50	14M + 30SM + 6 ST/A	94				ACN=50	Korea	L-77
Platysmacheilus	*exiguus*		F, M	50	24M + 14SM + 12ST	88	100			ACN=50	China (Guangxi)	Y-15
Pseudogobio	*esocinus*		F, M	50	16M + 28 SM/ST + 6A		94	2	3.1* FCM, 3.3 FD	ACN=50	Japan (Lake Biwa, Kobe)	T-4, O-18, O-48, S-141
Pseudogobio	*esocinus*		F, M	50	16M + 28SM + 6A	94				ACN=52	Korea (Ko-san)	L-77
Pseudogobio	*vaillanti guilinensis*		F, M	50	18M + 18SM + 12ST + 2A	86	98	2		ACN=50	China (Guilin)	L-43, Y-15

Table 6.11 Order CYPRINIFORMES (continued)

A	B	C	D	E	F	G	H	I	J	K	L
Current scientific name of taxon Superfamily/family/subfamily/species	Reported in karyotype paper	Sex	2n	Karyotype	NF_1	NF_2	Ag-NORs	Genome size (pg/cell)	Comments	Locality	Reference
Pseudogobio vaillanti vaillanti		F, M	50	18M + 22SM + 8ST + 2A	90	98			ACN=52	China (Guangdong)	H-22, Y-15
Pseudopungtungia nigra		F, M	50	14M + 36 SM/ST		100			ACN=50	Korea (Kum R.)	K-59
Pseudopungtungia tenuicorpus		M	50	20M + 30 SM/ST		100			ACN=50	Korea (Han R.)	K-59
Pseudorasbora parva			50	16M + 22SM + 12 ST/A	88		4	3.1* FCM	ACN=50	Japan (Kobe)	T-17, O-48
Pseudorasbora parva		F, M	50	14M + 36 SM/ST		100			ACN=50	Japan (Lake Biwa)	O-18
Pseudorasbora parva			50	14M + 36SM	100	100			ACN=50	Korea	L-15
Pseudorasbora parva		F, M	50	18M + 22SM + 10ST	90	100	2	2.6* FD	ACN=50	China (Hubei)	L-29, R-43, C-83
Pseudorasbora parva			50					2.5 FD		China (Hubei)	L-41
Pseudorasbora parva			50	20M + 26SM + 4ST	96	100			ACN=50	China (Kunming)	L-38
Pseudorasbora parva		F, M	50	18M + 22SM + 10 ST/A	90				ACN=50	China (Amur)	W-14
Pseudorasbora pumila pumila			50	26M + 20SM + 4ST	96	100			ACN=50	Japan (Akita)	A-78
Pseudorasbora pumila subsp.		F, M	50	14M + 36 SM/ST		100			ACN=50	Japan	M-49
Pungtungia helzi			50	18M + 18SM + 14 ST/A	86		4	3.0* FCM	ACN=50	Japan (Kobe)	T-17, O-48
Pungtungia helzi		F, M	50	20M + 30 SM/ST		100			ACN=50	Japan (Lake Biwa)	0-18, K-10
Pungtungia helzi		F, M	50	16M + 26SM + 8 ST/A	92				ACN=50	Korea (Ko-san)	L-77
Pungtungia helzi			50	16M + 22SM + 6ST + 6A	88	94			ACN=50	Korea	L-15
Rhinogobio cylindricus		F, M	50	14M + 22SM + 12ST + 2A	86	98			ACN=50	China (Wuhan)	H-22, Y-15
Rhinogobio typus		F, M	50	14M + 22SM + 12ST + 2A	86	98			ACN=50	China (Wuhan)	H-22, Y-15
Rhinogobio ventralis		F, M	50	12M + 24SM + 12ST + 2A	86	98			ACN=50	China (Sichuan)	L-43, Y-15
Romanogobio banaticus	*Gobio kessleri banaticus*		50	24M + 20SM + 4ST + 2A	94	98			ACN=50	Rumania (Uliuc village)	R-34
Romanogobio kessleri	*Gobio*		50	30M + 18 SM/ST + 2A		98		3.7 FCM	ACN=50	Czech	J-12, G-85
Romanogobio uranoscopus	*Gobio*		52	24M + 12SM + 12ST + 4A	88	100				Rumania (Buda village)	R-34
Romanogobio uranoscopus	*Gobio*	F, M	50	30M + 18SM + 2ST	98	100			ACN=50	Slovakia	R-22
Romanogobio vladykovi	*Gobio albipinnatus vladykovi*		50	28M + 20SM + 2A	98			3.4 FCM	ACN=50	Rumania (Uliuc village)	R-34, G-85
Rostrogobio amurensis			50	18M + 24SM + 6ST + 2A	92	98			ACN=50	China (Amur)	W-14
Sarcocheilichthys biwaensis			50	12M + 30SM + 8A	92				ACN=52	Japan (Lake Biwa)	K-12
Sarcocheilichthys czerskii		M	50	18M + 32 SM/ST		100			ACN=52	Korea (Geum R.)	U-33
Sarcocheilichthys kiangsiensis		M	50	18M + 22SM + 8ST + 2A	90	98			ACN=52	China (Guangdong)	H-22, Y-15
Sarcocheilichthys lacustris		F, M	50	18M + 22SM + 8ST + 2A	90	98			ACN=52	China (Amur)	W-14
Sarcocheilichthys nigripinnis czerskii		F, M	50	18M + 22SM + 10ST	90	100			ACN=52	China (Amur)	W-14
Sarcocheilichthys nigripinnis morii		F, M	50	16M + 26SM + 8 ST/A	92				ACN=52	Korea (Ko-san)	L-77
Sarcocheilichthys nigripinnis nigripinnis		F, M	50	18M + 22SM + 10ST	90	100	2	2.6 FD	ACN=52	China (Hubei)	H-22, R-43, R-105, Z-28
Sarcocheilichthys nigripinnis nigripinnis			50					2.5, 2.8* FD		China (Hubei)	L-41, C-83
Sarcocheilichthys parvus		F, M	50	18M + 22SM + 8ST + 2A	90	98			ACN=52	China (Guangdong)	H-22, Y-15
Sarcocheilichthys sinensis sinensis		F, M	50	18M + 22SM + 8ST + 2A	90	98			ACN=52	China (Wuhan)	H-22, Y-15
Sarcocheilichthys variegatus microoculus	*variegatus*	F, M	50	12M + 30 SM/ST + 8A		92	4	2.8* FCM, 2.4 FD	ACN=52	Japan (Lake Biwa)	T-4, T-12, O-18, O-48
Sarcocheilichthys variegatus variegatus	*variegatus*	M	50	18M + 32 SM/ST		100			ACN=52	Japan (Okayama)	U-33
Saurogobio dabryi		F, M	50	18M + 26SM + 6ST	94	100			ACN=50	China (Wuhan)	H-22, Y-15
Saurogobio dumerili		F, M	50	18M + 26SM + 6ST	94	100			ACN=50	China (Wuhan)	H-22, Y-15

Table 6.11 Order CYPRINIFORMES (continued)

A	B	C	D	E	F	G	H	I	J	K	L
Current scientific name of taxon Superfamily/family/subfamily/species	Reported in karyotype paper	Sex	2n	Karyotype	NF_1	NF_2	Ag-NORs	Genome size (pg/cell)	Comments	Locality	Reference
Saurogobio gymnocheilus		F, M	50	18M + 24SM + 8ST	92	100			ACN=50	China (Wuhan)	H-22, Y-15
Squalidus argentatus	*Gnathopogon*	F, M	50	22M + 26SM + 2ST	98	100			ACN=50	China (Wuhan)	K-29, Y-15
Squalidus biwae			50	16M + 34SM	100	100	2		ACN=50	Japan (Yodo R.)	T-17
Squalidus chankaensis	*Gnathopogon*	F, M	50	22M + 24SM + 4ST	96	100			ACN=52	China (Amur)	W-14
Squalidus chankaensis biwae		F, M	50	20M + 30 SM/ST		100	2		ACN=50	Japan (Lake Biwa)	U-40
Squalidus chankaensis subsp.		F, M	50	20M + 30 SM/ST		100	2		ACN=50	Japan (Hyogo, Okayama)	U-40
Squalidus gracilis gracilis	*gracilis*		50	16M + 34SM	100	100			ACN=50	Japan (Lake Biwa)	K-14
Squalidus gracilis gracilis		F, M	50	20M + 30 SM/ST		100	2	(2.5 FD)	ACN=50	Japan (Shiga, Okayama)	U-40, S-141
Squalidus gracilis majimae		F, M	50	14M + 28SM + 8 ST/A	92				ACN=50	Korea (Ko-san)	L-77
Squalidus japonicus japonicus		F, M	50	20M + 30 SM/ST		100	2		ACN=50	Japan (Lake Biwa)	U-40
Squalidus nitens	*Gnathopogon sihuensis*	F	50	22M + 24SM + 4ST	96	100			ACN=52	China (Hubei)	L-29, Y-15
Hypophthalmichthyinae											
Aristichthys nobilis			48	26M + 20SM + 2ST	94	96	6		ACN=48	China (Beijing)	L-59, R-43
Aristichthys nobilis		F, M	48	18M + 22SM + 8ST	88	96		1.8* FD	ACN=48	China (Wuhan)	R-105, Y-15, C-83
Aristichthys nobilis			48	14M + 24SM + 10 ST/A	86			(2.3* FCM), 2.1 FD	ACN=48	China	Z-2, Z-8, F-5
Aristichthys nobilis	*Hypophthalmichthys*		48	18M + 20SM + 10ST	86	96				Hungary	B-54
Hypophthalmichthys molitrix			48	24M + 16SM + 8ST	88	96			ACN=48	China (Beijing)	L-60
Hypophthalmichthys molitrix			48	14M + 24SM + 10A	86	86		(2.4* FCM), 2.0 FD	ACN=48	China	Z-2, Z-8, F-5
Hypophthalmichthys molitrix		F, M	48	18M + 22SM + 8ST	88	96		1.8*, 2.0 FD	ACN=48	China (Wuhan)	R-105, L-41, C-83
Hypophthalmichthys molitrix	*moritrix*	F, M	48	18M + 22SM + 8ST	88	96			ACN=48	China (Wuhan)	Z-26
Hypophthalmichthys molitrix		F, M	48	20M + 12SM + 6ST + 10A	80	86				India (WB)	M-27
Labeoninae											
Cirrhinus jullieni			50	26M + 14SM + 4ST + 6A	90	94	2		ACN=50	Thailand (Uthai Thani)	M-12
Cirrhinus molitorella		F	50	20M + 26SM + 2ST + 2A	96	98			ACN=50	China (Guangsi, Guangdong)	Z-16
Cirrhinus molitorella		F, M	50	16M + 24SM + 10ST	90	100	4		ACN=50	China (Guangdong)	G-69, R-43, R-105, Y-15
Cirrhinus mrigala			50	10M + 12SM + 10ST + 18A	72	82	2		ACN=50	Thailand (Ayuthaya)	M-12
Cirrhinus mrigala			50	4M + 18SM + 28ST	72	100			ACN=50	India	L-2
Cirrhinus mrigala	*Cirrhina*	F, M	50	8M + 6SM + 14ST + 22A	64	78			ACN=50	India (Haryana)	R-55
Cirrhinus mrigala			50	12M + 18SM + 6ST + 14A	80	86	2		ACN=50	India (near Lucknow)	N-4
Cirrhinus reba	*Cirrhina*		50	6M + 8SM + 14ST + 22A	64	78			ACN=50	India (Haryana)	R-55
Cirrhinus reba		M	48	18M + 20SM + 6ST + 4A	86	92			ACN=50	India	M-27
Crossocheilus latius latius		F	50	12M + 28SM + 10A	90	90			ACN=50	India (Assam)	K-46, C-108
Crossocheilus latius latius		F	50	8M + 12SM + 12ST + 18A	70	82			ACN=50	India (J & K)	T-52
Crossocheilus latius punjabensis			48	12M + 36A	60					India	L-1
Discogobio tetrabarbatus		F, M	50	10M + 18SM + 12ST + 10A	78	90			ACN=50	China (Guangdong)	G-69, R-43, Y-15
Garra cambodgigensis	*taeniata*		50	8M + 18 SM/ST + 24A		76		2.2 BFA	ACN=50	(Asia)	S-141, H-13
Garra dembeensis			50	14M + 18SM + 18A	82	82			ACN=52	Ethiopia	K-107
Garra gotyla gotyla		F	50	14M + 10SM + 10ST + 16A	74	84			ACN=50	India (Jammu)	K-38
Garra gotyla gotyla		F, M	50	14M + 26SM + 10A	90	90			ACN=50	India (Tamilnadu)	K-42

Table 6.11 Order CYPRINIFORMES (continued)

A		B	C	D	E	F	G	H	I	J	K	L
Current scientific name of taxon Superfamily/family/subfamily/species		Reported in karyotype paper	Sex	2n	Karyotype	NF_1	NF_2	Ag-NORs	Genome size (pg/cell)	Comments	Locality	Reference
Garra	*gotyla gotyla*			50	12M + 8SM + 8ST + 22A	70	78			ACN=50	India (A.P.)	S-171
Garra	*kempi*			50	14M + 14SM + 10ST + 12A	78	88			ACN=50	India (A.P.)	S-171
Garra	*lamta*		F, M	50	12M + 24SM + 2ST + 12A	86	88			ZW/ZZ	India (Bihar)	K-41
Garra	*lamta*		M	50	6M + 18SM + 12ST + 14A	74	86			ACN=50	India (Bihar)	K-32
Garra	*lissorhynchus*			50	16M + 16SM + 6ST + 12A	82	88			ACN=50	India (A.P.)	S-171
Garra	*makiensis*			50	14M + 20SM + 16A	84	84			ACN=52	Ethiopia	K-107
Garra	*mullya*			50	18M + 14SM + 10ST + 8A	82	92	2		ACN=50	India (Kerala)	N-73
Garra	*orientalis*		F, M	50	16M + 12SM + 14ST + 8A	78	92			ACN=50	China (Guangdong)	G-69, R-43, Y-15
Garra	*ornata*			50							W.C. Africa	G-84
Garra	*imberba*	*pingi pingi*		50	18M + 20SM + 12 ST/A	88				ACN=50	China (Yunnan)	Z-5
Garra	*imberba*	*pingi pingi*	F, M	50	14M + 20SM + 12ST + 4A	84	96			ACN=50	China (Sichuan)	L-43, Y-15
Garra	*quadrimaculata*			50	16M + 22SM + 12A	88	88			ACN=52	Ethiopia	K-107
Garra	*rufa obtusa*			44–52							Middle East	D-27
Garra	*rufa rufa*			44–52							Middle East	D-27
Garra	*rufa*			44	22M + 20SM + 2A	86					Turky	S-171
Garra	*surendranathanii*			50	14M + 20SM + 8ST + 8A	84	92			ACN=52	India (W. Ghats)	N-72
Garra	*variabilis*			74						3X	Middle East	D-27
Labeo	*alluaudi*			50							W. Africa	G-84
Labeo	*bata*			50	18M + 12SM + 8ST + 12A	80	88	2		ACN=50	India (Allahabad)	J-10
Labeo	*bata*		F, M	50	6M + 18SM + 16ST + 10A	74	90			ACN=50	India (WB)	M-27
Labeo	*behri*			50	12M + 8SM + 2ST + 28A	70	72			ACN=50	N.E. Thailand	A-84
Labeo	*calbasu*			50	10M + 10SM + 14ST + 16A	70	84	2		ACN=50	India (Allahabad)	J-10
Labeo	*calbasu*		F, M	50	6M + 8SM + 22ST + 14A	64	86			ACN=50	India (WB)	M-27
Labeo	*calbasu*		F, M	50	8M + 22 SM/ST + 20A		80	2		ACN=50	India (Haryana)	R-75, R-101
Labeo	*caeruleus*	*cerulaeus*	F, M	48	12M + 6SM + 6ST + 24A	66	72			ACN=48	India (Haryana)	R-55
Labeo	*coubie*			50							Mali, W. Africa	P-18
Labeo	*dero*		M	50	26M + 12SM + 2ST + 10A	88	90			ACN=50	India (Assam)	K-43
Labeo	*dero*		F, M	48	12M + 16SM + 20 ST/A	76				ACN=50	India (U.P.)	R-73
Labeo	*diplostomus*		F	50	10M + 6SM + 8ST + 26A	66	74			ACN=50	India (Jammu)	K-38
Labeo	*dussumieri*			50	12M + 12SM + 10ST + 16A	74	84	4		ACN=50	India (Kerala)	N-56
Labeo	*gonius*			54	54A	54		2			India	L-1
Labeo	*pangusia*		M	50	6M + 12SM + 16ST + 16A	68	84			ACN=50	India (J & K)	T-52
Labeo	*parvus*			50							WC. Africa	G-84
Labeo	*parvus*	*obscurus*		50							WC. Africa	G-84
Labeo	*rohita*			50	14M + 16SM + 8ST + 12A	80	88	2		ACN=50	Thailand (Ayuthaya)	M-12
Labeo	*rohita*			50	10M + 14SM + 6ST + 20A	74	80	2		ACN=50	India (Allahabad)	J-10
Labeo	*rohita*			50	10M + 20SM + 20A	80	80			ACN=50	India	L-2
Labeo	*rohita*			50	8M + 12SM + 6ST + 24A	70	76			ACN=50	India (Haryana)	R-67
Labeo	*rohita*			50	10M + 14SM + 8ST + 18A	74	82	4		ACN=50	India (near Lucknow)	N-4

Table 6.11 Order CYPRINIFORMES (continued)

A		B	C	D	E	F	G	H	I	J	K	L
Current scientific name of taxon Superfamily/family/subfamily/species		Reported in karyotype paper	Sex	2n	Karyotype	NF_1	NF_2	Ag-NORs	Genome size (pg/cell)	Comments	Locality	Reference
Labeo	*rohita*		F, M	50	10M + 16SM + 12ST + 12A	76	88			ACN=50	China (Guangdong)	G-69, R-43, Y-15
Labeo	*roseopunctatus*			50							Mali, W. Africa	P-18
Labeo	*rouaneti*			50							Guinea	G-84
Labeo	*senegalensis*			50							Mali, W. Africa	P-18
Labiobarbus	*lineatus*			50	20M + 10SM + 20A	80	80			ACN=50	Thailand (Ayuthaya)	M-10
Morulius	*chrysophekadion*	*Labeo*		50	14 M/SM + 18ST + 18A	64	82		2.8 FD, 2.2 FIA, 2.4 BFA	ACN=50	(Asia)	M-91, H-13, H-41
Osteochilichthys	*longidorsalis*			100	24M + 26SM + 22ST + 28A	150	172	2		4X, ACN=100	India (W. Ghats)	N-55
Osteochilus	*hasselti*			50	30M + 14SM + 6ST	94	100			ACN=50	Thailand (Ayuthaya)	M-10
Osteochilus	*vittatus*			50	16M + 30SM + 4ST	96	100			ACN=50	Thailand (Ayuthaya)	M-10
Osteochilus	*waandersi*			50	18M + 24SM + 4ST + 4A	92	96	2		ACN=50	Thailand (Kanchana Buri)	M-12
Parasinilabeo	*assimilis*		F, M	50	16M + 12SM + 18ST + 4A	78	96			ACN=50	China (Guangdong)	G-69, Y-15
Semilabeo	*notabilis*		F, M	50	8M + 10SM + 12ST + 20A	68	80			ACN=50	China (Guangdong)	G-69, Y-15
Semilabeo	*prochilus*		F	50	16M + 18SM + 14ST + 2A	84	98			ACN=52	China (Sichuan)	L-43, Y-15
Semilabeo	*prochilus*		F	50	16M + 20SM + 14ST	86	100	7		ACN=50	China (Yunnan)	W-12
Sinilabeo	*decoratus decoratus*	*decorus decorus*	F, M	50	10M + 18SM + 10ST + 12A	78	88			ACN=50	China (Guangdong)	G-69, Y-15
Sinilabeo	*decoratus tungting*		M	50	12M + 16SM + 10ST + 12A	78	88			ACN=50	China (Hunan)	Z-16
Sinilabeo	*rendahli rendahli*		F, M	50	10M + 14SM + 18ST + 8A	74	92			ACN=52	China (Sichuan)	L-43, Y-15
Leuciscinae												
Eurasian Leuciscinae												
Abramis	*brama*			50	16M + 14SM + 12ST + 8A	80	92		2.7 FCM	ACN=50	Russia	A-118, V-86
Abramis	*brama*		M	52	30 M/SM + 22 ST/A	82				ACN=52	Sweden	N-50
Abramis	*brama*		F, M	50	16M + 18SM + 16A	84	84	2		ACN=50	Poland	J-2
Abramis	*brama*			50	12M + 18 SM/ST + 20A		80		2.5 FD	ACN=50	France	H-2, H-4
Acanthobrama	*marmid*		F, M	50	16M + 26SM + 8 ST/A	92		4		ACN=50	Turkey (Dam Lake)	G-79
Achondrostoma	*arcasii*	*Rutilus*	F, M	50	16M + 30 SM/ST + 4A		96		2.8 FCM	ACN=50	Portugal	C-69, C-75
Achondrostoma	*oligolepis*			50	12M + 32SM + 6 ST/A	94		4		ACN=50	Portugal	P-76
Alburnoides	*bipunctatus*		F, M	50	38 M/SM + 12 ST/A	88				ACN=50	Bosnia-Herzegovina	S-82
Alburnus	*albidus*			50	16M + 26SM + 8 ST/A	92		2		ACN=50	Italy	B-36
Alburnus	*arborella*	*albidus arborella*		50	12M + 25SM + 13 ST/A	87				ACN=50	(Europe)	F-30
Alburnus	*alburnus*	*alburnus alborella*	F	50	16M + 10SM + 16ST + 8A	76	92		(2.6 FCM)	ACN=50	Italy	C-34, G-85
Alburnus	*alburnus*			50	16M + 20 SM/ST + 14A		86		2.7 FD	0-2 B	France	H-3, H-4, H-5
Alburnus	*alburnus*		F, M	50	14M + 14SM + 14ST + 8A	78	92		3.1 FCM	0-2 B	Germany	S-164, L-82
Alburnus	*alburnus*		F, M	50	32 M/SM + 18 ST/A	82				ACN=50	Bosnia-Herzegovina	S-82
Alburnus	*filippii*		F, M	50	16M + 16SM + 18A	82				ACN=50	Turkey	G-80
Anaecypris	*hispanica*		F, M	50	10M + 36 SM/ST + 4A		96		2.5 FCM	ACN=50	Portugal	C-69, C-75
Aspius	*aspius*		F, M	50	14M + 28SM + 8 ST/A	92		2	(2.5 FCM)	ACN=50	Danube R., Don R.	R-16, G-85
Aspius	*aspius*		F, M	52	42 M/SM + 10 ST/A	94				ACN=52	Sweden	N-50
Ballerus	*ballerus*	*Abramis*	F, M	50	16M + 20SM + 6ST + 8A	86	92		(2.3 FCM)	ACN=50	Russia	A-118, G-85
Ballerus	*ballerus*	*Abramis*		50	14M + 28 SM/ST + 8A		92			ACN=50	Russia	V-72

Table 6.11 Order CYPRINIFORMES (continued)

A		B	C	D	E	F	G	H	I	J	K	L
Current scientific name of taxon Superfamily/family/subfamily/species		Reported in karyotype paper	Sex	2n	Karyotype	NF_1	NF_2	Ag-NORs	Genome size (pg/cell)	Comments	Locality	Reference
Ballerus	*ballerus*	*Abramis*		52		70					Sweden	N-50
Ballerus	*sapa*	*Abramis*	F, M	50	10M + 22SM + 10ST + 8A	82	92		2.2 FCM	ACN=50	Russia	A-118, G-85
Blicca	*bjoerkna*			50	14M + 26 SM/ST + 10A		90		(2.5 FCM)	ACN=50	Europe	V-72, G-85
Blicca	*bjoerkna*			50	12M + 14SM + 12ST + 12A	76	88				(Europe)	S-194
Chalcalburnus	*mossulensis*			48	12M + 20SM + 16A	80					Turkey	G-71
Chalcalburnus	*mossulensis*			50	12M + 16SM + 10ST + 12A	78	88	4		ACN=50	Turkey	Y-23
Chondrostoma	*kneri*			50	30 M/SM + 20 ST/A	80				ACN=50	former Yugoslavia	B-17
Chondrostoma	*nasus*			50	38 M/SM + 12A	88	88		(2.8 FCM)	ACN=50	Bosnia	S-85, G-85
Chondrostoma	*nasus*			50	15M + 15SM + 12ST + 8A	80	92				(Europe)	S-194
Chondrostoma	*phoxinus*			50	36 M/SM + 14 ST/A	86				ACN=50	former Yugoslavia	B-17
Chondrostoma	*soetta*		F, M	50	16M + 14SM + 14ST + 6A	80	94			ACN=50	Italy	C-34
Delminichthys	*adspersus*	*Paraphoxinus*	F, M	50	38 M/SM + 12A	88	88			ACN=50	Croatia, Bosnia	B-19
Delminichthys	*ghetaldii*	*Paraphoxinus pstrossii*	F, M	50	30 M/SM + 20 ST/A	80				ACN=50	former Yugoslavia	S-75, B-19
Elopichthys	*bambusa*		F, M	48	10M + 24SM + 12ST + 2A	82	94	4		ACN=50	China (Hubei)	Z-25, Z-28, L-42, Y-15
Eupallasella	*perenurus*		F, M	50	8M + 32SM + 10 ST/A	90		2		ACN=50	Poland	B-62
Iberochondrostoma	*almacai*			50	14M + 30 SM/ST + 6A		94	2	2.6 FCM	ACN=50	Portugal	M-141
Iberochondrostoma	*lemmingii*	*Chondrostoma*		50	12M + 32 SM/ST+6A		94		2.8 FCM	ACN=50	Iberia	C-69, C-75
Iberochondrostoma	*lusitanicum*	*Chondrostoma*	F, M	50	16M + 28 SM/ST + 6A		94		2.8 FCM	ACN=50	Portugal	C-68, C-75
Iberochondrostoma	*lusitanicum*			50	14M + 30 SM/ST + 6A		94	1-4	2.9 FCM	ACN=50	Portugal	M-141
Iberocypris	*alburnoides*	*Rutilus*	F, M	50	12M + 34 SM/ST + 4A		96		2.4 FCM	ACN=50	Portugal	C-75, C-102
Iberocypris	*alburnoides*	*Rutilus*		75	18M + 51 SM/ST + 6A		144			3X	Portugal	C-102
Iberocypris	*alburnoides*	*Tropidophoxinellus*		50, 75						2X, 3X	Portugal	M-45
Iberocypris	*alburnoides*	*Tropidophoxinellus*	F, M	50						ACN=50	Portugal	A-111
Iberocypris	*alburnoides*	*Tropidophoxinellus*	F, M	75						3X	Portugal	A-111
Iberocypris	*alburnoides*	*Tropidophoxinellus*	F, M	100						4X	Portugal	A-111
Iberocypris	*palaciosi*			50						ACN=50	Spain	E-8
Iberocypris	*palaciosi*			75						3X	Spain	E-8
Iberocypris	*palaciosi*			100						4X	Spain	E-8
Leucaspius	*delineatus*			50	14M + 30 SM/ST + 6A		94	2		ACN=50	Russia	V-72
Leucaspius	*delineatus*		F, M	50	16M + 26SM + 8 ST/A	92				ACN=50	Elbe R.	R-22
Leucaspius	*delineatus*		F, M	50	16M + 26 SM/ST + 8A		92	2		ACN=50	Elbe R., Danube R.	M-54
Leuciscus	*bergi*			50	18M + 22SM + 10A	90				ACN=50	Kyrgyzstan	M-58
Leuciscus	*idus*		F, M	50	32M + 6SM + 10ST + 2A	88	98		2.6 FCM, 3.0 BFA	ACN=50	Bosnia-Herzegovina	S-79, G-85, H-13
Leuciscus	*idus*		F, M	50	10M + 26SM + 6ST + 8A	86	92	2		ACN=50	Poland	B-66
Leuciscus	*leuciscus*		F, M	50	20M + 20SM + 8ST + 2A	90	98			ACN=50	Bosnia-Herzegovina	S-79
Leuciscus	*leuciscus*			50	16M + 22 SM/ST + 12A		88		2.6 FD	ACN=50	France	H-2, H-4
Leuciscus	*leuciscus*		F, M	50	12M + 24SM + 8ST + 6A	86	94	1-4		ACN=50	Poland	B-66
Leuciscus	*leuciscus baicalensis*	*leuciscus kirgisorum*		50	18M + 22SM + 10A	90				ACN=50	Kyrgyzstan	M-58
Leuciscus	*schmidti*			50	16M + 24SM + 10A	90				ACN=50	Kyrgyzstan	M-58

Table 6.11 Order CYPRINIFORMES (continued)

A		B	C	D	E	F	G	H	I	J	K	L
Current scientific name of taxon Superfamily/family/subfamily/species		Reported in karyotype paper	Sex	2n	Karyotype	NF_1	NF_2	Ag-NORs	Genome size (pg/cell)	Comments	Locality	Reference
Leuciscus	*waleckii*		F	50	18M + 20SM + 6ST + 6A	88	94	2		ACN=50	China (Nei Mongol, Gansu)	Z-17, R-42
Luciobrama	*macrocephalus*		M	48	12M + 22SM + 12ST + 2A	82	94			ACN=50	China (Hubei)	Z-28, L-42, Y-15
Ochetobius	*elongatus*		F, M	48	10M + 16SM + 22ST	74	96		1.9* FD	ACN=50	China (Hubei)	Z-28, L-42, C-83
Pachychilon	*macedonicum*		F	50	8M + 34SM + 8 ST/A	92		2		ACN=50	Macedonia	R-25
Pachychilon	*pictum*			50	8M + 22SM + 20A	80				ACN=50	(S.E. Europe)	K-133
Pachychilon	*pictum*			50							former Yugoslavia	B-19
Parachondrostoma	*arrigonis*		F, M	50	14M + 30SM + 6 ST/A	94		2		ACN=50	Spain (Valencia)	K-119
Parachondrostoma	*toxostoma*	*Chondrostoma*	F, M	50	16M + 14SM + 14ST + 6A	80	94			ACN=50	Italy	C-34
Parachondrostoma	*toxostoma*	*Chondrostoma*		50	16M + 24 SM/ST + 10A		90		2.6 FD	ACN=50	France	H-2, H-4
Pelecus	*cultratus*			50	16M + 30 SM/ST + 4A		96			ACN=50	Russia	V-72
Petroleuciscus	*borysthenicus*	*Leuciscus*	M	50	16M + 28 SM/ST + 6A		94			ACN=50	Greece	R-23
Petroleuciscus	*persidis*			50	29M + 18SM + 3ST	97	100			ACN=50	Iran (Fars)	E-7
Phoxinellus	*alepidotus*	*Paraphoxinus*		50	26 M/SM + 24 ST/A	76				ACN=50	Bosnia-Herzegovina	B-16
Phoxinus	*keumkang*	*Moroco keumgang*	F, M	50	12M + 28SM + 10 ST/A	90				ACN=50	Korea (Muju, Togu-ri)	K-55, L-12
Phoxinus	*lagowskii steindachneri*	*Moroco steindachneri*	F, M	50	12M + 28SM + 10 ST/A	90			2.4* FCM	ACN=50	Korea (Gangreung)	K-55, L-12, P-69
Phoxinus	*lagowskii steindachneri*	*Moroco steindachneri*		50	14M + 22SM + 8ST + 6A	86	94			ACN=50	Japan	O-22
Phoxinus	*oxycephalus*	*Moroco*	F, M	50	12M + 28SM + 10 ST/A	90			(2.2 FD)	ACN=50	Korea (Ochon-gun)	K-4, K-5, K-55, L-12
Phoxinus	*oxycephalus*	*Moroco*	M	50	12M + 28SM + 10 ST/A	90			2.5* FCM	ACN=50	Korea (Imjin R.)	L-13, P-69
Phoxinus	*oxycephalus*	*Moroco*	F	50	16M + 26SM + 8 ST/A	92				ACN=50	Korea (Taejong R.)	L-13
Phoxinus	*oxycephalus jouyi*	*Moroco jouyi*	F, M	50	10M + 34 SM/ST + 6A		94			ACN=50	Japan (Kumamoto)	U-33
Phoxinus	*oxycephalus jouyi*	*Moroco jouyi*		50	14M + 26SM + 10A	90		4	3.1* FCM	ACN=50	Japan (Lake Biwa, Kobe)	T-4, O-48
Phoxinus	*oxycephalus jouyi*	*Moroco jouyi*	F, M	50	18M + 24SM + 8 ST/A	92				ACN=50	Japan (Sagami R.)	L-13
Phoxinus	*oxycephalus jouyi*	*Moroco jouyi*	F, M	50	12M + 26SM + 12 ST/A	88				ACN=50	Japan (Ihara R.)	L-13
Phoxinus	*oxycephalus jouyi*	*Moroco jouyi*	F, M	50	12M + 30SM + 8 ST/A	92				ACN=50	Japan (Kanagawa)	L-13
Phoxinus	*phoxinus*		F, M	50	16M + 22SM + 12 ST/A	88				ACN=50	Korea (Chongson-gun)	K-55, L-12
Phoxinus	*phoxinus*		F, M	50	10M + 34 SM/ST + 6A		94			ACN=50	Korea (Han R.)	U-33
Phoxinus	*phoxinus*		F, M	50	8M + 32SM + 10 ST/A	90				ACN=50	Poland	B-62
Phoxinus	*phoxinus*			50	14M + 26 SM/ST + 10A		90		2.3 FD	ACN=50	France	H-2, H-4
Phoxinus	*phoxinus*		F	50	14M + 12SM + 16ST + 8A	76	92			ACN=50	Italy	C-34
Phoxinus	*phoxinus*		F, M	50	32 M/SM + 18 ST/A	82				ACN=50	Bosnia-Herzegovina	B-21
Pseudaspius	*leptocephalus*		F	50	14M + 26SM + 10 ST/A	90				ACN=50	Mongolia	R-20
Pseudochondrostoma	*duriense*	*Chondrostoma*		50					2.7 FCM		Iberia	C-75
Pseudochondrostoma	*duriense*			50	14M + 30SM + 6 ST/A	94		2		ACN=50	Portugal	P-76
Pseudochondrostoma	*polylepis*			50	14M + 30SM + 6 ST/A	94		2		ACN=50	Portugal	P-76
Pseudochondrostoma	*willkommii*	*Chondrostoma*		50					2.7 FCM		Iberia	C-75
Rhynchocypris	*lagowskii*	*Moroco*	F, M	50	12M + 28SM + 10 ST/A	90			2.0 FD	ACN=50	Korea (Mt. Sorak)	K-4, K-5, L-13
Rhynchocypris	*percnurus*	*Eupallasella perenurus*		50	8M + 32SM + 10ST	90	100	4		ACN=50	Poland	B-62
Rutilus	*aula*			50	16M + 26SM + 8 ST/A	92		2		ACN=50	Italy	B-36
Rutilus	*aula*	*Leuciscus*		50	15M + 23SM + 12 ST/A	88				ACN=50	(Europe)	F-30

Table 6.11 Order CYPRINIFORMES (continued)

A		B	C	D	E	F	G	H	I	J	K	L
Current scientific name of taxon Superfamily/family/subfamily/species		Reported in karyotype paper	Sex	2n	Karyotype	NF_1	NF_2	Ag-NORs	Genome size (pg/cell)	Comments	Locality	Reference
Rutilus	*frisii kutum*			50	32 M/SM + 18 ST/A	82				ACN=50	(Russia)	V-72
Rutilus	*macrolepidotus*		F, M	50	14M + 32 SM/ST + 4A		96		2.8 FCM	ACN=50	Portugal	C-69, C-75
Rutilus	*pigus virgo*		F	50	16M + 24SM + 10 ST/A	90		2		ACN=50	Danube R.	R-15
Rutilus	*prespensis*			50	16M + 26SM + 8 ST/A	92				ACN=50	Greece	B-36
Rutilus	*rubilio*			50	16M + 26SM + 8 ST/A	92		2		ACN=50	Italy	B-36
Rutilus	*rubilio*		F, M	50	18M + 10SM + 16ST + 6A	78	94			ACN=50	Italy	C-34
Rutilus	*rutilus*		F, M	50	16M + 28SM + 6 ST/A	94		2	(2.6 FCM)	ACN=50	Danube R., Elbe R.	R-15, V-86
Rutilus	*rutilus*			50	16M + 28SM + 6 ST/A	94				0-1 B	Danube R.	R-13
Rutilus	*rutilus*			50	14M + 18SM/ST + 18A		82		1.9 FD	ACN=50	France	H-3, H-4
Rutilus	*rutilus*			50	26 M/SM + 24 ST/A	76				ACN=50	Sweden	N-50
Rutilus	*rutilus*		F, M	50	16M + 16SM + 10ST + 8A	82	92			ACN=50	Germany (Baltic Sea side)	K-76
Rutilus	*rutilus*		F, M	51	16M + 18SM + 10ST + 7A	85	95				Germany (Baltic Sea side)	K-76
Rutilus	*rutilus*		F, M	52	16M + 16SM + 11ST + 9A	84	95				Germany (Baltic Sea side)	K-76
Rutilus	*rutilus caspicus*			50	14M + 18SM + 12ST + 6A	82	94			ACN=50	(Caspian Sea basin)	V-72
Rutilus	*ylikiensis*			50	16M + 26SM + 8 ST/A	92		2		ACN=50	Greece	B-36
Scardinius	*acarnanicus*			50	16M + 26SM + 8 ST/A	92		2		ACN=50	Greece	B-36
Scardinius	*erythrophthalmus*			50	16M + 26SM + 8 ST/A	92		2	(2.8 FCM)	ACN=50	Italy	B-36, G-85
Scardinius	*erythrophthalmus*			50	14M + 24SM + 8ST + 4A	88	96			ACN=50	(Europe)	V-72
Scardinius	*erythrophthalmus*	*erythrophtalmus*		48	13M + 35 ST/A	61					(Europe)	F-30
Scardinius	*erythrophthalmus*			48	22 M/SM + 20ST + 6A	70	90			ACN=50	Sweden	N-50
Scardinius	*erythrophthalmus*		F, M	50	20M + 12SM + 12ST + 6A	82	94			ACN=50	Italy	C-34
Scardinius	*erythrophthalmus*			50	14M + 20 SM/ST + 16A		84		2.0 FD	ACN=50	France	H-2, H-4
Scardinius	*erythrophthalmus*		F, M	50	16M + 26 SM/ST + 8A		92	2		ACN=50	Elbe R., Danube R.	M-54
Scardinius	*erythrophthalmus*		F, M	50	30 M/SM + 20 ST/A	80				ACN=50	Bosnia	S-84
Scardinius	*graecus*			50	16M + 26SM + 8 ST/A	92		2		ACN=50	Greece	B-36
Scardinius	*scardafa*			50	16M + 26SM + 8 ST/A	92		2		ACN=50	Italy	B-36
Squalius	*albus*	*Leuciscus cephalus albus*		50	28M + 8SM + 14A	86				ACN=50	Bosnia-Herzegovina	S-79
Squalius	*aradensis*	*Leuciscus*		50					2.4 FCM		Portugal	C-75
Squalius	*carolitertii*	*Leuciscus*		50					2.4 FCM		Iberia	C-75
Squalius	*cephalus*	*Leuciscus*		50	18M + 20 SM/ST + 12A		88		2.7 FD	ACN=50	France	H-2, H-4
Squalius	*cephalus*	*Leuciscus*	F, M	50	10M + 22SM + 10ST + 8A	82	92	2		ACN=50	Poland	B-66
Squalius	*cephalus*	*Leuciscus*		50	16M + 26SM + 8 ST/A	92		2		ACN=50	Italy	B-36
Squalius	*cephalus*	*Leuciscus*		50	34 M/SM + 16 ST/A	84				ACN=50	Croatia	A-27, A-28
Squalius	*cephalus cephalus*	*Leuciscus cephalus*	F, M	50	22M + 14SM + 14A	86				ACN=50	Bosnia-Herzegovina	S-79
Squalius	*keadicus*	*Leuciscus*		50	14M + 12SM + 10ST + 14A	76	86			ACN=50	Greece	T-54
Squalius	*pyrenaicus*	*Leuciscus*		50					2.4 FCM		Iberia	C-75
Squalius	*squalus*	*Leuciscus cephalus cabeda*		50	16M + 12SM + 12ST + 10A	78	90			ACN=50	Portugal	C-71
Squalius	*squalus*	*Leuciscus cephalus cabeda*	F, M	50	16M + 12SM + 12ST + 10A	78	90			ACN=50	Italy	C-34
Squalius	*svallize*	*Leuciscus svallize svallize*	F, M	50	20M + 12SM + 16ST + 2A	82	98			ACN=50	Bosnia-Herzegovina	S-79

Table 6.11 Order CYPRINIFORMES (continued)

A	B	C	D	E	F	G	H	I	J	K	L
Current scientific name of taxon Superfamily/family/subfamily/species	Reported in karyotype paper	Sex	2n	Karyotype	NF_1	NF_2	Ag-NORs	Genome size (pg/cell)	Comments	Locality	Reference
Squalius tenellus	*Leuciscus turskyi tenellus*	F, M	50	18M + 16SM + 8ST + 8A	84	92			ACN=50	Bosnia-Herzegovina	S-79
Telestes croaticus	*Paraphoxinus*	F, M	50	36 M/SM + 14A	86				ACN=50	Croatia	B-19
Telestes metohiensis	*Paraphoxinus*	F, M	50	32 M/SM + 18A	82				ACN=50	Bosnia-Herzegovina	B-19
Telestes muticellus	*Leuciscus souffia muticellus*	M	50	20M + 12SM + 8ST + 10A	82	90			ACN=50	Italy	C-34
Telestes polylepis	*Chondrostoma*		50					2.7 FCM		Croatia	C-71, C-75
Telestes souffia	*Leuciscus souffia agassizi*		50	20M + 22SM + 2ST + 6A	92	94			ACN=50	Bosnia-Herzegovina	S-79
Tribolodon ezoe		F, M	50	10M + 20SM + 12ST + 8A	80	92			ACN=50	Japan (Hokkaido)	I-17
Tribolodon ezoe			50	14M + 22SM + 8ST + 6A	86	94			ACN=50	Japan	O-22
Tribolodon hakonensis		F, M	50	10M + 20SM + 12ST + 8A	80	92			ACN=50	Japan (Hokkaido)	I-17
Tribolodon hakonensis			50	14M + 26SM + 10A	90		4	1.8 FD	ACN=50	Japan	T-4, S-141
Tribolodon hakonensis		M	50	14M + 30SM + 6 ST/A	94				ACN=50	Korea (Hadong-gun)	K-55, L-12
Vimba elongatus	*vimba* natio *carinata*	F	50	14M + 20SM + 16 ST/A	84				XX, ACN=50	Poland (Baltic Sea basin)	R-95
Vimba elongatus	*vimba* natio *carinata*	M	50	13M + 20SM + 17 ST/A	83				XY, ACN=50	Poland (Baltic Sea basin)	R-95
Vimba vimba		F, M	52	24-28 M/SM + 28-24 ST/A	76-80				ACN=50	Russia	S-70
Vimba vimba vimba		F	50	14M + 20SM + 16 ST/A	84				XX, ACN=50	Poland (Baltic Sea basin)	R-95
Vimba vimba vimba		M	50	13M + 20SM + 17 ST/A	83				XY, ACN=50	Poland (Baltic Sea basin)	R-95
North American Leuciscinae											
Acrocheilus alutaceus			50				8			USA (OR)	J-6
Campostoma anomalum		F, M	50	12M + 16SM + 18ST + 4A	78	96	6	2.3 FD	ACN=50	USA (TX)	G-28, G-35, G-36, G-48
Couesius plumbeus	*Hybopsis plumbea*		50							Canada (Quebec)	L-16
Chrosomus cumberlandensis	*Phoxinus*	F, M	50	14M + 24SM + 12 ST/A	88				ACN=50	USA (KY)	J-11
Chrosomus eos			50					2.8 FCM		Canada (Quebec)	L-16, G-88
Chrosomus erythrogaster			50	10M + 36SM + 4A	96			2.6 FD	ACN=50	USA (IL)	G-38, G-60
Chrosomus neogaeus	*Phoxinus*	F, M	50	12M + 24SM + 14 ST/A	86			3.1 FCM, 2.8 FIA	ACN=50	USA (MI, WY)	J-11, G-88, H-41
Chrosomus oreas	*Phoxinus*	F, M	50	14M + 24SM + 12 ST/A	88			2.5 FCM	ACN=50	USA (VA)	J-11, G-88
Cyprinella callistia	*Notropis callistius*		50	24 M/SM + 26 ST/A	74			2.1 FD	ACN=50	USA (AL)	D-6, G-87
Cyprinella camura	*Notropis camurus*		50	18M + 28 SM/ST + 4A		96			ACN=50	USA (LA)	G-33
Cyprinella camura	*Notropis camurus*		50	10M + 26SM + 14 ST/A	86		2	2.4 FD	ACN=50	USA (LA)	G-43, G-87
Cyprinella formosa			50	10M + 32SM + 8ST	92		2		ACN=50	USA (NM)	G-39
Cyprinella galactura	*Notropis galacturus*		50	20M + 28SM + 2 ST/A	98		2		ACN=50	USA (AR)	G-43
Cyprinella gibbsi			50	10M + 34SM + 6ST	94	100	2		ACN=50	USA (AL)	G-39
Cyprinella lepida	*Notropis lutrensis lepidus*		50		94		2	2.5 FD		USA (TX)	A-39, A-40, G-87
Cyprinella lutrensis	*Notropis*		50	50 M/SM	100	100	2	2.4 FD	ACN=50	USA (TX)	G-31, G-35, G-48
Cyprinella lutrensis	*Notropis*	F, M	50	10M + 32 SM/ST + 8A		92			ACN=50	USA (TX)	L-50
Cyprinella lutrensis	*Notropis*		50	10M + 34SM + 6A	94	94	2		ACN=50	USA	G-36
Cyprinella proserpina	*Notropis proserpinus*	F, M	50	12M + 24SM + 14 ST/A	86		2	2.9 FD	ACN=50	USA (TX)	A-40, G-43, G-87
Cyprinella spiloptera	*Notropis spilopterus*		50	8M + 38SM + 4 ST/A	96		2		ACN=50	USA (OH)	G-43
Cyprinella venusta	*Notropis*		50	48 M/SM + 2 ST/A	98		2	2.4 FD	ACN=50	USA (TX)	G-31, G-35, G-48
Cyprinella venusta	*Notropis venustus*	F, M	50	2M + 38SM + 10A	90				ACN=50	USA (TX)	C-6

Table 6.11 Order CYPRINIFORMES (continued)

A		B	C	D	E	F	G	H	I	J	K	L
Current scientific name of taxon Superfamily/family/subfamily/species		Reported in karyotype paper	Sex	2n	Karyotype	NF_1	NF_2	Ag-NORs	Genome size (pg/cell)	Comments	Locality	Reference
Cyprinella	*whipplei*	*Notropis*		50	22M + 26SM + 2 ST/A	98		2	2.5 FD	ACN=50	USA (AR)	G-38, G-43
Dionda	*argentosa*			50				2	2.1 FCM	ACN=50	USA (TX)	G-86
Dionda	*diaboli*			50				2	2.0 FCM	ACN=50	USA (TX)	G-86
Dionda	*episcopa*			50				2	2.1 FCM, 2.1 FD	ACN=50	USA (TX, NM)	G-33, G-86
Dionda	*serena*			50				2	2.0 FCM	ACN=50	USA (TX)	G-86
Dionda	sp.			50				2	2.0 FCM	ACN=50	USA (TX)	G-86
Exoglossum	*maxillingua*			48							Canada (Quebec)	L-16
Gila	*bicolor*			50	44 M/SM + 6 ST/A	94				ACN=50	USA (CA)	G-27
Gila	*elegans*			50				4			USA (NM)	J-6
Gila	*nigrescens*			50				4			USA (NM)	J-6
Gila	*orcutti*			50	14M + 24ST + 12A	88		4		ACN=50	USA (CA)	G-61, J-6
Gila	*pandora*			50				4			USA (NM)	J-6
Gila	*purpurea*			50				4			USA (AZ)	J-6
Hemitremia	*flammea*			50		96		2	2.4 FD		USA (AL)	A-39, A-40, G-87
Hesperoleucus	*symmetricus*			50	44 M/SM + 6 ST/A	94				ACN=50	USA (CA)	G-27
Hesperoleucus	*symmetricus*			50	20M + 20SM + 10A	90				ACN=50	USA (CA)	G-61
Hybognathus	*hayi*		F, M	50	10M + 36SM + 4 ST/A	96				ACN=50	USA (TX)	G-28
Hybognathus	*nuchalis*			50							USA (LA)	G-33
Hybognathus	*placitus*			50	6M + 32SM + 12 ST/A	88		2	2.8 FD	ACN=50	USA (TX)	G-36, G-87, A-39
Hybopsis	*amblops*	*Notropis* (*Hybopsis*)		50	16M + 30SM + 4A	96	96			ACN=50	USA (LA)	G-33
Lavinia	*exilicauda*			50	42 M/SM + 8 ST/A	92		4	2.5 FCM	ACN=50	USA (CA)	G-27, G-44, G-45, J-6
Lepidomeda	*albivallis*			50	12M + 32 SM/ST + 6A		94			ACN=50	USA (NV)	U-69
Lepidomeda	*mollispinis*			50	12M + 30 SM/ST + 8A		92			ACN=50	USA (AZ)	U-69
Lepidomeda	*vittata*			50	14M + 32 SM/ST + 4A		96			ACN=50	USA (AZ)	U-69
Luxilus	*albeolus*			50	8M + 32SM + 10 ST/A	90		4		ACN=50	USA (VA)	P-47
Luxilus	*cardinalis*			50	12M + 34SM + 4 ST/A	96		2		ACN=50	USA (AR)	P-47
Luxilus	*cerasinus*			50	8M + 28SM + 14 ST/A	86		2		ACN=50	USA (VA)	P-47
Luxilus	*chrysocephalus*	*Notropis*		50	48 M/SM + 2 ST/A	98		4	2.3 FD	ACN=50	USA (LA)	G-31, G-38
Luxilus	*chrysocephalus chrysocephalus*			50	6M + 40SM + 4 ST/A	96		4		ACN=50	USA (AR)	P-47
Luxilus	*chrysocephalus isolepis*			50	10M + 28SM + 12 ST/A	88		4		ACN=50	USA (AL, LA, MS)	P-47
Luxilus	*chrysocephalus isolepis*			50	10M + 28SM + 12 ST/A	88		2		ACN=50	USA (OK)	P-47
Luxilus	*coccogenis*			50	12M + 30SM + 8 ST/A	92		4		ACN=50	USA (VA)	P-47
Luxilus	*cornutus*	*Notropis*		50	6M + 40SM + 4A	96		4		ACN=50	USA (IL, MI)	G-60, P-47
Luxilus	*pilsbryi*	*Notropis*		50	14M + 34SM + 2 ST/A	98		2	2.5 FD	ACN=50	USA (MS, AR)	G-43, G-38, P-47
Luxilus	*zonatus*			50	6M + 34SM + 10 ST/A	90		2		ACN=50	USA (AR)	P-47
Luxilus	*zonistius*			50	12M + 30SM + 8 ST/A	92		2		ACN=50	USA (AL)	P-47
Lythrurus	*ardens*	*Notropis*	F, M	50	8M + 36SM + 6 ST/A	94		4	2.5 FD	ACN=50	USA (AL)	G-36, G-87, A-39
Lythrurus	*bellus*	*Notropis*		50		92			2.6 FD		USA (AL)	A-39, G-87
Lythrurus	*fumeus*	*Notropis*		50	48 M/SM + 2 ST/A	98				ACN=50	USA (TX)	G-31

Table 6.11 Order CYPRINIFORMES (continued)

A		B	C	D	E	F	G	H	I	J	K	L
Current scientific name of taxon Superfamily/family/subfamily/species		Reported in karyotype paper	Sex	2n	Karyotype	NF_1	NF_2	Ag-NORs	Genome size (pg/cell)	Comments	Locality	Reference
Lythrurus	*roseipinnis*	*Notropis*		50	48 M/SM + 2 ST/A	98				ACN=50	USA (LA)	G-31
Lythrurus	*umbratilis*	*Notropis*		50	50 M/SM	100		4	2.7 FD	ACN=50	USA (TX, OK)	A-40, G-31, G-38
Macrhybopsis	*aestivalis*	*Hybopsis*	F, M	50	6M + 16SM + 22ST + 6A	72	94	4	2.5 FD	ACN=50	USA (AL, TX)	G-28, G-36, G-87
Margariscus	*margarita*	*Semotilus*		50							Canada (Quebec)	L-16
Meda	*fulgida*			50	18M + 24 SM/ST + 8A		92			ACN=50	USA (AR)	U-69
Mylopharodon	*conocephalus*			50	44 M/SM + 6 ST/A	94		4	2.7 FCM	ACN=50	USA (CA)	G-27, G-44, G-45, J-6
Nocomis	*leptocephalus*			50	14M + 28SM + 8A	92	92	4	2.5 FD	ACN=50	USA (LA)	A-40, G-33, G-87
Nocomis	*micropogon*			50							USA (WV)	M-90
Notemigonus	*crysoleucus*		F, M	50	16M + 24 SM/ST + 10A		90			ACN=50	USA (TX)	L-50
Notemigonus	*crysoleucus*			50	12M + 24SM + 8ST + 6A	86	94	2	2.3 FD	ACN=50	USA (CA)	G-27, G-35, G-36, G-38, G-48
Notropis	*amabilis*			50	14M + 34SM + 2 ST/A	98		2	2.5 FD	ACN=50	USA (TX)	A-40, G-33, G-87
Notropis	*atherinoides*			50	16M + 32SM + 2 ST/A	98			2.4 FD	ACN=50	USA (TX)	G-33, G-87
Notropis	*atrocaudalis*			50	10M + 36SM + 4A	96			2.9 FD	ACN=50	USA (TX)	G-33, G-87
Notropis	*baileyi*			50	4M + 28SM + 18ST	82	100	2	2.8 FD	ACN=50	USA (MS)	G-43, G-87
Notropis	*boops*			50	6M + 32SM + 12ST	88	100	2	2.2 FD	ACN=50	USA (OK)	G-38, G-39, C-75
Notropis	*braytoni*		F, M	50	14M + 20 SM + 10ST + 6A	84	94	2	2.6 FD		USA (TX)	G-36, G-87, A-39
Notropis	*buccatus*	*Ericymba buccata*		50					2.3 FD		USA (LA)	G-33, G-87
Notropis	*buchanani*			50	8M + 36SM + 6 ST/A	94		2		ACN=50	USA (TX)	A-41
Notropis	*chrosomus*			50	42 M/SM + 8A	92	92		2.5 FD	ACN=50	USA (AL)	A-39, G-87
Notropis	*dorsalis*			50	10M + 32SM + 8ST	92	100	2		ACN=50	USA (IL)	G-39
Notropis	*jemezanus*			50	42 M/SM + 8A	92	92	2		ACN=50	USA (TX)	A-39, A-40
Notropis	*longirostris*			50	10M + 28SM + 12ST	88	100	2		ACN=50	USA (LA)	G-31, G-36
Notropis	cf *longirostris*			50	8M + 30SM + 4ST + 8A	88	92	2		ACN=50	USA (AL)	G-36, A-39
Notropis	*maculatus*			50	10M + 34SM + 6 ST/A	94		2		ACN=50	USA (LA)	A-41
Notropis	*nubilus*			50	16M + 30SM + 4 ST/A	96		2	2.4 FD	ACN=50	USA (AR)	G-38, G-43
Notropis	*oxyrhynchus*			50	48 M/SM + 2A	98	98	2	2.1 FD	ACN=50	USA (TX)	A-40, G-31, G-87
Notropis	*petersoni*			50	8M + 30SM + 12ST	88	100	4		ACN=50	USA (FL)	G-39
Notropis	*potteri*			50	12M + 34SM + 2ST + 2A	96	98	2	2.4 FD	ACN=50	USA (TX)	G-33, G-36, G-87
Notropis	*sabinae*			50	50 M/SM	100	100	4		ACN=50	USA (TX)	G-31, G-39
Notropis	*shumardi*			50	14M + 34SM + 2 ST/A	98		2	2.7 FD	ACN=50	USA (TX)	G-31, G-36, G-38
Notropis	*stilbius*			50					2.5 FD		USA (AL)	D-6, G-87
Notropis	*stramineus*			50	8M + 34SM + 8 ST/A	92		2	2.5 FD	ACN=50	USA (TX, OK)	A-41, G-38
Notropis	*stramineus*			50	16M + 34SM	100	100			ACN=50	USA (TX)	G-33
Notropis	*texanus*			50	48 M/SM + 2A	98	98	2	(2.7 FD)	ACN=50	USA (TX)	A-40, G-31, G-87
Notropis	*texanus*		F, M	50	2M + 42SM + 6A	94	94			ACN=50	USA (TX)	C-6
Notropis	*volucellus*			50	14M + 28SM + 8 ST/A	92		2	2.5 FD	ACN=50	USA (TX, AL)	A-41, G-87
Notropis	*volucellus*			50	46 M/SM + 4 ST/A	96				ACN=50	USA (TX)	G-31
Opsopoeodus	*emiliae*		F, M	48	2M + 30SM + 16A	80				ACN=50	USA (TX)	C-6
Opsopoeodus	*emiliae*	*Notropis*		50	8M + 20SM + 16ST + 6A	78	94	2	2.1 FD	ACN=50	USA (LA)	G-35, G-36, A-39

Table 6.11 Order CYPRINIFORMES (continued)

A		B	C	D	E	F	G	H	I	J	K	L
Current scientific name of taxon Superfamily/family/subfamily/species		Reported in karyotype paper	Sex	2n	Karyotype	NF_1	NF_2	Ag-NORs	Genome size (pg/cell)	Comments	Locality	Reference
Orthodon	*microlepidotus*			50	44 M/SM + 6 ST/A	94		4		ACN=50	USA (CA)	G-27, J-6
Phenacobius	*mirabilis*			50	18M + 28SM + 4 ST/A	96		6	3.2 FD	ACN=50	USA (TX, IL)	G-28, G-39, G-87
Pimephales	*notatus*			50	46 M/SM + 4 ST/A	96		2	2.2 FD		USA (OK)	A-39, A-40, L-46, G-38
Pimephales	*notatus*			52							Canada (Quebec)	L-16
Pimephales	*promelas*			50	14M + 34SM + 2A	98	98	4	2.2 FD	ACN=50	USA (TX)	G-33, G-38, L-46
Pimephales	*vigilax*		F, M	50	8M + 24SM + 16ST + 2A	82	98	2	2.2 FD	ACN=50	USA (TX)	G-28, G-35, G-36, G-38, G-48
Plagopterus	*argentissimus*			50	18M + 26 SM/ST + 6A		94	4		ACN=50	USA (AZ, NM)	U-69, J-6
Pogonichthys	*macrolepidotus*			50	44 M/SM + 6 ST/A	94				ACN=50	USA (CA)	G-27
Pteronotropis	*hubbsi*			50	22M + 22SM + 6ST	94	100	2		ACN=50	USA (TX)	A-41
Pteronotropis	*signipinnis*			50	22M + 22SM + 6 ST/A	94		4		ACN=50	USA (LA)	A-41
Pteronotropis	*welaka*			50	22M + 22SM + 6 ST/A	94		4		ACN=50	USA (LA)	A-41
Ptychocheilus	*grandis*			50	42 M/SM + 8 ST/A	92		4	2.6 FCM	ACN=50	USA (CA)	G-27, G-44, J-6
Ptychocheilus	*lucius*			50				4	2.5 FCM		USA (NM)	G-44, J-6
Ptychocheilus	*oregonensis*			50				4	2.8 FCM		USA (OR)	G-44, J-6
Ptychocheilus	*umpquae*			50				4	2.5–2.8 FCM		USA (OR)	G-44, J-6
Ptychocheilus	cf. *umpquae*			50				4			USA (OR)	J-6
Rhinichthys	*atratulus*			50	16M + 28SM + 2ST + 4A	94	96		2.6 FD	ACN=50	USA (NY, AL)	H-30, K-73, G-87
Rhinichthys	*cataractae*			50	16M + 28SM + 2ST + 4A	94	96	4		ACN=50	USA (NY, NM)	H-30, J-6
Rhinichthys	*cataractae*			50							USA (WV)	M-90
Rhinichthys	*cobitis*	*cobitus*		50				6			USA (AZ)	J-6
Rhinichthys	*evermanni*			50							N. America	M-61
Rhinichthys	*osculus*			50				6			USA (AZ)	J-6
Rhinichthys	cf. *osculus*			50				6			USA (OR)	J-6
Richardsonius	*balteatus*			50				8	2.5 FCM		USA (OR)	G-45, J-6
Richardsonius	*egregius*			50	36 M/SM + 14 ST/A	86		6	2.7 FCM	ACN=50	USA (CA)	G-27, G-45
Semotilus	*atromaculatus*			50	22M + 24SM + 4A	96		4	2.5 FD	ACN=50	USA (TX, MS, IL)	G-33, G-38, G-39
Semotilus	*atromaculatus*			52							Canada (Quebec)	L-16
Semotilus	*corporalis*			52					2.5 FIA		Canada (Quebec)	L-16, H-41
Rasborinae (= Danioninae)												
Amblypharyngodon	*microlepis*		F, M	50	12M + 14SM + 6ST + 18A	76	82	2		ACN=50	India (WB)	K-136
Amblypharyngodon	*mola*		F	50	12M + 20SM + 8ST + 10A	82	90			ACN=50	India (WB)	M-27
Aspidoparia	*morar*		F	50	8M + 18SM + 14ST + 10A	76	90			ACN=50	India (J & K)	T-52
Aspidoparia	*morar*		F	48	8M + 6SM + 34A	62					India (Bihar)	K-42
Aspidoparia	*morar*		M	48	14M + 28SM + 6A	90	90				India (Assam)	C-108
Barilius	*bakeri*			50	24M + 14SM + 6ST + 6A	88	94	2		ACN=50	India (W. Ghats)	N-73
Barilius	*barila*		F	50	8M + 20SM + 8ST + 14A	78	86			ACN=50	India (Bihar)	K-35
Barilius	*barna*		F	50	8M + 16SM + 12ST + 14A	74	86			ACN=50	India (WB)	K-45
Barilius	*bendelisis*		F, M	50	6M + 6SM + 10 ST + 28A	62	72			ACN=50	India (Bihar)	K-30
Barilius	*bendelisis*		F, M	50	6M + 18SM + 20ST + 6A	74	94			ACN=50	India (Jammu)	S-50

Table 6.11 Order CYPRINIFORMES (continued)

A		B	C	D	E	F	G	H	I	J	K	L
Current scientific name of taxon Superfamily/family/subfamily/species		Reported in karyotype paper	Sex	2n	Karyotype	NF_1	NF_2	Ag-NORs	Genome size (pg/cell)	Comments	Locality	Reference
Barilius	*bendelisis*	*bendelisis* var. *chedra*		50	24M + 4SM + 22A	78	78			ACN=50	India (Assam)	K-41
Barilius	*bendelisis*	*bendelisis* var. *cocksa*		50							India (Tamil Nadu)	K-41
Barilius	*gatensis*			50	18M + 16SM + 10ST + 6A	84	94	2		ACN=50	India (W. Ghats)	N-72
Barilius	*naseeri*			50	24M + 24SM + 2A	98	98			ACN=50	Pakistan (Punjab)	R-31
Barilius	*pakistanicus*			50	12M + 32SM + 6A	94	94			ACN=50	Pakistan (Punjab)	R-31
Barilius	*shacra*		F	52	22M + 24SM + 6A	98	98				India (WB)	K-45
Barilius	*shacra*		M	52	22M + 23SM + 7A	97	97			XY	India (Assam)	K-45
Barilius	*tileo*		M	50	12M + 32SM + 6A	94	94			ACN=50	India (Assam)	K-45
Barilius	*vagra*		M	50	27M + 17SM + 6 ST/A	94				ACN=50	India (U.P.)	R-74
Barilius	*vagra*			50	22M + 22SM + 6A	94	94			ACN=50	Pakistan (Punjab)	R-31
Chela	*cachius*	*Perilampus atpar*	F	70	16M + 6SM + 16ST + 32A	92	108			ACN=72	India (J & K)	T-52
Chela	*caeruleostigmata*	*mouhoti*		48					3.2 BFA		(Asia)	H-13
Chela	*laubuca*			50							India	L-1
Chelaethiops	*bibie*			50							E. Africa	G-84
Danio	*albolineatus*			50	10M + 39SM + 1A	99	99		2.8 FD	ACN=50	(Asia)	F-30, S-141
Danio	*rerio*			50	10M + 39SM + 1A	99	99		(3.4–3.6, 4.6 FCM)	ACN=50	(Asia)	F-30, V-86, G-85
Danio	*rerio*		F, M	50	10M + 16 SM/ST + 24A		76			ACN=50	(India)	R-51
Danio	*rerio*			50	12M + 26SM + 12ST	88	100	4		ACN=50	(Asia)	P-35
Danio	*rerio*	*Brachydanio frankei*		50							S. China	W-32
Devario	*aequipinnatus*	*Danio*		50	14M + 32SM + 4A	96	96				India (Assam)	K-41
Devario	*aequipinnatus*	*Danio*		50	8M + 28SM + 10ST + 4A	86	96			ACN=50	Thailand (Nakhonphanom)	M-166
Devario	*devario*	*Danio*		50	12M + 24SM + 10ST + 4A	86	96			ACN=50	India (Orissa)	K-41
Devario	*devario*	*Danio*		50	10M + 40 ST/A	60				ACN=50	(Asia)	F-30
Devario	*malabaricus*	*Danio*		50	10M + 40ST	60	100		2.8 FIA, 4.4 BFA	ACN=50	(Asia)	F-30, H-13, H-40
Esomus	*danrica*	*danricus*	F, M	50	12M + 16SM + 10ST + 12A	78	88			ACN=50	India (WB)	M-27
Esomus	*danrica*	*danricus*		50	10M + 18SM + 18ST + 4A	78	96			ACN=50	India (Jammu)	S-56
Hemigrammocypris	*rasborella*			48	10M + 38 SM/ST		96		1.8 FD		Japan	S-141
Leptocypris	*niloticus*			50							Africa	G-84
Opsariichthys	*bidens*		F, M	76	4M + 6SM + 4ST + 62A	86	90			ACN=76	China (Guangdong)	L-42
Opsariichthys	*bidens*		F, M	74	6M + 6SM + 4ST + 58A	86	90			ACN=74	China (Sichuan)	L-44
Opsariichthys	*bidens*			76	4M + 72A	80				ACN=76	Korea	L-15
Opsariichthys	*uncirostris amurensis*		F	76	4M + 4SM + 68A	84				ACN=76	Korea (Pongdong)	K-55, L-12
Opsariichthys	*uncirostris bidens*			74	6M + 6SM + 4ST + 58A	86	90			ACN=74	China	Y-14
Opsariichthys	*uncirostris*		F, M	78	4M + 4SM + 70A	86		4	2.3* FCM, 3.3 FD	ACN=78	Japan (Osaka, Kobe)	T-4, O-18, O-48
Raiamas	*bola*	*Barilius*		50	6M + 14SM + 10ST + 20A	70	80			ACN=50	India (Bihar)	K-35
Raiamas	*nigeriensis*			50							W. Africa	G-84
Raiamas	*senegalensis*			50							W. Africa	G-84
Raiamas	*steindachneri*			50	16M + 30SM + 4ST	96	100	2		ACN=50	Guinea	R-26
Rasbora	*argyrotaenia*			50	24M + 22SM + 4 ST/A	96				ACN=50	Thailand	M-166

Table 6.11 Order CYPRINIFORMES (continued)

A		B	C	D	E	F	G	H	I	J	K	L
Current scientific name of taxon Superfamily/family/subfamily/species		Reported in karyotype paper	Sex	2n	Karyotype	NF_1	NF_2	Ag-NORs	Genome size (pg/cell)	Comments	Locality	Reference
Rasbora	*aurotaenia*			50	28M + 16SM + 4ST + 2A	94	98			ACN=50	Thailand (Chi Nat)	M-166
Rasbora	*daniconius*		F	50	18M + 6SM + 6ST + 20A	74	80			ACN=50	India (U.P.)	K-30
Rasbora	*rasbora*	*buchanani*	F	50	30M + 18SM + 2ST	98	100			ACN=50	India (WB)	M-27
Rasbora	*sumatrana*			50	22M + 18SM + 6ST + 4A	90	96			ACN=50	Thailand (Khanchanaburi)	M-166
Rasbora	*trilineata*			50	24M + 14SM + 12 ST/A	88				ACN=50	Thailand (Chi Nat)	M-166
Salmostoma	*bacaila*	*Chela*	F, M	50	10M + 12SM + 10ST + 18A	72	82			ACN=50	India (WB)	M-27
Securicula	*gora*			50	20M + 8SM + 22A	78				ACN=50	India	L-1
Tanichthys	*albonubes albonubes*	White cloud mountain type	F, M	50	24M + 14SM + 12 ST/A	88				ACN=50	China	A-81
Tanichthys	*albonubes* subsp.	Hong Kong type	F, M	50	22M + 22SM + 6 ST/A	94				ACN=50	China	A-81
Trigonostigma	*espei*	*Rasbora*		50	14M + 6SM + 30A	70	70			ACN=50	Thailand (Narativat)	M-166
Zacco	*platypus*		F, M	48	18M + 22 SM/ST + 8A		88	4	2.2* FCM, 2.2 FD	ACN=48	Japan (Osaka, Kobe)	T-4, O-18, O-48
Zacco	*platypus*		F, M	48	18M + 22SM + 8A	88	88			ACN=48	Korea (Yesan Pongdeng)	K-55, L-12
Zacco	*platypus*		F, M	78	4M + 4SM + 4ST + 66A	86	90			ACN=78	China (Guangdong)	L-42, C-85
Zacco	*sieboldi*	*Zacco* sp. type A	F, M	48	22M + 20SM + 6 ST/A	90				ACN=48	Japan (Yamaguchi)	N-54
Zacco	*temmincki*	*Zacco* sp. type B	F, M	48	18M + 18SM + 12 ST/A	84				ACN=48	Japan (Yamaguchi)	N-54
Zacco	*temminckii*			48	12M + 22SM + 14A	82	82	8	2.3* FCM		Japan	T-4, O-48
Zacco	*temminckii*		F, M	48	18M + 22 SM/ST + 8A		88		2.9 FD		Japan (Osaka)	O-18
Zacco	*temminckii*		F, M	48	18M + 22SM + 8A	88	88			ACN=48	Korea (Yemgye-gun)	K-55, L-12
Schizothoracinae												
Chuanchia	*labiosa*	*labioza*	F, M	92	32M + 26SM + 18ST + 16A	150	168			4X	China (Heihe R.)	Y-16
Diptychus	*maculatus*	*gymnogaster microcephalus*	F, M	100	70 M/SM + 30 ST/A	170				4X, ACN=100	Kirghiz Tan	M-60
Diptychus	*maculatus*	*gymnogaster oschanini*		100	80 M/SM + 20 ST/A	180				4X, ACN=100	Kirghiz Tan	M-60
Diptychus	*micromaculatus*		F, M	98	62 M/SM + 36 ST/A	160				4X	Kirghiz Tan	M-60
Diptychus	*sewerzowi*		F, M	98	80 M/SM + 18 ST/A	178				4X, ACN=98	Kirghiz Tan	M-60
Diptychus	sp.		F, M	98	28M + 32SM + 38 ST/A	158			4.9 FD	4X	China (Yunnan)	Z-6, Z-8
Gymnocypris	*eckloni*		F, M	94	26M + 28SM + 22ST + 18A	148	170	4		4X, ACN=100	China (Heihe R.)	Y-16, R-42
Gymnodiptychus	*dybowskii*	*Diptychus*		98	54 M/SM + 4–6 ST + 40–38 A	152	156–158			4X	Kirghiz Tan	M-60
Gymnodiptychus	*dybowskii lansdelli*	*Diptychus*	F, M	98	54 M/SM + 4ST + 40A	152	156			4X	Kirghiz Tan	M-60
Gymnodiptychus	*pachycheilus*			90–98						4X	(China)	C-74
Oxygymnocypris	*stewarti*		F, M	92	26M + 30SM + 22ST + 14A	148	170			ACN=98	China (Central Asia)	Y-16
Platypharodon	*extremus*		F, M	90	24M + 30SM + 20ST + 16A	144	164		2.7* FD	4X	China (Heihe R.)	Y-16, C-83
Ptychobarbus	*dipogon*	*Diptychus*	F, M	417–470					17.2* FD	18X	China (Tibet)	Y-17, C-83
Schizocypris	*oconnori*	*Schizothorax*	F, M	92	30M + 26SM + 20ST + 16A	148	168		2.7* FD	ACN=96	China (Central Asia)	Y-16, C-83
Schizopyge	*curvifrons*	*Schizothorax intermedius*		98–100						4X	(C. Asia)	C-74
Schizopygopsis	*pylzovi*		F, M	92	32M + 26SM + 20ST + 14A	150	170			4X, ACN=98	China (Heihe R.)	Y-16
Schizopygopsis	*younghusbandi younghusbandi*		F, M	90	26M + 28SM + 20ST + 16A	144	164			4X, ACN=98	China (Central Asia)	Y-16
Schizothoraichthys	*kumaonensis*	*Schizothorax*	F M	98	24M + 6SM + 68 ST/A	128				4X	India (U.P.)	R-74
Schizothoraichthys	*niger*	*Schizothorax*	F, M	98	22M + 26SM + 8ST + 42A	146	154			4X	India (Kashmir)	K-37
Schizothoraichthys	*progastus*		F, M	98	16M + 20SM + 12ST + 50A	134	146			4X	India (Haryana)	R-62

Table 6.11 Order CYPRINIFORMES (continued)

A		B	C	D	E	F	G	H	I	J	K	L
Current scientific name of taxon Superfamily/family/subfamily/species		Reported in karyotype paper	Sex	2n	Karyotype	NF_1	NF_2	Ag-NORs	Genome size (pg/cell)	Comments	Locality	Reference
Schizothoraichthys	*richardsonii*	*Schizothorax*		98	24M + 28SM + 24ST + 22A	150	174	2		4X, ACN=100	India (Beas R.)	B-5
Schizothoraichthys	*richardsonii*	*Schizothorax*		98	24M + 28SM + 24ST + 22A	150	174	4		4X, ACN=100	India (Kosi R.)	B-5
Schizothoraichthys	*richardsonii*	*Schizothorax*		98	66M + 16SM + 6ST + 10A	180	186			4X	India (Jammu)	B-5
Schizothorax	*argentatus*			98–150						4X, 6X	(Asia)	D-27
Schizothorax	*davidi*		F, M	98	20M + 34SM + 24ST + 20A	152	176			4X, ACN=104	China (Sichuan)	L-45, Y-14
Schizothorax	*grahami*		F, M	148	52M + 30SM + 66 ST/A	230			6.5 FD	6X	China (Yunnan)	Z-6, Z-7, Z-8
Schizothorax	*lissolabiatus*			148	38M + 32SM + 78 ST/A	218				6X	China (Yunnan)	Z-8
Schizothorax	*macropogon*			90–98						4X	(C. Asia)	C-74
Schizothorax	*prenanti*		F, M	148	28M + 40SM + 36ST + 44A	216	252			6X, ACN=156	China (Sichuan)	L-45, Y-14
Schizothorax	*pseudoaksaiensis issykkuli*			98–100						4X	Kazakhstan	C-74
Schizothorax	*taliensis*	*daliensis*	M	148	48M + 30SM + 70 ST/A	226			6.9 FD	6X	China (Yunnan)	Z-6, Z-7, Z-8
Schizothorax	*taliensis*	*yunnanensis daliensis*		148	48M + 28SM + 72 ST/A	224				6X	China (Yunnan)	Z-8
Schizothorax	*waltoni*		F, M	92	26M + 28SM + 22ST + 16A	146	168			4X	China (Central Asia)	Y-16
Schizothorax	sp. 1			148	50M + 28SM + 70 ST/A	226			7.0 FD	6X	China (Yunnan)	Z-6, Z-7, Z-8
Schizothorax	sp. 2			148	48M + 30SM + 70 ST/A	226				6X	China (Yunnan)	Z-8
Squaliobarbinae												
Ctenopharyngodon	*idellus*			48	16M + 20SM + 12ST	84	96		(2.4* FCM)	ACN=50	China (Beijing)	Y-11, F-5
Ctenopharyngodon	*idellus*			48	20M + 20SM + 8ST	88	96				China (Hunan)	L-63
Ctenopharyngodon	*idellus*		F, M	48	18M + 24SM + 6ST	90	96	6	2.1 FD	ACN=50	China (Hubei)	L-42, G-70, R-42, R-105
Ctenopharyngodon	*idellus*			48	18M + 22SM + 8 ST/A	88			2.0 FD	ACN=50	China (Yunnan)	Z-1, Z-8
Ctenopharyngodon	*idellus*			48	16M + 32SM	96	96		(2.0 FCM)		China (Beijing)	L-58, T-74
Ctenopharyngodon	*idellus*			48	18M + 20SM + 10ST	86	96	4	(1.9* FD)	ACN=50	China (Shashi)	Z-22, C-83
Ctenopharyngodon	*idellus*			48	14M + 24SM + 10 ST/A	86					Japan	O-16
Ctenopharyngodon	*idellus*		F, M	48	16M + 20 SM/ST + 12A		84		2.1 FD		Japan (Osaka)	O-18
Ctenopharyngodon	*idellus*		F, M	48	14M + 20SM + 8ST + 6A	82	90				India (WB)	M-27
Ctenopharyngodon	*idellus*			48	32 M/SM + 16 ST/A	80					former Yugoslavia	A-28
Ctenopharyngodon	*idellus*			48	28M + 20SM	96	96				Hungary	B-54
Mylopharyngodon	*piceus*			48	24M + 20SM + 4ST	92	96	4	(2.4* FCM)	ACN=50	China (Shashi)	Z-22, F-5
Mylopharyngodon	*piceus*			48	14M + 34 SM/ST		96			ACN=50	China (Shashi)	Z-27
Mylopharyngodon	*piceus*			48	16M + 28SM + 4ST	92	96				China (Shanghai)	L-68
Mylopharyngodon	*piceus*		M	48	18M + 24SM + 6ST	90	96		(1.9* FD)	ACN=50	China (Wuhan)	G-70, C-83
Squaliobarbus	*curriculus*		F, M	48	14M + 30SM + 4ST	92	96			ACN=50	China (Wuhan)	L-42, Y-15
Tincinae												
Tinca	*tinca*			48	14M + 12SM + 22ST	74	96				Bulgaria	P-37
Tinca	*tinca*			48	30 M/SM + 18 ST/A	78					Croatia	A-28
Tinca	*tinca*		F, M	48	36 M/SM + 12 ST/A	84				ACN=50	Bosnia-Herzegovina	B-23
Tinca	*tinca*			48	6M + 28SM + 14 ST/A	82			2.3 FCM		(Europe)	C-75, F-30
Tinca	*tinca*		M	48	8M + 12SM + 10ST + 18A	68	78			ACN=50	Italy	C-34
Tinca	*tinca*			48	20 M/SM + 16ST + 12A	68	84			ACN=50	Sweden	N-50

Table 6.11 Order CYPRINIFORMES (continued)

A		B	C	D	E	F	G	H	I	J	K	L
Current scientific name of taxon Superfamily/family/subfamily/species		Reported in karyotype paper	Sex	2n	Karyotype	NF_1	NF_2	Ag-NORs	Genome size (pg/cell)	Comments	Locality	Reference
Tinca	*tinca*		F, M	48	16M + 26 SM/ST + 6A		90	2			Elbe R., Danube R.	M-54
Tinca	*tinca*			48	18M + 18SM + 6ST + 6A	84	90				Hungary	B-54
Tinca	*tinca*			48	14M + 24 SM/ST + 10A		86		2.1 FD	ACN=50	France	H-2, H-4
Xenocyprinae												
Distoechodon	*tumirostris*		F, M	48	18M + 26SM + 4ST	92	96			ACN=50	China (Sichuan)	L-32, Y-15
Plagiognathops	*microlepis*		F, M	48	18M + 26SM + 4ST	92	96	4	1.8* FD	ACN=50	China (Hubei)	L-28, Z-25, C-83
Pseudobrama	*simoni*	*Acanthobrama*	F, M	48	18M + 26SM + 4ST	92	96			ACN=50	China (Hubei)	L-28, Y-15
Xenocypris	*argentea*		F, M	48	20M + 26SM + 2ST	94	96	4		ACN=50	China (Hubei)	L-28, Y-15, Z-25
Xenocypris	*davidi*		F, M	48	18M + 26SM + 4ST	92	96	4	2.5* FD	ACN=50	China (Hubei)	L-28, Z-25, C-83
Xenocypris	*fangi*		F, M	48	16M + 28SM + 4ST	92	96	4		ACN=50	China (Sichuan)	Z-25, Y-15
Xenocypris	*sechuanensis*		F, M	48	18M + 26SM + 4ST	92	96	4		ACN=50	China (Sichuan)	Z-25, Y-15
Psilorhynchidae												
Psilorhynchus	*balitora*	*Psylorhynchus bolitora*		50	24M + 16SM + 10A	90				ACN=50	India (Assam)	K-41
Psilorhynchus	*sucatio*		M	50	22M + 18SM + 10A	90				ACN=50	India (Assam)	K-43
Superfamily Cobitoidea												
Gyrinocheilidae												
Gyrinocheilus	*aymonieri*			48	4M + 4SM + 4ST + 36A	56	60	2	1.2* FCM, 1.0 FD, 1.3 FIA	ACN=50	S.E. Asia	A-82, O-48, H-41
Catostomidae												
Carpiodes	*carpio*			96–100						4X	USA (KS)	U-66
Carpiodes	*cyprinus*			100					4.5 FCM	4X, ACN=100	USA	F-23
Catostomus	*catostomus*			98	8M + 6SM + 84 ST/A	112			4.2 FIA	4X	Canada (Ont.)	B-11, H-41
Catostomus	*clarki*			96–100						4X	USA (AZ)	U-66
Catostomus	*commersoni*			98	12M + 14SM + 72 ST/A	124			(5.5 FCM)	4X	Canada (Ont.)	B-11, F-23
Catostomus	*commersoni*			96–98					(5.1 FIA)	4X	USA (MI)	U-66, H-41
Catostomus	*commersoni*		F, M	100	24 M/SM + 76 ST/A	124				4X, ACN=100	USA (ME)	B-64
Catostomus	*discobolus*			96–100						4X	USA (CO)	U-66
Catostomus	*latipinnis*			96–100						4X	USA (UT)	U-66
Cycleptus	*elongatus*			96–100						4X	USA (MO)	U-66
Erimyzon	*sucetta*			96–100						4X	USA (MI)	U-66
Hypentelium	*nigricans*			96–100						4X	USA (MI)	U-66
Ictiobus	sp.			96–100						4X	USA (MO)	U-66
Minytrema	*melanops*			<96						4X	USA (MI)	U-66
Moxostoma	*duquesnei*			96–100						4X	USA (MI)	U-66
Moxostoma	*erythrurum*			96–100					4.3 FCM	4X	USA (MI)	U-66, F-23
Moxostoma	*macrolepidotum*			96–100						4X	USA (MI)	U-66
Myxocyprinus	*asiaticus*		F, M	100	18 M/SM + 82 ST/A	118				4X, ACN=100	China (Sichuan)	L-37

Table 6.11 Order CYPRINIFORMES (continued)

A		B	C	D	E	F	G	H	I	J	K	L
Current scientific name of taxon Superfamily/family/subfamily/species		Reported in karyotype paper	Sex	2n	Karyotype	NF_1	NF_2	Ag-NORs	Genome size (pg/cell)	Comments	Locality	Reference
Myxocyprinus	*asiaticus*			100	6M + 8SM + 86 ST/A	114				4X, ACN=100	(China)	U-37
Myxocyprinus	*asiaticus*			100	4M + 14 SM/ST + 82A		118		4.0 FD	4X, ACN=100	(China)	S-138
Cobitidae												
Botiinae												
Botia	*almorhae*			100						4X, ACN=100	(Asia)	S-157
Botia	*birdi*		M	98	14M + 18SM + 4ST + 62A	130	134			4X	India (Kashmir)	K-37
Botia	*dario*			98	32M + 6SM + 60A	136	136			4X	India (Assam)	K-41
Botia	*dario*			90	12M + 2SM + 76A	104				4X	India (Assam)	R-65
Botia	*dayi*		F, M	98						4X	India (U.P.)	R-74
Botia	*histrionica*			100						4X	(Asia)	S-157
Botia	*lohachata*			100	28M + 26 SM/ST + 46A		154		1.9 FD	4X, ACN=100	(Asia)	S-144
Botia	*lohachata*		F, M	98	16M + 20SM + 62 ST/A	134				4X	India (U.P.)	R-74
Botia	*rostrata*			98	22M + 8SM + 2ST + 66A	128	130			4X	India (WB)	K-46
Botia	*striata*			100	20M + 26 SM/ST + 54A		146			4X, ACN=100	(Asia)	S-144
Chromobotia	*macracantha*	*Botia*		100	16M + 20 SM/ST + 64A		136		1.9 FD	4X, ACN=100	(Sumatra, Borneo)	S-144
Chromobotia	*macracantha*	*Botia*		98	28 M/SM + 70A	126			1.9 FD	4X	(Asia)	M-91
Sinibotia	*pulchra*	*Botia*		100	12M + 40 SM/ST + 48A		152			4X, ACN=100	(Asia)	S-144
Sinibotia	*pulchra*	*Botia*	F	100	10M + 12SM + 14ST + 64A	122	136			4X, ACN=100	China (Guilin)	Y-15
Syncrossus	*berdmorei*	*Botia*		100	8M + 14 SM/ST + 78A		122		2.0 FD	4X, ACN=100	(Myanmar)	S-144
Syncrossus	*helodes*	*Botia*		100	8M + 12 SM/ST + 80A		120		1.7 FD	4X, ACN=100	(Cambodia)	S-144
Syncrossus	*hymenophysa*	*Botia*		100	8M + 12 SM/ST + 80A		120			4X, ACN=100	(Asia)	S-144
Syncrossus	*hymenophysa*	*Botia*	F	90	4M + 86A	94				4X	India (Manipur)	R-65
Syncrossus	*reversa*			100						4X	(Indonesia)	S-157
Yasuhikotakia	*eos*			100						4X	(Laos)	S-157
Yasuhikotakia	*lecontei*	*Botia*		100	10M + 20 SM/ST + 70A		130			4X, ACN=100	(Thailand)	S-144
Yasuhikotakia	*modesta*	*Botia*		100	14M + 14 SM/ST + 72A		128		1.6* FCM, 1.2 FD	4X, ACN=100	(Thailand)	S-144, O-48
Yasuhikotakia	*morleti*	*Botia*		100	14M + 12 SM/ST + 74A		126		1.5 FD	4X, ACN=100	(Cambodia, Thailand)	S-144
Yasuhikotakia	*nigrolineata*			100						4X	(China, Yunnan)	S-157
Yasuhikotakia	*sidthimunki*	*Botia*		100	12M + 12 SM/ST + 76A		124			4X, ACN=100	(Thailand)	S-144
Leptobotiinae												
Leptobotia	*curta*			50	8M + 10SM + 32 ST/A	68		2	1.1 FD	ACN=50	Japan (Okayama)	S-17, S-144
Leptobotia	*elongata*		F, M	50	6M + 12SM + 18ST + 14A	68	86			ACN=50	China (Leshan)	Y-15
Leptobotia	*guilinensis*		F, M	50	6M + 8SM + 8ST + 28A	64	72			ACN=50	China (Guilin)	Y-15
Leptobotia	*pellegrini*		F, M	50	8M + 8SM + 12ST + 22A	66	78			ACN=50	China (Guilin)	Y-15
Leptobotia	*pellegrini*			50	6M + 22 SM/ST + 22A		78			ACN=50	(China)	S-144
Leptobotia	*taeniops*		F	50	6M + 10SM + 12ST + 22A	66	78			ACN=50	China (Nanchong)	Y-15
Leptobotia	*zebra*		F, M	50	6M + 10SM + 10ST + 24A	66	76			ACN=50	China (Guilin)	Y-15
Parabotia	*banarescui*			50	6M + 14 SM/ST + 30A		72		1.1 FD	2B, ACN=50	(China)	S-137

Table 6.11 Order CYPRINIFORMES (continued)

A		B	C	D	E	F	G	H	I	J	K	L
Current scientific name of taxon Superfamily/family/subfamily/species		Reported in karyotype paper	Sex	2n	Karyotype	NF_1	NF_2	Ag-NORs	Genome size (pg/cell)	Comments	Locality	Reference
Parabotia	*fasciata*		F, M	50	10M + 8SM + 14ST + 18A	68	82			ACN=50	China (Guangdong)	Y-15
Parabotia	*kwangsiensis*	*Botia*	F, M	50	10M + 6SM + 4ST + 30A	66	70			ACN=50	China (Guilin)	Y-15
Parabotia	*lijiangensis*		F, M	50	8M + 8SM + 10ST + 24A	66	76			ACN=50	China (Guilin)	Y-15
Parabotia	*maculosa*		F, M	50	8M + 8SM + 14ST + 20A	66	80			ACN=50	China (Guilin)	Y-15
Cobitinae												
Acanthopsis	*choiriorhynchos*			50	16M + 30 SM/ST + 4A		96			ACN=50	(Asia)	S-137
Acanthopsis	sp. 2			50	22M + 18SM + 4ST + 6A	90	94			ACN=50	Thailand	D-28
Canthophrys	*gongota*	*Somileptes*		50	8M + 4SM + 2ST + 36A	62	64			ACN=50	India (WB)	K-46
Cobitis	*biwae*		F, M	48	18M + 30SM	96	96			2X	Japan	K-61, K-91
Cobitis	*biwae*			48	20M + 24SM + 4A	92	92			2X	Japan (Tochigi)	U-30
Cobitis	*biwae*			46	20M + 22SM + 4A	88	88			2X	Japan (Toyama)	U-30
Cobitis	*biwae*		F, M	48	42 M/SM + 6 ST/A	90				2X	Japan (Shiga)	T-1
Cobitis	*biwae*			48	18M + 26 SM/ST + 4A		92		4.3 FD	2X	Japan (Okayama)	S-137
Cobitis	*biwae*			48	20M + 22 SM/ST + 6A		90			2X	Japan (Shiga)	U-27, U-29
Cobitis	*biwae*			48	16M + 24 SM/ST + 8A		88			2X	Japan (Kochi)	U-29
Cobitis	*biwae*			48	16M + 22 SM/ST + 10A		86	2		2X	Japan (Shimane)	U-29
Cobitis	*biwae*			96	32M + 54 SM/ST + 10A		182		(6.7* FCM)	4X	W. Japan	U-27, U-29, O-48
Cobitis	*biwae*		F, M	96	36M + 52SM + 8ST	184	192			4X	W. Japan	K-61, K-91
Cobitis	*biwae*		F, M	96	58 M/SM + 38 ST/A	154				4X	Japan (Hyogo)	H-16, O-14
Cobitis	*biwae*			96	34M + 50 SM/ST + 12A		180		7.2 FD	4X	Japan (Okayama)	S-137
Cobitis	*calderoni*		F, M	50	6M + 14SM + 30A	70				ACN=50	Portugal, Spain	M-4
Cobitis	*calderoni*			50	6M + 12SM + 32 ST/A	68				ACN=50	Portugal	V-25
Cobitis	*hankugensis*	*taenia taenia*		48	14M + 4SM + 30 ST/A	66				2X	Korea	K-56
Cobitis	*hankugensis*	*sinensis*		48	14M + 4SM + 30 ST/A	66			(2.8 FCM)	2X	Korea (Nagdong R.)	K-116, V-102
Cobitis	*koreensis pumilus*			50	10M + 12SM + 28A	72				ACN=50	Korea	K-56
Cobitis	cf. *laosensis*			48	14M + 4SM + 30 ST/A	66				2X	Korea	K-129
Cobitis	*lutheri*	*taenia lutheri*		50	10M + 6SM + 34 ST/A	66				ACN=50	Korea (Yongjin)	K-56
Cobitis	*lutheri*	*taenia lutheri*		49	11M + 6SM + 32A	66				2X	Korea (Tamjin R.)	K-129
Cobitis	*lutheri*	*taenia lutheri*		51	9M + 6SM + 36A	66				2X	Korea (Tamjin R.)	K-129
Cobitis	*lutheri*	*taenia lutheri*		50	12M + 4SM + 34A	66				2X, ACN=50	Korea (Han R.)	U-35
Cobitis	*lutheri*			50	12M + 8SM + 30 ST/A	70			3.5, 3.9 FCM	2X, ACN=50	Russia (Amur basin)	V-86, V-98, V-102
Cobitis	*maroccana*		F, M	50	6M + 12SM + 32A	68	68	2		ACN=50	Portugal, Spain	M-4, V-73
Cobitis	*melanoleuca*			50	8M + 16SM + 26 ST/A	74				2X, ACN=50	Mongolia	V-98
Cobitis	*melanoleuca*			50	6M + 16SM + 28 ST/A	72			(3.1 FCM)	2X, ACN=50	Russia (Amur basin)	V-86, V-98
Cobitis	*melanoleuca gladkovi*	*granoei*		50	8M + 18SM + 24 ST/A	76				2X, ACN=50	Russia	V-24, V-98
Cobitis	*melanoleuca gladkovi*			50	8M + 18SM + 24 ST/A	76				2X, ACN=50	Russia (Don basin)	V-98
Cobitis	*pacifica*	*taenia granoei*		50	24 M/SM + 26 ST/A	74				2X, ACN=50	Korea	K-128
Cobitis	*pacifica*	*granoei*		50	6M + 24SM + 20ST	80	100			2X, ACN=50	Korea	L-15
Cobitis	*sinensis*		F	40	20M + 8SM + 4ST + 8A	68	72			2X	China (Guilin)	Y-15

Table 6.11 Order CYPRINIFORMES (continued)

A		B	C	D	E	F	G	H	I	J	K	L
Current scientific name of taxon Superfamily/family/subfamily/species		Reported in karyotype paper	Sex	2n	Karyotype	NF_1	NF_2	Ag-NORs	Genome size (pg/cell)	Comments	Locality	Reference
Cobitis	*sinensis*		F	90	26M + 18SM + 16ST + 30A	134	150			4X	China (Guilin)	Y-15
Cobitis	*taenia*		F, M	48	12M + 18SM + 18 ST/A	78		2		2X	Poland	B-48, B-52
Cobitis	*taenia*		F	74	21M + 30SM + 23 ST/A	125				3X	Poland	B-48, B-52
Cobitis	*taenia*			50	8M + 16SM + 6ST + 20A	74	80			2X, ACN=50	Russia (Volga R.)	V-15
Cobitis	*taenia*			48	12M + 12SM + 10ST + 14A	72	82			2X	Russia (Volga R.)	V-15
Cobitis	*taenia*									3X, 4X, 5X	Russia (Volga R.)	V-15
Cobitis	*taenia*			48	10M + 18SM + 20 ST/A	76			(3.5 FCM)	ACN=50	Russia	V-24, V-102
Cobitis	*taenia*		F, M	50	12M + 6SM + 16ST + 16A	68	84			2X, ACN=50	Italy	C-34
Cobitis	sp.			49	9M + 24SM + 16 ST/A	82				2X	Russia (Dnepr R.)	V-19
Cobitis	sp.			50	8M + 24SM + 18 ST/A	82				2X, ACN=50	Russia (Dnepr R.)	V-19
Cobitis	sp.			74	23M + 25SM + 26 ST/A	122			6.0 FCM	3X	Russia (Dnepr R.)	V-19, V-102
Cobitis	sp.			98					7.6 FCM	4X	Russia	V-102
Cobitis	*taenia satunini*			50						2X	(Asia)	V-72
Cobitis	*taenia striata*	small race	F	50	16 M/SM + 34 ST/A	66				$X_1X_1X_2X_2$	Japan (Okayama)	S-3, S-4
Cobitis	*taenia striata*	small race	M	49	17 M/SM + 32 ST/A	66				X_1X_2Y, ACN=50	Japan (Okayama)	S-3, S-4
Cobitis	*taenia striata*	middle race	F, M	50	16 M/SM + 34A	66				2X, ACN=50	Japan (Okayama)	S-3
Cobitis	*taenia striata*	large race	F, M	98	42 M/SM + 56A	140				4X	Japan (Lake Biwa)	S-3
Cobitis	*taenia striata*		F	50	16 M/SM + 34 ST/A	66				2X, ACN=50	Japan (Yodo R.)	T-1
Cobitis	*taenia striata*			50	12M + 4SM + 34A	66	66		(3.9* FCM)	2X, ACN=50	W. Japan	U-27, U-29, O-48
Cobitis	*taenia striata*			98	20M + 22 SM/ST + 56A		140			4X	Japan (Shiga)	U-27, U-29
Cobitis	*taenia taenia*			50	12M + 4SM + 34 ST/A	66				2X, ACN=50	Japan (Fukuoka)	U-29
Cobitis	*taenia taenia*			86	32M + 32 SM/ST + 22A		150	2			Japan (Fukuoka, Nagasaki)	U-29
Cobitis	*taenia taenia*			94	26M + 32 SM/ST + 36A		152	2		4X	Japan (Fukuoka, Oita)	U-29
Cobitis	*taenia taenia*		M	50	38 M/SM + 12 ST/A	88				2X, ACN=50	Bosnia	S-81
Cobitis	*taenia taenia*		F	75	57 M/SM + 18 ST/A	132				3X	Bosnia	S-81
Cobitis	*taenia taenia*		F, M	86	64 M/SM + 22 ST/A	150					Japan (Saga)	T-1
Cobitis	*takatsuensis*			48	12M + 18 SM/ST + 18A		78			2X	W. Japan	K-60
Cobitis	*takatsuensis*			48	12M + 12SM + 24A	72				2X	Japan	K-117
Cobitis	*tetralineata*	*taenia striata*		50	10M + 6SM + 34 ST/A	66				2X, ACN=50	Korea	K-56
Cobitis	*vardarensis*			50	26M + 20SM + 4ST	96	100			2X, ACN=50	Macedonia	R-104
Iksookimia	*choii*			50	18 M/SM + 32 ST/A	68				2X, ACN=50	Korea	K-127
Iksookimia	*choii*	*Cobitis*		50	8M + 10SM + 8ST + 24A	68	76			2X, ACN=50	Amur basin	V-98
Iksookimia	*hugowolfeldi*			50	24 M/SM + 26 ST/A	74				2X, ACN=50	Korea	K-128
Iksookimia	*koreensis*	*Cobitis*		50	12M + 8SM + 30 ST/A	70				2X, ACN=50	Korea (Han R., Geum R.)	U-35
Iksookimia	*koreensis*	*Cobitis*		50	10M + 12SM + 28A	72				2X, ACN=50	Korea (Mankyong R.)	K-56
Iksookimia	*longicorpa*	*Cobitis longicorpus*		50	14M + 6SM + 30 ST/A	70				2X, ACN=50	Korea (Nakton R.)	K-116
Iksookimia	*longicorpa*	*Cobitis longicorpus*	F, M	50	12M + 8SM + 30 ST/A	70				2X, ACN=50	Korea (Nakton R.)	K-56, U-35
Iksookimia	*pumila*		F	50	22 M/SM + 28 ST/A	72				2X, ACN=50	Korea	V-98
Iksookimia	*yongdokensis*			100	44 M/SM + 56 ST/A	144				4X	Korea (Kyongsangbuk-do)	K-127

Table 6.11 Order CYPRINIFORMES (continued)

A	B	C	D	E	F	G	H	I	J	K	L
Current scientific name of taxon Superfamily/family/subfamily/species	Reported in karyotype paper	Sex	2n	Karyotype	NF_1	NF_2	Ag-NORs	Genome size (pg/cell)	Comments	Locality	Reference
Kichulchoia brevifasciata		F, M	48	6M + 10SM + 32 ST/A	64				2X, ACN=48	Korea (Jeollanam-do)	K-128
Koreocobitis rotundicaudata	*Cobitis*		50	10M + 4SM + 36 ST/A	64				2X, ACN=50	Korea (Han R.)	U-35
Lepidocephalichthys berdmori		M	62	24 M/SM + 38A	86					India (Portonovo)	N-13
Lepidocephalichthys guntea		F	50	18M + 18SM + 10ST + 4A	86	96	2		1B, 2X, ACN-50	India (Jammu)	S-57, K-136
Lepidocephalichthys guntea		M	50	18M + 18SM + 10ST + 4A	86	96	2		2B, 2X, ACN=50	India (Jammu)	S-57, K-136
Lepidocephalichthys guntea var. *balgara*		F, M	50	22M + 14SM + 2ST + 12A	86	88			2X, ACN=50	India (WB)	B-4
Misgurnus anguillicaudatus		F, M	50	14 M/SM + 36 ST/A	64			(2.8 BFA)	2X, ACN=50	Japan (Hokkaido)	H-13, H-15, H-16, O-14
Misgurnus anguillicaudatus		F, M	50	10M + 4SM + 36 ST/A	64				ACN=50	Japan	O-28, I-15
Misgurnus anguillicaudatus			48	12M + 4SM + 32 ST/A	64				ACN=50	Japan (Shiga)	O-28
Misgurnus anguillicaudatus		F	75	15M + 6SM + 54 ST/A	96				3X	Japan (Shiga)	O-28
Misgurnus anguillicaudatus		F, M	100	20M + 8SM + 72 ST/A	128				4X	Japan (Shiga)	O-28
Misgurnus anguillicaudatus			50	8M + 6 SM/ST + 36A		64		3.7 FD	2X, ACN=50	Japan	S-137
Misgurnus anguillicaudatus			50					2.4 FCM	2X	S. Korea	L-15, P-70
Misgurnus anguillicaudatus			50	10M + 4SM + 36 ST/A	64				2X, ACN=50	N. Korea	V-28
Misgurnus anguillicaudatus		F	50	8M + 6SM + 36A	64	64			2X, ACN=50	China (Sichuan)	L-44, Y-15
Misgurnus anguillicaudatus		F, M	100	16M + 12SM + 72A	128	128		4.6, 4.0* FD	4X	China (Hubei)	L-27, L-41, Y-15, C-83
Misgurnus bufoensis			48	10M + 2SM + 36 ST/A	60					N. Korea	V-98
Misgurnus fossilis			100	36 M/SM/ST + 64A		136			4X	Rumania	R-32
Misgurnus mohoity			50							Russia (Amur basin)	V-98
Misgurnus nikolskyi			50	8M + 6SM + 36 ST/A	64				2X, ACN=50	Russia (Far East)	V-28
Misgurnus nikolskyi			50	10M + 4SM + 36 ST/A	64				ACN=50	Russia (Amur basin)	V-98
Niwaella delicata	*Cobitis*	F, M	50	18 M/SM + 32 ST/A	68				2X, ACN=50	Japan (Shiga, Gifu)	O-14, T-1, H-16
Niwaella delicata		F, M	50	6M + 14 SM/ST + 30A		70			2X, ACN=50	W. Japan	K-60
Niwaella multifasciata			50	38 M/SM + 12 ST/A	88				2X, ACN=50	Korea	K-128
Pangio borneensis	*Acanthophthalmus*		50	10M + 18 SM/ST + 22A		78		2.0 FD	2X, ACN=50	(Asia)	S-137
Pangio khulii	*Acanthophthalmus*	F	50	14 M/SM + 12ST + 24A	64	76		2.0 FD	2X, ACN=50	(Asia)	M-91
Pangio khulii	*Acanthophthalmus*		50	14M + 4ST + 32A	64	68		(2.4 BFA)	2X, ACN=50	Thailand	D-20, H-13
Pangio pangia	*Acanthophthalmus*		50	16M + 10SM + 24A	76	76				India	L-1
Paramisgurnus dabryanus		F, M	48	12M + 4SM + 32A	64			2.2, 2.0* FD	0-1 B, ACN=50	China (Hubei)	L-27, L-41, Y-15, C-83
Paramisgurnus dabryanus	*Misgurnus mizolepis*		48	12M + 4SM + 32 ST/A	64				2X, ACN=50	Korea (Geum R.)	U-35
Sabanejewia balcanica	*aurata balcanica*	F, M	50	4M + 12SM + 34 ST/A	66		2		2X, ACN=50	Slovakia	R-19, V-26
Sabanejewia caspia			50	4M + 6SM + 22ST + 18A	60	82			2X		V-72
Sabanejewia caspia			50	4M + 18SM + 28 ST/A	72				2X	(Caspian Sea basin)	V-108
Sabanejewia kubanica	*aurata kubanica*		50	6M + 14SM + 30 ST/A	70				2X, ACN=50	Russia (Kuban R).	V-25
Sabanejewia kubanica?	*aurata*		50	6M + 14SM + 30 ST/A	70				2X, ACN=50		V-72
Sabanejewia larvata		F, M	50	4M + 6SM + 22ST + 18A	60	82			2X, ACN=50	Italy (near Torino)	L-65

Table 6.11 Order CYPRINIFORMES (continued)

A		B	C	D	E	F	G	H	I	J	K	L
Current scientific name of taxon Superfamily/family/subfamily/species		Reported in karyotype paper	Sex	2n	Karyotype	NF_1	NF_2	Ag-NORs	Genome size (pg/cell)	Comments	Locality	Reference
Balitoridae (= Homalopteridae)												
Nemacheilinae												
Acanthocobitis	*botia*	*Nemacheilus*	F, M	50	20M + 16 SM/ST + 14A		86			ACN=50	(India)	R-51
Acanthocobitis	*botia*	*Nemacheilus aureus*		50	28M + 4SM + 18A	82	82			ACN=50	India (Orissa)	K-41
Barbatula	*barbatula*	*Nemacheilus barbatulus*	F, M	50	8M + 20SM + 22A	78	78	2		ACN=50	Swiss, Spain	M-4
Barbatula	*barbatula*	*Nemacheilus barbatulus*	F, M	50	8M + 20SM + 22 ST/A	78		2		ACN=50	Poland	B-51
Barbatula	*barbatula*	*Nemacheilus barbatulus*	F, M	50	20 M/SM + 30 ST/A	70				ACN=50	Bosnia-Herzegovina	S-83
Barbatula	*barbatula*	*Nemacheilus barbatulus*		50	8M + 20 SM/ST + 22A		78			ACN=50	Slovakia	C-73
Barbatula	*barbatula*	*Nemacheilus barbatulus*		75	12M + 30 SM/ST + 33A		117			3X	Slovakia	C-73
Barbatula	*barbatula*	*Nemachilus barbatulus*		50	6M + 12 SM/ST + 32A		68			ACN=50		V-72
Barbatula	*toni*		F, M	50	16 M/SM + 34 ST/A	66				ACN=50	Japan (Hokkaido)	H-16
Barbatula	*toni*			50	6M + 24 SM/ST + 20A		80		1.1 FD	ACN=50	Japan (Hokkaido)	S-137
Barbatula	*toni*	*Nemacheilus*	M	50	8M + 12SM + 30 ST/A	70				ACN=50	Korea (Han R., Nakton R.)	K-58
Barbatula	*toni*	*Nemacheilus*	F	50	8M + 6SM + 36 ST/A	64				ACN=50	Korea (Maeubchon R.)	K-58
Lefua	*costata*		M	50	6M + 4SM + 40 ST/A	60				ACN=50	Korea	K-58
Lefua	*echigonia*			50	4M + 8 SM/ST + 38A		62		1.0 FD	ACN=50	Japan (Kanagawa)	S-124, S-137
Lefua	*echigonia*			50	4M + 8SM + 38 ST/A	62				ACN=50	(Japan)	U-30
Lefua	*nikkonis*		F, M	50	12 M/SM + 38 ST/A	62				ACN=50	Japan (Hokkaido)	H-16
Lefua	*nikkonis*			50	4M + 24 SM/ST + 22A		78	2	0.9 FD	ACN=50	Japan (Hokkaido)	S-145
Micronemacheilus	*pulcher*		F, M	50	10M + 12SM + 12ST + 16A	72	84			ACN=50	China (Guilin)	Y-15
Micronemacheilus	*pulcher*			50	8M + 30 SM/ST + 12A		88	2		ACN=50	(China)	S-145
Nemacheilus	*mooreh*	*sinuatus*	M	50	24M + 22SM + 4A	96	96			ACN=50	India (Assam)	K-46
Nemacheilus	*selangoricus*			40	6M + 2SM + 32A	48	48			ACN=42	(Asia)	S-137
Paracobitis	*potanini*	*Nemacheilus*	F, M	48	14M + 26SM + 6ST + 2A	88	94			ACN=48	China (Ya'an, Sichuan)	Y-15
Schistura	*fasciolata*	*Nemacheilus fasciolatus*	M	44	10M + 8SM + 10ST + 16A	62	72			ACN=48	China (Guilin)	Y-15
Schistura	*fasciolata*	*Nemacheilus fasciolatus*	F, M	50	12M + 14SM + 14ST + 10A	76	90			ACN=50	China (Guilin)	Y-15
Schistura	*incerta*	*Nemacheilus incertus*	F, M	50	8M + 8SM + 4ST + 30A	66	70			ACN=50	China (Guilin)	Y-15
Schistura	*prashadi*	*Nemacheilus*		50	10M + 10SM + 4ST + 26A	70	74			ACN=50	India	L-1
Schistura	*rupecula*	*Nemacheilus rupicola*	F, M	50	20M + 8SM + 22 ST/A	78				ACN=50	India (U.P.)	R-74
Schistura	*savona*	*Nemacheilus*	F, M	36	20M + 6SM + 2ST + 8A	62	64				India (Bihar)	K-41
Triplophysa	*kungessana*	*Nemacheilus dorsalis*		50	16 M/SM + 34 ST/A	66				ACN=50	Central Asia	M-57
Triplophysa	*siluroides*			48					2.0* FD		Tibet	C-83
Triplophysa	*stoliczkai*	*Nemacheilus*		50	12 M/SM + 38 ST/A	62				ACN=50	Central Asia	M-57
Triplophysa	*strauchii*	*Nemacheilus*		50	18 M/SM + 32 ST/A	68				ACN=50	Central Asia	M-57
Balitorinae												
Beaufortia	*kweichowensis kweichowensis*			48	10M + 26 SM/ST + 12A		84	2			China	S-145
Jinshaia	*abbreviata*	*Hemimyzon abbreviata*		50							China	Y-15
Pseudogastromyzon	*myersi*			50	6M + 16 SM/ST + 28A		72		0.9 FD	ACN=50	(Asia)	S-139
Sinohomaloptera	*kwangsiensis*	*Homaloptera hoffmanni*		50	6M + 18 SM/ST + 26A		74		0.9 FD	ACN=50	(Asia)	S-139
Vanmanenia	*pingchowensis*		F, M	50	8M + 6SM + 6ST + 30A	64	70			ACN=50	China (Guilin)	Y-15
Vaillantellidae												
Vaillantella	*maassi*			50	26 M/SM + 24ST	76	100	2		ACN=50	Thailand	B-72

Table 6.12 Order CHARACIFORMES

A		B	C	D	E	F	G	H	I	J	K	L
Current scientific name of taxon Suborder/family/subfamily/species		Reported in karyotype paper	Sex	2n	Karyotype	NF_1	NF_2	Ag-NORs	Genome size (pg/cell)	Comments	Locality	Reference
Suborder Characoidei												
Acestrorhynchidae												
Acestrorhynchus	*altus*		F	50	8M + 22SM + 14ST + 6A	80	94			ACN=52	Brazil (MS)	F-2
Acestrorhynchus	*lacustris*		F, M	50	12M + 32SM + 4ST + 2A	94	98			ACN=52	Brazil (SP)	F-2
Acestrorhynchus	*pantaneiro*			50	36 M/SM + 14 ST/A	86			1.7 FD		Argentina	F-20, C-94
Alestiidae												
Arnoldichthys	*spilopterus*			56					2.4 BFA		(Africa)	H-13
Anostomidae												
Abramites	*hypselonotus*			54	30M + 22SM + 2ST	106	108				Argentina	F-20, O-50
Abramites	*hypselonotus*	*solarii*	F, M	54	54 M/SM	108	108	2		ACN=56	Brazil (AM)	M-44
Anostomus	*anostomus?*	*anostomus*		54	54 M/SM	108	108		2.8 BFA		(S. America)	O-22, H-13
Anostomus	*ternetzi*		F, M	54	54 M/SM	108	108	2		ACN=56	Brazil (AM)	M-44, O-50
Leporellus	*vittatus*		F, M	54	30M + 24SM	108	108	2		ACN=56	Brazil (SP)	G-2, G-4
Leporinus	*acutidens*			54	28M + 26SM	108	108				Argentina	F-20
Leporinus	*affinis*			54	54 M/SM	108	108				(Brazil)	P-88
Leporinus	*amblyrhynchus*			54	54 M/SM	108	108				(S. America)	A-92
Leporinus	*brunneus*			54	54 M/SM	108	108				(Brazil, Venezuela)	A-92
Leporinus	*conirostris*		F	54	52 M/SM + 1SM + 1ST	107	108	2		ZW, ACN=54	Brazil (SP)	G-10
Leporinus	*conirostris*		M	54	52 M/SM + 2SM	108	108	2		ZZ, ACN=54	Brazil (SP)	G-10
Leporinus	*copelandii*			54	26M + 28SM	108	108				Brazil (SP)	B-32
Leporinus	*cylindriformis*			54	26M + 28SM	108	108				(Brazil)	A-92
Leporinus	*desmotes*		F	54	54 M/SM	108	108	2		ACN=56	Brazil (TO)	M-155

Table 6.12 Order CHARACIFORMES (continued)

A		B	C	D	E	F	G	H	I	J	K	L
Current scientific name of taxon Suborder/family/subfamily/species		Reported in karyotype paper	Sex	2n	Karyotype	NF_1	NF_2	Ag-NORs	Genome size (pg/cell)	Comments	Locality	Reference
Leporinus	sp		F	54	32M + 21SM + 1ST	107	108	2		ZW, ACN=56	Brazil (MS)	G-91, K-135
Leporinus	sp		M	54	32M + 22SM	108	108	2		ZZ, ACN=56	Brazil (MS)	G-91
Pseudanos	*trimaculatus*		F, M	54	54 M/SM	108	108	2		ACN=56	Brazil (AM)	M-44
Rhytiodus	*microlepis*			54							Brazil (AM)	B-32
Schizodon	*altoparanae*			54	54 M/SM	108	108	2		ACN=56	Brazil (SP)	M-43
Schizodon	*borellii*			54	54 M/SM	108	108				Argentina	F-20
Schizodon	*borellii*			54	32M + 22SM	108	108		(3.0 FD)		Brazil (SP)	B-32, C-94
Schizodon	*borellii*		F, M	54	54 M/SM	108	108	2		ACN=56	Brazil (Paraguay R.)	M-154
Schizodon	*fasciatus*			54	54 M/SM	108	108				Brazil	O-50
Schizodon	*fasciatus*			54	28M + 26SM	108	108	2			Brazil	P-88
Schizodon	*intermedius*		F, M	54	54 M/SM	108	108	2	2.9 FD	ACN=56	Brazil (SP)	M-43, C-93
Schizodon	*isognathus*	*isognathum*		54	54 M/SM	108	108	2		ACN=56	Brazil (Paraguay R.)	M-154
Schizodon	*knerii*			54	54 M/SM	108	108	2		ACN=56	Brazil (MG)	M-43
Schizodon	*nasutus*		F, M	54	32M + 22SM	108	108	2	3.1 FD	ACN=56	Brazil (SP)	G-2, G-4, M-43, C-93
Schizodon	*nasutus*			54	32M + 22SM	108	108	2		0-1 B	Argentina	P-81, M-43, F-20, C-98
Schizodon	*platae*			54	54 M/SM	108	108				Argentina	F-20
Schizodon	*vittatus*			54	54 M/SM	108	108	2		ACN=56	Brazil (MT)	M-43
Characidae												
Aphyocharacinae												
Aphyocharax	*alburnus*			50*						ACN=50	(S. America)	S-29
Aphyocharax	*anisitsi*	*rubropinnis*		50*					2.7 FD, 3.4 BFA	ACN=50	(S. America)	S-29, C-94, H-13
Aphyocharax	*dentatus*	*difficilis*	F, M	50	2M + 2SM + 2ST + 44A	54	56	4	(2.5 FD)	ACN=50	Brazil (Parana R.)	S-98, C-94
Inpaichthys	*kerri*		M	52	12M + 26SM + 14ST	90	104			ACN=54	(Brazil)	A-91
Bryconinae												
Brycon	*cephalus*		F, M	50	24M + 20SM + 6ST	94	100	2		ACN=52	Brazil (AM)	M-37
Brycon	cf. *cephalus*		F, M	50	24M + 26 SM/ST		100	2		ACN=52	Brazil (AM)	A-22
Brycon	cf. *cephalus*			50	20M + 22SM + 8ST	92	100	2			Brazil (AM)	P-88
Brycon	cf. *cephalus*	cf. *erythropterum*		50	20M + 22SM + 8ST	92	100	2			Brazil (AM)	P-88
Brycon	*falcatus*	*brevicauda*	F, M	50	20M + 24SM + 6ST	94	100	2		ACN=52	Brazil (MT)	M-37
Brycon	*hilarii*	*microlepis*	F, M	50	20M + 24SM + 6ST	94	100	2	2.4 FD	ACN=52	Brazil (MT)	M-37, M-151, C-94
Brycon	*insignis*		F, M	50	22M + 20SM + 8ST	92	100	2		ACN=52	Brazil (SP)	M-37
Brycon	*insignis*		F, M	50	26M + 24 SM/ST		100	2		ACN=52	Brazil (SP)	A-22
Brycon	cf. *nattereri*	cf. *reinhardtii*	M	50	22M + 28 SM/ST		100	2		ACN=52	Brazil (SP)	A-22
Brycon	*orbignyanus*		F, M	50	24M + 22SM + 4ST	96	100	2		ACN=52	Brazil (PR)	M-37

Table 6.12 Order CHARACIFORMES (continued)

A		B	C	D	E	F	G	H	I	J	K	L
Current scientific name of taxon Suborder/family/subfamily/species		Reported in karyotype paper	Sex	2n	Karyotype	NF_1	NF_2	Ag-NORs	Genome size (pg/cell)	Comments	Locality	Reference
Leporinus	sp		F	54	32M + 21SM + 1ST	107	108	2		ZW, ACN=56	Brazil (MS)	G-91, K-135
Leporinus	sp		M	54	32M + 22SM	108	108	2		ZZ, ACN=56	Brazil (MS)	G-91
Pseudanos	*trimaculatus*		F, M	54	54 M/SM	108	108	2		ACN=56	Brazil (AM)	M-44
Rhytiodus	*microlepis*			54							Brazil (AM)	B-32
Schizodon	*altoparanae*			54	54 M/SM	108	108	2		ACN=56	Brazil (SP)	M-43
Schizodon	*borellii*			54	54 M/SM	108	108				Argentina	F-20
Schizodon	*borellii*			54	32M + 22SM	108	108		(3.0 FD)		Brazil (SP)	B-32, C-94
Schizodon	*borellii*		F, M	54	54 M/SM	108	108	2		ACN=56	Brazil (Paraguay R.)	M-154
Schizodon	*fasciatus*			54	54 M/SM	108	108				Brazil	O-50
Schizodon	*fasciatus*			54	28M + 26SM	108	108	2			Brazil	P-88
Schizodon	*intermedius*		F, M	54	54 M/SM	108	108	2	2.9 FD	ACN=56	Brazil (SP)	M-43, C-93
Schizodon	*isognathus*	*isognathum*		54	54 M/SM	108	108	2		ACN=56	Brazil (Paraguay R.)	M-154
Schizodon	*knerii*			54	54 M/SM	108	108	2		ACN=56	Brazil (MG)	M-43
Schizodon	*nasutus*		F, M	54	32M + 22SM	108	108	2	3.1 FD	ACN=56	Brazil (SP)	G-2, G-4, M-43, C-93
Schizodon	*nasutus*			54	32M + 22SM	108	108	2		0-1 B	Argentina	P-81, M-43, F-20, C-98
Schizodon	*platae*			54	54 M/SM	108	108				Argentina	F-20
Schizodon	*vittatus*			54	54 M/SM	108	108	2		ACN=56	Brazil (MT)	M-43
Characidae												
Aphyocharacinae												
Aphyocharax	*alburnus*			50*						ACN=50	(S. America)	S-29
Aphyocharax	*anisitsi*	*rubropinnis*		50*					2.7 FD, 3.4 BFA	ACN=50	(S. America)	S-29, C-94, H-13
Aphyocharax	*dentatus*	*difficilis*	F, M	50	2M + 2SM + 2ST + 44A	54	56	4	(2.5 FD)	ACN=50	Brazil (Parana R.)	S-98, C-94
Inpaichthys	*kerri*		M	52	12M + 26SM + 14ST	90	104			ACN=54	(Brazil)	A-91
Bryconinae												
Brycon	*cephalus*		F, M	50	24M + 20SM + 6ST	94	100	2		ACN=52	Brazil (AM)	M-37
Brycon	cf. *cephalus*		F, M	50	24M + 26 SM/ST		100	2		ACN=52	Brazil (AM)	A-22
Brycon	cf. *cephalus*			50	20M + 22SM + 8ST	92	100	2			Brazil (AM)	P-88
Brycon	cf. *cephalus*	cf. *erythropterum*		50	20M + 22SM + 8ST	92	100	2			Brazil (AM)	P-88
Brycon	*falcatus*	*brevicauda*	F, M	50	20M + 24SM + 6ST	94	100	2		ACN=52	Brazil (MT)	M-37
Brycon	*hilarii*	*microlepis*	F, M	50	20M + 24SM + 6ST	94	100	2	2.4 FD	ACN=52	Brazil (MT)	M-37, M-151, C-94
Brycon	*insignis*		F, M	50	22M + 20SM + 8ST	92	100	2		ACN=52	Brazil (SP)	M-37
Brycon	*insignis*		F, M	50	26M + 24 SM/ST		100	2		ACN=52	Brazil (SP)	A-22
Brycon	cf. *nattereri*	cf. *reinhardtii*	M	50	22M + 28 SM/ST		100	2		ACN=52	Brazil (SP)	A-22
Brycon	*orbignyanus*		F, M	50	24M + 22SM + 4ST	96	100	2		ACN=52	Brazil (PR)	M-37

Table 6.12 Order CHARACIFORMES (continued)

A		B	C	D	E	F	G	H	I	J	K	L
Current scientific name of taxon Suborder/family/subfamily/species		Reported in karyotype paper	Sex	2n	Karyotype	NF_1	NF_2	Ag-NORs	Genome size (pg/cell)	Comments	Locality	Reference
Brycon	*orthotaenia*	*lundii*	F, M	50	22M + 24SM + 4ST	96	100	2		ACN=52	Brazil (MG)	M-37, M-151
Brycon	cf. *pesu*			50	20M + 22SM + 8ST	92	100	2			(Brazil)	P-88
Brycon	sp.		F	50	20M + 22SM + 8ST	92	100	2		ACN=52	Brazil (MT)	M-37
Brycon	sp.			50							Brazil (AM)	O-50
Salminus	*brasiliensis*	*maxillosus*	F, M	50	14M + 30SM + 6ST	94	100	2			Brazil (SP)	O-81
Salminus	*hilarii*		F, M	50	14M + 30SM + 6ST	94	100	2	(2.6 FD)		Brazil (SP)	O-81, C-93
Salminus	*hilarii*		F, M	50	12M + 18SM + 20ST	80	100			ACN=52	Brazil (SP)	M-151
Chalceinae												
Chalceus	*macrolepidotus*	Characinidae		54	32 M/SM + 22ST	86	108		2.1 FD	ACN=56	(S. America)	M-91
Chalceus	*macrolepidotus*			52	44 M/SM + 8ST	96	104		(2.2 BFA)		(S. America)	O-22, H-13
Characinae												
Charax	*leticiae*			52					2.9 FD		(S. America)	C-25, C-94
Charax	sp.		F	52	22 M/SM + 30 ST/A	74		2			Brazil (Acre)	C-25
Cynopotamus	cf. *kincaidi*			52	30 M/SM + 22 ST/A	82					Argentina	F-20
Exodon	*paradoxus*		F	52	2M + 4SM + 10ST + 36A	58	68		(3.4 BFA)	ACN=52	Brazil (MT)	V-32, H-13
Exodon	*paradoxus*			52*			(74)			ACN=52	(S. America)	S-152
Galeocharax	*gulo*		F, M	52	6M + 24SM + 22ST	82	104	2		ACN=52	Brazil (MT)	V-32
Galeocharax	*humeralis*			52	36 M/SM + 16 ST/A	88					Argentina	F-20
Galeocharax	*knerii*		F, M	52	6M + 26SM + 20ST	84	104		3.2 FD	ACN=52	Brazil (SP, MG)	F-2, C-93
Phenacogaster	aff. *microstictus*	cytotype A		52*							(S. America)	S-30
Phenacogaster	aff. *microstictus*	cytotype B		50*							(S. America)	S-30
Phenacogaster	cf. *pectinatus*		M	46	12M + 2ST + 32A	58	60	2			Brazil (Acre)	C-26
Roeboides	*descalvadensis*	*paranensis*		52	4M + 20SM + 8ST + 20A	76	84	2		ACN=52	Parana R.	V-32
Roeboides	*microlepis*	*bonariensis*		52	34 M/SM + 18 ST/A	86			2.2 FD		Argentina	F-20, C-94
Roeboides	*xenodon*		F, M	52	4M + 28SM + 16ST + 4A	84	100	2		ACN=52	Brazil (MG)	V-32
Roeboides	sp.		F, M	52	6M + 20SM + 12ST + 14A	78	90	2		ACN=52	Brazil (MT)	V-32
Cheirodontinae												
Cheirodon	sp.			52							(Brazil)	W-19
Grundulus	*bogotensis*		F, M	50	10M + 28SM + 12ST	88	100			ACN=52	Colombia	G-53
Microschemobrycon	*casiquiare*			42*							(S. America)	S-30
Nematobrycon	*palmeri*			50						ACN=52	(Colombia)	S-152
Nematobrycon	*palmeri*		F, M	50	8 M/SM + 10ST + 32A	58	68				(Colombia)	A-93
Odontostilbe	*claudiae*			52							Parana	W-19
Odontostilbe	*paranensis*		F	52	36 M/SM + 13ST + 3A	88	101			ZW, ACN=52	Brazil (SP)	W-19
Odontostilbe	*paranensis*		M	52	36 M/SM + 12ST + 4A	88	100			ZZ, ACN=52	Brazil (SP)	W-19

Table 6.12 Order CHARACIFORMES (continued)

A	B	C	D	E	F	G	H	I	J	K	L
Current scientific name of taxon Suborder/family/subfamily/species	Reported in karyotype paper	Sex	2n	Karyotype	NF_1	NF_2	Ag-NORs	Genome size (pg/cell)	Comments	Locality	Reference
Odontostilbe stenodon	*Cheirodon stenodon*		52					3.7 FD		(Brazil, PR)	C-93
Paracheirodon axelrodi	*Cheirodon*		52			(54)			ACN=52	(S. America)	S-153
Paracheirodon innesi			32			(64)			ACN=52	(S. America)	S-153
Paracheirodon innesi	*Hyphessobrycon*	M	36							Brazil, Colombia, Peru	L-75
Paracheirodon simulans	*Hyphessobrycon*		50			(96)			ACN=52	(Amazon R., Orinoco R.)	S-152
Pristella maxillaris	*riddlei*		52*			(74)			ACN=52	(S. America)	S-152
Probolodus heterostomus			50							Brazil (SP)	O-50
Serrapinnus heterodon	*Holoshestes*	F	52	37 M/SM + 12ST + 3A	89	101			ZW, ACN=52	Brazil (SP)	W-19
Serrapinnus heterodon			52	16M + 20SM + 14ST + 2A	88	102	2			Brazil (Sao Francisco R.)	P-77
Serrapinnus heterodon			52	15M + 20SM + 14ST + 3A	87	101				Brazil (Sao Francisco R.)	P-77
Serrapinnus heterodon			52	17M + 20SM + 14ST + 1A	89	103				Brazil (Sao Francisco R.)	P-77
Serrapinnus notomelas	*Cheirodon*		52						ZW/ZZ	(Brazil)	W-19
Serrapinnus notomelas		F	52	16M + 23SM + 10ST + 3A	91	101	2		ZW, ACN=52	Brazil (SP)	S-201
Serrapinnus notomelas		M	52	16M + 22SM + 10ST + 4A	90	100	2		ZZ, ACN=52	Brazil (SP)	S-201
Serrapinnus piaba			52	16M + 20SM + 14ST + 2A	88	102	3			Brazil (Sao Francisco R.)	P-77
Serrapinnus sp. 1		F	52	8M + 15SM + 4ST + 25A	75	79			ZW	Paraná R.	S-201
Serrapinnus sp. 1		M	52	8M + 16SM + 4ST + 24A	76	80	3		ZZ, ACN=52	Paraná R.	S-201
Glandulocaudinae											
Gephyrocharax caucanus			52*							(Colombia)	S-30
Gephyrocharax valencia			52*							(Venezuela)	S-30
Glandulocauda melanogenys			52	4M + 12SM + 22ST + 14A	68	90				Brazil (SP)	T-71
Glandulocauda melanopleura			54	8M + 18SM + 14ST + 14A	80	94				Brazil (PR)	T-71
Mimagoniates lateralis			52	6M + 20SM + 16ST + 10A	78	94				Brazil (SP)	T-71
Mimagoniates microlepis			52	6M + 20SM + 18ST + 8A	78	96		3.1 FD		Brazil (SP)	T-71, C-94
Mimagoniates microlepis		F, M	52	12M + 18SM + 14ST + 8A	82	96	2		ACN=52	Brazil (PR)	T-71
Iguanodectinae											
Iguanodectes spilurus			50*							(S. America)	S-30
Paragoniatinae											
Prionobrama filigera			52	12 M/SM + 40 ST/A	64				ACN=54	(S. America)	A-93
Serrasalminae											
Catoprion mento			60							(Brazil)	P-88
Colossoma macropomum		F, M	54	20M + 34SM	108	108		(3.0 FIA)	ACN=54	Brazil	A-129, H-41
Colossoma macropomum			54	18M + 36SM	108	108				S. America	K-113
Colossoma macropomum			54	28M + 26SM	108	108	4		ACN=54	Venezuela	N-25
Metynnis argenteus			62*							(Brazil)	S-30

Table 6.12 Order CHARACIFORMES (continued)

A		B	C	D	E	F	G	H	I	J	K	L
Current scientific name of taxon Suborder/family/subfamily/species		Reported in karyotype paper	Sex	2n	Karyotype	NF_1	NF_2	Ag-NORs	Genome size (pg/cell)	Comments	Locality	Reference
Metynnis	*hypsauchen*	*schreitmuelleri*		62	60 M/SM + 2A	122	122		3.4 BFA		(S. America)	O-22, H-13
Metynnis	*lippincottianus*		F, M	62	30M + 30 SM/ST + 2A		122		(3.4 BFA)		(Brazil, French Guiana)	A-91, A- 92, H-13
Metynnis	*lippincottianus*			62	30M + 24SM + 6ST + ZW	116	122			0-2 B	Brazil (AM)	C-98
Metynnis	*lippincottianus*	cf. *roosevelti*		62	26M + 20SM + 14ST + 2A	108	122				Brazil	M-167
Metynnis	*maculatus*			62	32M + 22SM + 4ST + 4A	116	120			1B	Brazil (SP)	C-98
Metynnis	*mola*			62	60 M/SM + 2A	122	122				Brazil	J-19
Mylesinus	*paraschomburgkii*			58				6-12			Brazil (Amazon R.)	P-88
Mylesinus	*schomburgkii*	*Myleus*		58	42 M/SM + 16 ST/A	100		5-8			(Guyana, Venezuela)	P-88
Myleus	*arnoldi*	*Myloplus*		58	34 M/SM + 24 ST/A	92					(Brazil)	O-22
Myleus	*micans*		F, M	58	26M + 18SM + 8ST + 6A	102	110	4			Brazil (São Francisco R.)	A-128
Myleus	*pacu*			58	40 M/SM + 18 ST/A	98		5-9			(Guyana)	P-88
Myleus	*tiete*			58							Argentina	F-20
Myloplus	*rubripinnis*			58				5-8			(Guyana)	P-88
Mylossoma	*aureum*			54	54 M/SM	108	108	6-14			(Brazil)	P-88
Mylossoma	*duriventre*	*duriventris*		54	54 M/SM	108	108	6-14			(S. America)	P-88
Mylossoma	*duriventre*	*duriventris*		54	18M + 34SM + 2A	106	106				(S. America)	K-113
Mylossoma	*duriventre*	*duriventris*		54	50 M/SM + 4A	104	104				(S. America)	O-22
Mylossoma	*duriventre*	*duriventris*		54							Argentina	F-20
Mylossoma	*duriventre*	*paraguayensis*		54					2.9 FD		Brazil (Miranda R., MG)	C-94
Piaractus	*brachypomus*			54	28M + 26SM	108	108	2		ACN=54	Venezuela	N-25
Piaractus	*mesopotamicus*			54							Argentina	F-20
Piaractus	*mesopotamicus*	*Colossoma mitrei*	F, M	54	20M + 34SM	108	108			ACN=54	Brazil	A-129
Pristobrycon	*striolatus*			62	46 M/SM + 16 ST/A	108					(S. America)	P-88
Pygocentrus	*nattereri*			60	50 M/SM + 10A	110					(S. America)	P-84
Pygocentrus	*nattereri*	*Serrasalmus*		60							Argentina	F-20
Pygocentrus	*piraya*			60	48 M/SM + 12A	108	108				(Brazil, Sao Francisco R.)	P-84
Serrasalmus	*altispinis*			60	24M + 20SM + 6ST + 10A	104	110	9		ACN=60	Brazil (Amazon basin)	N-8
Serrasalmus	*altispinis*		F, M	60	20M + 28SM + 2ST + 10A	108	110	5-12		ACN=60	Brazil (Amazon basin)	N-71
Serrasalmus	*altuvei*			60	46 M/SM + 14 ST/A	106					(Venezuela)	P-88
Serrasalmus	*brandtii*		F, M	60	18M + 24SM + 8ST + 10A	102	110	10	(3.3 FD)	ACN=60	Brazil (MG)	C-87, C-94
Serrasalmus	*compressus*		F, M	60	18M + 30SM + 2ST + 10A	108	110	5-12			Brazil (Amazon basin)	N-71
Serrasalmus	*eigenmanni*	*Pristobrycon*		60	44 M/SM + 16 ST/A	104					(S. America)	P-88
Serrasalmus	*elongatus*		F, M	60	22M + 22SM + 4ST + 12A	104	108	5-12			Brazil (Amazon basin)	N-71
Serrasalmus	*gouldingi*			60	22M + 22SM + 6ST + 10A	104	110	8		ACN=60	Brazil (Amazon basin)	N-8
Serrasalmus	*hollandi*	Characinidae	M	64	30 M/SM + 16ST + 18A	94	110		3.4 FD		(Brazil, Guyana)	M-91

Table 6.12 Order CHARACIFORMES (continued)

A	B	C	D	E	F	G	H	I	J	K	L
Current scientific name of taxon Suborder/family/subfamily/species	Reported in karyotype paper	Sex	2n	Karyotype	NF_1	NF_2	Ag-NORs	Genome size (pg/cell)	Comments	Locality	Reference
Serrasalmus humeralis			60	16M + 26SM + 6ST + 12A	102	108				Brazil (Amazon basin)	N-71
Serrasalmus manueli		F, M	60	22M + 24SM + 6ST + 8A	106	112	5-12			Brazil (Amazon basin)	N-71
Serrasalmus marginatus		F, M	60	16M + 26SM + 8ST + 10A	102	110	8		ACN=60	Brazil (Paraguay R.)	C-87
Serrasalmus marginatus			60	18M + 26SM + 4ST + 12A	104	108				Argentina	F-20
Serrasalmus rhombeus		F, M	60	30M + 16SM + 4ST + 10A	106	110	8		ACN=60	Venezuela (Bolivar)	N-60
Serrasalmus rhombeus	cytotype 1	F, M	60	20M + 24SM + 6ST + 10A	104	110	5-12		ACN=60	Brazil (Amazon basin)	N-9
Serrasalmus rhombeus	cytotype 2		58	22M + 24SM + 2ST + 10A	104	106	5-12		ACN=60	Brazil (Amazon basin)	N-9
Serrasalmus serrulatus			60	20M + 22SM + 8ST + 10A	102	110	12		ACN=60	Brazil (Amazon basin)	N-8
Serrasalmus serrulatus	*Pristobrycon*		60	44 M/SM + 16 ST/A	104		6-10			(S. America)	P-88
Serrasalmus spilopleura		F, M	60	50 M/SM + 10A	110	110	5-10	(2.9 FD)		Brazil (SP)	G-90, C-93
Serrasalmus spilopleura	cytotype A	F, M	60	24M + 20SM + 4ST + 12A	104	108	5-12	(3.2 FD)	ACN=60	Brazil (Amazon basin)	C-94, C-103, N-71
Serrasalmus spilopleura	cytotype C		60	23M + 21SM + 4ST + 12A	104	108			ACN=60	Brazil (Central Amazon)	C-103
Serrasalmus spilopleura	cytotype D		60	24M + 20SM + 4ST + 12A	104	108			ACN=60	Brazil (Central Amazon)	C-103
Serrasalmus spilopleura	cytotype a	F, M	60	20M + 26SM + 4ST + 10A	106	110	10		ACN=60	Upper Parana	C-41
Serrasalmus spilopleura	cytotype b	F, M	60	18M + 26SM + 4ST + 12A	104	108	12		ACN=60	Paraguay	C-41, F-20
Serrasalmus spilopleura	cytotype c	F, M	60	19M + 26SM + 4ST + 11A	105	109	11		ACN=60	Lower Parana, Argentina	C-41, F-20
Stethaprioninae											
Brachychalcinus copei		F, M	50	42 M/SM + 8A	92	92	2	3.5 FD	ACN=52	Brazil (Acre)	C-25, C-94
Orthospinus franciscensis			50	10M + 32SM + 8A	92	92	2		ACN=52	Brazil (MG)	P-25
Orthospinus franciscensis			50	22M + 20SM + 2ST + 6A	92	94	2		ACN=52	Brazil (Sao Francisco R.)	A-122
Poptella compressa			50*							(Argentina)	F-20, S-30
Poptella paraguayensis		F, M	50	10M + 26SM + 8ST + 6A	86	94	2	3.5 FD	ACN=52	Brazil (MS, MT)	F-39, C-94
Tetragonopterinae											
Astyanax abramis			50					3.2 FD		Brazil (Miranda R., MG)	C-94
Astyanax altiparanae	cytotype 1		50	10M + 26SM + 4ST + 10A	86	90	1-4		ACN=52	Brazil (PR)	P-1
Astyanax altiparanae	cytotype 2		50	10M + 24SM + 4ST + 12A	84	88	1-4		ACN=52	Brazil (PR)	P-1
Astyanax altiparanae	cytotype 3		50	10M + 22SM + 4ST + 14A	82	86	1-4		ACN=52	Brazil (PR)	P-1
Astyanax altiparanae			50	6M + 28SM + 8ST + 8A	84	92	4		ACN=52	Brazil (Tibagi, PR)	D-29
Astyanax altiparanae			50	6M + 30SM + 8ST + 6A	86	94	2		ACN=52	Brazil (Iguacu, PR)	D-29
Astyanax altiparanae			50						1B	Brazil (SP)	C-98
Astyanax asuncionensis			50					2.4 FD		Brazil (Miranda R., MG)	C-94
Astyanax bimaculatus		F, M	50	6M + 22SM + 12ST + 10A	78	90		(2.1 FD)	ACN=52	Brazil (SP)	M-86, C-93
Astyanax bimaculatus		F, M	50	10M + 18SM + 12ST + 10A	78	90	6		ACN=52	Argentina	A-14, F-20
Astyanax bimaculatus			50	28 M/SM + 22 ST/A	78		2			Brazil (PR)	T-80

Table 6.12 Order CHARACIFORMES (continued)

A		B	C	D	E	F	G	H	I	J	K	L
Current scientific name of taxon Suborder/family/subfamily/species		Reported in karyotype paper	Sex	2n	Karyotype	NF_1	NF_2	Ag-NORs	Genome size (pg/cell)	Comments	Locality	Reference
Astyanax	*bimaculatus*			50	34 M/SM + 16 ST/A	84		2			Brazil (PR)	T-80
Astyanax	*bimaculatus*			50	40 M/SM + 10 ST/A	90		2			Brazil (PR)	T-80
Astyanax	*bimaculatus*			50*			(96)				(Brazil)	S-152
Astyanax	*bimaculatus lacustris*			50					2.9 FD		Brazil (MG)	C-94
Astyanax	*bockmanni*		F, M	50	10M + 12SM + 12ST + 16A	72	84	1–4		ACN=52	Brazil	K-126
Astyanax	*eigenmanniorum*		F, M	50	6M + 20SM + 8ST + 16A	76	84			ACN=52	Brazil (MG)	F-6
Astyanax	*eigenmanniorum*		F, M	75	9M + 30SM + 12ST + 24A	114	126			3X	Brazil (MG)	F-6
Astyanax	*eigenmanniorum*		F, M	48	14M + 20SM + 4ST + 10A	82	86	5		0–1 B, ACN=50	Brazil (MG)	T-77, C-98
Astyanax	*fasciatus*		F, M	46	6M + 24SM + 8ST + 8A	76	84		(3.5 FD)	0–1 B	Brazil (MG)	M-85, C-93
Astyanax	*fasciatus*		F, M	48	8M + 22SM + 12ST + 6A	78	90			ACN=52	Brazil (SP)	C-39
Astyanax	*fasciatus*			48	8M + 22SM + 12ST + 6A	78	90	2		ACN=52	Brazil (SP)	P-79
Astyanax	*fasciatus*			46	12M + 20SM + 10ST + 4A	78	88	2, 4		0–1 B, ACN=52	Brazil (SP)	P-79, P-80, C-98
Astyanax	*fasciatus*			45	12M + 20SM + 10ST + 3A	77	87	2		ACN=52	Brazil (SP)	P-79
Astyanax	*fasciatus*			47	12M + 19SM + 10ST + 6A	78	88	2		ACN=52	Brazil (SP)	P-79
Astyanax	*fasciatus*			47	12M + 21SM + 10ST + 4A	80	90	2		ACN=52	Brazil (SP)	P-79
Astyanax	*fasciatus*			47	12M + 20SM + 10ST + 5A	79	89	2		ACN=52	Brazil (SP)	P-79
Astyanax	*fasciatus*		F, M	46	8M + 26SM + 10ST + 2A	80	90				Brazil (SP)	M-86
Astyanax	*fasciatus*		F, M	48	8M + 24SM + 14ST + 2A	80	94				Brazil (Juquia R., SP)	M-86
Astyanax	*fasciatus*		F, M	48	8M + 20SM + 16ST + 4A	76	92	2	(2.8 FD)	ACN=52	Brazil (MG)	P-91, C-94
Astyanax	*fasciatus*		F, M	48	8M + 20SM + 16ST + 4A	76	92	2–5		ACN=52	Brazil (Piumhi R.)	P-91
Astyanax	*giton*		F, M	50	6M + 8SM + 8ST + 28A	64	72	3			Brazil (RJ)	K-19
Astyanax	*hastatus*			50	4M + 8SM + 10ST + 28A	62	72	3		ACN=52	Brazil (RJ)	K-121
Astyanax	*hastatus*			50	8M + 10SM + 14ST + 18A	68	82	3		ACN=52	Brazil (RJ)	K-121
Astyanax	*hastatus*			50	6M + 8SM + 4ST + 32A	64	68	2		ACN=52	Brazil (RJ)	K-121
Astyanax	*intermedius*		F, M	50	6M + 8SM + 4ST + 32A	64	68	6			Brazil (RJ)	K-19
Astyanax	*janeiroensis*		F, M	50	6M + 14SM + 14ST + 16A	70	84	6		ACN=52	Brazil (SP)	C-26, C-94
Astyanax	*jordani*	*Anoptichthys*	F, M	50	40 M/SM + 10ST	90	100			ACN=52	(Mexico)	K-63
Astyanax	*jordani*	*Anoptichthys*		50*			(92)				(S. America)	S-152
Astyanax	*mexicanus*		F, M	50	40 M/SM + 10ST	90	100			ACN=52	USA (TX)	K-63
Astyanax	aff. *mexicanus*			50	8M + 18SM + 12ST + 12A	76	88			0–2 B	Mexico	C-98
Astyanax	*parahybae*		F, M	48	8M + 18SM + 12ST + 10A	74	86	6		ACN=52	Brazil (Paraiba basin)	C-39, K-19
Astyanax	*scabripinnis*		F, M	50	6M + 8SM + 36A	64		4–12		0–4 B	Brazil (ES)	S-106
Astyanax	*scabripinnis*		F	50	8M + 20SM + 6ST + 16A	78	84	3		ACN=52	Brazil (MG)	B-70
Astyanax	*scabripinnis*		F, M	50	6M + 28SM + 6ST + 10A	84	90	3		ACN=52	Brazil (MG)	B-70
Astyanax	*scabripinnis*		F, M	50	6M + 24SM + 8ST + 12A	80	88	5		ACN=52	Brazil (MG)	B-70

Table 6.12 Order CHARACIFORMES (continued)

A	B	C	D	E	F	G	H	I	J	K	L
Current scientific name of taxon Suborder/family/subfamily/species	Reported in karyotype paper	Sex	2n	Karyotype	NF_1	NF_2	Ag-NORs	Genome size (pg/cell)	Comments	Locality	Reference
Astyanax scabripinnis		F, M	50	6M + 22SM + 10ST + 12A	78	88		(3.7 FD)	0–2 B	Brazil (SP)	F-6, S-8, M-85, C-93
Astyanax scabripinnis		F	75	9M + 33SM + 15ST + 18A	117	132			0–2 B, 3X	Brazil (SP)	F-6
Astyanax scabripinnis		F, M	50	6M + 22SM + 10ST + 12A	78	88			0–2 B, ACN=52	Brazil (SP, 1720m alt.)	V-39
Astyanax scabripinnis		F, M	50	6M + 22SM + 10ST + 12A	78	88			0–2 B	Brazil (SP, 1920m alt.)	N-20
Astyanax scabripinnis		F, M	50	6M + 22SM + 10ST + 12A	78	88			0–1 B	Brazil (SP, 1800m alt.)	N-20
Astyanax scabripinnis		F, M	50	6M + 22SM + 10ST + 12A	78	88			0 B	Brazil (SP, 700m alt.)	N-20
Astyanax scabripinnis		F, M	50	6M + 24SM + 6ST + 14A	80	86	2		0–1 B, ACN=52	Brazil (SP, 1800m alt.)	S-99
Astyanax scabripinnis		F, M	50	4M + 10SM + 6ST + 30A	64	70	8		0 B, ACN=52	Brazil (SP, 780m alt.)	S-99
Astyanax scabripinnis	cytotype I	F, M	50	4M + 26SM + 4ST + 16A	80	84	1–6		ACN=52	Brazil (SP)	S-100, S-101, S-174
Astyanax scabripinnis	cytotype II	F, M	48	6M + 22SM + 8ST + 12A	76	84			ACN=52	Brazil (SP)	S-100
Astyanax scabripinnis		F, M	50	6M + 30SM + 4ST + 10A	86	90	1–3		0–1 B, ACN=52	Brazil (PR, Ivai basin)	M-75, M-76
Astyanax scabripinnis		F, M	50	6M + 28SM + 16A	84	84	1–4		0–2 B, ACN=52	Brazil (PR, Parana basin)	M-75, M-76
Astyanax scabripinnis		F, M	48	10M + 20SM + 8ST + 10A	78	86	3		ACN=50	Brazil (PR)	A-33
Astyanax scabripinnis		F, M	48	10M + 24SM + 6ST + 8A	82	88	7		ACN=50	Brazil (PR)	A-33
Astyanax scabripinnis		F, M	48	8M + 24SM + 4ST + 12A	80	84	8		ACN=50	Brazil (PR, Ivai basin)	F-21
Astyanax scabripinnis		F, M	48	8M + 20SM + 6ST + 14A	76	82	6		ACN=50	Brazil (PR, Ivai basin)	F-21
Astyanax scabripinnis	cytotype I	F, M	50	6M + 22SM + 6ST + 16A	78	84			1B, ACN=52	Brazil (PR, Ivai basin)	F-22
Astyanax scabripinnis	cytotype II	F, M	48	8M + 26SM + 6ST + 8A	82	88			3B, ACN=52	Brazil (PR, Ivai basin)	F-22
Astyanax scabripinnis	cytotype III	F, M	46	8M + 22SM + 6ST + 10A	76	82			2B, ACN=50	Brazil (PR, Ivai basin)	F-22
Astyanax scabripinnis paranae		F, M	50	8M + 20SM + 12ST + 10A	78	90			ACN=52	Brazil (SP)	M-86
Astyanax scabripinnis paranae			50	8M + 22SM + 10ST + 10A	80	90			0–1 B	Brazil (SP)	P-82, C-98
Astyanax scabripinnis paranae		F, M	50	4M + 30SM + 4ST + 12A	84	88	4		0–1 B, ACN=52	Brazil (SP)	M-135
Astyanax aff. *scabripinnis*		F, M	48	6M + 22SM + 4ST + 16A	76	80			ACN=50	Brazil (PR)	C-79
Astyanax aff. *scabripinnis*		F	72	9M + 33SM + 6ST + 24A	114	120	4		3X, ACN=75	Brazil (PR)	C-79
Astyanax schubarti		F, M	36	16M + 14SM + 4ST + 2A	66	70			ACN=52	Brazil (SP)	M-86, M-87
Astyanax schubarti			54	24M + 21SM + 6ST + 3A	99	105			3X	Brazil (SP)	M-87
Astyanax schubarti		F, M	36	12M + 14SM + 10ST	62	72			0–1 B, ACN=52	Argentina	M-85
Astyanax schubarti		F	36	14M + 14SM + 6ST + 2A	64	70	2		ACN=52	Argentina (Paraná)	A-14
Astyanax taeniatus			50	12M + 2SM + 24ST + 12A	64	88				Brazil	S-192
Astyanax sp. B		F, M	50	6M + 24SM + 6ST + 14A	80	86			0–2 B, ACN=52	Brazil (PR)	F-7
Astyanax sp. C		F, M	50	4M + 22SM + 8ST + 16A	76	84	2, 4		ACN=52	Brazil (PR)	K-7
Astyanax sp.		F, M	48	6M + 18SM + 14ST + 10A	72	86	2, 4		ACN=50	Brazil (PR, sinkhole)	M-47
Astyanax sp.			50							Argentina	F-20
Bryconamericus aff. *exodon*	cytotype I	F, M	52	16M + 12SM + 6ST + 18A	80	86			ACN=52	Brazil (PR)	P-2
Bryconamericus aff. *exodon*	cytotype II	F, M	52	10M + 24SM + 6ST + 12A	86	92			ACN=52	Brazil (PR)	P-2

Table 6.12 Order CHARACIFORMES (continued)

A		B	C	D	E	F	G	H	I	J	K	L
Current scientific name of taxon Suborder/family/subfamily/species		Reported in karyotype paper	Sex	2n	Karyotype	NF_1	NF_2	Ag-NORs	Genome size (pg/cell)	Comments	Locality	Reference
Bryconamericus	aff. *iheringii*	cytotype I	F, M	52	12M + 18SM + 8ST + 14A	82	90	4		ACN=54	Brazil (PR)	C-27, P-65
Bryconamericus	aff. *iheringii*	cytotype II	F, M	52	8M + 28SM + 6ST + 10A	88	94	4		ACN=54	Brazil (PR)	C-27, P-65
Bryconamericus	aff. *iheringii*	cytotype III	F, M	52	8M + 20SM + 8ST + 16A	80	88	2		ACN=54	Brazil (Ivai R. basin)	C-27
Bryconamericus	aff. *iheringii*		F, M	52	8M + 28SM + 6ST + 10A	88	94			ACN=54	Brazil (PR)	P-65
Bryconamericus	aff. *iheringii*		F, M	52	8M + 22SM + 10ST + 12A	82	92	2		ACN=54	Brazil (PR)	P-90
Bryconamericus	*stramineus*		F	52	26 M/SM + 26 ST/A	78			3.3 FD	ACN=52	Brazil (SP)	P-38, C-93
Bryconamericus	sp. A		F	52	6M + 30SM + 6ST + 10A	88	94	2–3		ACN=54	Brazil (SP)	W-17, W-18
Bryconamericus	sp. B		F, M	52	6M + 10SM + 20ST + 16A	68	88	3		ACN=54	Brazil (SP)	W-17, W-18
Bryconamericus	sp. C		F, M	52	6M + 18SM + 14ST + 14A	76	90	4		ACN=54	Brazil (PR)	W-17, W-18
Bryconamericus	sp. D		F	52	8M + 14SM + 16ST + 14A	74	90	4		ACN=54	Brazil (MG)	W-17, W-18
Bryconamericus	sp. E		F, M	54	10M + 16SM + 22ST + 6A	80	102			ACN=54	Brazil (MG)	W-17
Bryconella	*pallidifrons*	*palifrons*		50*							(Brazil, Peru)	S-30
Bryconops	*humeralis*	*Cynopotamus*		50							Brazil	O-50
Ctenobrycon	*hauxwellianus*		F, M	50	10M + 6SM + 34ST	66	100	2		ACN=52	Brazil (Acre)	C-26
Deuterodon	*pedri*		F, M	50	14 M/SM + 36 ST/A	64				ACN=52	Brazil (SP)	P-38
Gymnocorymbus	*ternetzi*		F, M	50	14M + 12SM + 6ST + 18A	76	82		3.7 FCM, 4.2 BFA	ACN=52	Argentina	A-14, V-86, H-13
Gymnocorymbus	*ternetzi*			50			(60)			ACN=52	(S. America)	S-152
Hasemania	*nana*	*marginata*	F, M	50	12M + 18SM + 10ST + 10A	80	90			ACN=52	(Brazil, Sao Francisco R.)	A-93
Hasemania	*nana*	*marginata*		50*			(100)			ACN=52	(Brazil, Sao Francisco R.)	S-152
Hasemania	*nana*			50	8M + 42SM	100	100	2		ACN=52	Brazil, Sao Francisco R.	A-122
Hemigrammus	*analis*			54*			(82)				(S. America)	S-152
Hemigrammus	*bellottii*	*Hyphessobrycon*		48*							(S. America)	S-30
Hemigrammus	*erythrozonus*		F	48	2SM + 46 ST/A	50			(4.0* FCM)		(Guyana)	L-75, O-48
Hemigrammus	*erythrozonus*			52*			(56)			ACN=52	(Guyana)	S-152
Hemigrammus	*hyanuary*		F, M	52	22 M/SM + 30 ST/A	74					(Brazil, Peru)	A-93
Hemigrammus	*hyanuary*			54*			(70)				(Brazil, Peru)	S-152
Hemigrammus	*marginatus*		F, M	50	10M + 34SM + 6A	94	94	2		ACN=52	Brazil (PR)	P-39
Hemigrammus	*marginatus*			50*			(100)			ACN=52	(S. America)	S-152
Hemigrammus	*matei*	*mattei*		36			(70)				(Argentina)	S-152
Hemigrammus	*micropterus*			52*							(Venezuela)	S-30
Hemigrammus	*ocellifer*			38*					3.0 FCM, 3.4 BFA		(Amazon R.)	S-152, V-86, H-13
Hemigrammus	*pulcher*			50*							(Peru)	S-30
Hemigrammus	*rhodostomus*			50*			(98)			ACN=52	(Lower Amazon R.)	S-152
Hemigrammus	*rodwayi*	*Hyphessobrycon*		50*						ACN=50	(S. America)	S-29
Hemigrammus	aff. *schmardae*			52			(88)			ACN=52	(Amazon R.)	S-152

Table 6.12 Order CHARACIFORMES (continued)

A		B	C	D	E	F	G	H	I	J	K	L
Current scientific name of taxon Suborder/family/subfamily/species		Reported in karyotype paper	Sex	2n	Karyotype	NF_1	NF_2	Ag-NORs	Genome size (pg/cell)	Comments	Locality	Reference
Hemigrammus	*stictus*	*Hyphessobrycon*		50			(100)			ACN=52	(Amazon R., Orinoco R.)	S-152
Hemigrammus	aff. *ulreyi*			52*			(76)			ACN=52	(Argentina, Brazil)	S-152
Hemigrammus	*unilineatus*			52*							(northern S. America)	S-30
Hemigrammus	*vorderwinkleri*			48*							(Brazil, Negro R.)	S-30
Hollandichthys	*multifasciatus*		F, M	50	10M + 12SM + 28ST	72	100	2		ACN=52	Brazil (SP)	C-26
Hyphessobrycon	aff. *agulha*			50*							(Brazil)	S-30
Hyphessobrycon	*anisitsi*		F, M	50	6M + 16SM + 12ST + 16A	72	84	0-4		ACN=52	Brazil (SP)	C-40
Hyphessobrycon	*anisitsi*	*Hemigrammus caudovittatus*		50	2M + 32SM + 16 ST/A	84			(3.8 FCM)	ACN=52	(S. America)	F-30, V-86
Hyphessobrycon	*anisitsi*	*Hemigrammus caudovittatus*		50*			(98)		(3.4 BFA)	ACN=52	(S. America)	S-152, H-13
Hyphessobrycon	*bentosi*			52*			(66)			ACN=52	(Amazon R.)	S-152
Hyphessobrycon	*bentosi*	*robertsi*		52*			(52)			ACN=52	(Amazon R.)	S-152
Hyphessobrycon	*bifasciatus*			50*			(98)			ACN=52	(Argentina, Brazil)	S-152
Hyphessobrycon	*copelandi*			52*			(100)			ACN=52	(Brazil)	S-152
Hyphessobrycon	aff. *eos*			48*							(S. America)	S-30
Hyphessobrycon	*eques*	*callistus*		52*			(90)		3.3 FCM	ACN=52	(S. America)	S-152, V-86
Hyphessobrycon	*eques*	*serpae*		52*			(82)			ACN=52	(S. America)	S-152
Hyphessobrycon	*erythrostigma*	*rubrostigma*		52*			(76)				(S. America)	S-152
Hyphessobrycon	*flammeus*		F, M	52	18 M/SM + 32ST + 2A	70	102		(2.5 FCM)		(Brazil)	A-93, V-86
Hyphessobrycon	*flammeus*			50*			(100)			ACN=52	(Brazil)	S-152
Hyphessobrycon	*griemi*			50*			(100)			ACN=52	(Brazil)	S-152
Hyphessobrycon	*griemi*			48					2.6 FD		Brazil (SP)	C-94
Hyphessobrycon	*herbertaxelrodi*			52	10 M/SM + 42 ST/A	62					(Brazil)	A-93
Hyphessobrycon	*herbertaxelrodi*			52*			(76)			ACN=52	(Brazil)	S-152
Hyphessobrycon	*heterorhabdus*			48*							(Brazil)	S-30
Hyphessobrycon	*megalopterus*	*Megalamphodus melanopterus*		52*			(94)			ACN=52	(Brazil, Bolivia)	S-152
Hyphessobrycon	*minimus*			52*							(Guyana)	S-30
Hyphessobrycon	*minor*	*'minor'*		52	14M + 12SM + 26 ST/A	78					(Guyana)	A-91
Hyphessobrycon	*peruvianus*			50*						ACN=50	(Amazon R., Peru)	S-29
Hyphessobrycon	*pulchripinnis*			50*			(82)		3.7 FCM	ACN=52	(Brazil)	S-152, V-86
Hyphessobrycon	*reticulatus*		F, M	50	14M + 20SM + 16ST	84	100		2.3 FD		Brazil (SP)	C-26, C-94
Hyphessobrycon	*rosaceus*	*ornatus*		52*							(Guyana, Suriname)	S-152
Hyphessobrycon	*rosaceus*	*Cheirodon troemneri*		50*						ACN=50	(Guyana, Suriname)	S-29, S-152, S-153
Hyphessobrycon	*santae*			50*			(90)			ACN=52	(Brazil, Sao Francisco R.)	S-152
Hyphessobrycon	*scholzei*		F, M	50	8M + 20SM + 8ST + 14A	78	86				(Lower Amazon R.)	A-93
Hyphessobrycon	*scholzei*			50*			(86)			ACN=52	(S. America)	S-152

Table 6.12 Order CHARACIFORMES (continued)

A		B	C	D	E	F	G	H	I	J	K	L
Current scientific name of taxon Suborder/family/subfamily/species		Reported in karyotype paper	Sex	2n	Karyotype	NF_1	NF_2	Ag-NORs	Genome size (pg/cell)	Comments	Locality	Reference
Hyphessobrycon	*simulatus*	*Pseudopristella simulata*		52*			(74)			ACN=52	(French Guiana)	S-152
Hyphessobrycon	*sweglesi*	*Megalamphodus*		52*			(98)			ACN=52	(Colombia)	S-152
Hyphessobrycon	*tropis*			46*							(Rio Negro)	S-30
Knodus	aff. *beta*			52*			(102)				(Colombia)	S-152
Markiana	*nigripinnis*		M	52					2.2 FD		Brazil (Miranda R.)	C-94
Moenkhausia	*collettii*			50*			(94)			ACN=52	(S. America)	S-152
Moenkhausia	*costae*		F	50	50 M/SM	100	100	1-2		ACN=52	Brazil (MG)	P-38
Moenkhausia	*dichroura*			50	32M + 14SM + 4ST	96	100		2.0 FD		Brazil (MT)	P-39, C-94
Moenkhausia	*gracilima*	cytotype A		50	14M + 26SM + 6ST + 4A	90	96				Brazil (AM)	P-39
Moenkhausia	*gracilima*	cytotype B		48	4M + 24SM + 12ST + 8A	76	88				Brazil (AM)	P-39
Moenkhausia	*intermedia*		F, M	50	50 M/SM	100	100	1-2		0-1 B, ACN=52	Brazil (SP)	P-38
Moenkhausia	*intermedia*		F, M	50	16M + 34SM	100	100	2		ACN=52	Brazil (PR)	P-39
Moenkhausia	*intermedia*		F, M	50	16M + 34SM	100	100	2		0 B, ACN=52	Brazil (SP)	D-31
Moenkhausia	*oligolepis*			50*			(100)		3.2 BFA	ACN=52	(Brazil, Venezuela)	S-152, H-13
Moenkhausia	*pittieri*			50*			(92)			ACN=52	(Venezuela)	S-152
Moenkhausia	*pittieri*			50	4M + 40 SM/ST + 6A		94				(Venezuela)	A-91
Moenkhausia	*pittieri*			49	4M + 39 SM/ST + 6A		92				(Venezuela)	A-91
Moenkhausia	*sanctaefilomenae*		F, M	50	48 M/SM + 2 ST/A	98				0-3 B, ACN=52	Argentina	A-14
Moenkhausia	*sanctaefilomenae*		F, M	50	48 M/SM + 2A	98	98		(2.4 FD)	1-8 B, ACN=52	Brazil (SP)	F-35, C-94
Moenkhausia	*sanctaefilomenae*		F, M	50	12M + 36SM + 2ST	98	100	3	(2.8 FD)	0-6 B, ACN=52	Brazil (SP)	D-31, C-93
Moenkhausia	*sanctaefilomenae*		F, M	50	12M + 36SM + 2ST	98	100	2		0-2 B, ACN=52	Brazil (PR)	P-39, P-73
Moenkhausia	sp.		F, M	50	16M + 34SM	100	100	2		0 B, ACN=52	Brazil (MT)	D-31
Oligosarcus	*hepsetus*			50	2M + 16SM + 16ST + 16A	68	84	3	3.3 FD	ACN=52	Brazil	H-38, C-94
Oligosarcus	*hepsetus*		F, M	50	6M + 10SM + 16ST + 18A	66	82	2		ACN=52	Brazil (Paraiba do Sul R.)	K-21
Oligosarcus	*jenynsii*			50	2M + 24SM + 10ST + 14A	76	86	2		ACN=52	Brazil	H-38
Oligosarcus	*macrolepis*		F, M	50	8M + 20SM + 6ST + 16A	78	84			ACN=52	Brazil (MG)	F-2
Oligosarcus	*paranensis*		F, M	50					3.3 FD		Brazil (SP)	C-93
Oligosarcus	*pintoi*			50	2M + 20SM + 12ST + 16A	72	84	2		ACN=52	Brazil (SP)	H-38
Oligosarcus	*pintoi*	*Paroligosarcus*	F, M	50	24 M/SM + 26 ST/A	74				0-1 B, ACN=52	Brazil (SP)	F-56
Piabina	*argentea*		F, M	52	26 M/SM + 26 ST/A	78		1-4		ACN=52	Brazil (SP)	P-38
Piabina	*argentea*			52	8M + 14SM +16ST + 14A	74	90	2		ACN=52	Brazil (MG)	A-122
Piabina	*argentea*			52	18 M/SM + 12ST + 22A	70	82		2.4 FD	0-1 B	Brazil (SP)	C-98, C-93
Schultzites	*axerlodi*			52*							(Colombia)	S-30
Tetragonopterus	*argenteus*		F	52	16 M/SM + 2ST + 34A	68	70	2	3.0 FD	ACN=52	Argentina (Parana R.)	A-14, C-94
Tetragonopterus	*chalceus*		F, M	52	26 M/SM + 26 ST/A	78		1-2	3.9 FD	ACN=52	Brazil (MG)	P-38, C-94

Table 6.12 Order CHARACIFORMES (continued)

A		B	C	D	E	F	G	H	I	J	K	L
Current scientific name of taxon Suborder/family/subfamily/species		Reported in karyotype paper	Sex	2n	Karyotype	NF_1	NF_2	Ag-NORs	Genome size (pg/cell)	Comments	Locality	Reference
Thayeria	*boehlkei*			50*			(100)			ACN=52	(Peru, Brazil)	S-152
Triporthinae												
Triportheus	*albus*		F, M	52	14M + 20SM + 14ST + 4A	86	100	1–4		ZW/ZZ	Amazon basin	S-12
Triportheus	*culter*			52	14M + 16SM + 16ST + 6A	82	98				Amazon basin	S-12
Triportheus	*auritus*	*elongatus*	F, M	52	22M + 12SM + 16ST + 2A	86	102	1–4		ZW/ZZ	Amazon basin	S-12
Triportheus	cf. *auritus*	cf. *elongatus*	F, M	52						ZW/ZZ	Brazil (MT)	A-117
Triportheus	*angulatus*	*flavus*	F, M	52	22M + 14SM + 12ST + 4A	88	100	1–4		ZW/ZZ	Amazon basin	S-12
Triportheus	*guentheri*		F	52	20 M/SM + 12 ST/A	72		1–4		ZW, ACN=53	Brazil (MG)	A-117, S-12
Triportheus	*guentheri*		M	52	20 M/SM + 12 ST/A	72		1–4		ZZ, ACN=54	Brazil (MG)	A-117, S-12
Triportheus	*nematurus*		F	52	13M + 23SM + 16ST	88	104	3		ZW, ACN=53	Brazil (SP)	D-30
Triportheus	*nematurus*		M	52	14M + 22SM + 16ST	88	104	2		ZZ, ACN=54	Brazil (SP)	D-30
Triportheus	*nematurus*	*paranensis*	F	52	25M + 23SM + 4ST	100	104	1–4		ZW, ACN=53	Argentina	S-12
Triportheus	*nematurus*	*paranensis*	M	52	26M + 22SM + 4ST	100	104	1–4	2.7 FD	ZZ, ACN=54	Argentina	S-12, C-94
Triportheus	*pictus*			52					3.5 FD		(Brazil, Peru)	P-88, C-94
Triportheus	*pictus*			50*			(100)			ACN=52	(S. America)	S-152
Triportheus	*pictus*			52*							(Brazil)	S-30
Triportheus	*rotundatus*			52*							(Brazil)	S-30
Triportheus	*rotundatus*			50*			(100)			ACN=52	(S. America)	S-152
Triportheus	*signatus*			52						ZW/ZZ	(Parana–Paraguai basin)	A-117
Triportheus	*venezuelensis*		F	52	20M + 16SM + 16ST	88	104	2–4		ZW, ACN=55	Venezuela (Bolivar)	N-69
Triportheus	*venezuelensis*		M	52	20M + 16SM + 16ST	88	104	2–4		ZZ, ACN=56	Venezuela (Bolivar)	N-69
Chilodontidae												
Caenotropus	*labyrinthicus*		F, M	54	54 M/SM	108	108	2		ACN=56	Brazil (MT)	M-44
Chilodus	*punctatus*		F, M	54	54 M/SM	108	108	2	3.2 BFA		Brazil (AM)	M-44, H-13
Crenuchidae												
Characidium	cf. *alipioi*		F, M	50	30M + 20SM	100	100	2		ZW/ZZ	Brazil (SP)	C-91
Characidium	cf. *fasciatum*			50	32M + 18SM	100	100		2.4 FD	0–4 B, ZW/ZZ	Brazil (SP)	C-93, C-98
Characidium	*gomesi*		F, M	50	32M + 18SM	100	100	2		1–2 B, ZW/ZZ	Brazil (SP)	M-113
Characidium	cf. *gomesi*		F	50	31M + 18SM + 1ST	99	100	2–7		ZW, ACN=52	Brazil (PR)	V-91
Characidium	cf. *gomesi*		M	50	32M + 18SM	100	100	2–7		ZZ, ACN=52	Brazil (PR)	V-91
Characidium	cf. *lagosantense*		F, M	50	32M + 18SM	100	100			ACN=52	Brazil (SP)	M-74
Characidium	*lauroi*		F, M	50	24M + 24SM + 2ST	98	100	2–3			Brazil (SP)	C-91
Characidium	*oiticicai*		F	50	30M + 18SM + ZW					0–3 B, ZW	Brazil (SP)	C-98

Table 6.12 Order CHARACIFORMES (continued)

A		B	C	D	E	F	G	H	I	J	K	L
Current scientific name of taxon Suborder/family/subfamily/species		Reported in karyotype paper	Sex	2n	Karyotype	NF_1	NF_2	Ag–NORs	Genome size (pg/cell)	Comments	Locality	Reference
Characidium	*pterostictum*		F, M	50	32M + 16SM + 2ST	98	100			ACN=52	Brazil (SP)	M–74
Characidium	cf. *zebra*		F, M	50	32M + 18SM	100	100	6		0–1 B, ACN=52	Brazil (SP)	M–74, V–94
Characidium	sp.		F	50	32M + 18SM	100	100			ACN=52	Brazil (SP)	M–74
Characidium	sp.			50	32M + 16SM + 2A	98	98			0–2 B	Brazil (SP)	C–98
Ctenoluciidae												
Boulengerella	sp.			36							(S. America)	P–88
Ctenolucius	*hujeta*	*hujeti*		36	26 M/SM + 10 ST/A	62					(S. America)	A–92
Curimatidae												
Curimata	*cyprinoides*		F, M	54	44M + 10SM	108	108	2		ACN=56	Brazil (AM)	F–12, V–92
Curimata	*inornata*		F, M	54	40M + 14SM	108	108	2		ACN=56	Brazil (AM)	F–12, V–92
Curimata	*knerii*		F, M	54	40M + 12SM + 2ST	106	108	2		ACN=56	Brazil (AM)	F–12
Curimata	*ocellata*		F	56	40M + 16SM	112	112	2		ACN=56	Brazil (AM)	F–12
Curimata	*vittata*		F, M	54	42M + 12SM	108	108	2	3.0 FD	ACN=56	Brazil (AM)	F–12, V–92, C–94
Curimatella	*alburna*		M	54	46M + 8SM	108	108	2		ACN=56	Brazil (AM)	F–12
Curimatella	*dorsalis*		F, M	54	46M + 8SM	108	108	2	2.8 FD	ACN=56	Brazil (MS)	N–14, C–94
Curimatella	*dorsalis*		F, M	54	54 M/SM	108	108	2		ACN=56	Argentina	B–74
Curimatella	*immaculata*		F, M	54	46M + 8SM	108	108	2			Brazil (GO)	V–92
Curimatella	*lepidura*		F, M	54	54 M/SM	108	108	2			Brazil (MG)	V–93
Curimatella	*meyeri*		F, M	54	46M + 8SM	108	108	2		ACN=56	Brazil (AM)	F–12
Curimatopsis	aff. *macrolepis*	cytotype A		52*							(S. America)	S–30
Curimatopsis	aff. *macrolepis*	cytotype B		46*							(S. America)	S–30
Curimatopsis	*myersi*		F	46	42M + 4SM	92	92	2			Brazil (MS)	N–14
Cyphocharax	*gilberti*		F, M	54	44M + 10SM	108	108	2			Brazil (SP)	V–92
Cyphocharax	cf. *gillii*		F	54	54 M/SM	108	108	2			Brazil (MT)	V–93
Cyphocharax	*gouldingi*		F, M	54	54M	108	108	2		1 B	Brazil (GO)	V–92
Cyphocharax	*modestus*	*Curimata modesta*	F, M	54	54 M/SM	108	108	2	(3.2 FD)	0–1 B	Brazil (SP)	V–30, C–93
Cyphocharax	*modestus*	*Curimata modesta*		82	82 M/SM	164	164	3		0–1 B, 3X	Brazil (SP)	V–30
Cyphocharax	*modestus*	*modesta*	F, M	54	54 M/SM	108	108	2		0–1 B, ACN=56	Brazil (PR)	M–42
Cyphocharax	*modestus*		F, M	54	54 M/SM	108	108	2		0–2 B, ACN=56	Brazil (PR)	G–95, T–82
Cyphocharax	*modestus*	*modesta*	F, M	54	54 M/SM	108	108	2			Brazil (Upper Parana)	V–93
Cyphocharax	*modestus*	*modesta*		54	54 M/SM					0–1 B	Brazil (SP)	V–94
Cyphocharax	*modestus*		F, M	54	50M + 4SM	108	108	2		1 B	Brazil (SP)	V–92
Cyphocharax	*nagelii*		F, M	54	54 M/SM	108	108	2		1 B	Brazil (Upper Parana)	V–93, C–98

Table 6.12 Order CHARACIFORMES (continued)

A		B	C	D	E	F	G	H	I	J	K	L
Current scientific name of taxon Suborder/family/subfamily/species		Reported in karyotype paper	Sex	2n	Karyotype	NF_1	NF_2	Ag-NORs	Genome size (pg/cell)	Comments	Locality	Reference
Cyphocharax	*nagelii*		F, M	54	46M + 8SM	108	108	2–9			Brazil (SP)	V-92
Cyphocharax	*platanus*		F, M	58	52 M/SM + 6ST	110	116	2			Argentina	B-74
Cyphocharax	*platanus*		F, M	58	48M + 4SM + 6ST	110	116	2			Argentina	V-92
Cyphocharax	*spilotus*		F, M	54	54 M/SM	108	108	2		0–1 B, ACN=56	Argentina	B-74, C-98
Cyphocharax	cf. *spilurus*	*spilura*	F, M	54	54 M/SM	108	108	2			Brazil (RO)	V-93
Cyphocharax	*vanderi*		F, M	54	54 M/SM	108	108	2			Brazil (SP)	V-93
Cyphocharax	*voga*		F, M	54	54 M/SM	108	108	2			Brazil (RS)	V-93
Cyphocharax	*voga*		F, M	54	54 M/SM	108	108	2		ACN=56	Argentina	B-74
Potamorhina	*altamazonica*		F, M	102	2M + 2SM + 98A	106	106	2			Brazil (AM)	F-13
Potamorhina	*latior*		F, M	56	52M + 2SM + 2ST	110	112	2		ACN=56	Brazil (AM)	F-13
Potamorhina	*pristigaster*		F, M	54	44M + 10SM	108	108	2		ACN=56	Brazil (AM)	F-13
Potamorhina	*squamoralevis*		F, M	102	14 M/SM + 88A	116	116	2	3.8 FD		Argentina	B-74, C-94
Psectrogaster	*amazonica*		F, M	54	44M + 10SM	108	108	2			Brazil (MT)	V-92
Psectrogaster	*curviventris*		F, M	54	42M + 12SM	108	108	2		ACN=56	Brazil (MS)	N-14
Psectrogaster	*curviventris*		F, M	54	54 M/SM	108	108	2		ACN=56	Argentina	B-74
Psectrogaster	*rutiloides*		F, M	54	42M + 12SM	108	108	2			Brazil (AM)	F-12
Steindachnerina	*amazonica*		F, M	54	42M + 12SM	108	108	3			Brazil (GO)	V-92
Steindachnerina	*brevipinna*		F, M	54	48M + 6SM	108	108	2		ACN=56	Brazil (MS)	N-14
Steindachnerina	*brevipinna*		F, M	54	54 M/SM	108	108	2		ACN=56	Argentina	B-74
Steindachnerina	*conspersa*		F, M	54	46M + 8SM	108	108	2		ACN=56	Argentina	B-74
Steindachnerina	*conspersa*		F, M	54	54 M/SM	108	108	2			Brazil (MS)	V-93
Steindachnerina	*elegans*		F, M	54	54 M/SM	108	108	2	3.5 FD		Brazil (MG)	V-93, C-94
Steindachnerina	*gracilis*		F, M	54	38M + 16SM	108	108	2–6			Brazil (MT)	V-92
Steindachnerina	cf. *guentheri*		F, M	54	54 M/SM	108	108	2	3.2 FD		Brazil (Acre)	C-25, C-94
Steindachnerina	*insculpta*		F, M	54	54 M/SM	108	108	2	(2.9 FD)	0–2 B, ACN=56	Brazil (PR)	V-93, C-93, G-95
Steindachnerina	*insculpta*		F, M	54	50M + 4SM	108	108	2			Brazil (SP)	V-92
Steindachnerina	*leucisca*	*leuciscus*	F	54	48M + 6SM	108	108	2		ACN=56	Brazil (AM)	F-12
Erythrinidae												
Erythrinus	*erythrinus*		F, M	54	6M + 2ST + 46A	60	62			4B, no sex chrom.	Brazil (Birigui)	B-35
Erythrinus	*erythrinus*		F, M	54	6M + 2ST + 46A	60	62			No sex chrom.	Brazil (Penapolis)	B-35
Erythrinus	*erythrinus*		F	54	6M + 2ST + 46A	60	62			2B, $X_1X_1X_2X_2$	Brazil (Guaira, PR)	B-35, C-110
Erythrinus	*erythrinus*		M	53	7M + 2ST + 44A	60	62			X_1X_2Y, ACN=54	Brazil (Guaira, PR)	B-35, C-110
Erythrinus	*erythrinus*		F	52	6M + 2SM + 6ST + 38A	60	66	3–8		$X_1X_1X_2X_2$, ACN=54	Brazil (Manaus, AM)	B-33, B-35, C-110
Erythrinus	*erythrinus*		M	51	7M + 2SM + 6ST + 36A	60	66			X_1X_2Y, ACN=54	Brazil (Manaus, AM)	B-35, C-110

Table 6.12 Order CHARACIFORMES (continued)

A		B	C	D	E	F	G	H	I	J	K	L
Current scientific name of taxon Suborder/family/subfamily/species		Reported in karyotype paper	Sex	2n	Karyotype	NF_1	NF_2	Ag-NORs	Genome size (pg/cell)	Comments	Locality	Reference
Erythrinus	*erythrinus*		F	52	4M + 2SM + 2ST + 44A	58	60			$X_1X_1X_2X_2$, ACN=54	Brazil (Natal)	B-35, C-110
Erythrinus	*erythrinus*		M	51	5M + 2SM + 2ST + 42A	58	60			X_1X_2Y, ACN=54	Brazil (Natal)	B-35, C-110
Erythrinus	*erythrinus*			54							Argentina	F-20
Hoplerythrinus	*unitaeniatus*		F, M	48	48 M/SM	96	96			ACN=50	Brazil (AM)	G-96
Hoplerythrinus	*unitaeniatus*		F, M	48	47 M/SM + 1 ST/A	95				ACN=48	Brazil (AM)	G-96
Hoplerythrinus	*unitaeniatus*		F, M	48	46 M/SM + 2 ST/A	94					Brazil (AM)	G-96
Hoplerythrinus	*unitaeniatus*	cytotype A	F, M	48	48 M/SM	96	96			ACN=50	Brazil (RO, MS, Argentina)	G-21
Hoplerythrinus	*unitaeniatus*	cytotype B	F, M	48	46 M/SM + 2A	94	94			ACN=48	Brazil (Su)	G-21
Hoplerythrinus	*unitaeniatus*		F, M	52	46 M/SM + 6A	98	98			ACN=52	Brazil (MG)	G-21
Hoplerythrinus	*unitaeniatus*		F, M	48	44 M/SM + 4ST	92	96	2–4		No sex chrom.	Brazil (SP, MS)	D-16
Hoplerythrinus	*unitaeniatus*		F, M	52	44 M/SM + 4ST + 4A	96	100	4–6		No sex chrom.	Brazil (MG)	D-16
Hoplerythrinus	*unitaeniatus*		F	72	69 M/SM + 3 ST/A	141		3		3X	Brazil (AM)	G-20
Hoplerythrinus	*unitaeniatus*			48							Argentina	F-20
Hoplias	*lacerdae*		F, M	50	50 M/SM	100	100	2		No sex chrom.	Brazil	B-78
Hoplias	*lacerdae*		F, M	50	36M + 14SM	100	100			XX/XY, ACN=50	Brazil (SP)	B-29
Hoplias	*malabaricus*	cytotype A	F, M	42	22M + 20SM	84	84	2–6		No sex chrom.	Brazil (RS, MG)	B-61, B- 77
Hoplias	*malabaricus*	cytotype A	F, M	42	24M + 18SM	84	84	3–8	(2.3 FD)	No sex chrom.	Brazil (SP, AM, PR)	B-61, B-77, P-21, C-93
Hoplias	*malabaricus*	cytotype A	F, M	42	20M + 22SM	84	84	3–7		No sex chrom.	Brazil (SP, MT)	B-61, B- 77
Hoplias	*malabaricus*	cytotype A	F, M	42	42 M/SM	84	84			No sex chrom.	Argentina	L-90
Hoplias	*malabaricus*	cytotype B	F	42	24M + 16SM + 2ST	82	84	4–7		XX, ACN=50	Brazil (MG)	B-61, B-76
Hoplias	*malabaricus*	cytotype B	M	42	24M + 17SM + 1ST	83	84	4–7		XY, ACN=50	Brazil (MG)	B-76
Hoplias	*malabaricus*	cytotype C	F, M	40	40 M/SM	80	80			No sex chrom.	Brazil (AM, PA, RO, MT, PR)	B-61, P-21
Hoplias	*malabaricus*	cytotype C	F, M	40	40 M/SM	80	80			No sex chrom.	Argentina	L-90
Hoplias	*malabaricus*	cytotype D	F	40	40 M/SM	80	80		(2.8 BFA)	$X_1X_1X_2X_2$	Brazil (SP, MG, PR)	B-31, B-34, B- 61, H-13
Hoplias	*malabaricus*	cytotype D	M	39	39 M/SM	78	78			X_1X_2Y, ACN=48	Brazil (SP, PR)	B-31, B-34, B- 61, P-21
Hoplias	*malabaricus*	cytotype E	M	42	40 M/SM + 2A	82	82			ACN=50	Brazil (PA)	B-61
Hoplias	*malabaricus*	cytotype F	F, M	40	40 M/SM	80	80			No sex chrom.	Brazil (PA, MA, RN, MG)	B-61
Hoplias	*malabaricus*	cytotype G	F	40	40 M/SM	80	80			XX, ACN=48	Brazil (PA, RO, MT)	B-61
Hoplias	*malabaricus*	cytotype G	M	41	40 M/SM + 1A	81	81			XY_1Y_2, ACN=48	Brazil (PA, RO, MT)	B-61
Gasteropelecidae												
Carnegiella	*marthae*		F	50	20M + 12SM + 4ST + 14A	82	86	1–3		ZW, ACN=55	Brazil (AM)	T-76
Carnegiella	*marthae*		M	50	20M + 12SM + 4ST + 14A	82	86	1–3		ZZ, ACN=56	Brazil (AM)	T-76
Carnegiella	*strigata*		F	50	4M + 4SM + 2ST + 40A	58	60	1–3		ZW, ACN=56	Brazil (AM)	T-76
Carnegiella	*strigata*		M	50	4M + 4SM + 2ST + 40A	58	60	1–3		ZZ, ACN=56	Brazil (AM)	T-76

Table 6.12 Order CHARACIFORMES (continued)

A		B	C	D	E	F	G	H	I	J	K	L
Current scientific name of taxon Suborder/family/subfamily/species		Reported in karyotype paper	Sex	2n	Karyotype	NF_1	NF_2	Ag-NORs	Genome size (pg/cell)	Comments	Locality	Reference
Carnegiella	*strigata*			50–52					2.8 BFA		(S. America)	H-13
Carnegiella	*strigata*			48*								S-30
Gasteropelecus	*sternicla*			54*							(S. America)	S-30
Thoracocharax	*stellatus*			54							Paraguai R.	C-101
Thoracocharax	*stellatus*		F, M	54	6M + 6SM + 6ST + 36A	66	72	4		ZW/ZZ	Brazil (MT)	V-90
Thoracocharax	cf. *stellatus*		F	52	8M + 16SM + 4ST + 24A	76	80	2	2.2 FD	ZW, ACN=54	Brazil (Rio Branco, Acre)	C-94, C-101
Thoracocharax	cf. *stellatus*		M	52	8M + 16SM + 4ST + 24A	76	80	2	2.2 FD	ZZ, ACN=54	Brazil (Rio Branco, Acre)	C-94, C-101
Hemiodontidae												
Anodus	*elongatus*		M	54	24M + 26SM + 4ST	104	108	2		ACN=56	Brazil (AM)	P-40
Anodus	*elongatus*	*steatops*		54	52 M/SM + 2A	106	106				(Brazil)	A-92
Anodus	*elongatus*	*melanopogon*	F, M	54	20M + 28SM + 6ST	102	108	2		ACN=56	Brazil (AM)	P-40
Anodus	sp.		F, M	54	24M + 24SM + 6ST	102	108	2		ACN=56	Brazil (AM)	P-40
Argonectes	*longiceps*	*scapularis*		54	50 M/SM + 4ST	104	108	2			(S. America)	P-88
Hemiodus	*argenteus*	*ocellatus*	F, M	54	26M + 24SM + 4ST	104	108	2		ACN=56	Brazil (AM)	P-40
Hemiodus	*immaculatus*		F, M	54	22M + 26SM + 6ST	102	108	2		ACN=56	Brazil (AM)	P-40
Hemiodus	*microlepis*			54	52 M/SM + 2A	106	106				(Brazil, Venezuela)	A-92
Hemiodus	cf. *microlepis*		F, M	54	20M + 30SM + 4ST	104	108	2		ACN=56	Brazil (AM)	P-40
Hemiodus	*unimaculatus*		F, M	54	26M + 24SM + 4ST	104	108	2		ACN=56	Brazil (AM)	P-40
Lebiasinidae												
Copeina	*guttata*			42*							(S. America)	S-30
Copella	*arnoldi*			44*							(S. America)	S-30
Copella	*nattereri*			36*							(S. America)	S-30
Nannostomus	*beckfordi*		M	42	2M + 40A	44	44			ACN=44	(S. America)	A-92
Nannostomus	*beckfordi*	cytotype A		44*							(S. America)	S-30
Nannostomus	*beckfordi*	cytotype B		36*							(S. America)	S-30
Nannostomus	*eques*	*Nannobrycon*		34	34A	34					(Brazil, Peru)	A-92
Nannostomus	*eques*	*Poecilobrycon*		36*							(S. America)	S-30
Nannostomus	*harrisoni*	*Poecilobrycon*		40*							(Guyana)	S-30
Nannostomus	*marginatus*			42*							(S. America)	S-30
Nannostomus	*trifasciatus*	*erythrurus*		46*							(S. America)	S-30
Nannostomus	*trifasciatus*	*erythrurus*		38*							(S. America)	S-30
Nannostomus	*trifasciatus*	*erythrurus*		30*							(S. America)	S-30
Nannostomus	*trifasciatus*	*erythrurus*		24*							(S. America)	S-30

Table 6.12 Order CHARACIFORMES (continued)

A		B	C	D	E	F	G	H	I	J	K	L
Current scientific name of taxon Suborder/family/subfamily/species		Reported in karyotype paper	Sex	2n	Karyotype	NF_1	NF_2	Ag-NORs	Genome size (pg/cell)	Comments	Locality	Reference
Nannostomus	*unifasciatus*	*Poecilobrycon*		22*							(S. America)	S-30
Pyrrhulina	cf. *australis*		F, M	40	6ST + 34A	40	46	8		ACN=42	Brazil (SP)	O-53
Pyrrhulina	*australis*	*rachoviana*		42					(2.0 FD), 2.4 BFA		(S. America)	C-93, H-13
Pyrrhulina	sp.			42	2M + 2SM + 38 ST/A	46					Brazil	O-82
Parodontidae												
Apareiodon	*affinis*		F	55	51 M/SM + 4ST	106	110	2		ZW_1W_2, ACN=56	Brazil (MG, SP)	M-82, M-83, L-85
Apareiodon	*affinis*		M	54	50 M/SM + 4ST	104	108	2	(2.0, 2.5 FD)	ZZ, ACN=56	Brazil (MG, SP)	M-82, M-83, L-85, C-93, C-94
Apareiodon	*affinis*		F	55	47 M/SM + 8ST	102	110	2		ACN=56	Argentina	C-96
Apareiodon	*affinis*		F	55	43 M/SM + 10ST + 2A	98	108	2		ACN=55, 56	Argentina	C-96
Apareiodon	*affinis*		F	54	45 M/SM + 8ST + 1A	99	107	2		ACN=55	Argentina	C-96
Apareiodon	*affinis*		M	54	40 M/SM + 12ST + 2A	94	106	2		ACN=54	Argentina	C-96
Apareiodon	*affinis*		M	54	42 M/SM + 8ST + 4A	96	104	2		ACN=55	Argentina	C-96
Apareiodon	*ibitiensis*			54	50 M/SM + 4ST	104	108	2		ACN=56	Brazil (PR, SP)	M-83, J-8
Apareiodon	*piracicabae*		F, M	54	52 M/SM + 2ST	106	108	4		0–1 B, ACN=56	Brazil (SP)	J-8, F-56
Apareiodon	*vittatus*			54	52 M/SM + 2ST	106	108	2		ACN=56	Brazil (Timbo R)	J-8
Apareiodon	sp. A			54	50 M/SM + 4ST	104	108	2		ACN=56	Brazil (MG)	J-8
Apareiodon	sp. B			54	50 M/SM + 4ST	104	108	2		ACN=56	Brazil (MG)	J-8
Apareiodon	sp. C			54	52 M/SM + 2ST	106	108	2		ACN=54	Brazil (Araguaia R., MT)	J-8
Parodon	*hilarii*		F	54	31M + 22SM + 1ST	107	108	2		ZW, ACN=55	Brazil (MG)	M-84, J-7
Parodon	*hilarii*		M	54	32M + 22SM	108	108	2		ZZ, ACN=54	Brazil (MG)	M-84, J-7
Parodon	*nasus*		F, M	54	48 M/SM + 6ST	102	108	2		ACN=56	Brazil (MT)	B-81
Parodon	*nasus*	*tortuosus*	F, M	54	48 M/SM + 6ST	102	108	2		ACN=56	Brazil (SP)	B-81
Parodon	*nasus*	*tortuosus*	F, M	54	48 M/SM + 6ST	102	108	2		ACN=56	Brazil (PR)	J-7
Parodon	sp.		F, M	54	50 M/SM + 4ST	104	108	2		ACN=54	Brazil (MT)	J-7
Prochilodontidae												
Prochilodus	*argenteus*		F, M	54	40M + 14SM	108	108	2	3.1 FD	ACN=56	Brazil (CE)	P-20, C-94
Prochilodus	*brevis*	*cearaensis*	F, M	54	40M + 14SM	108	108	2		0–2 B, ACN=56	Brazil (CE)	P-20
Prochilodus	*costatus*	*affinis*	F, M	54	40M + 14SM	108	108	2	3.1 FD	ACN=56	Brazil (MG)	P-20, C-94
Prochilodus	*lineatus*		F, M	54	40M + 14SM	108	108		(3.4 FD)	ACN=56	Brazil (MS)	P-20, C-93
Prochilodus	*lineatus*			54	54 M/SM	108	108				Argentina	F-20
Prochilodus	*lineatus*	*platensis*		54	54 M/SM	108	108				Argentina	F-20
Prochilodus	*lineatus*		F, M	54	40M + 14SM	108	108	2		0–7 B, ACN=56	Brazil (PR, SP)	D-12, A-119, V-107
Prochilodus	*lineatus*	*scrofa*	F, M	54	40M + 14SM	108	108	2		0–5 B, ACN=56	Brazil (SP)	P-19, P-20

Table 6.12 Order CHARACIFORMES (continued)

A	B	C	D	E	F	G	H	I	J	K	L
Current scientific name of taxon Suborder/family/subfamily/species	Reported in karyotype paper	Sex	2n	Karyotype	NF_1	NF_2	Ag-NORs	Genome size (pg/cell)	Comments	Locality	Reference
Prochilodus marggravii		F, M	54	40M + 14SM	108	108	2			Brazil (MG)	P-20
Prochilodus mariae			54	40M + 14SM	108	108			0-3 B, ACN=54	Venezuela (Bolivar)	C-77, C-98
Prochilodus nigricans		F, M	54	40M + 14SM	108	108	2		0-2 B, ACN=56	Brazil (AM.)	P-20, V-94
Prochilodus vimboides		F, M	54	40M + 14SM	108	108			ACN=56	Brazil (SP)	P-20
Semaprochilodus insignis		F, M	54	54 M/SM	108	108	2		ACN=56	Brazil (AM.)	F-11
Semaprochilodus kneri			54	40M + 14SM	108	108	2		ACN=54	Venezuela (Bolivar)	O-78
Semaprochilodus laticeps			54	40M + 14SM	108	108	2		ACN=54	Venezuela (Bolivar)	O-78
Semaprochilodus taeniurus		F	54	53 M/SM + 1large M	108	108	2		ZW, ACN=55	Brazil (AM.)	F-11
Semaprochilodus taeniurus		M	54	54 M/SM	108	108	2		ZZ, ACN=54	Brazil (AM.)	F-11

Table 6.13 Order SILURIFORMES

A	B	C	D	E	F	G	H	I	J	K	L
Current scientific name of taxon Family/subfamily/species	Reported in karyotype paper	Sex	2n	Karyotype	NF_1	NF_2	Ag-NORs	Genome size (pg/cell)	Comments	Locality	Reference
Amblycipitidae											
Amblyceps mangois		F, M	36	4M + 12SM + 16ST + 4A	52	68				India (J & K)	S-48
Liobagrus andersoni		F, M	28	18M + 10SM	56	56			ACN=30	Korea (Han R.)	S-94
Liobagrus andersoni		F, M	28	14M + 14SM	56	56			ACN=30	Korea (Gapyeong)	K-52
Liobagrus anguillicauda			34	20M + 12SM + 2ST	66	68				China (Zhejiang)	L-38
Liobagrus marginatoides	*Leiobagrus*	F	30	16M + 6SM + 6ST + 2A	52	58			XX, ACN=36	China (Nanchong)	Y-15
Liobagrus marginatoides	*Leiobagrus*	M	30	16M + 5SM + 7ST + 2A	51	58			XY, ACN=35	China (Nanchong)	Y-15
Liobagrus marginatus		F, M	24	20M + 2SM + 2ST	46	48				China (Sichuan)	L-36
Liobagrus marginatus	*Leiobagrus*	F	24	20M + 2SM + 2ST	46	48			XX, ACN=30	China (Sichuan)	L-30, Y-15
Liobagrus marginatus	*Leiobagrus*	M	24	19M + 2SM + 2ST + 1A	45	47			XY, ACN=30	China (Sichuan)	L-30, Y-15
Liobagrus mediadiposalis		F, M	42	26M + 16SM	84	84			ACN=42	Korea (Nakdong R.)	S-94
Liobagrus mediadiposalis		F, M	42	26M + 12SM + 4 ST/A	80		2		ACN=42	Korea (Hamyang)	K-52
Liobagrus mediadiposalis			28	18M + 6SM + 4ST	52	56			ACN=30	Korea (Gum R.)	U-34
Liobagrus nigricauda	*Leiobagrus*	F	30	16M + 6SM + 6ST + 2A	52	58			XX, ACN=36	China (Nanchong)	Y-15
Liobagrus nigricauda	*Leiobagrus*	M	30	16M + 5SM + 7ST + 2A	51	58			XY, ACN=35	China (Nanchong)	Y-15
Liobagrus obesus		F, M	20	20M	40	40				Korea (Kum R.)	S-94
Liobagrus reinii		F, M	38	26M + 8SM + 4ST	72	76			ACN=46	Japan (Gifu)	U-34
Liobagrus reinii			38	28M + 10SM	76	76				Japan	O-22

Table 6.13 Order SILURIFORMES (continued)

A		B	C	D	E	F	G	H	I	J	K	L
Current scientific name of taxon Family/subfamily/species		Reported in karyotype paper	Sex	2n	Karyotype	NF_1	NF_2	Ag-NORs	Genome size (pg/cell)	Comments	Locality	Reference
Ariidae												
Ariopsis	*felis*	*Arius*		54	26 M/SM + 28 ST/A	80			(4.5 FCM)	ACN=54	USA (LA)	L-19, T-73
Ariopsis	*felis*	*Arius*	F, M	54	16M + 12SM + 20ST + 6A	82	102			ACN=54	Mexico	G-74
Ariopsis	*guatemalensis*	*Galeichthys caerulescens*		52	16M + 24SM + 10ST + 2A	92	102				Mexico	G-74
Bagre	*bagre*			56	24M + 26SM + 6ST	106	112				Brazil (SP)	G-52
Bagre	*marinus*			54	12M + 8SM + 34 ST/A	74			4.8 BFA		USA (LA)	F-27, H-13
Cathorops	*melanopus*	*Arius*		52	16M + 30SM + 6ST	98	104				Mexico	G-74
Cathorops	*spixii*			54							Brazil	G-52
Cathorops	sp.			54	13M + 13SM + 28ST	80	108				Brazil (SP)	B-86
Genidens	*barbus*	*Netuma barba*		56	18M + 18SM + 18ST + 2A	92	110			XX/XY	Brazil	B-86, G-52
Genidens	*genidens*			56	12M + 20SM + 20ST + 4A	88	108				Brazil (SP)	B-86, G-52
Nemapteryx	*caelata*	*Arius caelatus*		54	6M + 20SM + 28A	80	80				India (Bombay)	K-101
Nemapteryx	*nenga*	*Arius*	F, M	54	16M + 36SM + 2ST	106	108			ACN=54	India (Orissa)	C-63
Netuma	*thalassina*	*Arius serratus*	F	56	8M + 24SM + 24ST	88	112			ACN=56	India (Orissa)	C-63
Notarius	*luniscutis*	*Sciadeichthys*		56							Brazil	G-52
Plicofollis	*dussumieri*	*Arius*	F, M	54	12M + 18SM + 12ST + 12A	84	96				India (Bombay)	R-59
Sciades	*parkeri*	*Arius*		56	16M + 16SM + 22ST + 2A	88	110				Brazil (SP)	B-86
Aspredinidae												
Bunocephalus	cf. *larai*			50	6M + 8SM + 8ST + 28A	64	72	2			S. America	F-20
Auchenipteridae												
Ageneiosus	*atronasus*	*atronases*	F, M	56	16M + 16SM + 12ST + 12A	88	100	2			Brazil (AM)	F-18
Ageneiosus	*inermis*	*brevifilis*	F, M	56	20M + 16SM + 10ST + 10A	92	102	2			Brazil (AM)	F-18
Auchenipterus	*nuchalis*		F, M	58	24M + 14SM + 10ST + 10A	96	106	2		ACN=58	Brazil (PR)	R-39
Glanidium	*ribeiroi*		F, M	58	28M + 16SM + 10ST + 4A	102	112	2		ACN=58	Brazil (Iguaçu, PR)	R-39
Trachelyopterus	*galeatus*	*Parauchenipterus*	F, M	58	22M + 12SM + 6ST + 18A	92	98	2		ACN=58	Brazil (PR)	R-39
Trachelyopterus	*galeatus*			58							Argentina	F-20
Bagridae												
Bagrichthys	*macracanthus*	*Bagroides*		50	16M + 26SM + 2ST + 6A	92	94				Thailand (Nakhon Phanom)	M-8
Coreobagrus	*brevicorpus*		F, M	44	22M + 14SM + 8 ST/A	80				ACN=48	Korea (Hamyang)	K-52
Coreobagrus	*ichikawai*		F, M	56	18M + 14SM + 24 ST/A	88				ACN=56	Japan (Gifu)	U-34
Hemibagrus	*guttatus*	*Mystus*	F, M	60	20M + 12SM + 16ST + 12A	92	108			ACN=62	China (Guangdong)	Y-15
Hemibagrus	*guttatus*	*Mystus elongatus*	F, M	60	20M + 12SM + 16ST + 12A	92	108			ACN=62	China (Yangshan)	Y-15
Hemibagrus	*macropterus*	*Mystus*	F, M	60	20M + 12SM + 16ST + 12A	92	108			ACN=62	China (Wuhan)	H-20, Y-15

Table 6.13 Order SILURIFORMES (continued)

A		B	C	D	E	F	G	H	I	J	K	L
Current scientific name of taxon Family/subfamily/species		Reported in karyotype paper	Sex	2n	Karyotype	NF_1	NF_2	Ag-NORs	Genome size (pg/cell)	Comments	Locality	Reference
Hemibagrus	*macropterus*	*Mystus*		60							China	C-86
Hemibagrus	*menoda*	*Mystus*		48	12M + 12ST + 24A	60	72				India	L-1
Hemibagrus	*menoda*			58	22M + 20SM + 16A	100	100			ACN=58	India (WB)	D-34
Hemibagrus	*menoda menoda*	*Mystus*	F	56	36M + 16SM + 4A	108					India (Assam)	K-46, C-108
Hemibagrus	*menoda menoda*	*Mystus corsula*	M	58	22M + 20SM + 16A	100		2			India (WB)	B-4
Hemibagrus	*nemurus*	*Mystus bleekeri*	F, M	56	20M + 14SM + 10ST + 12A	90	100				India (Jammu)	S-55
Hemibagrus	*nemurus*	*Mystus bleekeri*		56	32M + 14SM + 10A	102	102				India (Assam)	K-46, C-108
Hemibagrus	*wyckii*	*Mystus*		54	16M + 14SM + 24A	84				ACN=54	Thailand (Nakhon Phanom)	M-8
Horabagrus	*brachysoma*			60	28M + 20SM + 8ST+ 4A	108	116	2		ACN=60	India (Kerala)	N-56
Horabagrus	*nigricollaris*			60	26M + 20SM + 10ST + 4A	106	116	2			India (Kerala)	N-57
Mystus	*cavasius*			58	18M + 16SM + 10ST + 14A	92	102	2			India (Jammu)	S-55, R-122
Mystus	*cavasius*		F, M	58	18M + 22SM + 18A	98				ACN=58	India (Orissa)	T-49
Mystus	*cavasius*		F, M	58	14M + 26SM + 4ST + 14A	98	102				India (Bihar)	K-32
Mystus	*gulio*		F, M	58	30M + 12SM + 2ST + 14A	100	102	2		ACN=58	India (WB)	M-28, D-33
Mystus	*gulio*		F	58	12M + 34SM + 4ST + 8A	104	108			XX, ACN=58	India (Orissa)	C-63
Mystus	*gulio*		M	58	13M + 33SM + 4ST + 8A	104	108			XY, ACN=58	India (Orissa)	C-63
Mystus	*tengara*		F	54	9M + 38 SM/ST + 7A		101			ZW	India (Haryana)	R-46
Mystus	*tengara*		M	54	10M + 38 SM/ST + 6A		102			ZZ	India (Haryana)	R-46
Mystus	*tengara*		F	54	25M + 18 SM/ST + 11A		97			ZW	India (Haryana)	R-75
Mystus	*tengara*		M	54	26M + 18 SM/ST + 10A		98			ZZ	India (Haryana)	R-75
Mystus	*vittatus*		F, M	54	22M + 26SM + 6ST	102	108				India (Orissa)	T-49
Mystus	*vittatus*		F, M	54	22M + 20SM + 12ST	96	108				India (Jammu)	S-55
Mystus	*vittatus*		F, M	58	10M + 30SM + 12ST + 6A	98	110			ACN=58	India (Orissa)	C-63
Mystus	*vittatus*	var. A	F, M	54	20M + 24SM + 10ST	98	108			ACN=56	India (WB)	M-23
Mystus	*vittatus*	var. B	M	58	16M + 10SM + 20ST + 12A	84	104			ACN=58	India (WB)	M-23
Mystus	*vittatus*	*vittatus vittatus*	F, M	54	28M + 22SM + 2ST + 2A	104	106	2			India (Orissa)	K-42, J-9
Pelteobagrus	*eupogon*		M	50	20M + 14SM + 16ST	84	100			ACN=52	China (Wuhan)	H-20, Y-15
Pelteobagrus	*nudiceps*		F, M	56	18M + 12SM + 26 ST/A	86			(1.8* FCM)	ACN=56	Japan (Lake Biwa)	U-34, O-48
Pelteobagrus	*nudiceps*			56	44 M/SM + 12 ST/A	100					Japan (Shimane)	F-51
Pelteobagrus	*ussuriensis*	*Pseudobagrus*	M	52	24M + 18SM + 10ST	94	104			ACN=58	China (Wuhan)	H-20, Y-15
Pelteobagrus	*ussuriensis*	*Pseudobagrus*	F, M	52	18M + 16SM + 18 ST/A	86		2		ACN=58	Korea (Kangkyeong)	K-52
Pelteobagrus	*ussuriensis*	*Pseudobagrus emarginatus*	F, M	52	24M + 10SM + 18ST	86	104			ACN=54	China (Nanchong)	Y-15
Pelteobagrus	*ussuriensis*	*emarginatus*	F, M	52	20M + 20SM + 12 ST/A	92				ACN=54	Korea (Han R.)	U-34
Pseudobagrus	*aurantiacus*		F, M	48	20M + 12SM + 16 ST/A	80				ACN=54	Japan (Kyushu)	U-25
Pseudobagrus	*aurantiacus*		F, M	48	18M + 14SM + 16 ST/A	80				ACN=54	Japan (Kumamoto)	U-34
Pseudobagrus	*crassilabris*	*Leiocassis*	F, M	52	24M + 14SM + 14ST	90	104			ACN=56	China (Wuhan)	H-20, Y-15

Table 6.13 Order SILURIFORMES (continued)

A		B	C	D	E	F	G	H	I	J	K	L
Current scientific name of taxon Family/subfamily/species		Reported in karyotype paper	Sex	2n	Karyotype	NF_1	NF_2	Ag-NORs	Genome size (pg/cell)	Comments	Locality	Reference
Pseudobagrus	*koreanus*	sp.	F, M	48	18M + 14SM + 16 ST/A	80				ACN=54	Korea (Han R.)	U-34
Pseudobagrus	*koreanus*	sp.	F, M	48	20M + 14SM + 14 ST/A	82				ACN=54	Korea (Gapyeong)	K-52
Pseudobagrus	*longirostris*	*Leiocassis*	F, M	52	20M + 16SM + 16ST	88	104			ACN=56	China (Wuhan)	H-20, Y-15
Pseudobagrus	*nitidus*	*Pelteobagrus*	F, M	52	22M + 20 SM/ST + 10A		94			ACN=54	Korea (Kangkyeong)	K-52
Pseudobagrus	*nitidus*	*Pelteobagrus*	F, M	52	20M + 16SM + 16ST	88	104			ACN=56	China (Wuhan)	H-20, Y-15
Pseudobagrus	*pratti*		M	52	20M + 14SM + 8ST + 10A	86	94			ACN=58	China (Guilin)	Y-15
Pseudobagrus	*tenius*		F, M	52	22M + 16SM + 14ST	90	104			ACN=56	China (Wuhan)	H-20, Y-15
Pseudobagrus	*tokiensis*	*aurantiacus*	F, M	56	24M + 12SM + 20 ST/A	92				ACN=56	Japan (Tokyo, Chiba, Iwate)	U-25
Pseudobagrus	*truncatus*		F, M	52	26M + 14SM + 12ST	92	104			ACN=58	China (Nanchong)	Y-15
Pseudobagrus	*vachellii*	*Pelteobagrus*	F, M	52	22M + 16SM + 14ST	90	104			ACN=56	China (Wuhan)	H-20, Y-15
Pseudobagrus	*vachellii*	*Pelteobagrus*	M	52	16M + 20SM + 16 ST/A	88		2		ACN=56	Korea (Gum R.)	U-34
Rita	*chrysea*			54	28M + 20SM + 6ST	102	108				India (Orissa)	D-2
Rita	*chrysea*			52	8M + 22SM + 2ST + 20A	82	84			ACN=52	India (WB)	K-139
Rita	*rita*		F, M	54	14M + 34SM + 6ST	102	108	2		ACN=54	India (WB)	K-42, D-33
Rita	*rita*			54	14M + 24SM + 12ST + 4A	92	104				India	M-20
Sperata	*aor*	*Aorichthys*		52	20M + 14SM + 10ST + 8A	86	96				India	L-1
Sperata	*seenghala*	*Mystus*	F	54	28M + 12SM + 8ST + 6A	94	102	2			India (Jammu)	S-55, R-122
Sperata	*seenghala*			50	10M + 14SM + 6ST + 20A	74	80	2			India	D-34
Tachysurus	*adiposalis*	*Pseudobagrus*	F, M	50	20M + 14SM + 14ST + 2A	84	98			ACN=56	China (Yangshan)	Y-15
Tachysurus	*fulvidraco*	*Pelteobagrus*	F, M	52	24M + 14SM + 10ST + 4A	90	100		1.9*, 2.0 FD	ACN=54	China (Hubei)	H-20, L-41, C-83
Tachysurus	*fulvidraco*	*Pelteobagrus*		52	22M + 12SM + 14ST + 4A	86	100	4			China (Shasi)	Z-21
Tachysurus	*fulvidraco*	*Pelteobagrus*	F, M	52	18M + 26 SM/ST + 8A		96			ACN=54	Korea (Kangkyeong)	K-52
Tachysurus	*fulvidraco*	*Pseudobagrus*	F, M	52	28M + 12SM + 12ST	92	104				China (Amur)	S-63
Tachysurus	*fulvidraco*	*Pseudobagrus*		52	22 M + 24 SM/ST + 6A		98				(Hubei, China)	L-53
Tachysurus	*fulvidraco*	*Pseudobagrus*		52	24M + 24SM + 4A	100	100				Korea	L-15
Tachysurus	*fulvidraco*		F, M	52	18M + 26SM + 8A	96	96				Korea (Kum R.)	P-68
Callichthyidae												
Callichthyinae												
Callichthys	*callichthys*		F, M	58	22M + 22SM + 14ST	102	116	2	1.9 FD, (3.4 BFA)	0-5 B, ACN=58	Brazil (SP)	O-56, H-13
Callichthys	*callichthys*		F, M	56	14M + 10SM + 32 ST/A	80		2		0-2 B	Argentina	S-11
Callichthys	*callichthys*			54	46 M/SM + 8 ST/A	100		3			Brazil (Central Amazon)	P-89
Callichthys	*callichthys*			52	44 M/SM + 8 ST/A	96					Brazil (Central Amazon)	P-89
Dianema	*longibarbis*			60	6M + 54A	66	66	2			Brazil (Central Amazon)	M-148
Dianema	*urostriatum*			62	6M + 6ST + 50A	68	74	2			Brazil (Central Amazon)	M-148
Dianema	*urostriata*		F, M	62	8M + 4SM + 4ST + 46A	74	78	2	1.2 FD	ACN=64	Brazil (SP)	O-56

Table 6.13 Order SILURIFORMES (continued)

A		B	C	D	E	F	G	H	I	J	K	L
Current scientific name of taxon Family/subfamily/species		Reported in karyotype paper	Sex	2n	Karyotype	NF_1	NF_2	Ag–NORs	Genome size (pg/cell)	Comments	Locality	Reference
Hoplosternum	*littorale*		F, M	60	4M + 4SM + 52A	68	68	2		ACN=62	Brazil (Amazon R.)	O–56, P–85
Hoplosternum	sp.			60	6M + 2SM + 52A	68	68	2	1.4 FD	ACN=62	Brazil (SP)	O–56
Hoplosternum	sp.			62	8 M/SM + 54ST	70	124	4			Brazil (Amazon R.)	P–88
Megalechis	*thoracata*	*Hoplosternum*	F, M	64	8M + 12SM + 44 ST/A	84		2	3.2 FCM	ACN=66	Brazil (Amazon R.)	P–85, V–86
Corydoradinae												
Aspidoras	*fuscoguttatus*		F, M	44	28M + 12SM + 4ST	84	88	2	1.5 FD	ACN=62	Brazil (SP)	O–56
Aspidoras	cf. *fuscoguttatus*		F, M	46	32M + 10SM + 4ST	88	92	2		ACN=58	(Brazil)	S–190
Aspidoras	*poecilus*		F	46	30M + 10SM + 6ST	86	92	2		ACN=60	Brazil (MG)	S–190
Aspidoras	*taurus*		F, M	46	30M + 10SM + 6ST	86	92	1		ACN=60	Brazil (MG)	S–190
Brochis	*britskii*		F	90	4M + 10SM + 22ST + 54A	104	126	2			Brazil (SP)	O–56
Brochis	*britskii*	*Corydoras*	F, M	90	4M + 10SM + 22ST + 54A	104	126	4		ACN=90	Brazil (MG)	S–190
Brochis	*splendens*		M	100	18M + 18SM + 20ST + 44A	136	156	4	2.3 FD	4X, ACN=100	Brazil (SP)	O–56
Corydoras	*aeneus*		F	56	32 M/SM + 24 ST/A	88			3.6 FD	XX	Brazil (Belem, PA)	T–58
Corydoras	*aeneus*		M	56	33 M/SM + 23 ST/A	89				XY	Brazil (Belem, PA)	T–58
Corydoras	*aeneus*	*schultzei*		ca. 58	48 non–A + 10A		(106)		(3.2 FIA)		(S. America)	S–27, H–40
Corydoras	*aeneus*			60	20M + 20SM + 20 ST/A	100					(Brazil)	K–17
Corydoras	*aeneus*		F, M	60	26M + 26SM + 8ST	112	120	8	2.8 FD	0–3 B, ACN=64	Brazil (SP)	O–51, O–56
Corydoras	*aeneus*		F	61	25M + 26 SM + 8ST + 2A	112	120			0–1B, ACN=64	Brazil (SP)	O–51
Corydoras	*aeneus*			60	26M + 18SM + 16ST	104	120			2B	Brazil (SP)	C–98
Corydoras	*aeneus*			134					6.3 FD	4X	Peru, Guyana	T–58
Corydoras	*aeneus*			ca. 132	90 non–A + 42A		(222)		(8.4* FCM)	4X	(S. America)	S–27, O–48
Corydoras	*aeneus*			120					8.8 BFA		(S. America)	H–13
Corydoras	*agassizii*			98	82 non–A + 16A		(180)				(S. America)	S–27
Corydoras	*araguaiaensis*		F, M	94	46M + 40SM + 8ST	180	188	6			(Brazil)	S–69
Corydoras	*arcuatus*		M	46	28M + 18SM	92	92	2	(4.5 FD)		Brazil, Peru	O–54
Corydoras	*arcuatus*			46	46 non–A		(92)				(S. America)	S–27
Corydoras	*axelrodi*			46	46 non–A		(92)				(Colombia)	S–27
Corydoras	*bondi*			46	46 non–A		(92)				(northern S. America)	S–27
Corydoras	*delphax*			84	12M + 22SM + 50 ST/A	118					(Colombia)	K–16
Corydoras	*difluviatilis*		F, M	78	6M + 2SM + 20ST + 50A	86	106	2			Brazil (SP)	S–69
Corydoras	*ehrhardti*		F, M	44	22M + 22SM	88	88	2			Brazil (SC)	S–69
Corydoras	*ehrhardti*		F, M	44	18M + 26SM	88	88	4		ACN=58	Brazil (SC)	O–55
Corydoras	*elegans*			50	50 non–A		(100)				(Brazil, Colombia, Peru)	S–27
Corydoras	*elegans*			50					6.0 BFA		(S. America)	H–13
Corydoras	*flaveolus*		F, M	58	18M + 26SM + 14ST	102	116	4	3.0 FD	ACN=58	Brazil (SP)	O–54
Corydoras	*julii*			46	32M + 14SM	92	92				(Brazil)	K–16

Table 6.13 Order SILURIFORMES (continued)

A		B	C	D	E	F	G	H	I	J	K	L
Current scientific name of taxon Family/subfamily/species		Reported in karyotype paper	Sex	2n	Karyotype	NF_1	NF_2	Ag–NORs	Genome size (pg/cell)	Comments	Locality	Reference
Corydoras	*julii*			92	92 non–A		(184)		(8.4 BFA)		(S. America)	S-27, H-13
Corydoras	*melanistius*			46	32M + 14SM	92	92				(northern S. America)	K-16
Corydoras	*melanistius*			46	46 non–A		(92)				(S. America)	S-27
Corydoras	*melanistius*			48					6.0 BFA		(S. America)	H-13
Corydoras	*metae*		F, M	92	40M + 36SM + 6ST + 10A	168	174	6	(8.8 FD, 6.4 FIA)	4X, ACN=106	Brazil (PA)	O-54, H-41
Corydoras	*metae*			92	88 non–A + 4A		(180)				(S. America)	S-27
Corydoras	*nattereri*		M	44	18M + 26SM	88	88	2			Brazil (PR)	S-69
Corydoras	*nattereri*		F, M	44	18M + 26SM	88	88	2			Brazil (PR)	O-55
Corydoras	*nattereri*		F, M	44	20M + 24SM	88	88	2		ACN=58	Brazil (RJ)	O-52, O-55
Corydoras	*nattereri*		F, M	40	20M + 20SM	80	80	2	(3.6 FD)	ACN=58	Brazil (Bonito R., RJ)	O-52, O-55
Corydoras	*nattereri*		F, M	42	18M + 24SM	84	84	4		ACN=58	Brazil (Bigua R., SP)	O-52
Corydoras	cf. *nijsseni*		F, M	52	32M + 20SM	104	104	2			(Brazil)	S-69
Corydoras	aff. *osteocarus*			76	50 non–A + 26A		(126)				(Venezuela, Suriname)	S-27
Corydoras	*paleatus*		F, M	44	20M + 24SM	88	88	4			Brazil (RS)	S-69
Corydoras	*paleatus*		F, M	44	20M + 24SM	88	88	4–6		ACN=58	Brazil (PR, RS)	O-55
Corydoras	*paleatus*		F, M	44	22M + 22SM	88	88	4		ACN=58	Brazil (RS)	O-55
Corydoras	*panda*		F, M	46	24M + 22SM	92	92	2		ACN=58	Brazil (SP)	O-56
Corydoras	*pulcher*		M	102	10M + 14SM + 42ST + 36A	126	168	2		4X	(Brazil)	S-69
Corydoras	*punctatus*			44–46					5.8 BFA		(S. America)	H-13
Corydoras	aff. *punctatus*		F	102	10M + 14SM + 20ST + 58A	126	146	2		4X, ACN=106	Brazil (AM)	O-54
Corydoras	*rabauti*		F, M	58	20M + 22SM + 10ST + 6A	100	110	2		ACN=60	Brazil (SP)	O-56
Corydoras	*rabauti*	*myersi*		56					4.6 BFA		S. America	O-54, H-13
Corydoras	*reticulatus*		F, M	74	16M + 20SM + 12ST + 26A	110	122	2	(2.0 FD)	ACN=76	Brazil (AM)	O-54
Corydoras	*reticulatus*			74	20M + 10SM + 44 ST/A	104					(Brazil)	K-16
Corydoras	*robineae*		F, M	84	8M + 18SM + 16ST + 42A	110	126	2			(Brazil)	S-69
Corydoras	*schwartzi*		F, M	46	32M + 14SM	92	92	2	4.8 FD	ACN=62	Brazil (AM)	O-54
Corydoras	*simulatus*		M	62	32M + 18SM + 6ST + 6A	112	118	4	1.3 FD	ACN=62	Colombia	O-54
Corydoras	cf. *simulatus*		M	62	34M + 22SM + 6ST	118	124	2	1.0 FD	ACN=62	Colombia	O-54
Corydoras	*sodalis*		F, M	74	16M + 18SM + 10ST + 30A	108	118	2		ACN=84	(Brazil)	S-190
Corydoras	*trilineatus*		M	46	28M + 18SM	92	92	2	4.9 FD	ACN=58	Brazil (PA)	O-54
Corydoras	*undulatus*		F, M	52	24M + 14SM + 12ST + 2A	90	102	10			Brazil (RS)	S-69
Corydoras	*undulatus*			50					6.0 BFA			H-13
Corydoras	*zygatus*		F, M	56	18M + 20SM + 10ST + 8A	94	104	2			(Brazil)	S-69
Corydoras	sp.		M	60	38M + 16SM + 6ST	114	120	2	1.3 FD	ACN=62	Brazil (Caripi R., PA)	O-54
Corydoras	sp.		F, M	84	4M + 2SM + 18ST + 60A	90	108	10	2.4 FD	ACN=88	Brazil (Galheiro R., MG)	O-54
Corydoras	sp.			62							Argentina	F-20

Table 6.13 Order SILURIFORMES (continued)

A		B	C	D	E	F	G	H	I	J	K	L
Current scientific name of taxon Family/subfamily/species		Reported in karyotype paper	Sex	2n	Karyotype	NF_1	NF_2	Ag-NORs	Genome size (pg/cell)	Comments	Locality	Reference
Scleromystax	*barbatus*	*Corydoras*	F, M	66	38M + 22SM + 4ST + 2A	126	130	6			Brazil (SP)	S-69
Scleromystax	*barbatus*	*Corydoras*	F, M	66	38M + 22SM + 6ST	126	132	6			Brazil (PR)	S-69
Scleromystax	*barbatus*	*Corydoras*	F, M	64	38M + 20SM + 4ST + 2A	122	126	8	1.7, 1.9 FD	ACN=64	Brazil (SP)	O-55
Scleromystax	*barbatus*	*Corydoras*	F, M	66	38M + 22SM + 4ST + 2A	126	130	8		ACN=66	Brazil (PR, SC)	O-55
Scleromystax	*macropterus*	*Corydoras*	F, M	66	28M + 14SM + 16ST + 8A	108	124	6	1.4 FD	ACN=66	Brazil (SP)	O-55
Scleromystax	*prionotos*	*Corydoras*	F, M	68	14M + 12SM + 14ST + 28A	94	108	4	1.2 FD	ACN=68	Brazil (SP)	O-55
Scleromystax	*prionotos*	*Corydoras*	F	86	20M + 28SM + 20ST + 18A	134	154	4	1.6 FD	ACN=86	Brazil (RJ)	O-55
Clariidae												
Clarias	*batrachus*		F	100	4M + 6SM + 78A + 12 MC	110				XX, 4X	Thailand	W-30
Clarias	*batrachus*		M	100	4M + 7SM + 77A + 12 MC	111				XY, 4X	Thailand	W-30
Clarias	*batrachus*			56					1.6* FD		China	C-83, C-85
Clarias	*batrachus*	*Clarius*		52	6 M/SM + 46 ST/A	58					India (U.P.)	S-104
Clarias	*batrachus*			50	16M + 8SM + 14ST + 12A	74	88				India	R-52
Clarias	*batrachus*			50	18M + 20SM + 8ST + 4A	88	96				India (WB)	P-92
Clarias	*batrachus*			54					2.4 BFA			H-13
Clarias	*camerunensis*			54				2			(W. and W.C. Africa)	O-83
Clarias	*ebriensis*			50							(W. Africa)	O-83
Clarias	*fuscus*		F	56	18M + 24SM + 8ST + 6A	98	106			XX, ACN=56	China (Guangdong)	W-30
Clarias	*fuscus*		M	56	19M + 23SM + 8ST + 6A	98	106			XY, ACN=56	China (Guangdong)	W-30
Clarias	*fuscus*		F	56	20M + 22SM + 8ST + 6A	98	106			XX, ACN=56	China (Guangdong)	L-76
Clarias	*fuscus*		M	56	20M + 22SM + 8ST + 6A	98	106			XY, ACN=56	China (Guangdong)	L-76
Clarias	*fuscus*		F, M	56	18M + 14SM + 14ST + 10A	88	102				China (Guangdong)	Y-15
Clarias	*fuscus*			56	32 M/SM + 24 ST/A	88				ACN=56	Japan (Ishigaki)	A-56
Clarias	*gariepinus*		F	56	8M + 25SM + 23A	89		2		ZW, ACN=56	Africa, Israel	T-29, O-69
Clarias	*gariepinus*		M	56	8M + 24SM + 24A	88		2	(2.4 BFA)	ZZ, ACN=56	Africa, Israel	T-29, O-69, H-13
Clarias	*gariepinus*		F	56	14M + 17SM + 25A	87				ZW	Egypt	N-3
Clarias	*gariepinus*		M	56	14M + 18SM + 24A	88				ZZ	Egypt	N-3
Clarias	*gariepinus*			56	20M + 16SM + 10ST + 10A	92	102				India (introduced)	N-3
Clarias	*macrocephalus*		F, M	54	24M + 20SM + 6ST + 4A	98	104			XX/XY	Thailand	W-30
Clarias	*platycephalus*			54				2			(W. and W.C. Africa)	O-83
Heterobranchus	*longifilis*		F	52	6M + 25SM + 21A	83		2		ZW	Ivory Coast	T-29
Heterobranchus	*longifilis*		M	52	6M + 24SM + 22A	82		2		ZZ	Ivory Coast	T-29
Cranoglanididae												
Cranoglanis	*bouderius*	*sinensis*	F, M	74	8M + 16SM + 18ST + 32A	98	116			ACN=74	China	Y-15

Table 6.13 Order SILURIFORMES (continued)

A		B	C	D	E	F	G	H	I	J	K	L
Current scientific name of taxon Family/subfamily/species		Reported in karyotype paper	Sex	2n	Karyotype	NF_1	NF_2	Ag–NORs	Genome size (pg/cell)	Comments	Locality	Reference
Diplomystidae												
Diplomystes	*camposensis*		F, M	56	16M + 24SM + 8ST + 8A	96	104	2		ACN=58	Chile	C–7
Diplomystes	*nahuelbutaensis*		M	56	14M + 26SM + 8ST + 8A	96	104	2		ACN=58	Chile	C–7
Olivaichthys	*mesembrinus*	*Diplomystes*	F, M	56	22M + 24SM + 6ST + 4A	102	108	2	2.6 FD	ACN=56	Argentina (Chubut)	O–60
Doradidae												
Doras	*eigenmanni*			66							(S. America)	E–11
Hassar	*orestis*			58	42M + 14SM + 2A	114	114	2			Brazil (MT)	M–144
Hassar	*orestis*		F, M	58	32M + 20SM + 6ST	110	116	2		ACN=58	Brazil (Xingú R., PA)	M–144
Hassar	cf. *orestis*		F, M	58	32M + 18SM + 8ST	108	116	2		ACN=58	Brazil (Jari R., PA)	M–144
Hassar	*wilderi*		F	58	32M + 16SM + 10ST	106	116	2		ACN=58	Brazil (Araguaia R., MT)	E–11
Hassar	sp.		F	58	32M + 18SM + 8ST	108	116	2		ACN=58	Brazil (Jari R., PA)	M–144
Leptodoras	*acipenserinus*			58	24M + 16SM + 14ST + 4A	98	112	2		ACN=58	Brazil (MT)	M–144
Nemadoras	*humeralis*	*Opsodoras*		58							Brazil (Amazon)	D–32
Opsodoras	*ternetzi*		F, M	58	44M + 12SM + 2A	114	114	2		ACN=58	Brazil (Xingú R., PA)	M–144
Opsodoras	sp.		F	58	21M + 18SM + 12ST + 7A	97	109	2			Brazil (MT)	M–144
Opsodoras	sp.		M	58	20M + 18SM + 12ST + 8A	96	108				Brazil (MT)	M–144
Platydoras	cf. *costatus*		F	58	26M + 16SM + 4ST + 12A	100	104	2		ACN=58	Brazil (Xingú R., PA)	M–144
Oxydoras	*niger*	*Pseudodoras*	F, M	58	20M + 16SM + 8ST + 14A	94	102	2		ACN=58	Brazil (PA)	F–16
Pterodoras	*granulosus*			58							Brazil (Parana R., PR)	J–20
Rhinodoras	*dorbignyi*		F, M	58	20M + 20SM + 4ST + 14A	98	102	2	3.5 FD	ACN=58	Brazil (SP)	F–16, F–64
Rhinodoras	sp.			58	18M + 16SM + 12ST + 12A	92	104	2			Brazil (MT)	M–144
Trachydoras	*paraguayensis*	*paraguaiensis*	F, M	56	32M + 20SM + 4ST	108	112	2		ACN=56	Argentina	F–16
Wertheimeria	*maculata*			58	24M + 14SM + 8ST + 12A	96	104	2		ACN=58	Brazil (MG)	E–11
Erethistidae												
Pseudolaguvia	*ribeiroi*	*Laguvia rebeiroi*		50	6M + 28SM + 16A	84					India (Assam)	K–46, C–108
Heptapteridae												
Cetopsorhamdia	*iheringi*	*iheringhi*		58	22M + 16SM + 10ST + 10A	96	106	2	(1.8 FD)	ACN=58	Brazil (MG, PR)	F–19, F–64
Cetopsorhamdia	*iheringi*	Pimelodidae	F, M	58	28M + 24SM + 6ST	110	116	2			Brazil (SP)	V–46
Cetopsorhamdia	sp.		F, M	58	22M + 16SM + 10ST + 10A	96	106	2		ACN=58	Brazil (SP)	F–19
Imparfinis	*borodini*	*Heptapterus longicauda*	F, M	52	22M + 26SM + 4ST	100	104	4	2.2 FD		Brazil (SP)	V–46, F–64
Imparfinis	*hollandi*	*Pariolius*		42	22M + 10SM + 4ST + 6A	74	78				Brazil (PR)	S–180
Imparfinis	*hollandi*		F, M	42	22M + 10SM + 10ST	74	84	2			Brazil (PR)	M–129
Imparfinis	*mirini*		F	58	23M + 35SM	116	116	2	1.9–2.4 FD	ZW	Brazil (SP)	V–45, F–64

Table 6.13 Order SILURIFORMES (continued)

A		B	C	D	E	F	G	H	I	J	K	L
Current scientific name of taxon Family/subfamily/species		Reported in karyotype paper	Sex	2n	Karyotype	NF_1	NF_2	Ag-NORs	Genome size (pg/cell)	Comments	Locality	Reference
Imparfinis	*mirini*		M	58	24M + 34SM	116	116	2		ZZ	Brazil (SP)	V-45
Imparfinis	*piperatus*	cytotype A	F, M	58	32M + 26SM	116	116	2			Brazil (SP)	V-48
Imparfinis	*piperatus*	cytotype B	F, M	58	26M + 22SM + 8ST + 2A	106	114	2			Brazil (SP)	V-48
Imparfinis	cf. *piperatus*		M	56	22M + 26SM + 4ST + 4A	104	108	2			Brazil (SP)	V-48
Imparfinis	cf. *piperatus*		F, M	56	24M + 12SM + 20ST	92	112	2		ACN=58	Brazil (SP)	F-19
Imparfinis	aff. *mirini*	aff. *schubarti*	F, M	58	28M + 28SM + 2ST	114	116	2		ACN=58	Brazil (PR)	S-180, S-198
Imparfinis	aff. *mirini*	aff. *schubarti*	F, M	58	22M + 18SM + 10ST + 8A	98	108	2		ACN=58	Brazil (SP, PR)	F-19
Pariolius	cf. *longicaudus*			52	22M + 16SM + 4ST + 10A	90	94				Brazil (PR)	S-180
Pimelodella	*avanhandavae*	Pimelodidae	M	46	20M + 20SM + 6ST	86	92	2			Brazil (SP)	V-46
Pimelodella	aff. *avanhandavae*		F, M	52	30M + 22SM	104	104	2			Brazil (PR)	S-149
Pimelodella	*cristata*			52							Brazil (MT)	S-180
Pimelodella	*gracilis*			52					1.8 BFA		Brazil (MS)	S-180, H-13
Pimelodella	*kronei*		F, M	58	54 M/SM + 4ST	112	116	2		0-1 B	Brazil (SP)	A-20
Pimelodella	*meeki*	Pimelodidae	F, M	46	30M + 12SM + 4ST	88	92	2			Brazil (Tibagi R., PR)	V-95
Pimelodella	*transitoria*		F, M	58	54 M/SM + 4ST	112	116	2			Brazil (SP)	A-20, F-57
Pimelodella	sp.			46			84				S. America	L-21
Pimelodella	sp.			56							Argentina	F-20
Pimelodella	sp.			46							Argentina	F-20
Pimelodella	sp. 1		F, M	46	20M + 20SM + 6ST	86	92	2	1.1 FD		Brazil (PR)	V-84, F-64
Pimelodella	sp. 2			52	22M + 22SM + 8ST	96	104	8	2.0 FD		Brazil (PR)	V-84, F-64
Pimelodella	sp.	Rhamdiidae		46	34M + 12SM	92	92				Brazil (Tibagi R., PR)	S-166
Pimelodella	sp.		F	46	40 M/SM + 6 ST/A	86		2		XX	Brazil (MG)	D-11
Pimelodella	sp.		M	46	40 M/SM + 6 ST/A	86		2		XY	Brazil (MG)	D-11
Rhamdella	*microcephala*			56	18M + 30SM + 8ST	104	112	2			Brazil (MG)	F-29, S-180
Rhamdella	sp.	Rhamdiidae		56	26 M/SM + 30 ST/A	82					Brazil (BA)	S-166
Rhamdia	*laticauda*			58							C. America	L-21
Rhamdia	*quelen*	*branneri*		58	36M + 14SM + 4ST + 4A	108	112			0-2 B	Brazil (PR)	S-166
Rhamdia	*quelen*	*branneri*		58	30M + 10SM + 14ST + 4A	98	112			0-4 B	Brazil (PR)	C-98
Rhamdia	*quelen*	*branneri*	F, M	58	36M + 14SM + 4ST + 4A	108	112			0-4 B	Brazil (SC)	A-5
Rhamdia	*quelen*	*hilarii*	F, M	58	58 M/SM	116	116	2	(2.0-2.3 FD)	0-2 B, ACN=58	Brazil (SP)	S-166, M-13, F-64
Rhamdia	*quelen*	*hilarii*	F, M	58-63		>100		2		0-5 B	Brazil (SP)	F-15
Rhamdia	*quelen*	*hilarii*	F, M	58	26M + 16SM + 8ST + 8A	100	108	2		0 B, ACN=58	Argentina	F-19
Rhamdia	cf. *quelen*	cf. *hilarii*	F, M	58	30M + 18SM + 10ST	106	116	2		0-3 B	Brazil (SP)	V-47
Rhamdia	*quelen*		F, M	58	26M + 20SM + 6ST + 6A	104	110	2		0-4 B	Brazil (SP)	S-154
Rhamdia	*quelen*		F, M	58	26M + 22SM + 6ST + 4A	106	112	2		0 B	Brazil (PR)	S-154
Rhamdia	*quelen*			58	26M + 24SM + 8ST	108	116	2		0 B	Brazil (PR)	S-154

Table 6.13 Order SILURIFORMES (continued)

A		B	C	D	E	F	G	H	I	J	K	L
Current scientific name of taxon Family/subfamily/species		Reported in karyotype paper	Sex	2n	Karyotype	NF_1	NF_2	Ag-NORs	Genome size (pg/cell)	Comments	Locality	Reference
Rhamdia	*quelen*		F	58	26M + 16SM + 8ST + 8A	100	108	2		0 B, ACN=58	Argentina	F-19
Rhamdia	*quelen*		F, M	58	36M + 16SM + 6ST	110	116	2		0–3 B, ACN=58	Brazil (MS)	M-157
Rhamdia	*quelen*		F, M	58	36M + 16SM + 6ST	110	116			0–2 B, ACN=58	Brazil (SP, PR)	M-158
Rhamdia	*quelen*			58	36M + 16SM + 6ST	110	116			0–2 B	Brazil (SC)	M-158
Rhamdia	*quelen*	*sapo*	F, M	58	44 M/SM + 14 ST/A	102				0–1 B	Argentina	V-78
Rhamdia	*quelen*	*sapo*		56							Uruguai	S-166
Rhamdia	*quelen*	*voulezi*	F, M	58	36M + 14SM + 4ST + 4A	108	112			0–2 B	Brazil (PR)	A-5
Rhamdia	sp.		F, M	58	46 M/SM + 12ST	104	116	2		0–4 B, ACN=58	Brazil (SP)	G-13
Rhamdia	sp.		F	87	69 M/SM + 18ST (3n)	156	174	3		3X, ACN=87	Brazil (SP)	G-13
Heteropneustidae												
Heteropneustes	*fossilis*		F, M	56	14M + 26SM + 16ST	96	112				India	P-48
Heteropneustes	*fossilis*		F, M	56	18M + 10SM + 12ST + 16A	84	96				India (Jammu)	T-51
Heteropneustes	*fossilis*			56	22M + 16SM + 6ST + 12A	94	100	2			India	R-69
Heteropneustes	*fossilis*			56	18M + 18 SM/ST + 20A		92				India	R-52
Heteropneustes	*fossilis*			56	14M + 12SM + 30A	82					Thailand	D-28
Ictaluridae												
Ameiurus	*brunneus*	*Ictalurus*		62							(S.E. USA)	H-45
Ameiurus	*catus*	*Ictalurus*	F, M	48	16M + 20SM + 12ST	84	96			ACN=58	USA (AL)	L-22
Ameiurus	*melas*	*Ictalurus*	F, M	60	12M + 10SM + 12ST + 26A	82	94				USA (TN)	C-65
Ameiurus	*melas*	*Ictalurus*	M	60	16 M/SM + 44 ST/A	76				ACN=60	USA (OH)	L-21
Ameiurus	*natalis*	*Ictalurus*	F, M	62	12M + 10SM + 14ST + 26A	84	98				USA (TN)	C-65
Ameiurus	*natalis*	*Ictalurus*	M	62	22 M/SM + 40 ST/A	84				ACN=64	USA (OH, MO)	L-21
Ameiurus	*nebulosus*	*Ictalurus*	F, M	60	16 M/SM + 44 ST/A	76			1.9 FD, 2.4 BFA	ACN=60	USA (OH)	L-21, H-13
Ameiurus	*nebulosus*	*Ictalurus*	F, M	60	20 M/SM + 40 ST/A	80					Bosnia	B-22
Ameiurus	*nebulosus marmoratus*	*Ictalurus*	F, M	60	16 M/SM + 44 ST/A	76					Italy	B-10
Ameiurus	*platycephalus*	*Ictalurus*		54							(S.E. USA)	H-45
Ameiurus	*serracanthus*	*Ictalurus*	F	52	38 M/SM + 14 ST/A	90				ACN=58	USA (FL)	L-21
Ictalurus	*furcatus*			58	18M + 14SM + 26ST	90	116			ACN=58	USA (AL)	L-22
Ictalurus	*punctatus*			58	34 M/SM + 24 ST/A	92				ACN=58	USA (LA)	L-21
Ictalurus	*punctatus*		F, M	58	18M + 14SM + 26ST	90	116		(2.0 FCM, 2.1 FD)	ACN=58	USA	L-22, M-91, T-73
Noturus	*albater*			66–72		82					USA	L-21
Noturus	*elegans*		F, M	46	36 M/SM + 10 ST/A	82				ACN=54	USA (KY)	L-21
Noturus	*eleutherus*		F, M	42	24 M/SM + 18 ST/A	66				ACN=58	USA (AR, VA)	L-21
Noturus	*exilis*		F	54	14 M/SM + 40 ST/A	68				ACN=56	USA (MO)	L-21

Table 6.13 Order SILURIFORMES (continued)

A		B	C	D	E	F	G	H	I	J	K	L
Current scientific name of taxon Family/subfamily/species		Reported in karyotype paper	Sex	2n	Karyotype	NF_1	NF_2	Ag-NORs	Genome size (pg/cell)	Comments	Locality	Reference
Noturus	*flavater*		M	44	20 M/SM + 24 ST/A	64				ACN=58	USA (MO)	L-21
Noturus	*flavipinnis*		F	52	30 M/SM + 22 ST/A	82				ACN=56	USA (VA)	L-21
Noturus	*flavus*		F, M	48	22 M/SM + 26 ST/A	70				ACN=56	USA (OH, PA)	L-20, L-21
Noturus	*flavus*		F	50	20 M/SM + 30 ST/A	70				ACN=56	USA (VA)	L-20, L-21
Noturus	*flavus*			48							USA (IN)	H-17
Noturus	*funebris*		F	44	24 M/SM + 20 ST/A	68				ACN=58	USA (LA)	L-21
Noturus	*gilberti*		M	54	28 M/SM + 26 ST/A	82				ACN=58	USA (VA)	L-21
Noturus	*gyrinus*		F, M	42	30 M/SM + 12 ST/A	72				ACN=56	USA (LA, MO, OH)	L-21
Noturus	*hildebrandi hildebrandi*		F, M	46	34 M/SM + 12 ST/A	80				ACN=58	USA (MS)	L-21
Noturus	*hildebrandi lautus*		F, M	46	34 M/SM + 12 ST/A	80				ACN=58	USA (TN)	L-21
Noturus	*insignis*		F, M	54	20 M/SM + 34 ST/A	74				ACN=58	USA (PA, NC)	L-21
Noturus	*lachneri*		F, M	42	30 M/SM + 12 ST/A	72				ACN=54	USA (AR)	L-21
Noturus	*leptacanthus*		F, M	46	26 M/SM + 20 ST/A	72				ACN=56	USA (LA)	L-21
Noturus	*miurus*			50	24 M/SM + 26 ST/A	74				ACN=58	USA (LA, MO, OH)	L-21
Noturus	*munitus*		F	42	20 M/SM + 22 ST/A	62				ACN=58	USA (LA)	L-21
Noturus	*nocturnus*		F, M	48	24 M/SM + 24 ST/A	72				ACN=58	USA (LA, MS, MO)	L-21
Noturus	*phaeus*			42	26 M/SM + 16 ST/A	68				ACN=56	USA (MS, TN)	L-21
Noturus	*stigmosus*		F	42	20 M/SM + 22 ST/A	62				NAN=54	USA (OH)	L-21
Noturus	*taylori*		F, M	40	24 M/SM + 16 ST/A	64				ACN=56	USA (AR)	L-21
Prietella	*phreatophila*			50			ca. 80	2		ACN=52?	Mexico	A-37
Pylodictis	*olivaris*			56	26 M/SM + 30 ST/A	82				ACN=60	USA (OH)	L-21
Loricariidae												
Ancistrinae												
Ancistrus	cf. *dubius*		F, M	44	18M + 10SM + 8ST + 8A	72	80	2		ZW/ZZ	Brazil (MT)	M-164
Ancistrus	cf. *dubius*		F, M	42	24M + 10SM + 8ST	76	84	2		XX/XY	Brazil (Pantanal, MT)	M-165
Ancistrus	cf. *dubius*		F, M	42	24M + 10SM + 8ST	76	84	2		no sex chrom.	Brazil (Pantanal, MT)	M-165
Ancistrus	*multispinis*	Hypostominae, Ancistrini		52	28 M/SM + 24 ST/A	80				ACN=52	Brazil (SC)	A-34
Ancistrus	*ranunculus*		F	48	19M + 9SM + 6ST + 14A	76	82	4		ZW, ACN=51	Brazil (Amazon R.)	O-80
Ancistrus	*ranunculus*		M	48	20M + 8SM + 6ST + 14A	76	82	4		ZZ, ACN=52	Brazil (Amazon R.)	O-80
Ancistrus	sp.	Purus	F	34	20M + 12SM + 2ST	66	68	2		XX, ACN=42	Brazil (Purus, AM)	O-85
Ancistrus	sp.	Purus	M	34	21M + 11SM + 2ST	66	68	2		XY, ACN=42	Brazil (Purus, AM)	O-85
Ancistrus	sp.	Trombetas	M	38	22M + 8SM + 5ST + 3A	68	73	2		ACN=48	Brazil (Trombetas, PA)	O-85
Ancistrus	sp.	Vermelho	M	42	26M + 6SM + 4ST + 6A	74	78	2		ACN=42	Brazil (Demini, AM)	O-85
Ancistrus	sp.	Macoari	F	46	18M + 12SM + 6ST + 10A	76	82	2		XX, ACN=48	Brazil (Branco, RR)	O-85
Ancistrus	sp.	Macoari	M	46	18M + 11SM + 6ST + 11A	75	81	2		XY, ACN=48	Brazil (Branco, RR)	O-85

Table 6.13 Order SILURIFORMES (continued)

A		B	C	D	E	F	G	H	I	J	K	L
Current scientific name of taxon Family/subfamily/species		Reported in karyotype paper	Sex	2n	Karyotype	NF_1	NF_2	Ag-NORs	Genome size (pg/cell)	Comments	Locality	Reference
Ancistrus	sp.	Dimona	F, M	52	16M + 8SM + 2ST + 26A	76	78	2		ACN=56	Brazil (Dimona, AM)	O-85
Ancistrus	sp. 1	Balbina	F	38	26M + 10SM + 2ST	74	76	2		XX, ACN=48	Brazil (Barretinho, AM)	O-84
Ancistrus	sp. 1	Balbina	M	39	27M + 10SM + 2ST	76	78	2		XY_1Y_2, ACN=48	Brazil (Barretinho, AM)	O-84
Ancistrus	sp. 2	Barcelos	F	52	11M + 12SM + 4ST + 25A	75	79	2		$Z_1Z_2W_1W_2$	Brazil (Demeni, AM)	O-84
Ancistrus	sp. 2	Barcelos	M	52	12M + 12SM + 4ST + 24A	76	80	2		$Z_1Z_1Z_2Z_2$	Brazil (Demeni, AM)	O-84
Ancistrus	sp.	Piagaçu	F	52	16M + 9SM + 2ST + 25A	77	79	2		ZW, ACN=52	Brazil (Purus, AM)	O-80
Ancistrus	sp.	Piagaçu	M	52	16M + 8SM + 2ST + 26A	76	78	2		ZZ, ACN=52	Brazil (Purus, AM)	O-80
Ancistrus	sp. 1			38	30 M/SM + 8ST	68	76	2			Brazil (Acre)	A-34
Ancistrus	sp. 2			52	32 M/SM + 20 ST/A	84		2		ACN=52	Brazil (SP)	A-34
Ancistrus	n. sp. 1		F	40	34M + 6SM	80	80	2		XX	Brazil (GO)	A-112
Ancistrus	n. sp. 1		M	39	33M + 6SM	78	78	2		XO	Brazil (GO)	A-112
Ancistrus	n. sp. 2		F, M	52	10M + 16SM + 12ST + 14A	78	90	2			Brazil (SC)	A-112
Baryancistrus	aff. *niveatus*			52	16M + 32SM + 4A	100	100	2			Brazil (PA)	O-79
Hemiancistrus	*spilomma*		F	52	25M + 21SM + 6ST	98	104	6		ZW, ACN=57	Brazil (MT)	O-79
Hemiancistrus	*spilomma*		M	52	24M + 22SM + 6ST	98	104	6		ZZ, ACN=58	Brazil (MT)	O-79
Hemiancistrus	*spinosissimus*		F, M	52	26M + 22SM + 4ST	100	104	2		ACN=56	Brazil (MT)	O-79
Hemiancistrus	sp.		F	52	20M + 20SM + 12 ST/A	92				ZW	Brazil (MT)	A-109
Megalancistrus	*parananus*	*aculeatus*		52	26M + 26SM	104	104	2			Brazil (PR)	A-34
Panaque	cf. *nigrolineatus*		F, M	52	26M + 20SM + 6ST	98	104				Brazil (MT)	A-109
Hypoptopomatinae												
Corumbataia	*cuestae*		F, M	54	28M + 20SM + 6 ST/A	102		2			Brazil (SP)	C-3
Hisonotus	*depressicauda*	*Microlepidogaster*	F	54	14M + 28SM + 2ST + 10A	96	98	2			Brazil (SP)	A-49
Hisonotus	*leucofrenatus*	*Microlepidogaster*	F, M	54	24M + 26SM + 2ST + 2A	104	106	1-2		ZW, 0-2 B	Brazil (SP)	A-48
Hisonotus	*leucofrenatus*	*Microlepidogaster*	F	54	22M + 26SM + 4ST + 2A	102	106	1-2		ZW, 0-2 B	Brazil (PR)	A-48
Hisonotus	*leucofrenatus*		F	54	22M +24SM + 6ST + 2A	100	106	2		ACN=58	Brazil (SC)	A-108
Hisonotus	*nigricauda*		F	54	26M + 20SM + 8ST	100	108	2		ACN=58	Brazil (RS)	A-108
Hisonotus	sp. 1			54					2.7 FD	ZW	Brazil (SP)	F-64
Hisonotus	sp. 2			54					1.8 FD	ZW	Brazil (SP)	F-64
Hisonotus	sp. A	*Microlepidogaster*	F	54	30M + 20SM + 4ST	104	108	4			Brazil (SP)	A-49
Hisonotus	sp. A			54	26M + 26SM + 2ST	106	108	2		ACN=58	Brazil (SP)	A-108
Hisonotus	sp. B	*Microlepidogaster*	F	54	22M + 28SM + 4ST	104	108	2			Brazil (SP)	A-49
Hisonotus	sp. D			54	26M + 26SM + 2ST	106	108	2		ACN=58	Brazil (SP)	A-108
Lampiella	*gibbosa*	*Hisonotus gibbosos*		58							Brazil (SP)	K-20
Macrotocinclus	*affinis*	*Otocinclus*	F, M	54	46M + 8SM	108	108	2	(4.2 BFA)		Brazil (SP)	A-49, H-13
Macrotocinclus	*affinis*	*Otocinclus*	F, M	54	40M + 12SM + 2ST	106	108	2			Brazil (RJ)	A-49
Otocinclus	aff. *vestitus*		F	72	22M + 12SM + 4ST + 34A	106	110	2		ZW	Brazil (PA)	A-49

Table 6.13 Order SILURIFORMES (continued)

A		B	C	D	E	F	G	H	I	J	K	L
Current scientific name of taxon Family/subfamily/species		Reported in karyotype paper	Sex	2n	Karyotype	NF_1	NF_2	Ag-NORs	Genome size (pg/cell)	Comments	Locality	Reference
Parotocinclus	*maculicauda*		M	54	20M + 32SM + 2ST	106	108	2			Brazil (SP)	A-49
Pseudotocinclus	*tietensis*		F	54	26M + 22SM + 6ST	102	108	2		XX	Brazil (SP)	A-47
Pseudotocinclus	*tietensis*		M	54	27M + 21SM + 6ST	102	108	2		XY	Brazil (SP)	A-47
Pseudotothyris	*obtusa*		M	54	26M + 18SM + 4ST + 6A	98	102	2			Brazil (SP)	A-49
Hypostominae												
Corymbophanes	sp.	*Pareiorhina*		54	20M + 20SM + 14ST	94	108			ACN=54	Brazil (MG)	A-35
Hypostomus	*affinis*			66	14M + 14SM + 12ST + 26A	94	106	5			Brazil (Jacui Stream, SP)	K-20
Hypostomus	*albopunctatus*		F, M	74	10M + 20SM + 44 ST/A	104		6		ACN=76	Brazil (SP)	A-98
Hypostomus	*ancistroides*		F, M	68	16M + 18SM + 34 ST/A	102		6		ACN=76	Brazil (SP)	A-98
Hypostomus	*ancistroides*		F, M	68	18M + 10SM + 12ST + 28A	96	108	6			Brazil (SP)	A-112
Hypostomus	*ancistroides*	*Plecostomus*	F	68	10M + 28SM + 30 ST/A	106				XX	Brazil	M-66
Hypostomus	*ancistroides*	*Plecostomus*	M	68	10M + 27SM + 31 ST/A	105				XY	Brazil	M-66
Hypostomus	aff. *auroguttatus*		F	76	8M + 30SM + 38 ST/A	114		2		ACN=76	Brazil (SP)	A-98
Hypostomus	*commersoni*			68	14M + 14SM + 8ST + 32A	96	104				Argentina	F-20
Hypostomus	*goyazensis*		F, M	72	10M + 16SM + 10ST + 36A	98	108	2			Brazil (GO)	A-112
Hypostomus	*macrops*	*Plecostomus*	M	68	10M + 14SM + 44 ST/A	92					Brazil	M-66
Hypostomus	*nigromaculatus*			76	8M + 20SM + 48 ST/A	104		3		ACN=76	Brazil (SP)	R-119
Hypostomus	*nigromaculatus*			76	6M + 20SM + 50 ST/A	102		3		ACN=76	Brazil (PR)	R-119
Hypostomus	*paulinus*	*Plecostomus*	M	74	10M + 20SM + 44 ST/A	104					Brazil	M-66
Hypostomus	*plecostomus*			68					(3.2 FIA, 4.2 BFA)		(S. America)	H-13, H-41
Hypostomus	*plecostomus*?	*plecostomus*		54	24 M/SM + 12ST + 18A	78	90		3.6 FD		(S. America)	M-91
Hypostomus	*regani*		F, M	72	10M + 20SM + 42 ST/A	102				ACN=74	Brazil (SP)	A-98
Hypostomus	*regani*		F, M	72	12M + 18SM + 26ST + 16A	102	128	4			Brazil (SP)	A-112
Hypostomus	*strigaticeps*	*Plecostomus*	F, M	74	8M + 4SM + 62 ST/A	86					Brazil	M-66
Hypostomus	*tietensis*			68	14M + 12SM + 42 ST/A	94					Brazil (SP)	A-35
Hypostomus	sp.			67	15M + 12SM + 14ST + 26A	94	108			1B	Brazil (SP)	C-98
Hypostomus	sp. A		F, M	70	18M + 14SM + 38 ST/A	102		6		ACN=76	Brazil (SP)	A-98
Hypostomus	sp. B		F, M	72	12M + 18SM + 42 ST/A	102		2		ACN=76	Brazil (SP)	A-98, A-132
Hypostomus	sp. C		F, M	72	10M + 18SM + 44 ST/A	100		4	(4.3 FD)	ACN=74	Brazil (SP)	A-98, F-64
Hypostomus	sp. D_1		M	72	10M + 26SM + 36 ST/A	108		2	(4.7 FD)	ACN=74	Brazil (SP)	A-98, F-64
Hypostomus	sp. D_2		M	72	14M + 20SM + 38 ST/A	106		2	(3.7 FD)	ACN=78	Brazil (SP)	A-98, F-64
Hypostomus	sp. E		F, M	80	8M + 16SM + 56 ST/A	104		4		ACN=80	Brazil (SP)	A-98, A-132
Hypostomus	sp. 3			82	6M + 14SM + 62 ST/A	102				0-2 B	Brazil (MS)	C-98
Hypostomus	sp. 2	Rio Perdido	F, M	84	6M + 16SM + 62 ST/A	106		2		ACN=84	Brazil (MS)	C-107
Hypostomus	sp. 3	Córrego Salobrinha	F, M	82	6M + 12SM + 64 ST/A	100				ACN=82	Brazil (MS)	C-107
Hypostomus	sp. 3	Córrego Salobrinha	F, M	84	6M + 12SM + 66 ST/A	102				ACN=84	Brazil (MS)	C-107

Table 6.13 Order SILURIFORMES (continued)

A		B	C	D	E	F	G	H	I	J	K	L
Current scientific name of taxon Family/subfamily/species		Reported in karyotype paper	Sex	2n	Karyotype	NF_1	NF_2	Ag-NORs	Genome size (pg/cell)	Comments	Locality	Reference
Hypostomus	sp. F			76	10M + 16SM + 50 ST/A	102		2	(4.2, 4.9 FD)		Brazil (MG)	A-132, F-64
Hypostomus	sp. G			64	14M + 24SM + 26 ST/A	102				ZW	Brazil (MT)	A-109
Hypostomus	sp.		F	64	15M + 24SM + 25 ST/A	103		2	3.9 FD	ZW	Brazil (MT)	A-99, F-64
Hypostomus	sp.		M	64	14M + 24SM + 26 ST/A	102		2	3.9 FD	ZZ, ACN=82	Brazil (MT)	A-99, F-64
Hypostomus	sp. 1?			54	36 M/SM + 18 ST/A	90					Argentina	F-20
Hypostomus	sp. 2			72	28 M/SM + 44 ST/A	100					Argentina	F-20
Pterygoplichthys	*ambrosettii*	*Liposarcus anisitsi*	F, M	52	16M + 24SM + 8ST + 4A	92	100	2	(4.0 FD)		Brazil (SP)	A-133, F-64
Pterygoplichthys	*ambrosettii*	*Liposarcus anisitsi*	F, M	52	28M + 12SM + 8ST + 4A	92	100	2			Brazil (SP)	A-112
Pterygoplichthys	*ambrosettii*	*Liposarcus anisitsi*	F, M	52	8M + 14SM + 14ST + 16A	74	88	2			Brazil (MS).	A-112
Pterygoplichthys	*gibbiceps*	*Glyptoperichthys*	F, M	52	20M + 24SM + 8ST	96	104	2			Venezuela (Orinoco R.)	A-112
Pterygoplichthys	*joselimaianus*		F, M	52	28M + 16SM + 8 ST/A	96		2			Brazil (MG)	O-79
Pterygoplichthys	*multiradiatus*	*Liposarcus*	F, M	52	22M + 18SM + 12ST	92	104	2			Venezuela (Orinoco R.)	A-112
Pogonopoma	*wertheimeri*		F	54	20M + 30SM + 4ST	104	108			ACN=58	Brazil (Bahia)	A-109
Rhinelepis	*aspera*		F, M	54	20M + 26SM + 8ST	100	108			ACN=58	Brazil (PR)	A-109
Squaliforma	*emarginata*	*Hypostomus emarginatus*	F, M	52	16M + 30SM + 6ST	98	104			ACN=56	Brazil (MT)	A-109
Loricariinae												
Brochiloricaria	*macrodon*	*Loricaria*	M	58	18M + 2SM + 38 ST/A	78					Brazil	M-66
Harttia	*kronei*			58	40 M/SM + 18ST	98	116	2		ACN=58	Brazil (SP)	A-34
Harttia	*loricariformis*			52	32 M/SM + 20 ST/A	84		2		ACN=54	Brazil (SP)	A-34
Harttia	*loricariformis*			56	16M + 22SM + 10ST + 8A	94	104	2			Brazil (Paraitinga R., SP)	K-20
Harttia	sp.			56	14SM + 42A	70					Brazil (MG)	A-34
Loricaria	*cataphracta*	*carinata*		64	12 M/SM + 52 ST/A	76					Argentina	F-20
Loricaria	sp.			64	10M + 6SM + 4ST + 44A	80	84			1-3 B	Brazil (PR)	A-34, C-98
Loricaria	sp.			52							Brazil (PA)	A-34
Loricaria	sp.			62							Brazil (AM)	A-34
Loricariichthys	*maculatus*			56	22 M/SM + 34 ST/A	78					Argentina	F-20
Loricariichthys	*platymetopon*			54	7M + 20SM + 4ST + 23A	81	85				Argentina	F-20
Loricariichthys	*platymetopon*		F	54	7M + 20SM + 4ST + 23A	81	85	2		ZW	Brazil (Paraná R., PR)	S-186
Loricariichthys	*platymetopon*		M	54	6M + 20SM + 4ST + 24A	80	84	2		ZZ	Brazil (Paraná R., PR)	S-186
Loricariichthys	sp.			54	6M + 26SM + 4ST + 18A	86	90				Argentina	A-34
Loricariichthys	sp.			54	28M + 26A	82					Brazil (MG)	A-34
Proloricaria	*prolixa*	*Loricaria*		62	20M + 4SM + 38A	86	86			0-5 B	Brazil (PR)	A-34, C-98
Rineloricaria	*cadeae*	*cadee*		66	2M + 64 ST/A	68				ACN=66	Brazil (RS)	A-34
Rineloricaria	*cadeae*		F, M	64	2 M/SM + 62 ST/A	66		2		ACN=64	Brazil (RS)	M-168
Rineloricaria	*kronei*			64	6 M/SM + 58 ST/A	70		2		ACN=64	Brazil (SC)	A-34
Rineloricaria	*latirostris*			44	12M + 4SM + 28A	60	60				Brazil (Passa Cinco R., SP)	K-20

Table 6.13 Order SILURIFORMES (continued)

A		B	C	D	E	F	G	H	I	J	K	L
Current scientific name of taxon Family/subfamily/species		Reported in karyotype paper	Sex	2n	Karyotype	NF_1	NF_2	Ag-NORs	Genome size (pg/cell)	Comments	Locality	Reference
Rineloricaria	*latirostris*			44	10M + 4SM + 30A	58	58				Brazil (Passa Cinco R., SP)	K-20
Rineloricaria	*latirostris*			44	13M + 2SM + 29A	59	59				Brazil (Passa Cinco R., SP)	K-20
Rineloricaria	*latirostris*			44	13M + 4SM + 27A	61	61				Brazil (Passa Cinco R., SP)	K-20
Rineloricaria	*latirostris*			44	13M + 1SM + 30A	58	58				Brazil (Passa Cinco R., SP)	K-20
Rineloricaria	*latirostris*			44	10M + 3SM + 31A	57	57				Brazil (Passa Cinco R., SP)	K-20
Rineloricaria	*latirostris*			45	15 M/SM + 30 ST/A	60					Brazil (Passa Cinco R., SP)	K-20
Rineloricaria	*latirostris*			46	10M + 3SM + 33A	59	59				Brazil (Passa Cinco R., SP)	K-20
Rineloricaria	*latirostris*			46	14 M/SM + 32 ST/A	60					Brazil (Passa Cinco R., SP)	K-20
Rineloricaria	*latirostris*			47	13 M/SM + 34 ST/A	60					Brazil (Passa Cinco R., SP)	K-20
Rineloricaria	*latirostris*			36	24 M/SM + 12 ST/A	60					Brazil (Mogi-Guacu R., SP)	K-20
Rineloricaria	*latirostris*			37	23 M/SM + 14 ST/A	60					Brazil (Mogi-Guacu R., SP)	K-20
Rineloricaria	*latirostris*			38	22 M/SM + 16 ST/A	60					Brazil (Mogi-Guacu R., SP)	K-20
Rineloricaria	*latirostris*			39	21 M/SM + 18 ST/A	60					Brazil (Mogi-Guacu R., SP)	K-20
Rineloricaria	*latirostris*			40	20 M/SM + 20 ST/A	60					Brazil (Mogi-Guacu R., SP)	K-20
Rineloricaria	*latirostris*			43	17 M/SM + 26 ST/A	60					Brazil (Tres Bocas R., PR)	K-20
Rineloricaria	*latirostris*			44	16 M/SM + 28 ST/A	60					Brazil (Tres Bocas R., PR)	K-20
Rineloricaria	*latirostris*			46	14 M/SM + 32 ST/A	60					Brazil (Tres Bocas R., PR)	K-20
Rineloricaria	*latirostris*			47	13 M/SM + 34 ST/A	60					Brazil (Tres Bocas R., PR)	K-20
Rineloricaria	*latirostris*			48	12 M/SM + 36 ST/A	60					Brazil (Tres Bocas R., PR)	K-20
Rineloricaria	*parva*			48					3.2 BFA		(S. America)	G-73, H-13
Rineloricaria	*pentamaculata*		F, M	56	8 M/SM + 48 ST/A	64		2		ACN=60	Brazil (PR)	A-34, M-168
Rineloricaria	*strigilata*		F, M	68	6 M/SM + 62 ST/A	74		2		ACN=68	Brazil (RS)	M-168
Rineloricaria	sp.			70	2SM + 68A	72				ACN=70	Brazil (SP)	A-34
Sturisoma	cf. *nigrirostrum*		F, M	74	20M + 18SM + 36 ST/A	112					Brazil (MT)	A-109
Neoplecostominae												
Isbrueckerichthys	*duseni*			54	20M + 20SM + 14ST	94	108	2		ACN=56	Brazil (SP)	A-35
Kronichthys	*lacerta*			54	20M + 20SM + 14ST	94	108	2		ACN=54	Brazil (PR)	A-35
Kronichthys	*subteres*			54	20M + 20SM + 14ST	94	108	2		ACN=54	Brazil (SP)	A-35
Neoplecostomus	*microps*			54	20M + 20SM + 14ST	94	108	2		ACN=54	Brazil (SP)	A-35
Neoplecostomus	*microps*			54	24M + 20SM + 10ST	98	108	2			Brazil (Paraitinga R., SP)	K-20
Neoplecostomus	*paranensis*			54	20M + 20SM + 14ST	94	108	2	2.3 FD	0-2 B, ACN=54	Brazil (MG)	A-35, F-64, C-98
Neoplecostomus	*paranensis*			54	20M + 20SM + 14ST	94	108	2		ACN=54	Brazil (SP)	A-35
Pareiorhaphis	*splendens*	*Hemipsilichthys*		54	20M + 30SM + 4ST	104	108	2		ACN=54	Brazil (SC, PA)	A-35
Pareiorhaphis	*steindachneri*	*Hemipsilichthys*		54	20M + 20SM + 14ST	94	108	2		ACN=54	Brazil (SC)	A-35
Pareiorhaphis	*vestigipinnis*	*Hemipsilichthys*		54	20M + 20SM + 14ST	94	108	2		ACN=56	Brazil (SC)	A-35
Pareiorhaphis	sp.	*Hemipsilichthys*		54	20M + 20SM + 14ST	94	108	2		ACN=56	Brazil (PR)	A-35

Table 6.13 Order SILURIFORMES (continued)

A		B	C	D	E	F	G	H	I	J	K	L
Current scientific name of taxon Family/subfamily/species		Reported in karyotype paper	Sex	2n	Karyotype	NF_1	NF_2	Ag-NORs	Genome size (pg/cell)	Comments	Locality	Reference
Pareiorhina	*rudolphi*			54	26M + 16SM + 12ST	96	108	2		ACN=54	Brazil (SP)	A-35
Upsilodinae												
Upsilodus	sp.			96	16M + 8SM + 72A	120		2			Brazil (Paraitinga R., SP)	K-20
Mochokidae												
Hemisynodontis	*membranaceus*		M	54	30 M/SM + 14ST + 10A	84	98	2		ZZ, ACN=56	Africa	A-10, O-1
Hemisynodontis	*membranaceus*		F	54	31 M/SM + 13ST + 10A	85	98	2		ZW, ACN=56	Africa	A-10, O-1
Synodontis	*bastiani*		F	54	23 M/SM + 18ST + 13A	77	95			ZW, ACN=56	Ivory Coast	A-10
Synodontis	*budgetti*		M	54	30 M/SM + 18ST + 6A	84	102	2		ACN=56	Mali	A-10, O-1
Synodontis	*courteti*		F	54	28 M/SM + 19ST + 7A	82	101			ACN=56	Mali	A-10
Synodontis	*filamentosa*	*filamentosus*	F	56	24 M/SM + 22ST + 10A	80	102	2		ACN=56	Mali	A-10, O-1
Synodontis	*ocellifer*		F	54	22 M/SM + 20ST + 12A	76	96			ACN=56	Mali	A-10
Synodontis	*schall*		F	54	24 M/SM + 17ST + 13A	78	95	2	2.2 BFA	ACN=56	Mali	A-10, O-1, H-13
Synodontis	*sorex*		M	54	26 M/SM + 16ST + 12A	80	96	2		ZZ, ACN=56	Mali	A-10, O-1
Synodontis	*sorex*		F	54	27 M/SM + 15ST + 12A	81	96	2		ZW, ACN=56	Mali	A-10, O-1
Synodontis	*violacea*	*violaceus*	M	54	32 M/SM + 14ST + 8A	86	100	2		ACN=56	Mali	A-10, O-1
Pangasiidae												
Pangasianodon	*gigas*		F, M	60	10M + 26SM + 14ST + 10A	96	110			XX/XY	Thailand (Chiang Mai)	M-30
Pangasianodon	*gigas*			60	32M + 8SM + 12ST + 8A	100	112				Thailand	D-21
Pangasianodon	*hypophthalmus*	*Pangasius*		60	20M + 12SM + 4ST + 24A	92	96				Thailand	D-21
Pangasianodon	*hypophthalmus*	*Pangasius*	M	60	12M + 12SM + 6ST + 30A	84	90	2			(Thailand)	K-136
Pangasianodon	*hypophthalmus*	*Pangasius sutchi*		60	20M + 12SM + 4ST + 24A	92	96			ACN=60	Thailand (Nakhonsawan)	M-7
Pangasius	*larnaudii*			60	24M + 20SM + 4ST + 12A	104	108			ACN=60	Thailand (Nakhonsawan)	M-7
Pangasius	*pangasius*		F	58	14M + 20SM + 2ST + 22A	92	94				India (WB)	K-42
Pangasius	*pangasius*			62	14M + 6SM + 18ST + 24A	82	100				India	M-20
Pangasius	*sanitwongsei*			60	20M + 6SM + 18ST + 16A	86	104				Thailand	D-21
Pimelodidae												
Bergiaria	*westermanni*		F, M	56	42 M/SM + 14 ST/A	98		2		0-5 B	Brazil (MG)	D-11
Calophysus	*macropterus*			50	22M + 18SM + 10A	90		2			Brazil (R. Negro)	S-166
Hemisorubim	*platyrhynchos*		F, M	56	22M + 18SM + 6ST + 10A	96	102	2		ACN=56	Brazil (Parana basin)	M-46
Iheringichthys	*labrosus*		F	56	22M + 18SM + 10ST + 6A	96	106	2			Brazil (SP)	V-47
Iheringichthys	*labrosus*		F, M	56	32M + 8SM + 6ST + 10A	96	102			0-3 B	Brazil (Lower Tibagi R., PR)	C-97
Iheringichthys	*labrosus*		F, M	56	14M + 32SM + 4ST + 6A	102	106	2		0 B, ACN=56	Brazil (Upper Tibagi R., PR)	R-124
Iheringichthys	*labrosus*		F, M	56	26M + 12SM + 6ST + 12A	94	100	2		0-1 B, ACN=56	Brazil (PR)	C-104, C-105

Table 6.13 Order SILURIFORMES (continued)

A		B	C	D	E	F	G	H	I	J	K	L
Current scientific name of taxon Family/subfamily/species		Reported in karyotype paper	Sex	2n	Karyotype	NF_1	NF_2	Ag-NORs	Genome size (pg/cell)	Comments	Locality	Reference
Iheringichthys	*labrosus*			56							Argentina	F-20
Luciopimelodus	*pati*			50	16M + 14SM + 8ST + 12A	80	88				Argentina	S-180
Megalonema	*platanum*			54	14M + 18SM + 12ST + 10A	86	98				Argentina	S-180
Megalonema	*platanum*			54	26M + 10SM + 18ST	90	108			0-1 B	Brazil (PR)	C-98
Parapimelodus	*nigribarbis*		F, M	56	20M + 20SM + 4ST + 12A	96	100	2		ACN=56	Brazil (Porto Alegre, RS)	T-78
Parapimelodus	*valenciennis*			56							Brazil (RS)	S-166
Pimelodus	*absconditus*		F, M	56	24M + 18SM + 8ST + 6A	98	106	2			Brazil (PR)	B-46
Pimelodus	*albicans*			56							Argentina	F-20
Pimelodus	*argenteus*		F, M	56	24M + 16SM + 12ST + 4A	96	108	2			Brazil (Paraguai R.)	S-102, S-197
Pimelodus	*blochii*			56	36 M/SM + 20 ST/A	92					Brazil (MT)	B-47
Pimelodus	*clarias*			56					(2.4 BFA)		Argentina	F-20, H-13
Pimelodus	*fur*			56	30M + 14SM + 12A	100	100				S. America	L-21
Pimelodus	*fur*		F, M	54	32M + 8SM + 6ST + 8A	94	100	2		ACN=56	Brazil (MG)	G-81
Pimelodus	*heraldoi*		F, M	56	22M + 22SM + 6ST + 6A	100	106	2			Brazil (PR)	S-180, S-197
Pimelodus	*heraldoi*		F, M	56	18M + 24SM + 6ST + 8A	98	104	2		ACN=56	Brazil (PR)	T-81
Pimelodus	*maculatus*		F, M	56	22M + 16SM + 10ST + 8A	94	104	2	(2.7-2.8 FD)		Brazil (Paraguai R.)	S-102, S-197, F-64
Pimelodus	*maculatus*		F, M	56	40 M/SM + 16 ST/A	96		2			Brazil (MG)	D-11
Pimelodus	*maculatus*		M	56	41 M/SM + 15 ST/A	97		2			Brazil (MG)	D-11
Pimelodus	*maculatus*		F, M	56	20M + 20SM + 10ST + 6A	96	106	2		ACN=56	Brazil (PR)	B-46, M-161, V-46
Pimelodus	*maculatus*		F, M	56	32M + 12SM + 12ST	100	112	2		ACN=56	Brazil (MG)	G-81
Pimelodus	*maculatus*		F, M	56	24M + 20SM + 6ST + 6A	100	106	2		ACN=56	Brazil (Porto Alegre, RS)	T-78
Pimelodus	*mysteriosus*		F, M	56	26M + 20SM + 2ST + 8A	102	104	2			Brazil (Paraguai R.)	S-102, S-197
Pimelodus	*ornatus*		F, M	56	20M + 18SM + 8ST + 10A	94	102	2			Brazil (PR)	B-46
Pimelodus	*ornatus*			56							Argentina	F-20
Pimelodus	*ortmanni*		F, M	56	24M + 18SM + 8ST + 6A	98	106	2		0-4 B	Brazil (PR)	B-47
Pimelodus	*ortmanni*			56	20M + 12SM + 14ST + 10A	88	102				Brazil (PR)	B-47
Pimelodus	*paranaensis*		F, M	56	22M + 22SM + 4ST + 8A	100	104	2		ACN=56	Brazil (PR)	T-81
Pimelodus	sp.		F, M	56	41 M/SM + 15 ST/A	97		2			Brazil (MG)	D-11
Pimelodus	sp.		F, M	56	30M + 14SM + 8ST + 4A	100	108	2		0-4 B	Brazil (PR)	B-47
Pimelodus	sp.		F, M	56	32M + 12SM + 6ST + 6A	100	106	2		ACN=56	Brazil (MG)	G-81
Pinirampus	*pirinampu*		F	50	22M + 12SM + 4ST + 12A	84	88	2			Brazil (PR)	V-84
Pinirampus	*pirinampu*		F, M	50	26M + 12SM + 2ST + 10A	88	90	2			Brazil (PR)	S-146, S-147
Pseudoplatystoma	*corruscans*	*coruscans*	F, M	56	18M + 16SM + 10ST + 12A	90	100	2			Brazil (Parana basin)	M-46
Pseudoplatystoma	*corruscans*	*coruscans*		56	24M + 16SM + 8ST + 8A	96	104				Brazil	P-42
Pseudoplatystoma	*corruscans*	*coruscans*		56	20M + 16SM + 8ST + 12A	92	100	2			Brazil (MS, Paraguay R.)	S-150
Pseudoplatystoma	*corruscans*	*coruscans*		56	26M + 10SM + 6ST + 14A	92	98	2			Brazil (SP, Parana R.)	S-150

Table 6.13 Order SILURIFORMES (continued)

A		B	C	D	E	F	G	H	I	J	K	L
Current scientific name of taxon Family/subfamily/species		Reported in karyotype paper	Sex	2n	Karyotype	NF_1	NF_2	Ag-NORs	Genome size (pg/cell)	Comments	Locality	Reference
Pseudoplatystoma	*fasciatum*		F, M	56	18M + 14SM + 10ST + 14A	88	98	2		ACN=56	Brazil (AM)	F-17
Pseudoplatystoma	*fasciatum*		F, M	56	20M + 12SM + 12ST + 12A	88	100	2		ACN=56	Brazil (MS)	P-42
Pseudoplatystoma	*tigrinum*		F, M	56	18M + 16SM + 8ST + 14A	90	98	2		ACN=56	Brazil (AM)	F-17
Sorubim	*lima*		F, M	56	18M + 12SM + 14ST + 12A	86	100	2		ACN=56	Brazil (AM)	F-17
Sorubim	*lima*		F, M	56	20M + 14SM + 10ST + 12A	90	100	2		ACN=56	Brazil (Parana basin)	M-46
Sorubim	*lima*			56							Argentina	F-20
Steindachneridion	*scriptum*	*inscripta*		56							Argentina	F-20
Steindachneridion	*scriptum*	*scripta*	F, M	56	24M + 20SM + 4ST + 8A	100	104	2		ACN=56	Brazil (PR)	S-199, S-200
Steindachneridion	*melanodermatum*	sp.	F	56	20M + 24SM + 2ST + 10A	100	102	2		XX, ACN=56	Brazil (PR)	S-165, S-180, S-200
Steindachneridion	*melanodermatum*	sp.	M	56	21M + 23SM + 2ST + 10A	100	102	2		XY, ACN=56	Brazil (PR)	S-165, S-200
Zungaro	*zungaro*			56	32M + 6SM + 8ST + 10A	94	102	2			Brazil (SP)	S-148
Zungaro	*zungaro*	*Paulicea luetkeni*	F, M	56	26M + 10SM + 6ST + 14A	92	98	2		ACN=56	Brazil (Parana basin)	M-46
Plotosidae												
Plotosus	*lineatus*	*anguillaris*		48	12 M/SM + 36 ST/A	60				ACN=48	Japan (Okinawa)	A-58
Plotosus	*canius*		F, M	36	10M + 10SM + 16A	56				ACN=36	India (Orissa)	R-58
Plotosus	*canius*		F	36	14M + 12SM + 2ST + 8A	62	64			ACN=36	India (WB)	K-139
Plotosus	*canius*		F, M	36	20M + 8SM + 8 ST/A	64				ACN=36	India (Orissa)	T-49
Pseudopimelodidae												
Cephalosilurus	*apurensis*		M	54	6M + 28SM + 14ST + 6A	88	102	2		ACN=56	Venezuela (Orinoco R.)	M-142
Conorhynchos	*conirostris*	*Conorhynchus*	F, M	60	20M + 18SM + 10ST + 12A	98	108	2		ACN=60	Brazil (MG)	A-17
Lophiosilurus	*alexandri*			54	54 M/SM /ST/A			2			Brazil	M-128
Lophiosilurus	*alexandri*		F, M	54	16M + 18SM + 10ST + 10A	88	98	2		ACN=58	Brazil (MG)	A-17
Microglanis	*cottoides*		F, M	54	22M + 20SM + 12ST	96	108	2	2.5 FD		Brazil (SP)	V-46, F-64
Microglanis	aff. *cottoides*		F, M	54	10M + 32SM + 10ST + 2A	96	106	4		ACN=56	Brazil (SC)	M-142
Pseudopimelodus	*bufonius*			54	18M + 22SM + 6ST + 8A	94	100	2			Brazil (PR)	M-128
Pseudopimelodus	*bufonius*		F, M	54	12M + 30SM + 12ST	96	108	6		ACN=56	Brazil (AM)	M-142
Pseudopimelodus	*mangurus*		F, M	54	6M + 26SM + 12ST + 10A	86	98	2	2.2 FD	ACN=56	Brazil (SP)	M-128, F-64
Schilbeidae												
Ailia	*coila*			58	14M + 36SM + 8A	108	108			ACN=58	India (Assam)	K-46, C-108
Clupisoma	*garua*			56	18M + 34SM + 4A	108	108			ACN=56	India (Assam)	C-108
Clupisoma	*garua*			66	66A	66	66				India	L-1
Eutropiichthys	*vacha*		F, M	58	10M + 20SM + 12ST + 16A	88	100				India (WB)	M-28
Neotropius	*atherinoides*	*Pseudeutropius*		58	10M + 30SM + 18A	98	98			ACN=58	India (Assam)	K-46, C-108
Neotropius	*atherinoides*	*Pseudeutropius*	F, M	58	28M + 12SM + 2ST + 16A	98	100			ACN=58	India (Orissa)	R-58

Table 6.13 Order SILURIFORMES (continued)

A		B	C	D	E	F	G	H	I	J	K	L
Current scientific name of taxon Family/subfamily/species		Reported in karyotype paper	Sex	2n	Karyotype	NF_1	NF_2	Ag-NORs	Genome size (pg/cell)	Comments	Locality	Reference
Scoloplacidae												
Scoloplax	*distolothrix*		F, M	50	18M + 18SM + 10ST + 4A	86	96	2		ACN=54?	Brazil (MT)	O-72
Siluridae												
Kryptopterus	*bicirrhis*			60					1.8 BFA		(SE Asia)	H-13
Ompok	*bimaculatus*	*Callichrous*	F	42	18M + 12SM + 12A	72				XX, ACN=42	India (Haryana)	R-49
Ompok	*bimaculatus*	*Callichrous*	M	41	17M + 12SM + 12A	70				XY	India (Haryana)	R-49
Ompok	*bimaculatus*		F	42	6M + 24SM + 12A	72		2			India (WB)	K-136
Ompok	*bimaculatus*		M	41	5M + 24SM + 12A	70		2			India (WB)	K-136
Ompok	*pabda*			54	28M + 10SM + 8ST + 8A	92	100				India	D-2
Ompok	*pabo*			54	36M + 12SM + 6A	102				ACN=54	India (Assam)	K-43
Silurus	*aristotelis*	*Parasilurus*	F, M	58	30M + 20SM + 8ST	108	116				Greece	I-22
Silurus	*aristotelis*		F	58	20M + 24SM + 14 ST/A	102		2		ACN=58	Macedonia	R-121
Silurus	*asotus*	*Parasilurus*	M	58	38 M/SM + 8ST + 12A	96	104				Japan	M-92
Silurus	*asotus*	*Parasilurus*		58	46 M/SM + 12 ST/A	104		2		ACN=58	Japan (Yamaguchi)	F-51
Silurus	*asotus*	*Parasilurus*		58	44 M/SM + 14 ST/A	102			(2.3* FCM)	ACN=58	Japan (Tochigi)	A-58, O-48
Silurus	*asotus*	*Parasilurus*	F, M	58	24M + 24SM + 10 ST/A	106				ACN=58	Korea (Janghyeon)	K-52
Silurus	*asotus*		F, M	58	20M + 24SM + 10ST + 4A	102	112		(1.5* FD)	ACN=58	China (Wuhan)	H-18, Y-15, C-83
Silurus	*asotus*	*Parasilurus*		58	20M + 14SM + 6ST + 18A	92	98	2	2.9 FD		China (Shasi)	Z-21, L-41
Silurus	*asotus*	*Parasilurus*		58	20M + 24SM + 10ST + 4A	102	112				China (Amur)	Y-13
Silurus	*biwaensis*	*Parasilurus*		58	44 M/SM + 14 ST/A	102				ACN=58	Japan (Lake Biwa)	A-58
Silurus	*glanis*			48	30 M/SM + 18 ST/A	78					Yugoslavia	A-28
Silurus	*glanis*			60	28M + 26SM + 6ST	114	120			ACN=60	Czeck	R-1
Silurus	*glanis*			60	22M + 38 SM/ST		120	2		ACN=60	Czeck	R-18
Silurus	*glanis*			60	18M + 32 SM/ST + 10A		110				(Russia)	V-72
Silurus	*lithophilus*	*Parasilurus*		58	44 M/SM + 14 ST/A	102				ACN=58	Japan (Lake Biwa)	A-58
Silurus	*meridionalis*	*soldatovi meridionalis*	F, M	58	20M + 20SM + 14ST + 4A	98	112			ACN=58	China (Wuhan)	H-18, Y-15, C-83
Silurus	*microdorsalis*	*Parasilurus*	F, M	60	22M + 24SM + 14 ST/A	106				ACN=60	Korea (Janghyeon)	K-52
Silurus	*microdorsalis?*	*Parasilurus microdorsalis*		28	12M + 14SM + 2ST	54	56				Korea	L-15
Silurus	*soldatovi*			58	24M + 16SM + 14ST + 4A	98	112			ACN=58	China (Jilin)	H-34
Wallago	*attu*		F, M	86	12M + 6SM + 2ST + 66A	104	106			ACN=86	India (Haryana)	R-61
Wallago	*attu*		F, M	86	10M + 12SM + 8ST + 56A	108	116			ACN=86	India (Jammu)	S-202

Table 6.13 Order SILURIFORMES (continued)

A		B	C	D	E	F	G	H	I	J	K	L
Current scientific name of taxon Family/subfamily/species		Reported in karyotype paper	Sex	2n	Karyotype	NF_1	NF_2	Ag-NORs	Genome size (pg/cell)	Comments	Locality	Reference
Sisoridae												
Glyptosterninae												
Euchiloglanis	*davidi*		F, M	36	8M + 6SM + 22 ST/A	50				ACN=36	China (Sichuan)	L-35
Euchiloglanis	*kishinouyei*	*Coraglanis*	F, M	50	14M + 6SM + 30 ST/A	70				ACN=50	China (Sichuan)	L-35
Glyptosternon	*reticulatum*	*Exostoma stoliczkae*		42							India	R-74
Glyptothorax	*fokiensis*	*fukiensis*	F, M	52	20M + 18SM + 14ST	90	104			ACN=52	China (Guangdong)	Y-15
Glyptothorax	*telchitta*			56	18M + 26SM + 2ST + 10A	100	102			ACN=56	India (Bihar)	K-41
Glyptothorax	*trilineatus*			52	18M + 24SM + 10A	94				ACN=52	India (Assam)	K-46, C-108
Pseudecheneis	*sulcata*	*sulcatus*	F, M	52	8M + 14SM + 30 ST/A	74				ACN=52	India (U.P.)	R-74
Sisorinae												
Gagata	*cenia*			46	4M + 8SM + 8ST + 26A	58	66			ACN=46	India (U.P.)	M-162
Gogangra	*viridescens*	*Nangra punctata*		42	14M + 20SM + 8A	76				ACN=42	India (Assam)	K-46, C-108
Gogangra	*viridescens*	*Gagata*	M	48	12M + 22SM + 4ST + 10A	82	86			ACN=48	India (Jammu)	S-52
Trichomycteridae												
Bullockia	*maldonadoi*			60	46 M/SM + 14 ST/A	106					(Chile)	A-127, B-80
Eremophilus	*mutisii*		F, M	54	30M + 20SM + 4ST	104	108	2		ACN=56	Colombia	G-53
Hatcheria	*macraei*			52	30 M/SM + 22 ST/A	82					(Chile)	A-127, B-80
Trichogenes	*longipinnis*		F, M	54	36M + 12SM + 6ST	102	108	2		ACN=56	Brazil (SP)	L-51
Trichomycterus	*alternatus*	*florensis*		54	42M + 10SM + 2ST	106	108			ACN=56	Brazil (MG)	S-109
Trichomycterus	*areolatus*			56	56 M/SM	112	112				(Chile)	A-127
Trichomycterus	*areolatus*			54	44M + 8SM + 2ST	106	108	2	(5.0 FD)	ACN=56	Chile (Osomo)	C-66
Trichomycterus	*areolatus*			55	43M + 8SM + 2ST + 2A	106	108			ACN=56	Chile (Osomo)	C-66
Trichomycterus	*areolatus*			56	42M + 8SM + 2ST + 4A	106	108			ACN=56	Chile (Osomo)	C-66

Table 6.13 Order SILURIFORMES (continued)

A		B	C	D	E	F	G	H	I	J	K	L
Current scientific name of taxon Family/subfamily/species		Reported in karyotype paper	Sex	2n	Karyotype	NF_1	NF_2	Ag-NORs	Genome size (pg/cell)	Comments	Locality	Reference
Trichomycterus	*auroguttatus*			54	42M + 10SM + 2ST	106	108			ACN=56	Brazil (MG)	S-109
Trichomycterus	*chiltoni*			52	44 M/SM + 8 ST/A	96					(Chile)	A-127
Trichomycterus	*davisi*		F, M	54	40M + 12SM + 2ST	106	108	2		ACN=58	Brazil (Iguaçu R.)	B-67, B-79, B-80
Trichomycterus	*davisi*			55	41M + 12SM + 2ST	108	110			ACN=58	Brazil (Iguaçu R.)	B-79
Trichomycterus	*davisi*			56	40M + 12SM + 2ST + 2A	108	110			ACN=58	Brazil (Iguaçu R.)	B-79
Trichomycterus	*davisi*			54	42M + 10SM + 2ST	106	108			ACN=58	Brazil (PR)	S-109
Trichomycterus	*davisi*		M	81	60M + 18SM + 3ST	159	162	3		3X	Brazil (PR)	B-67
Trichomycterus	cf. *iheringi*		F, M	54					2.3 FD		Brazil (SP)	F-64
Trichomycterus	aff. *itatiayae*	aff. *itatiyae*		54	42M + 10SM + 2ST	106	108			ACN=58	Brazil (PR)	S-109
Trichomycterus	*laucaensis*			58	42 M/SM + 16 ST/A	100					(Chile)	A-127, B-80
Trichomycterus	*laucaensis*			62	62 M/SM/ST		124				Chile	B-80
Trichomycterus	*paolence*		F, M	54	46M + 6SM + 2ST	106	108	2			Brazil (Itatinga, SP)	T-48, T-69
Trichomycterus	*paolence*			54	46M + 6SM + 2ST	106	108			1B	Brazil (Itatinga, SP)	T-48
Trichomycterus	*paolence*			55	46M + 6SM + 3ST	107	110				Brazil (Itatinga, SP)	T-48
Trichomycterus	*paolence*			56	46M + 6SM + 4ST	108	112			1B	Brazil (Itatinga, SP)	T-48
Trichomycterus	*paolence*	cytotype A		54	44M + 8SM + 2ST	106	108	2		ACN=56	Brazil (Botucatu, SP)	T-69
Trichomycterus	*paolence*	cytotype B		54	40M + 14SM	108	108	2		ACN=56	Brazil (Bofete, SP)	T-69
Trichomycterus	*reinhardti*			54	42M + 10SM + 2ST	106	108			ACN=56	Brazil (MG)	S-109
Trichomycterus	*spegazzinii*			54	54 M/SM	108	108				Argentina	F-20
Trichomycterus	*stawiarski*			54	42M + 8SM + 4ST	104	108			ACN=58	Brazil (Iguaçu R.)	B-80
Trichomycterus	sp.			54	42M + 10SM + 2ST	106	108			ACN=58	Brazil (Iguaçu R.)	B-80
Trichomycterus	sp.		F, M	54					2.6 FD		Brazil	L-51, F-64
Trichomycterus	sp. B			54	42M + 8SM + 4ST	104	108			0-2 B	Brazil (PR)	C-98
Vandellia	*cirrhosa*			32							(S. America)	L-51

Table 6.14 Order GYMNOTIFORMES

A	B	C	D	E	F	G	H	I	J	K	L
Current scientific name of taxon Suborder/family/subfamily/species	Reported in karyotype paper	Sex	2n	Karyotype	NF_1	NF_2	Ag-NORs	Genome size (pg/cell)	Comments	Locality	Reference
Suborder Gymnotoidei											
Gymnotidae											
Electrophorus electricus			52	42 M/SM + 10A	94					Brazil (AM, GO)	A-125
Gymnotus carapo			54	54 M/SM	108	108			ACN=54	Brazil (SP)	A-125
Gymnotus carapo			54	44M + 8SM + 2ST	106	108	2		ACN=54	Brazil (SP)	F-71, F-72
Gymnotus carapo			54	52 M/SM + 2 ST/A	106				X_1X_2Y	Brazil (PR)	A-125
Gymnotus carapo		F, M	54	54 M/SM	108	108	2		ACN=54	Brazil (SP)	F-34
Gymnotus carapo		F, M	52	50 M/SM + 2 ST/A	102		2			Brazil (SP)	F-34
Gymnotus carapo			48	34 M/SM + 14 ST/A	82					Brazil (AM)	F-34
Gymnotus carapo		F, M	42	32 M/SM + 10 ST/A	74					Brazil (PA)	F-34
Gymnotus carapo			40	36 M/SM + 4 ST/A	76					Brazil (PR)	A-125
Gymnotus carapo			54	54 M/SM	108	108	2		ACN=54	Argentina	F-20
Gymnotus carapo			38					2.0 BFA		(S. America)	H-13
Gymnotus carapo			81	78 M/SM + 3 ST/A	159				3X	Brazil (SP)	A-125
Gymnotus inaequilabiatus			52	40M + 10SM + 2 ST/A	102		2		ACN=52	Brazil (SP)	F-71, F-72
Gymnotus pantanal			40	14 M/SM + 26 ST/A	54					Brazil (Pantanal)	A-125
Gymnotus pantherinus			52	38M + 8SM + 6 ST/A	98					East basin	A-125
Gymnotus pantherinus			52	38M + 8SM + 6 ST/A	98		2		ACN=52	Brazil (SP)	F-71, F-72
Gymnotus paraguensis		F, M	54	50 M/SM + 4 ST/A	104		3		ACN=54	Brazil (MG)	V-85
Gymnotus sylvius			40	30 M/SM + 10 ST/A	70		2			Brazil	A-15
Gymnotus sylvius			40	30 M/SM + 10 ST/A	70					Parana River	A-125
Gymnotus sylvius			40	30 M/SM + 10 ST/A	70					Brazil (SP)	A-125
Gymnotus sylvius			40	28M + 10SM + 2 ST/A	78		2		ACN=50	Brazil (SP)	F-71, F-72
Gymnotus sylvius		F, M	40	36 M/SM + 4 ST/A	76		2			Brazil (MG)	V-85
Gymnotus sp.		M	50	26 M/SM + 24 ST/A	76		2		ACN=52	Brazil (MG)	V-85
Gymnotus sp.		F	52	50 M/SM + 2 ST/A	102					Brazil (SP)	F-34
Gymnotus sp.		F	40	14 M/SM + 26 ST/A	54				$X_1X_1X_2X_2$	Brazil (PR)	A-125
Gymnotus sp.		M	39	14 M/SM + 25 ST/A	53				X_1X_2Y	Brazil (PR)	A-125
Gymnotus sp.		F	40	14 M/SM + 26 ST/A	54		2		$X_1X_1X_2X_2$	Argentina	S-155
Gymnotus sp.		M	39	15 M/SM + 24 ST/A	54		2		X_1X_2Y	Argentina	S-155

Table 6.14 Order GYMNOTIFORMES (continued)

A	B	C	D	E	F	G	H	I	J	K	L
Current scientific name of taxon Suborder/family/subfamily/species	Reported in karyotype paper	Sex	2n	Karyotype	NF_1	NF_2	Ag–NORs	Genome size (pg/cell)	Comments	Locality	Reference
Suborder Sternopygoidei											
Apteronotidae											
Apteronotus albifrons		F, M	24	14M + 2SM +2ST + 6A	40	42				(Brazil)	H-29
Apteronotus albifrons			24	14M + 2SM + 2ST + 6A	40	42			0–4 B	Brazil (PR)	C-98
Apteronotus albifrons		F, M	24	12M + 4SM + 2ST + 6A	40	42	2			Brazil (PA)	A-18
Apteronotus albifrons			22					1.4 BFA		(Brazil)	H-13
Apteronotus sp.			52	46 M/SM + 6 ST/A	98					Brazil (SP)	A-125
Parapteronotus bonapartii	*Apteronotus anas*		52	30M + 12SM + 10A	94					Brazil (Manaus)	A-125
Parapteronotus hasemani	*Apteronotus*		52	26M + 16SM + 10A	94					Brazil (Manaus)	A-125
Hypopomidae											
Brachyhypopomus brevirostris			36	4M + 2SM + 8ST + 22A	42	50				Brazil (AM)	A-125
Brachyhypopomus pinnicaudatus		F	42	42A	42	42			$X_1X_1X_2X_2$	Brazil (SP)	A-113
Brachyhypopomus pinnicaudatus		M	41	1M + 40A	42	42			X_1X_2Y, ACN=42	Brazil (SP)	A-113
Hypopomus artedi			38	32 M/SM + 6 ST/A	70					Brazil (AM)	A-125
Hypopygus lepturus			50	16M + 20SM + 10ST + 4A	86	96				Brazil (PA)	A-125
Rhamphichthyidae											
Rhamphichthys cf. *marmoratus*			52	38M + 10SM + 4ST	100	104				Brazil (AM)	A-125
Sternopygidae											
Eigenmannia humboldtii			40	8 M/SM + 32 ST/A	48					Brazil (PR)	A-125
Eigenmannia virescens			40							Argentina	F-20
Eigenmannia virescens		F, M	38	16 M/SM + 22 ST/A	54		2		no sex chrom.	Brazil (SP)	A-114
Eigenmannia virescens		F	38	16 M/SM + 22 ST/A	54		2		XX, ACN=44	Brazil (SP)	A-114
Eigenmannia virescens		M	38	16 M/SM + 22 ST/A	54		2		XY, ACN=44	Brazil (SP)	A-114
Eigenmannia virescens		F	38	23 M/SM + 15 ST/A	61		2		ZW, ACN=44	Brazil (MG)	A-115
Eigenmannia virescens		M	38	22 M/SM + 16 ST/A	60		2		ZZ, ACN=44	Brazil (MG)	A-115
Eigenmannia virescens		F	38	15 M/SM + 23 ST/A	53		2		ZW, ACN=44	Brazil (PA)	A-115, S-206
Eigenmannia virescens		M	38	14 M/SM + 24 ST/A	52		2		ZZ, ACN=44	Brazil (PA)	A-115, S-206
Eigenmannia sp.		F	31	13 M/SM + 18 ST/A	44		2		ACN=32	Brazil (PA)	A-19
Eigenmannia sp.		F, M	32	12 M/SM + 20 ST/A	44		2		ACN=32	Brazil (PA)	A-19
Eigenmannia sp.			46	20 M/SM + 26 ST/A	66		3		3X	Brazil (PA)	A-19
Eigenmannia sp.		F	30	6M + 24 ST/A	36		2			Brazil (PA)	M-169
Eigenmannia sp.		M	29	7 M/SM + 22 ST/A	36					Brazil (PA)	A-125
Eigenmannia sp.	cytotype A		36	14 M/SM + 22ST/A	50		2			Brazil (SP)	M-169

Table 6.14 Order GYMNOTIFORMES (continued)

A	B	C	D	E	F	G	H	I	J	K	L
Current scientific name of taxon Suborder/family/subfamily/species	Reported in karyotype paper	Sex	2n	Karyotype	NF_1	NF_2	Ag-NORs	Genome size (pg/cell)	Comments	Locality	Reference
Eigenmannia sp.	cytotype B		36	8 M/SM + 28 ST/A	44		2			Brazil (SP)	M-169
Eigenmannia sp.	cytotype C		36	16 M/SM + 20 ST/A	52		2		ACN=44	Brazil (MG)	M-169
Eigenmannia sp.			34	24 M/SM + 10 ST/A	58		2		ACN=44	Brazil (MG)	M-169
Eigenmannia sp.			34					2.0 BFA		(S. America)	H-13
Eigenmannia sp. 1		F, M	28	14 M/SM + 14 ST/A	42		2		ACN=32	Brazil (SP)	A-125, A-131
Eigenmannia sp. 2		F	32	8 M/SM + 24A	40	40	2		$X_1X_1X_2X_2$	Brazil (SP)	A-116, A-130
Eigenmannia sp. 2		M	31	9 M/SM + 22A	40	40	2		X_1X_2Y, ACN=44	Brazil (SP)	A-116, A-130
Sternopygus macrurus		M	46	30M + 16SM	92	92	2		ACN=46	Brazil (AM)	A-21
Sternopygus macrurus		F	46	32M + 14SM	92	92	2		ACN=46	Brazil (MG)	A-21
Sternopygus macrurus		F, M	46	28M + 18SM	92	92	2		ACN=46	Brazil (SP)	A-21
Sternopygus macrurus			48					2.0 BFA		(S. America)	H-13

Table 6.15 Order ARGENTINIFORMES

A	B	C	D	E	F	G	H	I	J	K	L
Current scientific name of taxon Family/subfamily/species	Reported in karyotype paper	Sex	2n	Karyotype	NF_1	NF_2	Ag-NORs	Genome size (pg/cell)	Comments	Locality	Reference
Argentinidae											
Argentina silus		M	44	17 M/SM + 27A	61	61		1.7 FD	ACN=47	(N. Atlantic)	E-2
Microstomatidae (= Bathylagidae)											
Bathylaginae											
Bathylagoides wesethi	*Bathylagus*	F	36	20M + 14SM + 2 satellited chrom.					XX	USA (off CA)	C-46, C-48
Bathylagoides wesethi	*Bathylagus*	M	36	19M + 14SM + 1A + 2 satellited chrom.				3.5 FD	XY	USA (off CA)	C-45, C-48, E-2
Leuroglossus stilbius	*Bathylagus*	M	62	14M + 9SM + 1A + 2 satellited chrom. + 36 MC				3.4 FD	XY	USA (off CA)	C-46, C-48, E-2
Lipolagus ochotensis	*Bathylagus*	M	54	9M + 2SM + 43A	65				XY, ACN=60	USA (off CA)	C-46, C-48, E-2
Pseudobathylagus milleri	*Bathylagus*	F	60	12M + 4SM + 2A + 2 satellited chrom. + 40 MC				6.3 FD	XX	USA (off CA)	E-2, E-6, C-46
Pseudobathylagus milleri	*Bathylagus*	M	60	11M + 4SM + 2A + 2 satellited chrom + 41 MC					XY	USA (off CA)	C-48

Table 6.16 Order OSMERIFORMES

A		B	C	D	E	F	G	H	I	J	K	L
Current scientific name of taxon Family/subfamily/species		Reported in karyotype paper	Sex	2n	Karyotype	NF_1	NF_2	Ag-NORs	Genome size (pg/cell)	Comments	Locality	Reference
Galaxiidae												
Brachygalaxias	*bullocki*		F, M	38	10M + 16SM + 12A	64	64			ACN=44	Chile	C-5
Brachygalaxias	*bullocki*		F, M	40	18M + 8SM + 14 ST/A	66		4			Chile	C-82
Brachygalaxias	*gothei*		F, M	40	12M + 6SM + 22 ST/A	58					Chile	C-82
Galaxias	*auratus*		F, M	32	10M + 2SM + 6ST + 14A	44	50			ACN=44	Tasmania	J-14
Galaxias	*brevipinnis*		F, M	44	6M + 8SM + 2ST + 28A	58	60			ACN=44	Tasmania	J-14
Galaxias	*fontanus*		F, M	44	10M + 8SM + 12ST + 14A	62	74			ACN=44	Tasmania	J-14
Galaxias	*johnstoni*		F, M	44	4M + 10SM + 8ST + 22A	58	66			ACN=44	Tasmania	J-14
Galaxias	*maculatus*		F, M	22	8M + 12SM + 2A	42	42			ACN=42	Chile, Australia	C-5
Galaxias	*maculatus*		F, M	22	8M + 10SM + 4ST	40	44			ACN=44	New Zealand	M-62
Galaxias	*maculatus*		F, M	22	18M + 4SM	44	44		(2.1* FD)	ACN=44	Tasmania	J-14, J-17
Galaxias	*platei*		F	30	2M + 16SM + 12A	48	48			Sex chrom.	Chile	C-5
Galaxias	*platei*		M	30	1M + 18SM + 11A	49	49		(1.8* FD)	Y chrom., ACN=42	Chile	C-5, J-18
Galaxias	*tanycephalus*		F, M	32	10M + 2SM + 4ST + 16A	44	48			ACN=44	Tasmania	J-14
Galaxias	*truttaceus*		F, M	32	10M + 2SM + 2ST + 18A	44	46			ACN=44	Tasmania	J-14
Osmeridae												
Hypomesinae												
Hypomesus	*olidus*			56	22 M/SM + 34 ST/A	78					Russia	V-72
Hypomesus	*olidus*			56	4M + 12SM + 40A	72	72				China	Z-23
Hypomesus	*pretiosus*			50±					1.5 FD	Figure absent	USA	O-8, O-11
Hypomesus	*transpacificus nipponensis*		F, M	56	26SM + 30A	82	82	2		ACN=56	Japan (Shimane)	K-69
Osmerinae												
Mallotus	*villosus*		F, M	54	26 M/SM + 28 ST/A	80				ACN=56	Barents Sea	G-89
Osmerus	*eperlanus*			54	16 M/SM + 38 ST/A	70			(1.2 FCM)	ACN=56	Sweden	N-42, V-86
Osmerus	*eperlanus*		F, M	56	10M + 18SM + 28 ST/A	84		2		ACN=56	Poland (Lake Galadus)	O-77
Osmerus	*eperlanus*			56		68					Russia (White Sea)	L-88
Spirinchus	*starksi*			50±					1.7 FD	Figure absent	USA	O-8, O-11
Plecoglossinae												
Plecoglossus	*altivelis*			56	12 M/SM/ST + 44A		68			ACN=56	Japan (Lake Biwa)	Y-8
Salanginae												
Neosalanx	*taihuensis*			56	50M + 6SM	112	112			ACN=56	China (Lake Taihu)	S-121
Salangichthys	*microdon*			56							Japan	N-37

Table 6.17 Order SALMONIFORMES

A	B	C	D	E	F	G	H	I	J	K	L
Current scientific name of taxon Family/subfamily/species	Reported in karyotype paper	Sex	2n	Karyotype	NF_1	NF_2	Ag-NORs	Genome size (pg/cell)	Comments	Locality	Reference
Salmonidae											
Coreginae											
Coregonus albula		M	80	14M + 2SM + 64 ST/A	96					Sweden	N-41, V-41
Coregonus albula			81	14M + 2SM + 4ST + 61A	97	101	4		0-2 B	Finland	J-3
Coregonus albula			80	14M + 2SM + 4ST + 60A	96	100			ACN=100	Finland	J-3
Coregonus artedi			80	14M + 2SM + 4ST + 60A	96	100			ACN=100	USA (Lake Huron)	P-33
Coregonus artedii			80	16M + 10SM + 54A	106	106		6.5 FD		USA (Lake Superior)	B-43
Coregonus autumnalis			78-80		96-98			6.0, 6.4 FCM		Europe, N. America	P-34
Coregonus chadary			80-84		98-100					Russia	P-34
Coregonus clupeaformis			80	20M + 8SM + 52A	108	108		6.9 FD		USA (Lake Superior)	B-43
Coregonus clupeaformis			80	20 M/SM + 60A	100	100	2	(4.9 FIA)	ACN=100	Canada (Ontario)	P-31, P-33, H-41
Coregonus hoyi			80	10M + 8SM + 62A	98	98		5.5 FD		USA (Lake Superior)	B-43
Coregonus hoyi			80	14M + 2SM + 4ST + 60A	96	100			ACN=100	USA (Lake Michigan)	P-33
Coregonus lavaretus			80	12 M/SM + 68 ST/A	92				ACN=100?	Sweden	N-45
Coregonus lavaretus		M	80	18 M/SM + 62 ST/A	98					Sweden	N-45
Coregonus lavaretus ludoga			80	22M + 58 ST/A	102					Russia (Siberia)	V-43
Coregonus lavaretus maraenoides			80	22M + 58 ST/A	102					Russia (Siberia)	V-43
Coregonus muksun			78		100					Russia	P-34
Coregonus nasus			80	10 M/SM + 2ST + 68A	90	92			ACN=100	Sweden	N-45
Coregonus nasus			60	30M + 8SM + 22 ST/A	98			(7.1 FCM)	3B, ACN=100	Russia (Anadyr River)	F-41, L-34
Coregonus nasus			58-60	22-24 M + 10SM + 24-28 ST/A	92					Russia (E. Siberia Sea)	V-42
Coregonus nigripinnis			80	14M + 2SM + 4ST + 60A	96	100			ACN=98	Canada (Lake Nipigon)	P-33
Coregonus peled			80	12 M/SM + 68 ST/A	92					Sweden	N-45
Coregonus peled			80	18 M/SM/ST + 62A		98			ACN=100	Sweden	N-45
Coregonus peled			74	22 S/SM + 52 ST/A	96					Russia	V-43
Coregonus peled			76							(Europe)	J-4
Coregonus pidschian		M	80	18 M/SM + 62 ST/A	98				ACN=100	Sweden	N-45
Coregonus pidschian	*lavaretus pidschian*		80	22M + 58 ST/A	102					Russia (Siberia)	V-43
Coregonus pidschian	*lavaretus pidschian*		80-82	12-14 M + 8SM + 62-58 ST/A	102					Russia (E. Siberia Sea)	V-42
Coregonus reighardi			80	12M + 12SM + 56A	104	104		6.0 FD		USA (Lake Huron)	B-43
Coregonus sardinella		F	80	8M + 10SM + 62 ST/A	98				XX, 1-6 B	Russia (Anadyr River)	F-45
Coregonus sardinella		M	81	9M + 10SM + 62 ST/A	100				XY_1Y_2, 1-6 B	Russia (Anadyr River)	F-42, F-45
Coregonus tugun			86	12M + 8SM + 66A	106	106			ACN=100	Russia (Siberia)	V-43
Coregonus ussuriensis			80	20 M/SM + 60 ST/A	100				ACN=100	Russia	V-43
Coregonus ussuriensis			80-82	10-12 M + 8SM + 64-60 A	100					Russia (Far East)	V-41

Table 6.17 Order SALMONIFORMES (continued)

A		B	C	D	E	F	G	H	I	J	K	L
Current scientific name of taxon Family/subfamily/species		Reported in karyotype paper	Sex	2n	Karyotype	NF_1	NF_2	Ag-NORs	Genome size (pg/cell)	Comments	Locality	Reference
Coregonus	*widegreni*	*lavaretus baeri*		80	12M + 4SM + 64 ST/A	96					Russia	V-41
Coregonus	*zenithicus*			80	10M + 8SM + 62A	98	98		5.3 FD		USA (Lake Superior)	B-43
Coregonus	*zenithicus*			80	14M + 2SM + 4ST + 60A	96	100	2		ACN=98	Canada (Lake Nipigon)	P-33
Coregonus	sp.			80	12M + 6SM + 62A	98	98			ACN=98	Russia (Anadyr R.)	V-43
Coregonus	sp.		F, M	80	14M + 16 SM/ST + 50A		110	5		ACN=99	Italy (Bolsena)	S-89
Coregonus	sp.		F, M	80	14M + 15 SM/ST + 51A		109	3		ACN=99	Italy (Bracciano)	S-89
Prosopium	*abyssicola*			72	12M + 16SM + 44A	100	100				USA (Bear Lake)	B-44
Prosopium	*coulteri*			82	10M + 8SM + 64A	100	100		5.1 FD	ACN=100?	USA (Lake Superior)	B-43
Prosopium	*cylindraceum*			78	12M + 10SM + 56A	100	100		5.0 FD, 4.0 FIA		USA (Lake Superior)	B-43, H-41
Prosopium	*gemmifer*	*gemmiferum*		64	24M + 12SM + 28A	100	100				USA (Bear Lake)	B-44
Prosopium	*spilonotus*			74	12M + 14SM + 48A	100	100				USA (Bear Lake)	B-44
Prosopium	*williamsoni*			78	8M + 14SM + 56A	100	100				USA	B-44
Stenodus	*leucichthys*			74	20M + 14SM + 40A	108	108		(6.5 FD)		USA (AK)	B-45, B-43
Stenodus	*leucichthys nelma*			76	22 M/SM/ST + 54A		98			ACN=100	Russia (Anadyr R.)	F-46
Thymallinae												
Thymallus	*arcticus*			100	58 SM/M + 10ST + 32A	158	168		(3.9–4.0 FIA)	ACN=102	Russia (upper Ob R.)	S-44, H-40
Thymallus	*arcticus*			102	56 SM/M + 10ST + 36A	158	168			ACN=102	Russia (upper Ob R.)	S-44
Thymallus	*arcticus*			98	52 M/SM + 10ST + 36A	150	160				Russia (lower Ob R.)	S-44
Thymallus	*grubei*			98–100	46–50 M/SM + 10–18 ST + 34–44A					ACN=100?	Russia (Far East)	M-16
Thymallus	*grubei*			98	32M + 18SM + 14ST + 34A	148	162				Russia (Far East)	M-16
Thymallus	*thymallus*			102	68 M/SM + 34A	170	170		(4.3 FCM)	ACN=102	Russia (Sylva R.)	S-43, V-86
Thymallus	*thymallus*			104	66 M/SM + 38A	170	170				Russia (Sylva R.)	S-43
Thymallus	*thymallus*			102	58 M/SM + 10ST + 34A	160	170			short-cycle type	Russia (Kama basin)	S-45
Thymallus	*thymallus*			102	58 M/SM + 10ST + 34A	160	170			middle-cycle type	Russia (Kama basin)	S-45
Thymallus	*thymallus*			104	56 M/SM + 10ST + 38A	160	170			long-cycle type	Russia (Kama basin)	S-45
Thymallus	*thymallus*		M	102	68 M/SM/ST + 34A		170				Sweden	N-41
Thymallus	*thymallus*							2			Europe	K-2
Thymallus	*thymallus*							1–4			Poland	J-5
Salmoninae												
Brachymystax	*lenok*			92	16M + 12SM + 64 ST/A	120				ACN=100	Russia	V-44
Brachymystax	*lenok*		F, M	90	26 M/SM + 64 ST/A	116				ACN=100	Korea	K-3
Hucho	*hucho*			82	26M + 6SM + 12ST + 38A	114	126			ACN=100	former Yugoslavia	R-5
Hucho	*hucho*			82	26M + 4SM + 12ST + 40A	112	124			ACN=100	Slovakia	R-5
Hucho	*taimen*			84	20M + 12SM + 52 ST/A	116				ACN=100	Russia	V-44
Hucho	*taimen*			82	22M + 8SM + 12ST + 40A	112	124				Russia (Amur R.)	F-50

Table 6.17 Order SALMONIFORMES (continued)

A		B	C	D	E	F	G	H	I	J	K	L
Current scientific name of taxon Family/subfamily/species		Reported in karyotype paper	Sex	2n	Karyotype	NF_1	NF_2	Ag-NORs	Genome size (pg/cell)	Comments	Locality	Reference
Hucho	*taimen*			83	22M + 7SM + 12ST + 42A	112	124				Russia (Amur R.)	F-50
Parahucho	*perryi*			62	42 M/SM + 4ST + 16A	104	108	2		ACN=102	Japan (Hokkaido)	C-38, U-72
Parahucho	*perryi*			62	38 M/SM + 8ST + 16A	100	108				Russia	V-72
Oncorhynchus	*aguabonita*	*Salmo*		58	44M + 2SM + 2ST + 10A	104	106		5.0* FCM	ACN=102	USA (CA)	M-71, G-25, J-15
Oncorhynchus	*apache*	*Salmo*		56	50 M/SM + 6 ST/A	106				ACN=102	USA (AZ)	M-71, G-26
Oncorhynchus	*chrysogaster*	*Salmo*		60		102					Mexico	P-34
Oncorhynchus	*clarkii alvordensi*	*Salmo*		64	40 M/SM + 24 ST/A	104					USA (OR)	G-26
Oncorhynchus	*clarkii bouvieri*	*Salmo*		64	38M + 2SM + 24A	104	104		5.0* FCM		USA (WY)	L-70, J-15
Oncorhynchus	*clarki clarki*	*Salmo*		68	36 M/SM + 32 ST/A	104			(5.2* FCM)	ACN=100	USA (CA)	G-26, G-97, J-15
Oncorhynchus	*clarki clarki*	*Salmo*		70	36 M/SM + 34 ST/A	106		2			USA (WA)	G-26, P-31
Oncorhynchus	*clarki henshawi*	*Salmo*		64	40 M/SM + 24 ST/A	104				ACN=100	USA (CA)	G-26, G-97
Oncorhynchus	*clarki henshawi*	*Salmo*		64	42 M/SM + 22 ST/A	106					USA (NV)	G-26
Oncorhynchus	*clarki lewisi*	*Salmo*		66	34M + 4SM + 12ST + 16A	104	116			ACN=100	USA (MT)	L-70
Oncorhynchus	*clarki lewisi*	*Salmo*		64	42 M/SM + 22 ST/A	106					USA (MT, WY)	G-26
Oncorhynchus	*clarki* subsp.	*Salmo*		64	42 M/SM + 22 ST/A	106					USA (NV)	G-26, M-71
Oncorhynchus	*gilae*	*Salmo*		56	49 M/SM + 7A	105	105			ACN=100	USA (NM)	B-13
Oncorhynchus	*gilae*	*Salmo*		56	48 M/SM + 8A	104	104				USA (NM)	B-13
Oncorhynchus	*gorbuscha*		F, M	52	48 M/SM + 4ST	100	104	2	(5.1* FCM, 4.5 FD)	ACN=100	Japan (Hokkaido)	M-95, J-15, P-93
Oncorhynchus	*gorbuscha*			52	52 M/SM	104	104			XX/XY	USA (AK)	S-71, P-62
Oncorhynchus	*gorbuscha*			52			104				Russia (Kamchatka)	G-56
Oncorhynchus	*keta*			74	26 M/SM + 6ST + 42A	100	106		(5.3* FCM, 5.5 FD)	ACN=100	Japan (Hokkaido)	S-14, J-15, O-33
Oncorhynchus	*keta*			74	28 M/SM + 46 ST/A	102			(5.0 FIA)		Japan	U-72, H-41
Oncorhynchus	*keta*			74	32 M/SM/ST + 42A		106			ACN=100	Japan (Hokkaido)	I-11
Oncorhynchus	*keta*			74	28 M/SM + 46 ST/A	102		2		XX/XY	USA (WA)	S-71, P-31, P-62
Oncorhynchus	*keta*			74	26M + 6SM + 42A	106	106				Russia (Kamchatka)	G-55
Oncorhynchus	*kisutch*			60	48 M/SM + 12ST	108	120		(5.2* FCM, 6.1 FD)		Japan	U-72, J-15, O-8
Oncorhynchus	*kisutch*			60	38M + 8SM + 14 ST/A	106					Korea?	K-118
Oncorhynchus	*kisutch*			60	44M + 14-16 ST + 2-0 A	104	118-120			ACN=102	Russia (Kamchatka)	G-55
Oncorhynchus	*kisutch*		F, M	60	34M + 8SM + 14ST + 4A	102	116				Russia (Pacific)	S-61
Oncorhynchus	*kisutch*			60	38M + 12SM + 10 ST/A	110		2		ACN=102	USA	U-67, P-93
Oncorhynchus	*kisutch*			60	52 M/SM + 8 ST/A	112		2			USA (WA)	S-71, P-31
Oncorhynchus	*masou*	Sakuramasu		66	38 M/SM + 28 ST/A	104			(5.1* FCM, 4.1 FD)		Japan	U-72, J-15, O-8
Oncorhynchus	*masou*			66	36 M/SM + 22ST + 8A	102	124	2		ACN=100	Japan (Hokkaido)	C-38
Oncorhynchus	*masou*			66	34 M/SM + 32 ST/A	100					Russia	V-72
Oncorhynchus	*masou ishikawae*	*rhodurus* var. *macrostomus*		66	34 M/SM + 24ST + 8A	100	124		5.9* FCM	ACN=100	Japan	M-95, O-48

Table 6.17 Order SALMONIFORMES (continued)

A		B	C	D	E	F	G	H	I	J	K	L
Current scientific name of taxon Family/subfamily/species		Reported in karyotype paper	Sex	2n	Karyotype	NF_1	NF_2	Ag-NORs	Genome size (pg/cell)	Comments	Locality	Reference
Oncorhynchus	*masou ishikawae*	*masou*, Amago		66	38 M/SM + 28 ST/A	104					Japan	U-72, F-52
Oncorhynchus	*masou ishikawae*	*masou* var. *ishikawae*		66	34 M/SM + 24ST + 8A	100	124			ACN=100	Japan	M-95
Oncorhynchus	*masou masou*	*masou* var. *masou*	F, M	66	34 M/SM + 24ST + 8A	100	124		6.6 FD	ACN=100	Japan	M-95, O-33
Oncorhynchus	*masou* subsp.	*rhodurus* var. *rhoduru*	F, M	66	34 M/SM + 24ST + 8A	100	124		4.3 FD	ACN=100	Japan	M-95, O-33
Oncorhynchus	*masou* subsp.	Biwamasu		66							Japan	F-52
Oncorhynchus	*mykiss*	*Parasalmo*		58	46M + 2ST + 10A	104	106		(5.2 FD)	XY, ACN=100	Russia (Kamchatka)	F-40, F-43, G-85
Oncorhynchus	*mykiss*	*Salmo*		58	46 M/SM + 2ST + 10A	104	106		(5.2, 5.4 FCM)	ACN=100	Russia (Kamchatka)	G-58, V-86
Oncorhynchus	*mykiss*	*Salmo*		60-62		106-108				diadromous	Russia (Kamchatka)	V-2
Oncorhynchus	*mykiss*	*Salmo*		58-60		104				lacustrine	Russia (Kamchatka)	V-2
Oncorhynchus	*mykiss*	*Salmo*		58	46M + 2ST + 10A	104	106				USA, Canada, Russia	O-64
Oncorhynchus	*mykiss*	*Salmo*		60	44M + 4ST + 12A	104	108				USA (WA), Canada	O-64
Oncorhynchus	*mykiss*	*Salmo*, Nijimasu		60							culture pond, Japan	F-52
Oncorhynchus	*mykiss*	*Salmo gairdnerii*		64							Mexico	N-21
Oncorhynchus	*mykiss*	*Salmo gairdnerii*		58-60		104		2	(4.9* FCM)		USA (AK-CA)	P-54, P-93, T-47, J-15
Oncorhynchus	*mykiss*	*Salmo gairdnerii*		58							USA (AK, ID)	T-47, P-78
Oncorhynchus	*mykiss*	*Salmo gairdnerii*		58-63		104					UK (Scotland)	H-6
Oncorhynchus	*mykiss*	*Salmo gairdnerii*		60	44 M/SM + 16 ST/A	104			(4.8 FIA)		USA (MI)	M-71, G-26, H-41
Oncorhynchus	*mykiss*	*Salmo gairdnerii*		58-60		104					USA (WA)	T-44
Oncorhynchus	*mykiss*	*Salmo gairdnerii*	F, M	58	46M + 2ST + 10A	104	106			XX/XY	USA (WA)	T-45
Oncorhynchus	*mykiss*	*Salmo gairdnerii*		60	44 M/SM + 2ST + 14A	104	106				USA (MI)	C-81
Oncorhynchus	*mykiss*	*Salmo gairdnerii*		90	66 M/SM + 3ST + 21A	156	159			3X	USA (MI)	C-81
Oncorhynchus	*mykiss*	*Salmo gairdnerii*	F, M	60	44 M/SM + 2ST + 14A	104	106	2		ACN=100	Japan	M-95, U-6
Oncorhynchus	*mykiss*	*Salmo gairdnerii*	F, M	60	44 M/SM + 2ST + 14A	104	106	2		XX/XY, ACN=100	Japan	U-6
Oncorhynchus	*mykiss*	*Salmo gairdnerii*	F, M	62	42 M/SM + 2ST + 18A	104	106	2		XX/XY, ACN=100	Japan (Tochigi)	U-6
Oncorhynchus	*mykiss*	*Salmo gairdnerii*	F, M	61	43 M/SM + 1ST + 17A	104	105	2		XX/XY, ACN=100	Japan (Tochigi)	U-6
Oncorhynchus	*mykiss*	*Salmo gairdnerii*		60	44 M/SM + 16 ST/A	104					Japan	U-78
Oncorhynchus	*mykiss*	*Salmo gairdnerii*		60	44 M/SM + 16 ST/A	104				ACN=100	Croatia	A-26
Oncorhynchus	*mykiss*	*Salmo gairdnerii*		62		104					Russia	V-2
Oncorhynchus	*mykiss*	*Salmo gairdneri aquilarum*		58	46 M/SM + 2ST + 10A	104	106				USA (CA)	B-60
Oncorhynchus	*mykiss irideus*	*Salmo irideus*		60	44 M/SM + 16A	104	104			ACN=100	Rumania (Carpathians)	R-35
Oncorhynchus	*mykiss irideus*	*Salmo irideus*		62	44 M/SM + 18A	106	106			ACN=102	Rumania (Carpathians)	R-35
Oncorhynchus	*mykiss irideus*	*Salmo irideus*		61	42 M/SM + 19A	103	103			ACN=101	Rumania (Carpathians)	R-35
Oncorhynchus	*mykiss irideus*	*Salmo irideus*		58	42 M/SM + 16A	100	100				Rumania (Carpathians)	R-35
Oncorhynchus	*mykiss*			58, 60							USA (CA, ID)	P-78
Oncorhynchus	*mykiss*	PdD 66		58	46 M/SM + 2ST + 10A	104	106				former Czechoslovakia	F-28

Table 6.17 Order SALMONIFORMES (continued)

A		B	C	D	E	F	G	H	I	J	K	L
Current scientific name of taxon Family/subfamily/species		Reported in karyotype paper	Sex	2n	Karyotype	NF_1	NF_2	Ag-NORs	Genome size (pg/cell)	Comments	Locality	Reference
Oncorhynchus	*mykiss*	PdD 66		60	44 M/SM + 2ST + 14A	104	106				former Czechoslovakia	F-28
Oncorhynchus	*nerka*			58	44M + 2SM + 12A	104	104		(6.1 FD)	ACN=102	Russia (Kamchatka)	G-55, O-33
Oncorhynchus	*nerka*		F	58	44 M/SM + 2ST + 12A	102	104				Russia (Kamchatka)	S-60, S-61
Oncorhynchus	*nerka*		M	57	45 M/SM + 2ST + 10A	102	104				Russia (Kamchatka)	S-60, S-61
Oncorhynchus	*nerka*		F	58		104				Sex chrom.	Russia	F-44
Oncorhynchus	*nerka*		M	57		104				Sex chrom.	Russia	F-44
Oncorhynchus	*nerka*			56	46 M/SM + 10A	102	102				Canada (B.C.)	S-71
Oncorhynchus	*nerka*		F	58	44 M/SM + 2ST + 12A	104	104	2	(5.5–5.9* FCM)	$X_1X_1X_2X_2$	USA (WA)	T-46, J-15, P-31, P-93
Oncorhynchus	*nerka*		M	57	45 M/SM + 2ST + 10A	102	104			X_1X_2Y	USA (WA)	T-46
Oncorhynchus	*nerka*	Himemasu	F	58	46 M/SM + 12A	104	104		(6.1 FD)	$X_1X_1X_2X_2$	Japan (Tochigi)	U-5, U-72, O-33
Oncorhynchus	*nerka*	Himemasu	M	57	47 M/SM + 10A	104	104			X_1X_2Y	Japan (Tochigi)	U-5, U-72
Oncorhynchus	*nerka*	Himemasu		58	44 M/SM + 2ST + 12A	102	104			ACN=102	Japan	S-14, M-95
Oncorhynchus	*penshinensis*	*Salmo*	F, M	58	46 M/SM + 2ST + 10A	104	106			ACN=102	Russia (Kamchatka)	G-58
Oncorhynchus	*tshawytscha*			68	32M + 4SM + 32A	104	104	2	(5.2* FCM, 4.9 FIA)	ACN=100	USA, E. Pacific	P-28, P-93, J-15, H-41
Oncorhynchus	*tshawytscha*			68	36 M/SM + 32A	104	104				USA (WA)	S-71
Oncorhynchus	*tshawytscha*		M	68	32 M/SM + 28ST + 8A	100	128			ACN=100	W. Pacific	M-95
Oncorhynchus	*tshawytscha*			68			120				Russia (Kamchatka)	G-56
Oncorhynchus	sp.	*Salmo*		58		104					USA (CA., OR)	G-26
Salmo	*carpio*			80		98					Italy	P-34
Salmo	*ischchan aestivalis*			82	18M + 64A	100	100			ACN=100	Armenia (Sevan basin)	D-22, R-98
Salmo	*ischchan danilewskii*			82	16M + 66A	98	98			ACN=100	Armenia (Sevan basin)	D-22, R-98
Salmo	*ischchan gegarkuni*			80	18M + 62A	98	98			ACN=100	Armenia (Sevan basin)	D-22, R-98
Salmo	*ischchan ischchan*			80	16M + 64A	96	96			ACN=98	Armenia (Sevan basin)	D-22, R-98
Salmo	*letnica*			80		104					Macedonia	P-34
Salmo	*marmoratus*			80	22 M/SM + 58A	102	102			ACN=100	Croatia	A-26
Salmo	*obtusirostris*	*Salmothymus o. oxyrhynchus*		82	12 M/SM + 70A	94	94			ACN=98	former Yugoslavia	B-18
Salmo	*salar*			54	18 M/SM + 36A	72	72	2		ACN=92	Atlantic	U-73
Salmo	*salar*		F, M	58	16 M/SM + 42A	74	74		(6.2 FCM)	ACN=98	Russia (Lake Onega)	Z-11, V-86
Salmo	*salar*			56	14 M/SM + 42A	70	70			ACN=94	Russia (Lake Onega)	Z-11
Salmo	*salar*			56	18 M/SM + 38A	74	74			ACN=98	Russia (White Sea basin)	Z-33
Salmo	*salar*			57	16 M/SM + 41A	73	73			ACN=97	Russia (White Sea basin)	Z-33
Salmo	*salar*			58	14M + 2SM + 42A	74	74			ACN=98	Russia (Lake Ladoga)	Z-10
Salmo	*salar*			58	14 M/SM + 44A	72	72			ACN=98	Russia (Neva)	B-7, B-8
Salmo	*salar*			58	16 M/SM + 42A	74	74				Russia (Neva)	B-7, B-8
Salmo	*salar*			58	16 M/SM + 42A	74	74			ACN=98	Sweden	N-39, N-48

Table 6.17 Order SALMONIFORMES (continued)

A		B	C	D	E	F	G	H	I	J	K	L
Current scientific name of taxon Family/subfamily/species		Reported in karyotype paper	Sex	2n	Karyotype	NF_1	NF_2	Ag–NORs	Genome size (pg/cell)	Comments	Locality	Reference
Salmo	*salar*			56	16 M/SM + 40A	72	72				Sweden	N-39
Salmo	*salar*			58	10M + 6SM + 42A	74	74			ACN=98	Norway	B-42
Salmo	*salar*			57–59							Norway	G-23
Salmo	*salar*			58	16 M/SM + 42A	74	74		(6.0* FCM)	ACN=98	UK (Scotland)	H-7, J-15
Salmo	*salar*			58	16M + 42 ST/A	74					UK (Scotland)	H-10
Salmo	*salar*			57	17 M/SM + 40A	74	74			ACN=98	UK (Scotland)	H-7
Salmo	*salar*			56	18 M/SM + 38A	74	74			ACN=98	UK (Scotland)	H-7
Salmo	*salar*			58	16 M/SM + 42A	74	74				Canada (Chaleur Bay)	N-48
Salmo	*salar*			56				2	(6.2–6.5 FIA)		Atlantic	P-31, P-54, H-40
Salmo	*salar salar*			54–56		72						B-63
Salmo	*salar salar*			54	18 M/SM + 36 ST/A	72						B-63
Salmo	*salar sebago*			56–57		72					(USA, Lake Sebago)	B-63
Salmo	*salar sebago*			56	16 M/SM + 40 ST/A	72					(USA, Lake Sebago)	B-63
Salmo	*trutta*			82	16M + 66A	98	98		(6.1 FCM)		Armenia (Marmarik R.)	R-99, V-86
Salmo	*trutta*			84	16M + 68A	100	100		(5.9 FCM)	ACN=100	Armenia (Argichi R.)	R-99, T-73
Salmo	*trutta*			80	14M + 8SM + 58A	102	102	2			Poland	W-29
Salmo	*trutta*			77–82	20–25 M/SM + 52–62 ST/A	102					Germany	Z-14
Salmo	*trutta*		F, M	80	20 M/SM + 60 ST/A	100				ACN=100	Sweden (Morrum)	N-44, N-51
Salmo	*trutta*			80						migratory	Norway	G-23
Salmo	*trutta*			80	22M/SM + 58A	102	102	4	(5.2* FCM)	ACN=100	UK (Scotland)	H-6, J-15
Salmo	*trutta*			80	21M/SM + 59A	101	101				UK (Scotland)	H-6
Salmo	*trutta*			78	20 M/SM + 58A	98	98				Armenia (Veda R.)	D-22
Salmo	*trutta*		F	80	18 M/SM + 62 ST/A	98				ACN=100	Japan (Tochigi)	U-77
Salmo	*trutta*	*trutta fario*		80	24 M/SM + 56A	104	104		(6.3 FCM, 5.8 FD)	ACN=100	Rumania (Carpathians)	R-35, G-85
Salmo	*trutta*	*trutta fario*		80	22 M/SM/ST + 58A		102			ACN=100	Spain	M-81
Salmo	*trutta*	*trutta fario*		79	23 M/SM/ST + 56A		102				Spain	M-81
Salmo	*trutta*	*trutta* m. *fario*		80	20 M/SM + 60 ST/A	100				ACN=100	Croatia	A-26
Salmo	*trutta*	*trutta* m. *fario*		80	22 M/SM + 58 ST/A	102					Poland	W-28
Salmo	*trutta*	*trutta* m. *trutta*		80	14M + 8SM + 58A	102	102	2		migratory	Poland	W-29
Salmo	*trutta*	*trutta* m. *trutta*		80	14M + 8SM + 58A	102	102	2–3	(5.9 FCM, 5.5 FD)		Spain (Galicia)	M-40, M-137, G-85
Salmo	*trutta oxianus*			80	18 M/SM + 62 ST/A	98					Kyrgyzstan (Kyzylsu R.)	M-59
Salvelinus	*albus*			78–80		98					Russia	P-34
Salvelinus	*alpinus*			82–84	16–18 M/SM + 64–68 ST/A	100					Russia (Kamchatka)	V-74
Salvelinus	*alpinus*			82	18 M/SM + 64 ST/A	100				ACN=98	Russia (Kamchatka)	V-74
Salvelinus	*alpinus*			80	20 M/SM + 60 ST/A	100				ACN=98	Russia (Kamchatka)	V-74

Table 6.17 Order SALMONIFORMES (continued)

A	B	C	D	E	F	G	H	I	J	K	L
Current scientific name of taxon Family/subfamily/species	Reported in karyotype paper	Sex	2n	Karyotype	NF_1	NF_2	Ag-NORs	Genome size (pg/cell)	Comments	Locality	Reference
Salvelinus alpinus			78	20 M/SM + 58A	98	98		6.9 FD	ACN=98	UK (Scotland)	H-9, H-28
Salvelinus alpinus			82				4–8	(5.8, 6.7 FIA)		(N. Europe)	P-54, P-93, H-41
Salvelinus alpinus			78–80							Norway	G-23
Salvelinus alpinus	*Salmo*	M	80	20 M/SM + 60 ST/A	100					Sweden	N-44
Salvelinus boganidae			76–78		98					Russia (Lake Elgygytgyn)	P-34
Salvelinus confluentus			78	24 M/SM + 54 ST/A	102		2		ACN=98	USA (MT)	C-38
Salvelinus elgyticus		F, M	78	18M + 2SM + 58 ST/A	98		2		ACN=96	Russia (Lake Elgygytgyn)	F-48
Salvelinus elgyticus		F, M	77	19M + 2SM + 56 ST/A	98		2		ACN=96	Russia (Lake Elgygytgyn)	F-48
Salvelinus elgyticus		M	76	20M + 2SM + 54 ST/A	98				ACN=96	Russia (Lake Elgygytgyn)	F-48
Salvelinus fontinalis			84				8–12	(5.7* FCM)		USA	K-73, P-54, J-15
Salvelinus fontinalis		F, M	84	18 M/SM + 66 ST/A	102			(6.5 FD)	ACN=98	Japan	U-2, O-33
Salvelinus fontinalis		F, M	84	16 M/SM + 6ST + 62A	100	106				Japan	M-95
Salvelinus fontinalis			84	16 M/SM + 68 ST/A	100			(6.5 FIA, 7.0 BFA)		N. America	U-67, B-63, H-39, H-40
Salvelinus kronocius	*kronicus*		78–82		100					Russia (Kamchatka)	P-34
Salvelinus leucomaenis			84	16 M/SM + 2ST + 66A	100	102				Russia	V-72
Salvelinus leucomaenis		F, M	84	14M + 4SM + 66 ST/A	102				ACN=98	Russia (Prymorye)	F-47
Salvelinus leucomaenis imbrius	Gogi		84	16 M/SM + 2 SM/ST + 66 ST/A	100/102					Japan (Hyogo)	U-72
Salvelinus leucomaenis imbrius	Gogi		84	16 M/SM + 68 ST/A	100		2		ACN=98	Japan (Shimane)	C-38
Salvelinus leucomaenis leucomaenis	Amemasu	F, M	84	16 M/SM + 2ST + 66A	100	102		(7.5 FD)	ACN=98	Japan (Hokkaido)	A-2, U-72, O-33
Salvelinus leucomaenis leucomaenis	Amemasu		84	16 M/SM + 2 SM/ST + 66 ST/A	100/102					Japan (Hokkaido)	U-72
Salvelinus leucomaenis leucomaenis	Amemasu		84	16 M/SM + 68 ST/A	100		2		ACN=98	Japan (Hokkaido)	C-38
Salvelinus leucomaenis pluvius	Nikkoiwana	F, M	84	16 M/SM + 2 SM/ST + 66 ST/A	100/102					Japan (Tochigi)	U-72, U-77
Salvelinus leucomaenis pluvius	Nikkoiwana		84	16 M/SM + 68 ST/A	100		2		ACN=98	Japan (Toyama)	C-38
Salvelinus levanidovi			78–80		98					Russia	P-34
Salvelinus malma		F, M	82	14M + 4SM + 64 ST/A	100				ACN=98	Russia (Prymorye)	F-47
Salvelinus malma krascheninnikovi			82	16 M/SM + 66 ST/A	98		2		ACN=98	Japan (Hokkaido)	C-38
Salvelinus malma lordi			82	16 M/SM + 66 ST/A	98		2		ACN=98	USA (AK)	C-38, U-14
Salvelinus malma malma	Oshorokoma	F, M	82	16 M/SM + 4ST + 62A	98	102			ACN=98	Japan (Hokkaido)	A-2, M-95
Salvelinus malma malma	Oshorokoma	F, M	82	16 M/SM + 2 SM/ST + 64 ST/A	98/100				ACN=98	Japan (Hokkaido)	U-2, U-72
Salvelinus malma miyabei	Miyabeiwana		82	16 M/SM + 66 ST/A	98		2		ACN=98	Japan (Hokkaido)	C-38
Salvelinus malma miyabei	Miyabeiwana		82	18 M/SM + 64 ST/A	100				ACN=98	Japan (Hokkaido)	U-2, U-72
Salvelinus namaycush			84	16 M/SM + 68 ST/A	100		8–12	5.7–6.3 FIA	XX/XY	N. America	B-63, P-34, P-54, H-40
Salvelinus taranetzi			76–78		98–100					Russia	P-34
Salvethymus svetovidovi		F, M	56	38M + 4SM + 14A	98	98			ACN=98	Russia (Lake Elgygytgyn)	F-49

Table 6.18 Order ESOCIFORMES

A		B	C	D	E	F	G	H	I	J	K	L
Current scientific name of taxon Family/species		Reported in karyotype paper	Sex	2n	Karyotype	NF_1	NF_2	Ag-NORs	Genome size (pg/cell)	Comments	Locality	Reference
Esocidae												
Esox	*a. americanus*			50	50A	50	50	2	2.4 FD	ACN=50	Canada, USA	B-12, R-115
Esox	*a. vermiculatus*			50	50A	50	50	2	2.3 FD, 2.1 FIA	ACN=50	Canada (Ontario)	B-12, R-115, H-40
Esox	*lucius*			50	50A	50	50		(2.3 FCM, 2.2 FIA)		USA	A-32, V-86, H-41
Esox	*lucius*			50	50A	50	50	2	2.7 FD	ACN=50	Canada, UK	B-12, R-115
Esox	*lucius*		M	50	50A	50	50			ACN=50	Sweden	N-40
Esox	*masquinongy*			50	50A	50	50		2.6 FD	ACN=50	Canada (Ontario)	B-12
Esox	*masquinongy*			50	50A	50	50	2			N. USA, Canada	R-115
Esox	*niger*			50	50 ST/A	50			(1.8 FIA)		N. America	B-63, H-41
Esox	*niger*			50		50	50	2			USA (NC)	R-115
Esox	*niger*			50					2.4 FD	ACN=50	USA (NC)	B-12
Esox	*reichertii*			50	50 ST/A	50			2.6 FD	ACN=50	Russia (Amur R.)	B-12, B-63
Umbridae												
Dallia	*pectoralis*			78	18M/SM + 60 ST/A	96		2	2.5 FD		USA (AK)	B-12, R-115
Novumbra	*hubbsi*		F, M	48	4M + 10SM + 14ST + 20A	62	76	4	2.1 FD	ACN=48	USA (WA)	B-12, C-80
Umbra	*krameri*			44	44A	44	44	2		ACN=44	Slovakia	R-2, R-12
Umbra	*limi*			22	22M	44	44		(5.0 FD, 5.1 FIA)	ACN=44	Canada (Ontario)	B-12, H-40
Umbra	*limi*			22	18M + 4SM	44	44	2	(5.4 BFA)		USA (NY)	K-72, K-73, H-13, H-26
Umbra	*pygmaea*			22	22M	44	44	2	4.8 FD	ACN=44	USA (NC)	B-12, R-115

Table 6.19 Order STOMIIFORMES

A		B	C	D	E	F	G	H	I	J	K	L
Current scientific name of taxon Suborder/family/subfamily/species		Reported in karyotype paper	Sex	2n	Karyotype	NF_1	NF_2	Ag-NORs	Genome size (pg/cell)	Comments	Locality	Reference
Suborder Gonostomatoidei												
Gonostomatidae												
Bonapartia	*pedaliota*			48*							Atlantic	P-46
Gonostoma	*elongatum*			ca. 48*							Atlantic	P-46
Sigmops	*bathyphilum*	*Gonostoma*		12*							Atlantic	P-46
Sternoptychidae												
Sternoptychinae												
Argyropelecus	*affinis*	*pacificus*		50	20 M/SM + 28A + 2 satellited chrom.					ACN=52	USA (off CA)	C-46
Argyropelecus	*hemigymnus*	*intermedius*		52	26M + 10SM + 10A + 6 satellited chrom.					ACN=52	USA (off CA)	C-46
Argyropelecus	*lychnus*			48	20M + 10SM + 12A + 6 satellited chrom.					ACN=52	USA (off CA)	C-46
Sternoptyx	*diaphana*		M	35	16M + 8SM + 5A + 6 satellited chrom.					XO	USA (off CA)	C-46, C-48

Table 6.20 Order AULOPIFORMES

A		B	C	D	E	F	G	H	I	J	K	L
Current scientific name of taxon Suborder/family/species		Reported in karyotype paper	Sex	2n	Karyotype	NF_1	NF_2	Ag-NORs	Genome size (pg/cell)	Comments	Locality	Reference
Suborder Synodontoidei												
Synodontidae												
Saurida	*elongata*		F	48	47A + 1 MC	48				ZW	Japan (Yamaguchi)	N-35
Saurida	*elongata*		M	48	48A	48	48			ZZ, ACN=48	Japan (Yamaguchi)	N-35
Saurida	*elongata*		F	48	47A + 1 MC	48				ZW	Japan (Wakayama)	U-44
Saurida	*elongata*		M	48	48A	48	48		2.8* FCM	ZZ, ACN=48	Japan (Wakayama)	U-44
Saurida	sp. 2	*undosquamis*, Ma-eso	F	48	47A + 1 MC	48				ZW	Japan (Yamaguchi)	N-35
Saurida	sp. 2	*undosquamis*, Ma-eso	M	48	48A	48	48			ZZ, ACN=48	Japan (Yamaguchi)	N-35
Synodus	*fuscus*			48	48A	48	48			ACN=48	Japan	I-21
Synodus	*hoshinonis*		F	48	2ST + 45A + 1 MC	48	50	1		ZW	Japan (Wakayama)	U-44
Synodus	*hoshinonis*		M	48	2ST + 46A	48	50		2.2* FCM	ZZ, ACN=48	Japan (Wakayama)	U-44
Synodus	*lucioceps*		F	48	28 M/SM + 20 ST/A	76			2.5 FD	ACN=48	USA (CA)	C-54, E-2
Synodus	*ulae*		F	48	47A + 1 MC	48		1		ZW	Japan (Wakayama)	U-44
Synodus	*ulae*		M	48	48A	48	48		2.9* FCM	ZZ, ACN=48	Japan (Wakayama)	U-44
Trachinocephalus	*myops*		F	27	23M + 1ST + 2A + 1 MC	50	51	2		ZW_1W_2	Japan (Wakayama)	U-44
Trachinocephalus	*myops*		M	26	24M + 2A	50	50		2.7* FCM	ZZ, ACN=48	Japan (Wakayama)	U-44
Suborder Chlorophthalmoidei												
Chlorophthalmidae												
Chlorophthalmus	*albatrossis*			36				2			Japan (Suruga Bay)	O-65
Chlorophthalmus	sp.	Bake-aome-eso		24	24M	48	48			ACN=48	Japan	I-21

Table 6.21 Order MYCTOPHIFORMES

A		B	C	D	E	F	G	H	I	J	K	L
Current scientific name of taxon Family/species		Reported in karyotype paper	Sex	2n	Karyotype	NF_1	NF_2	Ag–NORs	Genome size (pg/cell)	Comments	Locality	Reference
Neoscopelidae												
Scopelengys	*tristis*		M, F	48	2SM + 46A	50	50		2.6 FD	ACN=48	USA (off CA)	C–54, E–2
Myctophidae												
Bolinichthys	*photothorax*	*Lepidophanes*		48*							Atlantic	P–45
Bolinichthys	*supralateralis*	*Lepidophanes*		48*							Atlantic	P–45
Ceratoscopelus	*townsendi*			48	48 ST/A	48				ACN=48	USA (off CA)	C–54
Ceratoscopelus	*warmingii*	*warmingi*		48*							Atlantic	P–45
Diaphus	*brachycephalus*			48*							Atlantic	P–45
Diaphus	*dumerilii*	*dumerili*		48*							Atlantic	P–45
Diaphus	*fragilis*			48*							Atlantic	P–45
Diaphus	*rafinesquii*	*rafinesquei*		48*							Atlantic	P–45
Diaphus	*theta*			48						ACN=48	USA (off CA)	C–54
Electrona	*risso*	*rissoi*		48*							Atlantic	P–45
Hygophum	*hygomii*	*hygomi*		48*							Atlantic	P–45
Lampadena	*chavesi*			48*							Atlantic	P–45
Lampadena	*urophaos*			44	44 ST/A	44				ACN=48	USA (off CA)	C–54
Lepidophanes	*guentheri*			48*							Atlantic	P–45
Lobianchia	*gemellarii*			48*							Atlantic	P–45
Myctophum	*nitidulum*			48						ACN=48	USA (off CA)	C–54
Myctophum	*phengodes*	*Ctenoscopelus*		48*							Atlantic	P–45
Nannobrachium	sp. *"ater"*	*Lampanyctus*		48*							Atlantic	P–45
Nannobrachium	*nigrum*	*Lampanyctus niger*		50*							Atlantic	P–45
Nannobrachium	*regale*	*Lampanyctus regalis*	M	48	48A	48	48			ACN=48	USA (off CA)	C–54
Nannobrachium	*ritteri*	*Lampanyctus*	F	48	48A	48	48		4.0 FD	XX, ACN=48	USA (off CA)	C–48, C–54, E–2
Nannobrachium	*ritteri*	*Lampanyctus*	M	47	1SM + 46A	48	48		4.0 FD	XO, ACN=48	USA (off CA)	C–48, C–54, E–2
Notoscopelus	*resplendens*			48							Atlantic	P–45
Parvilux	*ingens*		M	49	1 M/SM + 48 ST/A	50			(3.9 FD)		USA (off CA)	C–46, C–48, E–2
Parvilux	*ingens*	*Lampanyctus*		50?							USA (off CA)	C–54
Protomyctophum	*crockeri*			48	48A	48	48			ACN=48	USA (off CA)	C–54
Stenobrachius	*leucopsarus*			48	48A	48	48			ACN=48	USA (off CA)	C–54
Symbolophorus	*californiensis*			48	2SM + 46A	50	50			ACN=48	USA (off CA)	C–46, C–54
Triphoturus	*mexicanus*		M	48	48A	48	48		3.8 FD	ACN=48	USA (off CA)	C–54, E–2

Table 6.22 Order PERCOPSIFORMES

A		B	C	D	E	F	G	H	I	J	K	L
Current scientific name of taxon Family/species		Reported in karyotype paper	Sex	2n	Karyotype	NF_1	NF_2	Ag-NORs	Genome size (pg/cell)	Comments	Locality	Reference
Aphredoderidae												
Aphredoderus	*sayanus*			48	2M + 18 SM/ST + 28A		68		1.2 FD	ACN=50	USA (TX)	G-82

Table 6.23 Order GADIFORMES

A		B	C	D	E	F	G	H	I	J	K	L
Current scientific name of taxon Family/subfamily/species		Reported in karyotype paper	Sex	2n	Karyotype	NF_1	NF_2	Ag-NORs	Genome size (pg/cell)	Comments	Locality	Reference
Gadidae												
Gadinae												
Eleginus	*gracilis*		F	26	22M + 4A	48	48				Japan (Hokkaido)	I-14
Eleginus	*navaga*			26	20M + 6A	46	46				Russia	V-72
Eleginus	*navaga*		F, M	26	22M + 4A	48	48			ACN=46	Russia (White Sea)	K-83
Eleginus	*navaga*			27		49					Russia (White Sea)	L-88
Gadus	*macrocephalus*		F	44	10M + 10ST + 24A	54	64		1.7 FIA	ACN=46	Japan (Hokkaido)	I-14, H-40
Gadus	*morhua*			46					(1.9 FIA)		Sweden	J-1, N-49, H-41
Gadus	*morhua*			46	10 M/SM + 36 ST/A	56				ACN=46	UK	F-4
Gadus	*morhua*			45	11 M/SM + 34 ST/A	56				ACN=46	UK	F-4
Gadus	*morhua*			46		52					Russia (White Sea)	L-88
Gadus	*ogac*	*morhua marisalbi*		46	8 M/SM + 38 ST/A	54					Russia	V-72
Gadus	*ogac*	*morhua marisalbi*	F, M	46	6M + 2SM + 10ST + 28A	54	64				Russia (White Sea)	K-83
Micromesistius	*poutassou*			44							Sweden	N-49
Pollachius	*pollachius*			38	10M + 28 ST/A	48				ACN=44	Sweden	N-49
Pollachius	*virens*			40	10M + 30 ST/A	50				ACN=46	Sweden	N-49
Theragra	*chalcogramma*		F, M	44	10M + 4ST + 30A	54	58		1.8 FIA	ACN=46	Japan (Hokkaido)	I-14, H-40
Trisopterus	*minutus*			48							Sweden	N-49
Lotinae												
Lota	*lota*		F, M	48	12M + 18SM + 14ST + 4A	78	92				Europe	R-9, K-83
Ranicipitinae												
Raniceps	*raninus*			48	30 M/SM + 18 ST/A	78				ACN=48	Sweden	N-49
Muraenolepididae												
Muraenolepis	*marmoratus*		M	48	4M + 2SM + 42 ST/A	54				ACN=48	Antarctica	D-4

Table 6.23 Order GADIFORMES (continued)

A		B	C	D	E	F	G	H	I	J	K	L
Current scientific name of taxon Family/subfamily/species		Reported in karyotype paper	Sex	2n	Karyotype	NF_1	NF_2	Ag-NORs	Genome size (pg/cell)	Comments	Locality	Reference
Phycidae												
Gaidropsarus	*mediterraneus*			28	24 M/SM + 4A	52	52			ACN=48	Black Sea	V-5
Gaidropsarus	*mediterraneus*	A karyotype	F, M	28	24 M/SM + 4A	52	52				Italy (Palermo)	V-50
Gaidropsarus	*mediterraneus*	B karyotype	F, M	28	18M + 5SM + 5A	51	51				Italy (Palermo)	V-50
Gaidropsarus	*mediterraneus*	C karyotype	F, M	28	22 M/SM + 6A	50	50	2			Italy (Palermo)	V-50
Phycis	*phycis*		F	48	48A	48	48			ACN=48	Spain	A-30

Table 6.24 Order OPHIDIIFORMES

A		B	C	D	E	F	G	H	I	J	K	L
Current scientific name of taxon Suborder/family/species		Reported in karyotype paper	Sex	2n	Karyotype	NF_1	NF_2	Ag-NORs	Genome size (pg/cell)	Comments	Locality	Reference
Suborder Ophidioidei												
Carapidae												
Carapus	*acus*		F, M	36	2SM + 34A	38	38	2		ACN=36	Italy (Palermo)	V-55
Ophidiidae												
Ophidion	*barbatum*	A-type	F, M	44	44A	44	44			ACN=44	Italy (Palermo)	V-75
Ophidion	*barbatum*	B-type	F, M	43	1M + 42A	44	44			ACN=44	Italy (Palermo)	V-75
Parophidion	*vassali*	A-type	F, M	44	44A	44	44			ACN=44	Italy (Palermo)	V-75
Parophidion	*vassali*	B-type	F, M	43	1M + 42A	44	44			ACN=44	Italy (Palermo)	V-75

Table 6.25 Order MUGILIFORMES

A		B	C	D	E	F	G	H	I	J	K	L
Current scientific name of taxon Family/species		Reported in karyotype paper	Sex	2n	Karyotype	NF_1	NF_2	Ag-NORs	Genome size (pg/cell)	Comments	Locality	Reference
Mugilidae												
Agonostomus	*monticola*		F, M	48	2ST + 46A	48	50	2		ACN=48	Panama, Venezuela	N-64
Chelon	*labrosus*		F, M	48	2ST + 46A	48	50			ACN=48	Italy (Tyrrhenian Sea)	C-30, C-32
Liza	*aurata*		F, M	48	2ST + 46A	48	50		2.3 FD	ACN=48	Mediterranean	C-9, C-32
Liza	*haematocheila*			48	48A	48	48			ACN=48	China (Shandong)	Y-20, Z-37
Liza	*parsia*	*Mugil*	M	48	48A	48	48			ACN=48	India (WB)	K-28
Liza	*ramada*		F, M	48	2ST + 46A	48	50		1.6 FCM	ACN=48	Italy (Tyrrhenian Sea)	C-30, C-32, G-85
Liza	*saliens*		F, M	48	2ST + 46A	48	50	2		ACN=48	Italy	C-32, G-75
Mugil	*cephalus*		F, M	48	48A	48	48			ACN=48	Italy (Rome)	C-32
Mugil	*cephalus*			48	48A	48	48	1-2	(2.4 FD)	ACN=48	Europe, USA, Australia, Taiwan	C-9, C-30, R-92
Mugil	*cephalus*			48	48A	48	48		(1.6 FIA)	ACN=48	USA (LA)	L-18, H-40
Mugil	*cephalus*			48	48A	48	48			ACN=48	India (A.P.)	L-3
Mugil	*cephalus*		F, M	48							India (Portonovo)	N-13
Mugil	*cephalus*		F, M	48	48A	48	48		(1.6* FCM)	ACN=48	China	L-54, Y-21, O-48
Mugil	*curema*			28	20M + 4ST + 4A	48	52		(1.4 FD)	ACN=48	USA (LA)	L-18, G-85
Mugil	*curema*			28	20M + 4ST + 4A	48	52	2		ACN=48	Brazil (PR)	N-74
Mugil	*curema*			24	22M + 2SM	48	48	2		ACN=48	Venezuela (Margarita Is.)	N-52, N-74, R-94
Mugil	*gaimardianus*		F, M	48	48A	48	48			ACN=48	Panama	N-52
Mugil	*gaimardianus*			48	48A	48	48			ACN=48	Venezuela (Margarita Is.)	N-52
Mugil	*liza*			48	48A	48	48	2		ACN=48	Venezuela (Margarita Is.)	R-94
Mugil	*platanus*			48	48A	48	48	2		ACN=48	Brazil (SP)	J-13
Mugil	*rubrioculus*			48	48A	48	48	2		ACN=48	Venezuela (Margarita Is.)	N-66
Mugil	*trichodon*		F, M	48	48A	48	48	2		ACN=48	Venezuela (Margarita Is.)	N-53
Oedalechilus	*labeo*		F, M	48	2ST + 46A	48	50	2		ACN=48	Italy (Tyrrhenian Sea)	C-32, R-93
Paramugil	*parmatus*	*Liza macrolepis*	F, M	48	48A	48	48			ACN=48	India (Orissa)	C-61
Paramugil	*parmatus*	*Liza oligolepis*	F, M	48	48A	48	48			ACN=48	India (Orissa)	C-61
Rhinomugil	*corsula*	*Mugil*		48	48A	48	48			ACN=48	India (WB)	N-18, K-26
Valamugil	*speigleri*	*Mugil*	F, M	48	48A	48	48			ACN=48	India (Orissa)	R-57
Valamugil	*speigleri*		M	48	48A	48	48			ACN=48	India (WB)	K-136

Table 6.26 Order ATHERINIFORMES

A		B	C	D	E	F	G	H	I	J	K	L
Current scientific name of taxon Suborder/family/subfamily/species		Reported in karyotype paper	Sex	2n	Karyotype	NF_1	NF_2	Ag–NORs	Genome size (pg/cell)	Comments	Locality	Reference
Suborder Atherinopsoidei												
Atherinopsidae												
Atherinopsinae												
Basilichthys	*australis*		F, M	48	4M + 4SM + 40 ST/A	56				ACN=48	Chile (Augostura)	G-1
Basilichthys	*microlepidotus*		F, M	46	2M + 8SM + 36 ST/A	56				ACN=46	Chile (Petorca)	G-1
Odontesthes	*bonariensis*			48							Argentina	F-20
Odontesthes	*bonariensis*	Atherinidae	F, M	48	4 M/SM + 44 ST/A	52		1–4			Italy (Lake Nemi)	S-170
Odontesthes	*bonariensis*	Atherinidae		48	3 M/SM + 45 ST/A	51					Italy (Lake Nemi)	S-170
Odontesthes	*bonariensis*	Atherinidae		48	2SM + 46 ST/A	50					Italy (Lake Nemi)	S-170
Odontesthes	*bonariensis*	Atherinidae		48	2M + 46 ST/A	50					Italy (Lake Nemi)	S-170
Odontesthes	*bonariensis*	*Basilichthys*		48	4SM + 44 ST/A	52				ACN=48	(introduced, Japan)	A-76
Menidiinae												
Chirostoma	*attenuatum*			48	4M + 24SM + 2ST + 18A	76	78				Mexico	U-50
Chirostoma	*estor*			48	12M + 8SM + 12ST + 16A	68	80			ACN=48	Mexico (Michoacán)	U-50
Chirostoma	*grandocule*			48							Mexico	U-50
Chirostoma	*jordani*			48	8M + 12SM + 10ST + 18A	68	78			ACN=48	Mexico (Lake Chapultepec)	U-50
Chirostoma	*patzcuaro*			44	12ST + 32A	44	56				Mexico (Michoacán)	U-50
Labidesthes	*sicculus*		F, M	48	12M + 22SM + 14 ST/A	82			1.3 FIA		USA	K-100, H-40
Membras	*martinica*		F, M	48	18M + 18SM + 12 ST/A	84					USA (LA)	K-100
Menidia	*beryllina*		F, M	48	8M + 18SM + 22 ST/A	74				freshwater	USA (LA)	K-100
Menidia	*beryllina*		F, M	48	10M + 24SM + 14 ST/A	82				marine	USA (LA)	K-100
Menidia	*menidia*			48	4M + 14SM + 12ST + 18A	66	78				USA (Long Island)	W-16
Suborder Atherinoidei												
Atherinidae												
Atherina	*boyeri*	*mochon caspia*		48	6SM + 42 ST/A	54					Russia	V-72
Atherina	*boyeri*	*mochon pontica*		48	6SM + 42 ST/A	54					Russia	V-72
Hypoatherina	*valenciennei*	*bleekeri*		48	4M + 44SM	96	96	2		ACN=48	Japan (Wakayama)	K-50
Atherionidae												
Atherion	*elymus*			48	2SM + 10ST + 36A	50	60			ACN=48	Japan (Chiba)	A-71
Melanotaeniidae												
Bedotia	*geayi*			48			(72)				Madagascar	S-24
Marosatherina	*ladigesi*	*Telmaterina*		48			(86)				(Sulawesi)	S-24
Melanotaenia	*fluviatilis*			48					2.6 BFA		(Australia)	H-13
Melanotaenia	*goldiei*	*nigricans*		48					2.6 BFA		(Papua New Guinea)	H-13
Melanotaenia	*maccullochi*			46	46A	46	46				Australia	S-24
Melanotaenia	cf. *splendida*	*Nematocentris* cf. *rubrostriatu*	M	48	48A	48	48			ACN=48	(Australia)	A-71

Table 6.27 Order BELONIFORMES

A		B	C	D	E	F	G	H	I	J	K	L
Current scientific name of taxon Suborder/family/subfamily/species		Reported in karyotype paper	Sex	2n	Karyotype	NF_1	NF_2	Ag-NORs	Genome size (pg/cell)	Comments	Locality	Reference
Suborder Adrianichthyoidei												
Adrianichthyidae												
Oryziinae												
Oryzias	*celebensis*		F, M	36	8M + 4SM + 24A	48	48	2	1.7 FD	ACN=44	Sulawesi	U-53, U-59
Oryzias	*curvinotus*			48	2M + 14SM + 32 ST/A	64		2	1.5 FD	XX/XY, ACN=48	China (Hainan Is.)	U-57, T-85
Oryzias	*dancena*	*melastigma*	F, M	48	48A	48	48	2	1.8 FD	XX/XY, ACN=48	India (Chidambaram)	U-58, U-59, T-85
Oryzias	*hubbsi*	*javanicus*		48	48A	48	48		1.8 FD	ZZ/ZW, ACN=48	Indonesia (Jakarta)	U-59, T-85
Oryzias	*javanicus*		M	48	2ST + 46A	48	50	2	1.7 FD	ZZ/ZW, ACN=48	Thailand (Phuket), Singapore	M-6, U-55, U-59, T-85
Oryzias	*latipes*		M	48	20 M/SM + 28A	68			(2.2 BFA)	ACN=48	Japan	O-15, H-13
Oryzias	*latipes*		F, M	48	4M + 16SM + 2ST + 26A	68	70	2	(2.1* FCM), 1.7 FD	XX/XY, ACN=48	Japan (Ehime)	U-54, O-48, T-85
Oryzias	*latipes*			48	4M + 18SM + 4ST + 22A	70	74	2	1.7 FD	ACN=48	Japan (Aomori)	U-59
Oryzias	*latipes*	Himedaka	F, M	48	20 M/SM + 2ST + 26A	68	70			ACN=48	Japan	A-53
Oryzias	*latipes*			48	22 M/SM + 26 ST/A	70				ACN=48	Korea	P-71
Oryzias	*latipes*			48	10M + 10SM + 2ST + 26A	68	70			ACN=48	Korea (Japan Sea side)	U-61, K-57
Oryzias	*luzonensis*		F, M	48	14M + 34SM	96	96	2	1.9 FD	XX/XY, ACN=48	Philippines (Solsona)	F-38, T-85
Oryzias	*marmoratus*			42	2M + 4SM + 36A	48	48	2		ACN=48	Sulawesi	U-75
Oryzias	*matanensis*			42	2M + 4SM + 2ST + 34A	48	50	2		ACN=48	Sulawesi	U-75
Oryzias	*mekongensis*			48	2M + 8SM + 24ST + 14A	58	82		1.5 FD	XX/XY, ACN=48	N.E. Thailand	U-60, M-6, T-85
Oryzias	*minutillus*			42	42A	42	42	2		ACN=42	S. and NE Thailand	M-11
Oryzias	*minutillus*			42	2SM + 40A	44	44	2		ACN=42	SE Thailand	M-11
Oryzias	*minutillus*			40	2M + 2SM + 36A	44	44	2		ACN=42	SE Thailand	M-11
Oryzias	*minutillus*			34	8M + 2SM + 24A	44	44	2	1.5 FD	XX/XY, ACN=42	C. Thailand (Bangkok)	M-5, M-6, M-11, T-85
Oryzias	*minutillus*			32	10M + 2SM + 20A	44	44	2		ACN=42	N. Thailand	M-11
Oryzias	*minutillus*			30	12M + 2SM + 16A	44	44	2	1.5 FD	XX/XY, ACN=42	C. and N. Thailand (Chiang Mai)	M-5, M-11, U-59, T-85
Oryzias	*minutillus*			28	14M + 2SM + 12A	44	44	2		ACN=42	C. Thailand	M-11
Oryzias	*nigrimas*			38	6M + 4SM + 28A	48	48	2		ACN=44	Sulawesi	U-75
Oryzias	*sinensis*	*latipes*		46	12M + 12SM + 22A	70	70			ACN=48	Korea (Yellow Sea side)	K-57
Oryzias	*sinensis*			46	6M + 16SM + 2ST + 22A	68	70			ACN=48	Korea	U-61
Oryzias	*sinensis*			46	22 M/SM + 24 ST/A	68				ACN=48	Korea	P-71
Oryzias	*sinensis*	*latipes*		46	6M + 18SM + 4ST + 18A	70	74			ACN=48	China (Shanghai)	U-59, C-58

Table 6.27 Order BELONIFORMES (continued)

A	B	C	D	E	F	G	H	I	J	K	L
Current scientific name of taxon Suborder/family/subfamily/species	Reported in karyotype paper	Sex	2n	Karyotype	NF_1	NF_2	Ag–NORs	Genome size (pg/cell)	Comments	Locality	Reference
Suborder Belonoidei (= Exocoetoidei)											
Exocoetidae											
Cheilopogon agoo agoo	*Cypselurus*		48					1.9* FCM		Japan	O-48
Hemiramphidae											
Dermogenys pusilla	*pusillus*		48			(58)		1.5 BFA		E. Asia	S-24, H-13
Hemiramphus brasiliensis	*marginatus*		52	52 ST/A	52					India (Goa)	R-45
Hyporhamphus kurumeus	*Hemiramphus*	F, M	40	2M + 38A	42			1.3* FD	ACN=48	China (Hubei)	Z-28, H-32, C-83
Rhynchorhamphus georgii	*Hemiramphus*		46	6 M/SM + 40 ST/A	52					India (Goa)	R-45
Belonidae											
Potamorrhaphis cf. *eigenmanni*		F, M	54	6 M/SM + 48 ST/A	60		2			Argentina (Misiones)	P-15
Potamorrhaphis guianensis			48					2.4 BFA		(S. America)	H-13
Pseudotylosurus microps	*Strongylura*	F, M	50	50 ST/A	50		2		0–2 B, ACN=50	Argentina (Misiones)	P-15
Strongylura incisa			48					2.2* FCM		Japan	O-48
Strongylura leiura	*Tylosurus leiurus*		48	48 ST/A	48					India (Goa)	R-45
Strongylura strongylura		F, M	50	16M + 14SM + 20A	80	80			ACN=50	India (Orissa)	R-57
Strongylura strongylura	*Tylosurus strongylurus*		48	48 ST/A	48					India (Tamil Nadu)	R-45
Xenentodon cancila		F, M	48	16M + 18SM + 6ST + 8A	82	88				India (Jammu)	S-51
Xenentodon cancila		F	48	30M + 2SM + 2ST + 14A	80	82			ACN=48	India (WB)	N-16
Xenentodon cancila			48	24M + 8SM + 4ST + 12A	80	84			ACN=48	Thailand	D-20
Scomberesocidae											
Cololabias saira		M	42							N. Pacific	Y-2

Table 6.28 Order CYPRINODONTIFORMES

A	B	C	D	E	F	G	H	I	J	K	L
Current scientific name of taxon Suborder/family/subfamily/species	Reported in karyotype paper	Sex	2n	Karyotype	NF_1	NF_2	Ag-NORs	Genome size (pg/cell)	Comments	Locality	Reference
Suborder Aplocheiloidei											
Aplocheilidae (Asian rivulines)											
Aplocheilus blockii			48	2 M/SM + 46 ST/A	50	(70)				S. India	S-24, S-158
Aplocheilus dayi			48	6 M/SM + 42 ST/A	54	(86)			ACN=48	Ceylon	S-24, S-158
Aplocheilus lineatus			48	48 ST/A	48	(60)				S. India	S-24, S-158
Aplocheilus panchax			38	12 M/SM + 26 ST/A	50	(56)				India, Thailand	S-24, S-158
Aplocheilus panchax		F	38	13M + 9SM + 4ST + 12A	60	64			ZW	India (WB)	K-31
Aplocheilus panchax		M	38	14M + 8SM + 4ST + 12A	60	64		(1.5 FD, 1.4 BFA)	ZZ	India (WB)	K-31, U-59, H-13
Aplocheilus werneri			48			(88)				Ceylon	S-24
Pachypanchax playfairii			48	22 M/SM + 26 ST/A	70	(96)		1.5 FIA	ACN=48	Seychelles	S-24, S-158, H-41
Nothobranchiidae (African rivulines)											
Aphyosemion ahli	AHL-NN		40	12SM + 6ST + 22A	52	58			ACN=44	E. Cameroon (Nyong)	S-158
Aphyosemion ahli	AHL-NS		38	2M + 2SM + 12ST + 22A	42	54			ACN=42	E. Cameroon (Nyong)	S-158
Aphyosemion ahli	AHL-ED		38	4M + 2SM + 6ST + 26A	44	50			ACN=42	E. Cameroon	S-158
Aphyosemion ahli	AHL-MB		36	8 M/SM + 28A	44	44			ACN=44	E. Cameroon	S-158
Aphyosemion ahli	AHL-SA		34	6M + 2SM + 2ST + 24A	42	44			ACN=44	E. Cameroon	S-158
Aphyosemion ahli	AHL-BE		26	10M + 2SM + 2ST + 12A	38	40			ACN=40	E. Cameroon	S-158
Aphyosemion ahli	AHL-CO		20	20 M/SM	40	40			ACN=40	E. Cameroon	S-158
Aphyosemion ahli	AHL-BN, KI, LO		22	22 M/SM	44	44			ACN=44	E. Cameroon	S-158
Aphyosemion alpha	*Chromaphyosemion*	F, M	38	2SM + 12ST+ 24A	40	52	2		ACN=42	N.W. Gabon	V-71
Aphyosemion amoenum	AMO-TY		34			(44)				E. Cameroon	S-158
Aphyosemion australe	AUS-CA		30	6M + 2SM + 22A	38	38			ACN=42	Gabon	S-158
Aphyosemion australe	AUS-LI		34	2M + 2SM + 30A	38	38			ACN=42	Gabon	S-158
Aphyosemion bamilekorum	BAM-TY		34	6M + 6SM + 22 ST/A	46	(66)			ACN=40	E. Cameroon	S-158
Aphyosemion batesii	BAT-AK		34	34 M/SM	68	68			ACN=44	E. Cameroon	S-158
Aphyosemion batesii	*splendidum* SPL-DJ		32			(60)				Congo	S-158
Aphyosemion bitaeniatum	BIT-BI		34	2M + 4SM + 8ST + 20A	40	48			ACN=42	Cameroon	S-158
Aphyosemion bitaeniatum	BIT-LO		36	6SM + 2ST + 28A	42	44			ACN=40	Cameroon (Kribi)	S-158
Aphyosemion bitaeniatum	BIT-EC		26	14 M/SM + 2ST + 10A	40	42			ACN=40	Cameroon (Rio Muni)	S-158
Aphyosemion bitaeniatum	BIT-KA		36	2M + 2ST + 32A	38	40			ACN=44	Cameroon	S-158
Aphyosemion bitaeniatum	BIT-DI		38	6ST + 32A	38	44			ACN=42	E. Cameroon	S-158
Aphyosemion bitaeniatum	BIT-BA, EK, MB, NN, TK		38	2ST + 36A	38	40			ACN=42	Cameroon	S-158
Aphyosemion bitaeniatum	BIT-IJ		40	4M + 2ST + 34A	44	46			ACN=42	S.W. Nigeria	S-158
Aphyosemion bitaeniatum	*Chromaphyosemion*, Afanyangan	F, M	40	10 M/SM/ST + 30A		50	2		ACN=44	Cameroon	V-97
Aphyosemion bivittatum	BIV-ES		34	4M + 4SM + 2ST + 24A	42	44			ACN=42	E. Cameroon	S-24
Aphyosemion bivittatum	BIV-NE		36	2M + 2ST + 32A	38	40			ACN=44	W. Cameroon	S-158
Aphyosemion bivittatum	BIV-NW		32	8SM + 4ST + 20A	40	44			ACN=44	W. Cameroon	S-158

Table 6.28 Order CYPRINODONTIFORMES (continued)

A		B	C	D	E	F	G	H	I	J	K	L
Current scientific name of taxon Suborder/family/subfamily/species		Reported in karyotype paper	Sex	2n	Karyotype	NF_1	NF_2	Ag-NORs	Genome size (pg/cell)	Comments	Locality	Reference
Aphyosemion	*bivittatum*	*Chromaphyosemion* 'Funge'		36	3 M/SM + 1ST + 32A	39	40	2		ACN=40	Cameroon	V-81
Aphyosemion	*bivittatum*	*Chromaphyosemion* CO3/4	F, M	34	3 M/SM + 3ST + 28A	37	40	2		ACN=40	Cameroon	V-81
Aphyosemion	*bivittatum*	*Chromaphyosemion* CO3/6		34	4 M/SM + 6ST + 24A	38	44	2		ACN=42	Cameroon	V-81
Aphyosemion	*bivittatum*	*Chromaphyosemion* CO3/6		33	7 M/SM + 4ST + 22A	40	44	2		ACN=42	Cameroon	V-81
Aphyosemion	*bivittatum*	*Chromaphyosemion* CO3/6		31	9 M/SM + 4ST + 18A	40	44	2		ACN=42	Cameroon	V-81
Aphyosemion	*bivittatum*	*Chromaphyosemion* CO3/6		30	8 M/SM + 6ST + 16A	38	44	2		ACN=42	Cameroon	V-81
Aphyosemion	*bivittatum*	*Chromaphyosemion* CO3/6		29	11 M/SM + 4ST + 14A	40	44	2		ACN=42	Cameroon	V-81
Aphyosemion	*bivittatum*	*Chromaphyosemion* CO3/9	F, M	32	6 M/SM + 4ST + 22A	38	42	2		ACN=40	Cameroon	V-81
Aphyosemion	*bivittatum*	*Chromaphyosemion* KVO3/34	F, M	32	6 M/SM + 4ST + 22A	38	42	2		ACN=40	Cameroon	V-81
Aphyosemion	*bualanum*	BUA-BA		40	4M + 4SM + 4ST + 28A	48	52			ACN=44	E. Cameroon	S-158
Aphyosemion	*bualanum*	BUA-ND		38	2M + 4SM + 10ST + 22A	44	54			ACN=44	W. Cameroon	S-158
Aphyosemion	*bualanum kekemense*	BUA-KE		36	4SM + 32 ST/A	40	(72)			ACN=42	S.W. Cameroon	R-30
Aphyosemion	*buytaerti*	BUY-AQ		38			(58)				Congo	S-158
Aphyosemion	*calliurum*	CAL-IJ		32	2M + 4SM + 26A	38	38			ACN=42	S.W. Nigeria	S-158
Aphyosemion	*calliurum*	CAL-DM, ME		20	20 M/SM	40	40			6B, ACN=44	E. Cameroon (Nyong)	S-158
Aphyosemion	*calliurum*	CAL-CM		24	16 M/SM + 8A	40	40			ACN=44	E. Cameroon (Nyong)	S-158
Aphyosemion	*calliurum*	CAL-BK, MO, MU		20	20 M/SM	40	40			ACN=44	E. & W. Cameroon	S-158
Aphyosemion	*calliurum*	CAL-NK		20	18 M/SM + 2ST	38	40			ACN=44	E. Cameroon	S-158
Aphyosemion	*calliurum*	CAL-WC		20	20 M/SM	40	40				Cameroon	S-158
Aphyosemion	*calliurum*	CAL-NT		18	18 M/SM	36	36			ACN=44	E. Cameroon	S-158
Aphyosemion	*cameronense*	CAM-MI		34	4M + 2SM + 6ST + 22A	40	46			ACN=44	E. Cameroon	S-158
Aphyosemion	*cameronense*	CAM-LN		34	4M + 2SM + 4ST + 24A	40	44			ACN=44	E. Cameroon	S-158
Aphyosemion	*cameronense*	CAM-BE, EC, YA		34							E. Cameroon	S-158
Aphyosemion	*cameronense*	CAM-MA		32			(46)				W. Cameroon	S-24
Aphyosemion	*cameronense*	CAM-KO		30			(46)				E. Cameroon	S-24
Aphyosemion	*cameronense*	CAM-NG		28	14 M/SM + 2ST + 12A	42	44				E. Cameroon	S-158
Aphyosemion	*cameronense*	CAM-LW		24	14 M + 6ST + 4A	38	44			ACN=44	E. Cameroon	S-158
Aphyosemion	*celiae*	CEL-BA, BU, MA		20	20 M/SM	40	40			ACN=44	W. Cameroon	S-158
Aphyosemion	*celiae*	*celiae winifredae* WIN-TY		20	20 M/SM	40	40			ACN=40	W. Cameroon	S-158
Aphyosemion	*chauchei*	CHO		24	12 M/SM + 12A	36	36			ACN=40	Congo	H-33
Aphyosemion	*christyi*	CHR-SA		36	36A	36	36				Zaire	H-33, S-158
Aphyosemion	*christyi*	CHR		18	16M + 2SM	36	36			ACN=40	Zaire	S-28, S-158
Aphyosemion	*christyi*	*shoutedeni* SCH-AQ		22	10M + 4SM + 8A	36	36			ACN=40	Zaire	H-33, S-28, S-158
Aphyosemion	*coeleste*	COL-TY		32	4 M/SM + 28 ST/A	36					Gabon	S-158
Aphyosemion	*coeleste*	COL-TI		30	6 M/SM + 24 ST/A	36					Congo	S-158
Aphyosemion	*cognatum*	COG-AQ		28	6M + 2SM + 20A	36	36			ACN=40	Congo	S-158
Aphyosemion	*cognatum*	COG		30	6 M/SM + 24A	36	36				Congo	S-158
Aphyosemion	*congicum*	*melanopteron* MET		30	6M + 24A	36	36			ACN=40	Zaire	S-24, S-28, H-33

Table 6.28 Order CYPRINODONTIFORMES (continued)

A		B	C	D	E	F	G	H	I	J	K	L
Current scientific name of taxon Suborder/family/subfamily/species		Reported in karyotype paper	Sex	2n	Karyotype	NF_1	NF_2	Ag-NORs	Genome size (pg/cell)	Comments	Locality	Reference
Aphyosemion	*cyanostictum*	CYA-AQ		34	14 M/SM + 6ST + 14A	48	54			ACN=38	N. Gabon	S-158
Aphyosemion	*decorsei*	DEC-KCF 5		24	18 M/SM + 6ST	42	48			ACN=42	Zaire	H-33
Aphyosemion	*elberti*	*rubrifascium*		38			(40)				Cameroon	S-159
Aphyosemion	*elberti kekemense*	*kekemense* KEK-TY		36	6 M/SM + 30 ST/A	42	(72)			ACN=42	E. Cameroon	S-158
Aphyosemion	*elegans*	ELE-SA		20	12M + 4SM + 4A	36	36			ACN=38	Congo	S-28, S-158, H-33
Aphyosemion	*escherichi*	*simulans* SIM-AQ		40			(70)				Gabon	S-158
Aphyosemion	*exigoideum*	EXO-AQ		22			(42)				Gabon	S-158
Aphyosemion	*exiguum*	EXI-NG		36	20 M/SM + 16ST	56	72			ACN=42	E. Cameroon	S-158
Aphyosemion	*franzwerneri*	FRA-TY		22	22 M/SM	44	44			ACN=44	E. Cameroon	S-158
Aphyosemion	*fulgens*	FUL-AQ		38	38A	38	38				N. Gabon	S-158
Aphyosemion	*gabunense*	MRG-AQ		36			(64)				Gabon	S-158
Aphyosemion	*gabunense*	GAB-AQ		40			(54)				Gabon	S-158
Aphyosemion	*georgiae*	GEO-AQ		36	4SM + 32A	40	40				Gabon	S-158
Aphyosemion	*georgiae*	GEO-AQ		34	4 M/SM + 30A	38	38				Gabon	S-158
Aphyosemion	*herzogi*	HEZ-AQ		34			(68)				N. Gabon	S-158
Aphyosemion	*kouamense*	*Chromaphyosemion*	F, M	38	10ST + 28A	38	48	2		ACN=42	N. W. Gabon	V-71
Aphyosemion	*labarrei*	LAB-AQ		28	8M + 12SM + 2ST + 6A	48	50			ACN=44	Congo	S-158
Aphyosemion	*labarrei*	LAB-NSC-4		26	20 M/SM + 4ST + 2A	46	50			ACN=44	Congo	S-158
Aphyosemion	*lamberti*	LAI-AQ		36	2M + 6SM + 4ST + 24A	44	48			ACN=44	Gabon	H-33, S-158
Aphyosemion	*loennbergii*	LOE-NS		38	2M + 4SM + 6ST + 26A	44	50			ACN=42	E. Cameroon	S-158
Aphyosemion	*loennbergii*	LOE-KI		34	2M + 6SM + 6ST + 20A	42	48			ACN=42	E. Cameroon	S-158
Aphyosemion	*loennbergii*	LOE-MO		32	4M + 6SM + 4ST + 18A	42	46			ACN=42	E. Cameroon	S-158
Aphyosemion	*loennbergii*	*pappenheimi* PAP-BI		32			(52)				E. Cameroon	S-158
Aphyosemion	*loennbergii*	*Chromaphyosemion* KV03/38	F	34	20 M/SM/ST + 14A		54	2		XX	Cameroon	V-97
Aphyosemion	*loennbergii*	*Chromaphyosemion* KV03/38	M	34	19 M/SM/ST + 15A		53	2		XY	Cameroon	V-97
Aphyosemion	*loennbergii*	*Chromaphyosemion* C03/29	F, M	34	8 M/SM/ST + 26A		42			XX/XY	Cameroon	V-97
Aphyosemion	*louessense*	LOU-AQ		40	2 M/SM + 38 ST/A	42	(74)				Congo	H-33
Aphyosemion	*louessense*	LOU-MI		20	16M + 4SM	40	40			ACN=40	Congo	S-158
Aphyosemion	*lugens*	*Chromaphyosemion*	F, M	36	2M + 34A	38	38	2		ACN=42	S. W. Cameroon	V-71
Aphyosemion	cf. *lugens*	*Chromaphyosemion* CMM36		28	12 M/SM/ST + 16A		40	2			Cameroon	V-97
Aphyosemion	*lujae*		M	40*							(Zaire)	G-65
Aphyosemion	*maculatum*	MAL-AQ		34			(44)				N. Gabon	S-158
Aphyosemion	*malumbresi*	*Chromaphyosemion* GEMHS00/31	F, M	24	20 M/SM/ST + 4A		44	2		XX/XY	Cameroon	V-97
Aphyosemion	*melanogaster*	*Chromaphyosemion* KV03/41	F, M	36	14 M/SM/ST + 22A		50	2		XX/XY	Cameroon	V-97
Aphyosemion	*microphtalmum*	MIP-AQ		38	16 M/SM + 22 ST/A	54	(76)			ACN=42	Congo, Gabon	S-158
Aphyosemion	*mimbon*	MIM-AQ		30			(50)				Gabon	S-158
Aphyosemion	*obscurum*	*Panchax obscurus*		34			(40)				Cameroon	S-159
Aphyosemion	*ocellatum*	OCE-AQ		30			(54)				Congo, Gabon	S-158

Table 6.28 Order CYPRINODONTIFORMES (continued)

A	B	C	D	E	F	G	H	I	J	K	L
Current scientific name of taxon Suborder/family/subfamily/species	Reported in karyotype paper	Sex	2n	Karyotype	NF_1	NF_2	Ag-NORs	Genome size (pg/cell)	Comments	Locality	Reference
Aphyosemion ogoense	OGO-AQ		40	18 M/SM + 22 ST/A	58	(72)			ACN=40	Gabon	S-158
Aphyosemion ogoense	OGO-BA		32			(56)				C. Congo	S-158
Aphyosemion ogoense caudofasciatum	CAU-AQ		38			(76)				S. Congo	S-158
Aphyosemion ogoense pyrophore	PYR-AQ		38			(64)				S. Congo	S-158
Aphyosemion ogoense ottogartneri	OTT-AQ		40	12 M/SM + 28 ST/A	52	(76)			ACN=44	C. Congo	S-24, S-158
Aphyosemion pascheni	PAS-TY		24	18 M/SM + 4ST + 2A	42	46			ACN=42	E. Cameroon	S-158
Aphyosemion pascheni	PAS-AQ		24			(48)			ACN=42	E. Cameroon	S-158
Aphyosemion primigenium	PRI-AQ		22			(42)				Gabon	S-158
Aphyosemion punctatum	PUC-OV		24	22 M/SM + 2ST	46	48				Gabon	H-33
Aphyosemion punctulatum	*Chromaphyosemion* CMM22	F, M	36	8 M/SM/ST + 28A		44	2		ACN=42	Cameroon	V-97
Aphyosemion raddai raddai	RAD-TY		32	12 M/SM + 20A	44	44			ACN=44	E. Cameroon	S-158
Aphyosemion rectogoense	REC		18	18 M/SM	36	36			ACN=44	Gabon	H-33
Aphyosemion riggenbachi	RIG-YA		38	38 ST/A	38	(42)			ACN=42	E. Cameroon (Dibonba)	S-158
Aphyosemion riggenbachi	RIG-GI		34	4SM + 30A	38	38			ACN=40	E. Cameroon (Vuri)	S-158
Aphyosemion riggenbachi	RIG-SA		30	2M + 8SM + 20A	40	40			ACN=36	E. Cameroon (Sabaga)	S-158
Aphyosemion riggenbachi	RIG-WN		20	12M + 8SM	40	40			ACN=42	E. Cameroon (Wuri)	S-158
Aphyosemion riggenbachi	*Chromaphyosemion* KVO3/16	M	36	2SM + 2ST + 32A	38	40	6		ACN=40	E. Cameroon	V-80
Aphyosemion riggenbachi	*Chromaphyosemion* KVO3/27	F, M	30	8 M/SM + 8ST + 14A	38	46	2		ACN=38	E. Cameroon	V-80
Aphyosemion riggenbachi	*Chromaphyosemion* KVO3/25	F, M	24	12 M/SM + 8ST + 4A	36	44	2		ACN=38	E. Cameroon	V-80
Aphyosemion riggenbachi	*Chromaphyosemion* KVO3/28	F, M	20	16 M/SM + 4ST	36	40	2, 4		ACN=36	E. Cameroon	V-80
Aphyosemion riggenbachi	*Chromaphyosemion* KVO3/29	F, M	20	16 M/SM + 4ST	36	40	2		ACN=36	E. Cameroon	V-80
Aphyosemion schioetzi	SIO-AQ		18	18 M/SM	36	36			ACN=36	C. Africa	H-33
Aphyosemion schluppi	SLU-AQ		28	12 M/SM + 16 ST/A	40	(54)			ACN=40	S. Gabon	S-158
Aphyosemion splendopleure	*Chromaphyosemion* Bioko		38	4ST + 34A	38	42	2		ACN=42	Cameroon	V-97
Aphyosemion striatum	STR-AQ		40	4 M/SM + 36 ST/A	44	(66)			ACN=44	N. Gabon	S-158
Aphyosemion thysi	THY-AQ		28			(52)				S. Congo	S-158
Aphyosemion volcanum	VOL-TY		36	2M + 2ST + 32A	38	40			ACN=44	W. Cameroon	S-158
Aphyosemion volcanum	*Chromaphyosemion*	F, M	38	2ST + 36A	38	40	2		ACN=44	W. Cameroon	V-79
Aphyosemion wachtersi	WAC-AQ		34			(68)				Congo	S-158
Aphyosemion wildekampi	WIL-TY		30	4M + 2SM + 24A	36	36			ACN=40	E. Cameroon	H-33, S-158
Aphyosemion zygaima	ZYG-MI		20			(40)				Congo	S-158
Archiaphyosemion guineense	*Aphyosemion* GUI-AQ		38	2M + 6SM + 2ST + 28A	46	48			ACN=44	Upper Niger	S-158
Callopanchax monroviae	*Aphyosemion*		46						annual	Liberia	S-158
Callopanchax occidentalis	*Roloffia o. occidentalis*	M	46*							Sierra Leone	G-65
Callopanchax occidentalis	*Aphyosemion* OCC-AQ		46	32 M/SM + 14 ST/A	78	(92)			ACN=46	Sierra Leone	S-24, S-158
Callopanchax toddi	*Roloffia occidentalis toddi*	M	46*							Sierra Leone	G-65
Callopanchax toddi	*Aphyosemion* TOD-TY		46	32 M/SM + 14 ST/A	78	(86)			ACN=46	Sierra Leone	S-24, S-158
Chromaphyosemion poliaki		F, M	38	2ST + 36A	38	40	2		ACN=44	W. Cameroon	V-79

Table 6.28 Order CYPRINODONTIFORMES (continued)

A		B	C	D	E	F	G	H	I	J	K	L
Current scientific name of taxon Suborder/family/subfamily/species		Reported in karyotype paper	Sex	2n	Karyotype	NF_1	NF_2	Ag-NORs	Genome size (pg/cell)	Comments	Locality	Reference
Chromaphyosemion	sp.	GEMHS00/41	F, M	36	8 M/SM/ST + 28A		44	2		XY	Rio Muni	V-97
Epiplatys	*annulatus*	*Aplocheilus*		50	20 M/SM + 30 ST/A	70	(86)				Guinee, Liberia	S-24, S-158, S-159
Epiplatys	*ansorgii*	*Aplocheilus*		46	4 M/SM + 42 ST/A	50	(88)				Gabon	S-158
Epiplatys	*barmoiensis*	*Aplocheilus* BAR		34	14 M/SM + 20 ST/A	48	(54)			ACN=46	Sierra Leone, W. Liberia	S-24, S-158
Epiplatys	*berkenkampi*	*Aplocheilus*		48	2 M/SM + 46A	50	50				Gabon	S-158
Epiplatys	*bifasciatus*		M	40							Volta	S-18
Epiplatys	*bifasciatus*	*Aplocheilus*		40	8M + 32A	48	48			ACN=48	W. Africa to Nile	S-24, S-159
Epiplatys	*chaperi*	*Aplocheilus*		50	50 ST/A	50	(56)				Ghana, Ivory Coast	S-24, S-159
Epiplatys	*chaperi schreiberi*	*Aplocheilus*		50			(52)				Ghana	S-158
Epiplatys	*chaperi sheljuzhkoi*	*Aplocheilus*		50			(52)				Ivory Coast	S-158
Epiplatys	*dageti*		M	50							Monrovia	S-18
Epiplatys	*dageti*	*Aplocheilus*		50	32 M/SM + 18 ST/A	82	(92)				Ghana to Liberia	S-24, S-159
Epiplatys	*dageti monroviae*	*Aplocheilus*		50			(92)				Liberia	S-158
Epiplatys	*duboisi*	*Aplocheilus*		48*		48	(52)				Congo	S-24, S-159
Epiplatys	*esekanus*	*Aplocheilus*		42	8M + 34A	50	50				E. Cameroon	S-24, S-159
Epiplatys	*fasciolatus fasciolatus*	*Aplocheilus*		40	10 M/SM + 30 ST/A	50	(76)			ACN=48	Guinee, Liberia	S-24, S-158, S-159
Epiplatys	*fasciolatus*		M	38							Sierra Leone	S-18
Epiplatys	*fasciolatus*	*Aplocheilus*		38			(76)				Guinee, Liberia	S-24, S-159
Epiplatys	*fasciolatus*	*Aplocheilus*		36			(76?)				Guinee, Liberia	S-24, S-159
Epiplatys	*grahami*	*Aplocheilus*		48	2SM + 46A	50	50			ACN=48	Nigeria to Congo	S-24
Epiplatys	*grahami*	*Aplocheilus nigromarginatus*		48			(50)				Cameroon	S-158
Epiplatys	*huberi*	*Aplocheilus*		46	6 M/SM + 40A	52	52				S.W. Gabon	S-158
Epiplatys	*lamottei*	*Aplocheilus* LAM		48	4 M/SM + 44A	52	52			ACN=48	Liberia	S-158
Epiplatys	*maeseni*	*Aphyosemion* MAE-AQ		42			(52)				Ivory Coast	S-158
Epiplatys	*mesogramma*	*Aplocheilus*		48	48A	48	48				C. Afrin Rep.	S-158
Epiplatys	*multifasciatus*	*Aplocheilus*		46	14 M/SM + 32 ST/A	60	(92)				Zaire	S-158
Epiplatys	*multifasciatus*	*Aplocheilus boulengeri*		46	14 M/SM + 32 ST/A	60	(92)				Congo	S-158
Epiplatys	*olbrechtsi*	*Aplocheilus*		38	12 M/SM + 26 ST/A	50	(76)				Ivory Coast	S-158
Epiplatys	*roloffi*	*Aplocheilus*		46			(92)				Liberia	S-158
Epiplatys	*sangmelinensis*	*Aplocheilus*		48*		48	48				E. Cameroon	S-24, S-159
Epiplatys	*sexfasciatus*	*Aplocheilus*		48	48 ST/A	48	(50)				Ghana to Gabon	S-24, S-159
Epiplatys	*sexfasciatus rathkei*	*Aplocheilus*		48			(50)				W. Cameroon	S-158
Epiplatys	*singa*	*Aplocheilus*		42			(58)				Zaire	S-158
Epiplatys	*spilargyreius*		M	34							Nigeria	S-18
Epiplatys	*spilargyreius*	*Aplocheilus*		34*			(48)				W. Africa to Nile	S-24, S-159
Fenerbahce	*formosus*	*Adamas* FOR-AQ		24	24A	24	24				Zaire	S-158
Foerschichthys	*flavipinnis*	FLA-AQ		40			52				Nigeria	S-158
Fundulopanchax	*arnoldi*	*Aphyosemion*		38	10 M/SM + 28 ST/A	48	(62)			ACN=42	Nigeria	S-24, S-159

Table 6.28 Order CYPRINODONTIFORMES (continued)

A		B	C	D	E	F	G	H	I	J	K	L
Current scientific name of taxon Suborder/family/subfamily/species		Reported in karyotype paper	Sex	2n	Karyotype	NF_1	NF_2	Ag-NORs	Genome size (pg/cell)	Comments	Locality	Reference
Fundulopanchax	*cinnamomeum*	*Aphyosemion* CIN-TY		40	4M + 4SM + 32 ST/A	48	(64)			ACN=44	Cameroon	S-24, S-158, R-29
Fundulopanchax	*fallax*	*Aphyosemion schwoiseri*		44			(48)				W. Cameroon	S-158
Fundulopanchax	*filamentosum*	*Aphyosemion* FIL-IJ		36	4 M/SM + 2ST + 30A	40	42			ACN=40	Nigeria	S-24, S-158
Fundulopanchax	*filamentosum*	*Aphyosemion* FIL-AQ		30	8M + 2ST + 20A	38	40			ACN=38	Nigeria	S-24, S-158
Fundulopanchax	*filamentosum*	*Aphyosemion* FIL-AQ		30	6M + 10ST + 14A	36	46			ACN=38	Nigeria	S-24, S-158
Fundulopanchax	*gardneri*	*Aphyosemion* GAR-GE		40	4M + 6SM + 30 ST/A	50	(68)				W. Cameroon	S-24, S-158, S-159
Fundulopanchax	*gardneri*	*Aphyosemion* GAR-MI		38			(68)				Nigeria	S-158
Fundulopanchax	*gardneri*	*Aphyosemion* GAR-EG		36	4 M/SM + 32 ST/A	40	(68)				Nigeria	S-24
Fundulopanchax	*gardneri*	*Aphyosemion* GAR-OW		36	4 M/SM + 32 ST/A	40	(66)				Nigeria (Owo)	S-158
Fundulopanchax	*gardneri clausen*	*Aphyosemion* CLA		36	4 M/SM + 32 ST/A	40	(66)				Nigeria	S-158
Fundulopanchax	*gardneri lacustre*	*Aphyosemion*	F	39	3M + 4SM + 12ST + 20A	46	58			ACN=42	W. Cameroon	R-29
Fundulopanchax	*gardneri lacustre*	*Aphyosemion*	M	37	5M + 4SM + 12ST + 16A	46	58			ACN=42	W. Cameroon	R-29
Fundulopanchax	*gardneri mamfense*	*Aphyosemion*	F	40	16 M/SM + 2ST + 22A	56	58			ACN=42	W. Cameroon	R-29
Fundulopanchax	*gularis*	*Aphyosemion gulare* GUL-AQ		32	32A	32	32			ACN=40	Nigeria	S-24, S-158
Fundulopanchax	*gularis*	*Aphyosemion gulare*	M	36							Nigeria	G-65
Fundulopanchax	*intermittens*	*Aphyosemion*		33, 36							Cameroon	S-158
Fundulopanchax	*kribianus*	*Aphyosemion* KRI-TY		44			(48)				E. Cameroon	S-158
Fundulopanchax	*marmoratum*	*Aphyosemion* MAM-TY		40	6 M/SM + 10ST + 24A	46	56			ACN=42	W. Cameroon	S-158, S-159
Fundulopanchax	*mirabilis*	*Aphyosemion mirabile*		32–38			(60)				W. Cameroon	S-158
Fundulopanchax	*mirabilis*	*Aphyosemion mirabile* MIR-TY		30	30 M/SM/ST		(60)				W. Cameroon	S-158
Fundulopanchax	*mirabilis*	*Aphyosemion mirabile*		38	20 M/SM + 18 ST/A	58	(60)				W. Cameroon	S-158
Fundulopanchax	*mirabilis*	*Aphyosemion mirabile mirabile*	F	37	19 M/SM + 18A	56	56			ACN=42	W. Cameroon	R-29
Fundulopanchax	*moensis*	*Aphyosemion mirabile moense* MOE-IN		32			(60)				W. Cameroon	S-158
Fundulopanchax	*moensis*	*Aphyosemion mirabile moense* MOE-TY		38			(66)				W. Cameroon	S-158
Fundulopanchax	*ndianus*	*Aphyosemion ndianum* NDI-AQ		40	40A	40	40			ACN=44	E. Nigeria	S-158
Fundulopanchax	*oeseri*	*Aphyosemion s. santaisabellae*		40	6 M/SM + 10ST + 24A	46	56			ACN=42	Fernando Po	S-24, S-158
Fundulopanchax	*puerzli*	*Aphyosemion* PUE-TY		38	2SM + 2ST + 34A	40	42			ACN=40	E. Cameroon	R-28, S-158
Fundulopanchax	*robertsoni*	*Aphyosemion* ROS-TY		42	20 M/SM + 22 ST/A	62	(78)			ACN=44	W. Cameroon	R-28, S-158
Fundulopanchax	*rubrolabialis*	*Aphyosemion rubrolabiale* RUL-MB		44	6M + 6SM + 2ST + 30A	56	58			ACN=44	W. Cameroon	R-28, S-158
Fundulopanchax	*rubrolabialis*	*Aphyosemion rubrolabiale* RUL-YO		40	10 M/SM + 8ST + 22A	50	58			ACN=44	W. Cameroon	S-158
Fundulopanchax	*rubrolabialis*	*Aphyosemion rubrolabiale*		40	8SM + 2ST + 30A	48	50				W. Cameroon	R-28
Fundulopanchax	*scheeli*	*Aphyosemion* SCE-AQ		40	10 M/SM + 30 ST/A	50	(76)			ACN=42	S.E. Nigeria, Cameroon	S-24
Fundulopanchax	*schwoiseri*	*Aphyosemion*		44			(48)				W. Cameroon	S-158
Fundulopanchax	*sjostedti*	*Aphyosemion* SJO-AQ		40	40A	40	40			ACN=42	W. Cameroon	S-158
Fundulopanchax	*traudeae*	*Aphyosemion mirabile traudeae*		37							W. Cameroon	R-29
Fundulopanchax	*walkeri*	*Aphyosemion* WAL-AQ		36	12 SM + 24 ST/A	48	(70)			ACN=40	Ivory Coast	S-158
Nimbapanchax	*petersii*	*Aphyosemion* PET-AQ		40	10M + 30A	50	50			ACN=44	Ivory Coast	S-24, S-158

Table 6.28 Order CYPRINODONTIFORMES (continued)

A		B	C	D	E	F	G	H	I	J	K	L
Current scientific name of taxon Suborder/family/subfamily/species		Reported in karyotype paper	Sex	2n	Karyotype	NF_1	NF_2	Ag-NORs	Genome size (pg/cell)	Comments	Locality	Reference
Nimbapanchax	*viridis*	*Aphyosemion viride* VIR-AQ		40	10M + 30A	50	50			ACN=48	Liberia	S-158
Nothobranchius	*eggersi*			36			(66)				S. Tanzania	S-158
Nothobranchius	*elongatus*			38			(76)				Kenya	S-158
Nothobranchius	*foerschi*		F, M	34	12 M/SM + 22A	46	46			ACN=42	Tanzania	E-5
Nothobranchius	*foerschi*	FOE-AQ		34			(46)				Tanzania	S-158
Nothobranchius	*furzeri*	FUR-TY		38			(76)				Rhodesia	S-158
Nothobranchius	*guentheri*		F	36	2M + 34A	38	38			$X_1X_1X_2X_2$, ACN=44	Zanzibar	E-5
Nothobranchius	*guentheri*		M	35	2M + 1SM + 32A	38	38			X_1X_2Y, ACN=44	Zanzibar	E-5
Nothobranchius	*guentheri*	GUE-ZA		36	4 M/SM + 32A	40	40				Zanzibar	S-158
Nothobranchius	*hengstleri*			38	6 M/SM + 32 ST/A	44				ACN=42	Mozambique	W-35
Nothobranchius	*janpapi*	JAN-TY		38			(70)				Tanzania	S-158
Nothobranchius	*jubbi cyaneus*	CYN-WA		34			(68)				Kenya	S-158
Nothobranchius	*jubbi jubbi*	JUB-TY		34			(66)				Kenya	S-158
Nothobranchius	*kirki*	KIR-TY		36			(58)				Malawi	S-24, S-158
Nothobranchius	cf. *kirki*	"KIR"		44			(84)				Malawi	S-158
Nothobranchius	*kiyawensis*	*Aphyosemion seymouri* SEM-VO		28	2M + 26 ST/A	30	(52)			ACN=38	Ghana	S-158
Nothobranchius	*korthausae*	KOR-TY		36			(42)				Tanzania	S-158
Nothobranchius	*kuhntae*	KUH-AQ		38			(66)				Mozambique	S-158
Nothobranchius	*lourensi*	LOR-AQ		32	4M + 28A	36	36				Tanzania	S-158
Nothobranchius	*lucius*			36	22 M/SM + 14 ST/A	58				ACN=42	Tanzania	W-35
Nothobranchius	*makondorum*			38	18 M/SM + 20 ST/A	56				ACN=42	Tanzania, Mozambique	W-35
Nothobranchius	*melanospilus*			38	14 M/SM + 24 ST/A	52				ACN=42	Tanzania	W-35
Nothobranchius	*melanospilus*	MEP-DS		38			(76)				Tanzania	S-158
Nothobranchius	*melanospilus*		F, M	36	2M + 34 ST/A	38	(76)			ACN=44	Tanzania	E-5
Nothobranchius	*microlepis*	MIL-WA		24			(26)				Kenya, Somalia	S-158
Nothobranchius	*palmqvisti*		F, M	36	4M + 32 ST/A	40				ACN=42	Kenya, Tanzania	E-5
Nothobranchius	*palmqvisti*	PAL 1956 strain		34			(40)				Kenya, Tanzania	S-24, S-158
Nothobranchius	*palmqvisti*	PAL 1957 strain		34	12M + 22A	46	46			ACN=38	Kenya, Tanzania	S-24, S-158
Nothobranchius	*patrizii*		F, M	36	12 M/SM + 24 ST/A	48				ACN=44	Kenya, Somalia	E-5
Nothobranchius	*patrizii*			36			(72)				Somali	S-158
Nothobranchius	*polli*	PLL-ZA		36			(72)				Zaire	S-158
Nothobranchius	*rachovii*		F, M	16	8M + 6SM + 2A	30	30			ACN=42	Somalia	K-106, E-5
Nothobranchius	*rachovii*	RAC-AQ		18	12 M/SM + 6A	30	30				Mozambique	S-24, S-158, S-159
Nothobranchius	*steinforti*	STI-AQ		36			(72)				Tanzania	S-158
Nothobranchius	*thierryi*			42			(48)			ACN=48	W. Africa	S-24
Nothobranchius	*thierryi*	*Fundulosoma* THI-AQ	M	43	4 M/SM + 39A	47	47				Ghana	S-158
Scriptaphyosemion	*bertholdi*			42			(>44)				Sierra Leone	S-158
Scriptaphyosemion	*bertholdi*	*Roloffia*	M	42							Sierra Leone	G-65

Table 6.28 Order CYPRINODONTIFORMES (continued)

A	B	C	D	E	F	G	H	I	J	K	L
Current scientific name of taxon Suborder/family/subfamily/species	Reported in karyotype paper	Sex	2n	Karyotype	NF_1	NF_2	Ag-NORs	Genome size (pg/cell)	Comments	Locality	Reference
*Scriptaphyosemion bruening*i		M	42			ca. 44				Sierra Leone	S-158
Scriptaphyosemion brueningi	*Roloffia*	M	42*							Sierra Leone	G-65
Scriptaphyosemion chaytori	*Roloffia*	M	42							Sierra Leone	G-65
Scriptaphyosemion fredrodi	*Aphyosemion* FRE-TY		42	2 M/SM + 40A	44	44				Sierra Leone	S-158
Scriptaphyosemion geryi	*Aphyosemion* GER-AQ		40	2M + 38A	42	42			ACN=44	Guinee, Sierra Leone	S-158
Scriptaphyosemion geryi	*Roloffia*	M	40	2M + 38A	42	42				Sierra Leone	G-65
Scriptaphyosemion geryi	*Aphyosemion guineense* GUI-AQ		38	2M + 6SM + 2ST + 28A	46	48			ACN=46	Guinee, Sierra Leone	S-158
Scriptaphyosemion geryi	*Roloffia guineensis*	M	40							Sierra Leone	G-65
Scriptaphyosemion guignardi	*Aphyosemion* GUG-TY		40	4 M/SM + 36A	44	44				W. Guinea	S-158
Scriptaphyosemion liberiense	*Aphyosemion*		42			(44)				Liberia	S-24
Scriptaphyosemion liberiense	*Roloffia calabaricus*		42*							(Africa)	G-93
Scriptaphyosemion liberiense	*Aphyosemion melantereon* MEL-AQ		48	2M + 46A	50	50				Liberia	S-158
Scriptaphyosemion roloffi	*Aphyosemion* ROL-AQ		42	2M + 40A	44	44			ACN=44	Sierra Leone	S-158
Scriptaphyosemion roloffi	*Roloffia*	M	42							Sierra Leone	G-65
Scriptaphyosemion roloffi	*Roloffia caldal*		42							(Africa)	G-93
Scriptaphyosemion schmitti	*Aphyosemion*		40			(60)				Liberia	S-158
Rivulidae (New World rivulines)											
Aphyolebias peruensis	*Pterolebias*		54*			(90)				Upper Amazon	S-26
Austrofundulus limnaeus		F, M	44	12M + 16SM + 16A	72					Venezuela	E-4
Austrofundulus transilis		F, M	44	12M + 16SM + 16A	72				ACN=46	Venezuela	E-4
Austrofundulus transilis			44			(80)			ACN=46	Venezuela	S-24
Austrolebias adloffi	*Cynolebias*	F, M	48	2SM + 46A	50		2		ACN=48	Brazil (RS)	G-15
Austrolebias cf. *adloffi*	*Cynolebias* sp.		48	4 M/SM + 44 ST/A	52		6			Uruguay	G-14
Austrolebias cf. *adloffi*	*Cynolebias* sp.		46	6 M/SM + 40 ST/A	52					Uruguay	G-14
Austrolebias cf. *adloffi*	*Cynolebias* sp.		48	8 M/SM + 40 ST/A	56					Uruguay	G-14
Austrolebias affinis	*Cynolebias*	F	48	2M + 2SM + 44A	52	52	4		ACN=48	Uruguay (Tacuarembo)	G-15
Austrolebias affinis	*Cynolebias*	F	49	2M + 2SM + 45A	53	53	4			Uruguay (Tacuarembo)	G-15
Austrolebias alexandrii	*Cynolebias*	F, M	48	4SM + 6ST + 38A	52	58	2		ACN=48	Uruguay (Salto)	G-15
Austrolebias alexandrii	*Cynolebias*	F, M	48	4SM + 10ST + 34A	52	62			ACN=48	Uruguay (Salto)	G-15
Austrolebias arachan	*Cynolebias uruguayensis*		48	4 M/SM + 44 ST/A	52		3			Uruguay	G-92
Austrolebias arachan	*Cynolebias uruguayensis*		48	2 M/SM + 46 ST/A	50		3			Uruguay	G-92
Austrolebias bellotti	*Cynolebias*	M	48	4 M/SM + 44 ST/A	52		5-6			Uruguay	G-14
Austrolebias bellotti	*Cynolebias*	M	48	6 M/SM + 42 ST/A	54					Argentina	G-14
Austrolebias bellotti	*Cynolebias*		48			52				La Plata	S-24
Austrolebias charrua	*Cynolebias*		48	2 M/SM + 46 ST/A	50				ACN=48	Uruguay	G-17
Austrolebias charrua	*Cynolebias*		48	4 M/SM + 44 ST/A	52				ACN=48	Uruguay	G-17
Austrolebias charrua	*Cynolebias*		48	6 M/SM + 42 ST/A	54				ACN=48	Uruguay	G-17

Table 6.28 Order CYPRINODONTIFORMES (continued)

A		B	C	D	E	F	G	H	I	J	K	L
Current scientific name of taxon Suborder/family/subfamily/species		Reported in karyotype paper	Sex	2n	Karyotype	NF_1	NF_2	Ag-NORs	Genome size (pg/cell)	Comments	Locality	Reference
Austrolebias	*charrua*	*Cynolebias*		48	8 M/SM + 40 ST/A	56				ACN=48	Uruguay	G-17
Austrolebias	*cheradophilus*	*Cynolebias*		40	4 M/SM + 36 ST/A	44		4			Uruguay	H-43
Austrolebias	*cinereus*	*Cynolebias*	F	44	4M + 2SM + 14ST + 24A	50	64	2		ACN=46	Uruguay (Colonia)	G-16
Austrolebias	*duraznensis*	*Cynolebias*	M	48	4M + 6ST + 38A	52	58	3			Uruguay (Durazno)	G-15
Austrolebias	*duraznensis*	*Cynolebias*	F, M	48	2M + 2SM + 6ST + 38A	52	58	3		ACN=48	Uruguay (Durazno)	G-15
Austrolebias	*gymnoventris*	*Cynolebias*	F, M	48	2M + 2SM + 10ST + 34A	52	62	2		ACN=48	Uruguay (Rocha)	G-15
Austrolebias	*luteoflammulatus*	*Cynolebias*	F, M	34	16 M/SM + 6ST + 12A	50	56	4		ACN=48	Uruguay (Rocha)	G-14
Austrolebias	*melanoorus*	*Cynolebias*	F, M	48	2SM + 22ST + 24A	50	72	2		ACN=48	Uruguay (Tacuarembo)	G-16
Austrolebias	*nigripinnis*	*Cynolebias*	F, M	48	8 M/SM + 40 ST/A	56		5–6		ACN=48	Uruguay (Salto)	G-14
Austrolebias	*nigripinnis*	*Cynolebias*	M	48	6 M/SM + 8ST + 34A	54	62			ACN=48	Uruguay (Artigas)	G-14
Austrolebias	*nigripinnis*	*Cynolebias*		48			(74)				La Plata	S-24
Austrolebias	*nioni*	*Cynolebias*	F, M	46	2M + 6SM + 34ST + 4A	54	88	2		ACN=48	Uruguay (Tacuarembo)	G-16
Austrolebias	*vazferreirai*	*Cynolebias*		46	2M + 2SM + 30ST + 12A	50	80	2		ACN=48	Uruguay (Cerro Largo)	G-16
Austrolebias	*vazferreirai*	*Cynolebias*		46	2M + 4SM + 24ST + 16A	52	76			ACN=48	Uruguay (Cerro Largo)	G-16
Austrolebias	*viarius*	*Cynolebias*	M	46	2M + 2ST + 42A	48	50	6		ACN=48	Uruguay	G-14
Austrolebias	*viarius*	*Cynolebias*	M	48	2 M/SM + 46 ST/A	50				ACN=48	Uruguay	G-14
Cynopoecilus	*melanotaenia*			44			(52)				Brazil	S-24
Cynopoecilus	*melanotaenia*	*Cynolebias*	F, M	44	14 M/SM + 30 ST/A	58		5			Uruguay	G-14
Gnatholebias	*zonatus*	*Pterolebias*	F, M	42	12 M/SM + 30A	54	54			ACN=46	Venezuela	E-3
Kryptolebias	*brasiliensis*	*Rivulus dorni*		48*			(70)				Brazil (RJ)	S-24
Kryptolebias	*marmoratus*	*Rivulus*		48			(52)			hermaphrodite	USA (FL), Cuba	S-24
Kryptolebias	*marmoratus*	*Rivulus*		48	4 M/SM + 46 ST/A	52		2		ACN=48	America	S-92
Kryptolebias	*ocellatus*	*Rivulus*		48			(54)				Brazil	S-24
Megalebias	*prognatus*	*Cynolebias*	F, M	36	12M + 6ST + 18A	48	54	3			Uruguay (Rocha)	G-14
Megalebias	*prognatus*	*Cynolebias*	M	36	10 M/SM + 6ST + 20A	46	52			ACN=48	Uruguay (Rocha)	G-14
Megalebias	*wolterstorffi*	*Cynolebias*	M	46	4 M/SM + 42 ST/A	50		5			Uruguay	G-14
Nematolebias	*whitei*	*Cynolebias*		46			(92)				Brazil (RJ)	S-24
Pterolebias	*hoignei*		F	46	6M + 40A	52	52			ACN=46	Venezuela	E-3
Pterolebias	*hoignei*		M	46	6M + 3SM + 37A	55	55			Y chrom.	Venezuela	E-3
Pterolebias	*hoignei*		F, M	46	6M + 4SM + 36A	56	56			ACN=46	Venezuela	E-4
Pterolebias	*longipinnis*			20	20A	20	20				(Lower Amazon)	S-24, K-106
Rachovia	*brevis*		F, M	44	12M + 14SM + 18A	70	70			ACN=46	Venezuela	E-4
Rachovia	*hummelincki*		F, M	44	10M + 10SM + 24A	64	64			ACN=46	Venezuela	E-4
Rachovia	*maculipinnis*	*Pterolebias*		44			(80)				Venezuela	S-24
Rachovia	*maculipinnis*	*maculipinna*	F, M	44	20M + 12SM + 12A	76	76			ACN=48	Venezuela	E-4
Rachovia	*pyropunctata*		F, M	44	10M + 8SM + 26A	62	62			ACN=46	Venezuela	E-4
Rachovia	*stellifer*	*Rivulus*	F, M	48	18M + 8SM + 22A	74	74			ACN=48	Venezuela	E-4
Rivulus	*agilae*			44	8M + 2SM + 14ST + 20A	54	68	4		ACN=46	French Guyana	Z-41

Table 6.28 Order CYPRINODONTIFORMES (continued)

A		B	C	D	E	F	G	H	I	J	K	L
Current scientific name of taxon Suborder/family/subfamily/species		Reported in karyotype paper	Sex	2n	Karyotype	NF_1	NF_2	Ag-NORs	Genome size (pg/cell)	Comments	Locality	Reference
Rivulus	*cylindraceus*			48			(58)				Cuba	S-24
Rivulus	*hartii*			44			(56)				S. America	S-24
Rivulus	*holmiae*			44			(72)				S. America	S-24
Rivulus	*magdalenae*	*milesi*		46			(92)			ACN=48	Colombia	S-24
Rivulus	*ornatus*			40			(66)			ACN=46	Lower Amazon	S-24
Rivulus	*strigatus*			46			(88)				Amazon	S-24
Rivulus	*urophthalmus*			46			(86)				Lower Amazon	S-24
Rivulus	*urophthalmus*			44					3.0 BFA		(S. America)	H-13
Terranatos	*dolichopterus*	*Austrofundulus*		44			(54)				Venezuela	S-24
Suborder Cyprinodontoidei												
Anablepidae												
Anableps	*dowi*		F, M	46	46 ST/A	46				ACN=46	Mexico (Tehuantepec)	M-69
Cyprinodontidae (pupfishes)												
Cyprinodontinae												
Aphanius	*asquamatus*	*Kosswigichthys*		48*							Turkey	K-9
Aphanius	*dispar*			48*							Middle East	K-9
Aphanius	*fasciatus*		F, M	48	48 ST/A	48		1–8		ACN=48	Italy (Sicily)	V-69
Aphanius	*iberus*			48*							Spain	K-9
Aphanius	*mento*	*cypris*		48							Iraq, Turkey	K-9
Aphanius	*persicus*			48	22SM + 26ST	70	96			ACN=48	Iran (Fars)	E-12
Aphanius	*sophiae*			48	28SM + 20ST	76	96			ACN=48	Iran (Fars)	E-12
Aphanius	*sophiae*			48*							Iran	K-9
Cyprinodon	*alvarezi*		F	48						X_1X_1	Mexico	H-1
Cyprinodon	*atrorus*		F, M	48	2M + 14SM + 32ST	64	96			ACN=48	Mexico	S-108
Cyprinodon	*beltrani*		F, M	48	2M + 14SM + 32ST	64	96			ACN=48	Mexico	S-108
Cyprinodon	*bifasciatus*		F, M	48	2M + 14SM + 32ST	64	96			ACN=48	Mexico	S-108
Cyprinodon	*bovinus*		F, M	48	2M + 14SM + 32ST	64	96			ACN=48	USA (TX)	S-108
Cyprinodon	*dearborni*		F, M	48	2M + 10SM + 36 ST/A	60		2		ACN=48	Venezuela (Margarita Is.)	N-61
Cyprinodon	*elegans*		F, M	48	2M + 14SM + 32ST	64	96			ACN=48	USA (TX)	S-108
Cyprinodon	*eximius*		F, M	48	2M + 14SM + 32ST	64	96			ACN=48	USA (TX)	S-108
Cyprinodon	*fontinalis*			48	2M + 6ST + 40A	50	56				Mexico	S-160
Cyprinodon	*hubbsi*		F, M	48	2M + 14SM + 32ST	64	96			ACN=48	USA (FL)	S-108
Cyprinodon	*macularius*			48	2M + 46A	50	50				USA (Death Valley)	T-62
Cyprinodon	*macularius*			48	2M/SM + 46 ST/A	50					USA (CA)	T-56
Cyprinodon	*nevadensis amargosae*			48	2M + 46A	50	50				USA (Death Valley)	T-62
Cyprinodon	*pecosensis*		F, M	48	2M + 14SM + 32ST	64	96			ACN=48	USA (TX)	S-108

Table 6.28 Order CYPRINODONTIFORMES (continued)

A		B	C	D	E	F	G	H	I	J	K	L
Current scientific name of taxon Suborder/family/subfamily/species		Reported in karyotype paper	Sex	2n	Karyotype	NF_1	NF_2	Ag–NORs	Genome size (pg/cell)	Comments	Locality	Reference
Cyprinodon	*radiosus*			48	2M + 46A	50	50				USA (Death Valley)	T–62
Cyprinodon	*rubrofluviatilis*		F, M	48	2M + 14SM + 32ST	64	96			ACN=48	USA (OK)	S–108
Cyprinodon	*salinus*			48	2M + 46A	50	50				USA (Death Valley)	T–62
Cyprinodon	*tularosa*			48	2M + 14SM + 32ST	64	96				USA (NM)	M–150
Cyprinodon	*variegatus*		F, M	48	2M + 14SM + 32ST	64	96		(3.2 BFA)	ACN=48	USA (LA)	S–59, H–13
Cyprinodon	*variegatus*		F, M	48	2M + 14SM + 32ST	64	96			ACN=48	USA (TX, FL), Mexico	S–108
Garmanella	*pulchra*		F	48	2M + 46 ST/A	50				$X_1X_1X_2X_2$, ACN=48	Yucatan Peninsula	L–26
Garmanella	*pulchra*		M	47	3 M/SM + 44 ST/A	50				X_1X_2Y, ACN=48	Yucatan Peninsula	L–26
Jordanella	*floridae*		F, M	48	2M + 46 ST/A	50					Mexico	L–26
Megupsilon	*aporus*		F	48	2M + 6SM + 40 ST/A	56				$X_1X_1X_2X_2$, ACN=48	Mexico (Nuevo Leon)	M–68, H–1
Megupsilon	*aporus*		M	47	3M + 6SM + 38 ST/A	56				X_1X_2Y, ACN=48	Mexico (Nuevo Leon)	M–68, H–1
Orestias	*agassii*		F, M	48	2M + 4SM + 14ST + 28A	54	68			ACN=48	Chile (Altiplano)	L–74, V–76, V–111
Orestias	*ascotanensis*		F, M	48	2M + 4SM + 4ST + 38A	54	58			ACN=48	Chile (Altiplano)	V–76, V–111
Orestias	*chungarensis*		F, M	50	2M + 2SM + 20ST + 26A	54	74			5B, ACN=48	Chile (Altiplano)	V–76, V–111
Orestias	*laucaensis*		F	50							Chile	V–76
Orestias	*laucaensis*		M	51							Chile	V–76
Orestias	*luteus*			48*							Peru (Lake Titicaca)	L–74
Orestias	*parinacotensis*			48	2M + 4SM + 10ST + 32A	54	64			ACN=48	Chile (Altiplano)	V–76, V–111
Orestias	*piacotensis*		F, M	50	2M + 2SM + 12ST + 34A	54	66			2B, ACN=48	Chile (Altiplano)	V–76, V–111
Fundulidae (topminnows and killifishes)												
Adinia	*xenica*	*multifasciatus*		32			(64)				USA (TX)	S–24
Adinia	*xenica*		F, M	32	16M + 6SM + 10ST	54	64			ACN=48	USA (LA)	S–59
Fundulus	*catenatus*			46						ACN=48	USA	C–52
Fundulus	*chrysotus*			34	14M + 20 ST/A	48		4		ACN=48	USA (FL)	C–52
Fundulus	*cingulatus*			46	2M + 44 ST/A	48		2		ACN=48	USA (FL)	C–52
Fundulus	*confluentus*			48	48 ST/A	48		2		ACN=48	USA (FL)	C–52
Fundulus	*diaphanus*		F, M	48	4SM + 44A	52	52		3.0 FCM	ACN=48	USA (CT)	C–52, A–87, G–85
Fundulus	*diaphanus*		F	48	4SM + 44A	52	52	2		ACN=48	USA (CT, NY)	C–49, H–27
Fundulus	*diaphanus*		M	48	2M + 2SM + 44A	52	52	2		ACN=48	USA (CT, NY)	C–49, H–27
Fundulus	*grandis*		F, M	48	4ST + 44A	48	52			ACN=48	USA (LA)	S–59
Fundulus	*grandis*			48	2SM + 46 ST/A	50		4		ACN=48	USA (FL)	C–52
Fundulus	*heteroclitus*			48	48 ST/A	48		2	(2.7 FCM, 2.6 FIA)	ACN=48	USA (CT)	C–49, C– 52, G–85, H–40
Fundulus	*heteroclitus*		M	48	48 ST/A	48				ACN=48	USA (ME)	K–99
Fundulus	*heteroclitus*	*Valencia lozanoi*	F, M	48	2SM + 10ST + 36A	50	60			ACN=48	S. W. Spain	B–27
Fundulus	*kansae*			48	48A	48	48	2		ACN=48	USA (MO)	C–52
Fundulus	*lineolatus*			46	2M + 44 ST/A	48		2		ACN=48	USA (FL)	C–52
Fundulus	*luciae*			32	16M + 4SM + 12 ST/A	52		2		ACN=48	USA (CT)	C–52

Table 6.28 Order CYPRINODONTIFORMES (continued)

A		B	C	D	E	F	G	H	I	J	K	L
Current scientific name of taxon Suborder/family/subfamily/species		Reported in karyotype paper	Sex	2n	Karyotype	NF_1	NF_2	Ag–NORs	Genome size (pg/cell)	Comments	Locality	Reference
Fundulus	*majalis*		F, M	48	2SM + 46 ST/A	50		2	(2.8 BFA)	ACN=48	USA (CT)	C-49, C-52, H-13
Fundulus	*notatus*			44	6M + 12ST + 26A	50	62			ACN=48	USA (AL)	B-38
Fundulus	*notatus*		F, M	40	10M + 12ST + 18A	50	62			ACN=48	USA (MS, TN, TX)	B-38, S-42
Fundulus	*notatus*			40	10M + 2SM + 28 ST/A	52		2		ACN=48	USA (MO)	C-52
Fundulus	*notti*			46	2M + 44 ST/A	48		2		ACN=48	USA (FL)	C-52
Fundulus	*notti*			48							USA (AL)	D-6
Fundulus	*olivaceus*			48	2M + 2SM + 44 ST/A	52		2		ACN=48	USA (MO)	C-52
Fundulus	*olivaceus*		F, M	48	2M + 12SM + 34A	62	62				USA (TX)	C-6
Fundulus	*olivaceus*		F, M	48	2M + 12ST + 34A	50	62			ACN=48	USA (AL, MS, TX)	B-38, S-42
Fundulus	*parvipinnis*		F, M	48	2SM + 46A	50	50				USA (CA)	K-99
Fundulus	*parvipinnis*		F	48	2SM + 46A	50	50			XX, ACN=48	USA (CA)	C-49
Fundulus	*parvipinnis*		M	48	1SM + 47A	49	49			XY, ACN=48	USA (CA)	C-49
Fundulus	*parvipinnis*			48	48A	48	48	2		ACN=48	USA (CA)	C-52
Fundulus	*pulvereus*			48	48 ST/A	48		2		ACN=48	USA (AL)	C-52
Fundulus	*rathbuni*			48	2SM + 46 ST/A	50		2		ACN=48	USA (NC)	C-52
Fundulus	*sciadicus*			44	4M + 2SM + 38 ST/A	50		2		ACN=48	USA (MO)	C-52
Fundulus	*seminolis*			48	48A	48	48	2		ACN=48	USA (FL)	C-52
Fundulus	*similis*		F, M	48	2SM + 46A	50	50	2		ACN=48	USA (FL)	C-52, S-59
Fundulus	*stellifer*			48						ACN=48	USA	C-52, D-6
Fundulus	*waccamensis*			48	4SM + 44 ST/A	52		2		ACN=48	USA (NC)	C-52
Fundulus	*zebrinus*			48						ACN=48	USA	C-52
Goodeidae												
Goodeinae												
Allodontichthys	*hubbsi*		F	42	8M + 34 SM/ST/A		54			ACN=48	Mexico (Jalisco)	M-70, U-65, U-70
Allodontichthys	*hubbsi*		M	41	9M + 32 SM/ST/A		54			ACN=48	Mexico (Jalisco)	M-70, U-65, U-70
Allodontichthys	*tamazulae*			48	2M + 2SM + 44 ST/A	52				ACN=48	Mexico	U-70
Allodontichthys	*zonistius*			48	2M + 2SM + 44 ST/A	52				ACN=48	Mexico	U-70
Allodontichthys	sp.			48	2M + 2SM + 44 ST/A	52				ACN=48	Mexico	U-70
Alloophorus	*robustus*			30	20M + 2ST + 8 ST/A	50				ACN=48	Mexico	U-70
Allotoca	*catarinae*	*Neoophorus*		46	2M + 4ST + 40A	48	52			ACN=48	Mexico	U-70
Allotoca	*diazi*	*Neoophorus*		46	2M + 4ST + 40A	48	52			ACN=48	Mexico	U-70
Allotoca	*dugesi*			26	22M + 4 SM/ST		52			ACN=48	Mexico	U-70
Allotoca	*dugesi*			26	22M + 4ST	48	52				W. Mexico	B-63, S-160
Allotoca	*goslinei*		F, M	48	6ST + 42A	48	54			ACN=48	Mexico (Jalisco)	S-73
Allotoca	*maculata*			48	4ST + 44A	48	52			ACN=48	W. Mexico	S-160, U-70
Allotoca	*meeki*	*Neoophorus*		46	2M + 6ST + 38A	48	54			ACN=48	Mexico	U-70
Ameca	*splendens*		F, M	26	22M + 2SM + 2A	50	50			ACN=48	Mexico (Pacific side)	M-72, U-70

Table 6.28 Order CYPRINODONTIFORMES (continued)

A		B	C	D	E	F	G	H	I	J	K	L
Current scientific name of taxon Suborder/family/subfamily/species		Reported in karyotype paper	Sex	2n	Karyotype	NF_1	NF_2	Ag-NORs	Genome size (pg/cell)	Comments	Locality	Reference
Ataeniobius	*toweri*			48	2SM + 46 ST/A	50				ACN=48	Mexico	U-70
Chapalichthys	*encaustus*			36	12M + 4SM + 16ST + 4A	52	68			ACN=48	Mexico	U-70
Chapalichthys	*pardalis*			36	12M + 2SM + 8ST + 14A	50	58			ACN=48	Mexico	U-70
Characodon	*lateralis*			24	24M	48	48			ACN=48	Mexico	F-59, U-70
Girardinichthys	*multiradiatus*			48	10ST + 38A	48	58			ACN=48	Mexico	U-70
Girardinichthys	*viviparus*			48	10ST + 38A	48	58			ACN=48	Mexico	U-70
Goodea	*atripinnis*			48	2SM + 46 ST/A	50				ACN=48	Mexico	U-70
Goodea	*gracilis*			48	2SM + 46 ST/A	50				ACN=48	Mexico	U-70
Goodea	*luitpoldi*			48	2SM + 46 ST/A	50				ACN=48	Mexico	U-70
Hubbsina	*turneri*			48	48 ST/A	48				ACN=48	Mexico	U-70
Ilyodon	*furcidens*		F, M	48	8ST + 40A	48	56			ACN=48	Mexico	U-70, M-163
Ilyodon	*furcidens*			48	7ST + 41A	48	55			ACN=48	Mexico (Pacific side)	T-57
Ilyodon	*furcidens*			48	2 M/SM + 11ST + 35A	50	61			ACN=48	Mexico (Pacific side)	T-57
Ilyodon	*furcidens*			48	4 M/SM + 12ST + 32A	52	64			ACN=48	Mexico (Pacific side)	T-57
Ilyodon	*furcidens*			48	6M + 3SM + 10ST + 29A	57	67			ACN=48	Mexico (Pacific side)	T-57
Ilyodon	*furcidens*			48	12 M/SM + 17ST + 19A	60	77			ACN=48	Mexico (Pacific side)	T-57
Ilyodon	*lennoni*		F	48	2SM + 46 ST/A	50				ACN=48	Mexico (Guerrero)	M-163
Ilyodon	*lennoni*		M	48	48 ST/A	48				ACN=48	Mexico (Guerrero)	M-163
Ilyodon	*whitei*		F, M	48	8ST + 40A	48	56			ACN=48	Mexico	U-70
Ilyodon	*whitei*		F	48	1M + 47 ST/A	49				ZW, ACN=48	Mexico (Guerrero)	M-163
Ilyodon	*whitei*		M	48	48 ST/A	48				ZZ, ACN=48	Mexico (Guerrero)	M-163
Ilyodon	*xantusi*		F, M	48	8ST + 40A	48	56			ACN=48	Mexico	U-70, M-163
Skiffia	*bilineata*			48	4M + 2SM + 34ST + 8A	54	88			ACN=48	Mexico	U-70
Skiffia	*francesae*			48	2M + 6SM + 40A	56	56			ACN=48	Mexico	U-70
Skiffia	*lermae*			26	22M + 4A	48	48			ACN=48	Mexico	U-70
Skiffia	*multipunctata*			46	4M + 4ST + 38A	50	54			ACN=48	Mexico	U-70
Xenoophorus	*captivus*		F	48	2SM + 46 ST/A	50				ACN=48	Mexico	U-70
Xenotaenia	*resolanae*		F	48	48 ST/A	48				ACN=48	Mexico	U-70
Xenotoca	*eiseni*		M	48	6ST + 42A	48	54			ACN=48	Mexico	U-70
Xenotoca	*melanosoma*			48	8ST + 40A	48	56			ACN=48	Mexico	U-70
Xenotoca	*variata*			48	4ST + 44A	48	52			ACN=48	Mexico	U-70
Zoogoneticus	*quitzeoensis*			28	22M + 6 ST/A	50					Mexico (Pacific side)	B-63

Table 6.28 Order CYPRINODONTIFORMES (continued)

A		B	C	D	E	F	G	H	I	J	K	L
Current scientific name of taxon Suborder/family/subfamily/species		Reported in karyotype paper	Sex	2n	Karyotype	NF_1	NF_2	Ag-NORs	Genome size (pg/cell)	Comments	Locality	Reference
Poeciliidae												
Aplocheilichthyinae												
Aplocheilichthys	*hutereaui*	*schalleri*		48	28ST + 20A	48	76				Mozambique	S-31
Aplocheilichthys	*spilauchen*			48			(94)				Africa	S-24
Procatopodinae												
Micropanchax	*macrophthalmus*	*Aplocheilichthys*		48		48	(48)			ACN=48	Nigeria	S-24
Poropanchax	*normani*	*Aplocheilichthys*		48		48	(48)			ACN=48	Africa	S-24
Procatopus	*aberrans*			48			(96)				Nigeria, Cameroon	S-24
Procatopus	*similis*			48*			(96)				Nigeria, Cameroon	S-24
Poeciliinae												
Gambusia	*affinis affinis*	*affinis*	F	48	1M + 47A	49	49		(1.8* FCM)	ZW, ACN=49	Japan (Kochi)	I-16, O-48
Gambusia	*affinis affinis*	*affinis*	M	48	48A	48	48			ZZ, ACN=48	Japan (Kochi)	I-16
Gambusia	*affinis affinis*	*affinis*	F	48	1M + 2SM + 45A	51	51		(1.9 FCM)	ZW, ACN=49	USA (CA, TX)	C-47, G-85
Gambusia	*affinis affinis*	*affinis*	M	48	2SM + 46A	50	50			ZZ, ACN=48	USA (CA, TX)	C-47
Gambusia	*affinis affinis*		F	48	1M + 47A	49	49			ZW, ACN=49	USA (AL, MS, AR, MO)	B-39
Gambusia	*affinis affinis*		M	48	48A	48	48			ZZ, ACN=48	USA (AL, MS, AR, MO)	B-39
Gambusia	*affinis holbrooki*		F, M	48	48A	48	48		(1.5 FCM, 1.7 BFA)	ACN=48	USA (NC)	R-81, T-73, H-13
Gambusia	*affinis holbrooki*		F, M	48	48A	48	48			ACN=48	USA (NC, SC, FL, AL)	B-39
Gambusia	*affinis holbrooki*		F, M	48	48A	48	48			ACN=48	India (J & K)	S-53, K-103
Gambusia	*affinis holbrooki*		F, M	48	48A	48	48	4		ACN=48	Italy, Cyprus	L-66, V-68, R-100
Gambusia	*affinis holbrooki*		F, M	48	1SM + 47A	49	49			ACN=48	Italy (Palermo)	V-68
Gambusia	*affinis holbrooki*		F, M	48	2SM + 46A	50	50			ACN=48	Italy (Palermo)	V-68
Gambusia	*gaigei*		F	48	1M + 47A	49	49			ZW, ACN=48	USA (TX)	C-4
Gambusia	*gaigei*		M	48	48A	48	48			ZZ, ACN=48	USA (TX)	C-4
Gambusia	*hurtadoi*		F	48	1M + 47A	49	49			ZW, ACN=48	Mexico	C-4
Gambusia	*hurtadoi*		M	48	48A	48	48			ZZ, ACN=48	Mexico	C-4
Gambusia	*luma*		F, M	48	48A	48	48			ACN=48	Belize	W-26
Gambusia	*marshi*		F, M	42	42A	42	42			ACN=48	Mexico	C-4
Gambusia	*nobilis*		F	48	1M + 47A	49	49			ZW, ACN=48	USA (TX)	C-4
Gambusia	*nobilis*		M	48	48A	48	48			ZZ, ACN=48	USA (TX)	C-4
Gambusia	*puncticulata puncticulata*		F	48	5 M/SM + 43A	53	53			ACN=48	Cuba	R-7
Gambusia	*puncticulata puncticulata*		M	48	6 M/SM + 42A	54	54				Cuba	R-7
Gambusia	*regani*		F, M	48	48A	48	48			ACN=48	Mexico	C-4
Gambusia	*rhizophorae*		F, M	48	2M + 4SM + 42A	54	54			ACN=48	USA (FL)	W-26
Gambusia	*sexradiata*		F	48	48A	48	48			ACN=48	Belize	W-26
Gambusia	*vittata*		F, M	48	48A	48	48			ACN=48	Mexico	C-4
Gambusia	*xanthosoma*		F, M	48	2M + 4SM + 42A	54	54			ACN=48	Cayman Islands	W-26
Gambusia	sp.			48	48A	48	48				Tenerife	S-24
Girardinus	*falcatus*	*Glaridichthys*	F, M	48	48A	48	48			ACN=48	Cuba	R-7
Girardinus	*metallicus*			48	2SM + 46A	50	50				(Cuba)	F-67
Limia	*vittata*		M	46	46A	46	46		1.9 FD	ACN=48	Cuba	R-7, G-85
Phalloceros	*caudimaculatus*			48							Argentina	F-20

Table 6.28 Order CYPRINODONTIFORMES (continued)

A		B	C	D	E	F	G	H	I	J	K	L
Current scientific name of taxon Suborder/family/subfamily/species		Reported in karyotype paper	Sex	2n	Karyotype	NF_1	NF_2	Ag-NORs	Genome size (pg/cell)	Comments	Locality	Reference
Poecilia	*formosa*			46	46A	46	46	2	1.9 FCM	ACN=48	Mexico (Tamaulipas)	G-8, L-82
Poecilia	*formosa*		F	69	69A	69	69	4	3.1 FCM	3X	Mexico (Tamaulipas)	G-8, L-82
Poecilia	*latipinna*			46		46			2.0 FCM, 1.9 FD		Mexico	K-73, L-82, G-85
Poecilia	*latipinna*		F	46	46A	46	46	1-4		ZW	Mexico	S-188
Poecilia	*latipinna*		M	46	46A	46	46	1-2		ZZ	Mexico	S-188
Poecilia	*latipunctata*		F, M	46	46 ST/A	46	46	8	1.8 FD	ACN=48	Mexico (Tamps)	G-7, G-85
Poecilia	*mexicana*			46					2.0 FCM, 1.5-1.8 FD		Mexico	G-7, G-85, L-82
Poecilia	*mexicana mexicana*		F, M	46	46A	46	46	4			Mexico	S-187
Poecilia	*mexicana mexicana*			69	69A	69	69		2.2 FD	3X	Mexico	S-187
Poecilia	*reticulata*	*Lebistes reticulatus*		46	46A	46	46		2.0 FCM, 1.5 FD	ACN=48	(Mexico)	I-16, C-109, V-86
Poecilia	*sphenops*	*Mollienesia*	F	46	1M + 45A	47	47			ZW	India	R-48
Poecilia	*sphenops*	*Mollienesia*	M	46	46A	46	46			ZZ	India	R-48
Poecilia	*sphenops*	*Mollienesia*	F, M	46	46A	46	46		(1.9 FD)	ACN=48	Japan	I-16, G-85
Poecilia	*velifera*			46							Mexico	G-7
Poecilia	*vivipara*			48	48A	48	48				Brazil	O-50
Poeciliopsis	*baenschi*			48	48A	48	48				(Mexico, Pacific)	F-67
Poeciliopsis	*latidens*			48					1.3 FD		Mexico	S-161, C-109
Poeciliopsis	*latidens*			72						3X	Mexico	S-161
Poeciliopsis	*lucida*			48					1.3 FCM, 1.4 FD	ACN=48	Mexico	S-161, C-109, G-85
Poeciliopsis	*monacha*			48					1.3 FCM, 1.3 FD		Mexico	S-183, C-109, G-85
Poeciliopsis	*occidentalis*			48					1.4 FD		(S. USA, N. Mexico)	S-207, C-109
Poeciliopsis	*viriosa*			48					1.4 FD		Mexico	S-183, C-109
Quintana	*atrizona*		F, M	48	48A	48	48			ACN=48	Cuba	R-7
Xenophallus	*umbratilis*			46	2SM + 44A	48	48				Costa Rica (Atlantic)	F-67
Xiphophorus	*helleri*		F, M	48	48A	48	48		1.9* FCM, 1.9 BFA	ACN=48	(C. America)	I-16, O-48, H-13
Xiphophorus	*helleri*		M	48	48A	48	48		1.5 FCM, 1.6 FD		(C. America)	O-4, G-85, T-72
Xiphophorus	*maculatus*		F, M	48	48A	48	48		(1.8* FCM, 1.9 BFA)	ACN=48	(C. America)	I-16, O-48, H-13
Xiphophorus	*maculatus*			48	48A	48	48		1.5 FCM	XY/XX	(C. America)	C-55, S-24, T-72
Xiphophorus	*montezumae*		F	48	48A	48	48		1.5 FCM	ACN=48	(Mexico)	L-75, T-72
Xiphophorus	*xiphidium*		M	48	48A	48	48			ACN=48	(Mexico)	L-75
Valenciidae												
Valencia	*hispanica*			48			(92)				(Spain)	S-24

Table 6.29 Order STEPHANOBERYCIFORMES

A		B	C	D	E	F	G	H	I	J	K	L
Current scientific name of taxon Family/species		Reported in karyotype paper	Sex	2n	Karyotype	NF_1	NF_2	Ag-NORs	Genome size (pg/cell)	Comments	Locality	Reference
Melamphaidae												
Melamphaes	*acanthomus*		M	48	48A	48	48			ACN=48	USA (off CA)	C-46
Melamphaes	*parvus*		M	50	50A	50	50			XY, ACN=50	USA (off CA)	C-48
Poromitra	*crassiceps*		M	58	56 M/SM/ST/A + 2 satellited chrom.						USA (off CA)	C-46
Scopeloberyx	*robustus*		M	42	40A + 2 satellited chrom.					XY, ACN=44	USA (off CA)	C-46, C-48
Scopelogadus	*mizolepis bispinosus*		M	46	46A	46	46			XY, ACN=48	USA (off CA)	C-46, C-48

Table 6.30 Order BERYCIFORMES

A		B	C	D	E	F	G	H	I	J	K	L
Current scientific name of taxon Suborder/family/species		Reported in karyotype paper	Sex	2n	Karyotype	NF_1	NF_2	Ag-NORs	Genome size (pg/cell)	Comments	Locality	Reference
Suborder Trachichthyoidei												
Anoplogasteridae												
Anoplogaster	*cornuta*		M	48	2SM + 46A	50	50			ACN=50	USA (off CA)	C-46
Diretmidae												
Diretmus	*argenteus*		M	44–46	2 macro. + 42–44 normal						Atlantic	P-74
Diretmus	sp. C		M	70±							Atlantic	P-74
Monocentridae												
Monocentris	*japonica*			48	48A	48	48			ACN=48	Japan (Chiba)	A-67
Monocentris	*japonica*			48	2ST + 46A	48	50			ACN=48	Japan (Suruga Bay)	M-104
Trachichthyidae												
Hoplostethus	*mediterraneus*			48	2ST + 46A	48	50			ACN=48	Japan (Suruga Bay)	M-104
Suborder Berycoidei												
Berycidae												
Beryx	*splendens*		F	48	4M + 8SM + 36A	60	60	2	1.7* FCM	$X_1X_1X_2X_2$, ACN=48	Japan (Izu Peninsula)	O-43, O-48
Beryx	*splendens*		M	47	5M + 8SM + 34A	60	60			X_1X_2Y, ACN=48	Japan (Izu Peninsula)	O-43

Table 6.30 Order BERYCIFORMES (continued)

A		B	C	D	E	F	G	H	I	J	K	L
Current scientific name of taxon Suborder/family/species		Reported in karyotype paper	Sex	2n	Karyotype	NF_1	NF_2	Ag-NORs	Genome size (pg/cell)	Comments	Locality	Reference
Suborder Holocentroidei												
Holocentridae												
Holocentrus	*adscensionis*			50			78		1.3 FD, 1.8 BFA		Brazil (CE, RN)	G-12, G-85, H-13
Myripristis	*jacobus*			48		48	48				Brazil (RN, SPR)	G-12
Sargocentron	*rubrum*	*Adioryx ruber*		48	48A	48	48			ACN=48	Japan (Yakushima Is.)	A-67

Table 6.31 Order ZEIFORMES

A		B	C	D	E	F	G	H	I	J	K	L
Current scientific name of taxon Suborder/family/species		Reported in karyotype paper	Sex	2n	Karyotype	NF_1	NF_2	Ag-NORs	Genome size (pg/cell)	Comments	Locality	Reference
Suborder Zeioidei												
Oreosomatidae												
Allocyttus	*verrucosus*			42	42A	42	42			ACN=48	Japan (Hokkaido)	S-140
Zeidae												
Zeus	*faber*		F	44	44A	44	44	2		ACN=46	Italy (Palermo)	V-57, V-64
Zeus	*faber*		M	42	1ST + 41A	42	43	2		Y chrom., ACN=46	Italy (Palermo)	V-64

Table 6.33 Order BATRACHOIDIFORMES

A		B	C	D	E	F	G	H	I	J	K	L
Current scientific name of taxon Family/species		Reported in karyotype paper	Sex	2n	Karyotype	NF_1	NF_2	Ag-NORs	Genome size (pg/cell)	Comments	Locality	Reference
Batrachoididae												
Amphichthys	*cryptocentrus*			46	4M + 2SM + 40A	52	52			ACN=48	Venezuela	N-24
Batrachoides	*manglae*			46	6M + 6SM + 34A	58	58			ACN=48	Venezuela	N-24
Batrachoides	*pacifici*		F, M	46	6M + 6SM + 34A	58	58			ACN=48	Panama (Pacific coast)	N-23
Halobatrachus	*didactylus*		F, M	46	8M + 12SM + 26 ST/A	66		2		ACN=48	Spain (Bay of Cadiz)	P-3
Porichthys	*notatus*		F	48	10M + 20SM + 18A	78	78		4.4 BFA		E. Pacific	C-46, H-13
Porichthys	*plectrodon*		F, M	44	8M + 10SM + 6ST + 20A	62	68	2		ACN=48	Venezuela	N-26
Porichthys	*porosissimus*		F, M	44	14 M/SM + 30 ST/A	58		1-3	3.4 BFA	ACN=48	Brazil (RJ)	B-58, H-13
Thalassophryne	*maculosa*		F, M	46	8M + 6SM + 32 ST/A	60				ACN=48	Venezuela	N-24
Thalassophryne	*maculosa*		F, M	46	12M + 6SM + 20ST + 8A	64	84	2		ACN=48	Venezuela	N-27
Thalassophryne	*nattereri*		F, M	46	8M + 8SM + 24ST + 6A	62	86	2		ACN=48	Brazil (RN)	C-106

Table 6.34 Order SYNBRANCHIFORMES

A		B	C	D	E	F	G	H	I	J	K	L
Current scientific name of taxon Suborder/family/species		Reported in karyotype paper	Sex	2n	Karyotype	NF_1	NF_2	Ag-NORs	Genome size (pg/cell)	Comments	Locality	Reference
Suborder Mastacembeloidei												
Mastacembelidae												
Macrognathus	*aculeatum*		M	48	8M + 2SM + 38A	58	58			ACN=48	India (WB)	M-26
Macrognathus	*aculeatum*		F, M	48	8M + 2SM + 38A	58	58			ACN=48	India (Bihar)	K-42
Macrognathus	*aculeatum*		F, M	48	10M + 38A	58	58	2			India (Haryana)	R-76
Macrognathus	*aculeatum*		F	48							India (Portonovo)	N-13
Macrognathus	*pancalus*	*Mastacembelus*	F, M	48	14M + 12SM + 14ST + 8A	74	88			ACN=48	India (Bihar)	K-42
Macrognathus	*pancalus*	*Mastacembelus*	F	48	16M + 6SM + 8ST + 18A	70	78			ACN=48	India (WB)	M-26
Mastacembelus	*armatus*		F, M	48	14M + 2SM + 4ST + 28A	64	68	2		ACN=48	China (Guangdong, Guilin)	Y-15
Mastacembelus	*armatus*		F	48	10M + 4SM + 2ST + 32A	62	64	2			India (WB)	M-28, D-34
Mastacembelus	*armatus*		F, M	48	10M + 6SM + 4ST + 28A	64	68	2	1.4 FD	ACN=48	(Asia)	O-57
Sinobdella	*sinensis*	*Mastacembelus aculeatus*	F	48	16M + 4SM + 2ST + 26A	68	70		(1.8* FD)	XX, ACN=48	China (Guilin)	Y-15, C-83
Sinobdella	*sinensis*	*Mastacembelus aculeatus*	M	48	15M + 4SM + 3ST + 26A	67	70			XY, ACN=48	China (Guilin)	Y-15
Sinobdella	*sinensis*	*Mastacembelus aculeatus*	F, M	48	16M + 4SM + 28A	68	68			XX/XY	China (Hubei)	L-55

Table 6.33 Order BATRACHOIDIFORMES

A	B	C	D	E	F	G	H	I	J	K	L
Current scientific name of taxon Family/species	Reported in karyotype paper	Sex	2n	Karyotype	NF_1	NF_2	Ag-NORs	Genome size (pg/cell)	Comments	Locality	Reference
Batrachoididae											
Amphichthys cryptocentrus			46	4M + 2SM + 40A	52	52			ACN=48	Venezuela	N-24
Batrachoides manglae			46	6M + 6SM + 34A	58	58			ACN=48	Venezuela	N-24
Batrachoides pacifici		F, M	46	6M + 6SM + 34A	58	58			ACN=48	Panama (Pacific coast)	N-23
Halobatrachus didactylus		F, M	46	8M + 12SM + 26 ST/A	66		2		ACN=48	Spain (Bay of Cadiz)	P-3
Porichthys notatus		F	48	10M + 20SM + 18A	78	78		4.4 BFA		E. Pacific	C-46, H-13
Porichthys plectrodon		F, M	44	8M + 10SM + 6ST + 20A	62	68	2		ACN=48	Venezuela	N-26
Porichthys porosissimus		F, M	44	14 M/SM + 30 ST/A	58		1-3	3.4 BFA	ACN=48	Brazil (RJ)	B-58, H-13
Thalassophryne maculosa		F, M	46	8M + 6SM + 32 ST/A	60				ACN=48	Venezuela	N-24
Thalassophryne maculosa		F, M	46	12M + 6SM + 20ST + 8A	64	84	2		ACN=48	Venezuela	N-27
Thalassophryne nattereri		F, M	46	8M + 8SM + 24ST + 6A	62	86	2		ACN=48	Brazil (RN)	C-106

Table 6.34 Order SYNBRANCHIFORMES

A	B	C	D	E	F	G	H	I	J	K	L
Current scientific name of taxon Suborder/family/species	Reported in karyotype paper	Sex	2n	Karyotype	NF_1	NF_2	Ag-NORs	Genome size (pg/cell)	Comments	Locality	Reference
Suborder Mastacembeloidei											
Mastacembelidae											
Macrognathus aculeatum		M	48	8M + 2SM + 38A	58	58			ACN=48	India (WB)	M-26
Macrognathus aculeatum		F, M	48	8M + 2SM + 38A	58	58			ACN=48	India (Bihar)	K-42
Macrognathus aculeatum		F, M	48	10M + 38A	58	58	2			India (Haryana)	R-76
Macrognathus aculeatum		F	48							India (Portonovo)	N-13
Macrognathus pancalus	*Mastacembelus*	F, M	48	14M + 12SM + 14ST + 8A	74	88			ACN=48	India (Bihar)	K-42
Macrognathus pancalus	*Mastacembelus*	F	48	16M + 6SM + 8ST + 18A	70	78			ACN=48	India (WB)	M-26
Mastacembelus armatus		F, M	48	14M + 2SM + 4ST + 28A	64	68	2		ACN=48	China (Guangdong, Guilin)	Y-15
Mastacembelus armatus		F	48	10M + 4SM + 2ST + 32A	62	64	2			India (WB)	M-28, D-34
Mastacembelus armatus		F, M	48	10M + 6SM + 4ST + 28A	64	68	2	1.4 FD	ACN=48	(Asia)	O-57
Sinobdella sinensis	*Mastacembelus aculeatus*	F	48	16M + 4SM + 2ST + 26A	68	70		(1.8* FD)	XX, ACN=48	China (Guilin)	Y-15, C-83
Sinobdella sinensis	*Mastacembelus aculeatus*	M	48	15M + 4SM + 3ST + 26A	67	70			XY, ACN=48	China (Guilin)	Y-15
Sinobdella sinensis	*Mastacembelus aculeatus*	F, M	48	16M + 4SM + 28A	68	68			XX/XY	China (Hubei)	L-55

Table 6.34 Order SYNBRANCHIFORMES (continued)

A		B	C	D	E	F	G	H	I	J	K	L
Current scientific name of taxon Suborder/family/species		Reported in karyotype paper	Sex	2n	Karyotype	NF_1	NF_2	Ag-NORs	Genome size (pg/cell)	Comments	Locality	Reference
Suborder Synbranchoidei												
Synbranchidae												
Monopterus	*albus*	*Fluta alba*	F, M	24	24A	24	24			ACN=24	Japan (Nara)	K-65
Monopterus	*albus*			24	24A	24	24		(1.3*, 1.6 FD)	ACN=24	China (Hubei)	L-39, C-83
Monopterus	*albus*			24	24A	24	24			ACN=24	China	L-61, Y-12
Monopterus	*albus*	*alba*	F	24	24A	24	24			ACN=24	India (Manipur)	R-56, R-64
Monopterus	*cuchia*	*Amphipnous*		42	2M + 4SM + 14ST + 22A	48	62				India (WB)	K-46
Monopterus	*cuchia*	*Amphipnous*	F, M	42	4SM + 38A	46	46			ACN=42	India (Assam)	R-64
Ophisternon	*bengalense*	*Synbranchus bengalensis*	F	46	6 M/SM + 40A	52	52				India (Portonovo)	N-13
Synbranchus	*marmoratus*		F, M	44	4 M/SM + 40 ST/A	48		4		0–2 B	Argentina	S-10, C-98
Synbranchus	*marmoratus*		F	44	4 M/SM + 40 ST/A	48		2		ACN=48	Brazil (CE, SP)	F-57
Synbranchus	*marmoratus*		F, M	42	4 M/SM + 38 ST/A	46		2		ACN=48	Brazil (MS, GO, SP)	F-57
Synbranchus	*marmoratus*			46	6 M/SM + 40 ST/A	52		2			Brazil (SP)	F-57
Synbranchus	*marmoratus*			42	4M + 6SM + 8ST + 24A	52	60		7.6–8.5 FD	ACN=48	Brazil (SP)	T-75
Synbranchus	*marmoratus*			42	4M + 2SM + 8ST + 28A	48	56		6.6–7.4 FD	ACN=48	Brazil (PR)	T-75
Synbranchus	*marmoratus*			42	4M + 2SM + 8ST + 28A	48	56		6.6 FD	ACN=48	Brazil (MS)	T-75
Synbranchus	*marmoratus*			44	4M + 2SM + 8ST + 30A	50	58		7.9 FD	ACN=48	Brazil (SP)	T-75
Synbranchus	*marmoratus*			46	4M + 2SM + 8ST + 32A	52	60		6.4 FD	ACN=48	Brazil (PR)	T-75
Synbranchus	*marmoratus*			46	6M + 2SM + 6ST + 32A	54	60		5.6 FD	ACN=48	Brazil (MS)	T-75

Table 6.35 Order SCORPAENIFORMES

A		B	C	D	E	F	G	H	I	J	K	L
Current scientific name of taxon Suborder/family/subfamily/species		Reported in karyotype paper	Sex	2n	Karyotype	NF_1	NF_2	Ag–NORs	Genome size (pg/cell)	Comments	Locality	Reference
Suborder Dactylopteroidei												
Dactylopteridae												
Dactylopterus	*volitans*			48	16M + 14SM + 6ST + 12A	78	84				Brazil (RJ)	B-86
Suborder Scorpaenoidei												
Scorpaenidae												
Sebastinae												
Helicolenus	*dactylopterus*		F, M	48	2M + 46A	50	50			ACN=48	Italy (Palermo)	V-54
Helicolenus	*dactylopterus*		F, M	47	3M + 44A	50	50			ACN=48	Italy (Palermo)	V-54
Helicolenus	*hilgendorfi*			48			52				Japan	Y-22
Sebastes	*hubbsi*		M	46	4M + 42A	50	50			ACN=48	Japan (Sanriku)	I-4, N-33
Sebastes	*hubbsi*			48	3M + 2ST + 43A	51	53				China	Z-23
Sebastes	*inermis*			48	2M + 46A	50	50			ACN=48	Japan	N-33
Sebastes	*inermis*			48			50		(1.9* FCM)		Japan	Y-22, O-48
Sebastes	*iracundus*		M	48							W.N. Pacific	I-12
Sebastes	*joyneri*			48	2M + 46A	50	50			ACN=48	Japan (Izu)	I-4
Sebastes	*longispinis*			48	2M + 46A	50	50			ACN=48	Japan	N-33
Sebastes	*matsubarae*			48			50				Japan	Y-22
Sebastes	*oblongus*		F, M	48	2SM + 46A	50	50			ACN=48	Japan (Sanriku)	I-4
Sebastes	*pachycephalus nudus*		F, M	48	2M + 46A	50	50			ACN=48	Japan (Sanriku)	I-4
Sebastes	*schlegeli*			48	2M + 2ST + 44A	50	52			ACN=48	Japan	N-33
Sebastes	*schlegeli*		M	48	2M + 46A	50	50			ACN=48	Japan (Sanriku)	I-4
Sebastes	*schlegeli*			48	2M + 2SM + 44A	52	52			ACN=48	China (Liaoning)	Z-15
Sebastes	*schlegeli*		F, M	48	2M + 46A	50	50	2		ACN=48	China (Shandong)	Y-18, W-8
Sebastes	*taczanowskii*		M	48	2M + 46A	50	50			ACN=48	Japan (Hokkaido)	S-15
Sebastes	*thompsoni*		F	48	2SM + 46A	50	50			ACN=48	Japan (Sanriku)	I-4
Sebastes	*trivittatus*		M	48	2M + 46A	50	50			ACN=48	Japan (Hokkaido)	I-4
Sebastes	*vulpes*			48	2M + 46A	50	50			ACN=48	Japan (Hokkaido)	I-4
Sebastiscus	*albofasciatus*			48			52				Japan	Y-22
Sebastiscus	*marmoratus*			48	2M + 46A	50	50		(1.8* FCM)	ACN=48	Japan	N-33, Y-22, O-48
Sebastiscus	*marmoratus*			48	2SM + 46A	50	50			ACN=48	China	P-4
Sebastiscus	*marmoratus*		F, M	48	2M + 46A	50	50		1.6* FCM	ACN=48	Korea (Busan)	P-70
Sebastolobus	*macrochir*		M	46							Okhotsk Sea	I-12

Table 6.35 Order SCORPAENIFORMES (continued)

A		B	C	D	E	F	G	H	I	J	K	L
Current scientific name of taxon Suborder/family/subfamily/species		Reported in karyotype paper	Sex	2n	Karyotype	NF_1	NF_2	Ag-NORs	Genome size (pg/cell)	Comments	Locality	Reference
Scorpaeninae												
Dendrochirus	*zebra*		F, M	48	4M + 6ST + 38A	52	58	2		ACN=48	Indonesia (Java)	C-23
Pterois	*lunulata*			48	2M + 10SM + 18 ST + 18A	60	78			ACN=48	Japan	N-33
Pterois	*lunulata*		F, M	48	2M + 12ST + 34A	50	62	2		ACN=48	Indonesia (Java)	C-23
Pterois	*radiata*		F, M	48	4M + 8ST + 36A	52	60	2		ACN=48	Indonesia (Java)	C-23
Pterois	*volitans*			48	2M + 10ST + 36A	50	60	2	2.0 FIA	ACN=48	Indonesia (Java)	C-23, H-41
Scorpaena	*brasiliensis*		F, M	46	2M + 12SM + 32 ST/A	60		2	2.8 BFA	ACN=48	Brazil (RJ)	C-78, H-13
Scorpaena	*isthmensis*		F, M	40	6M + 8SM + 26 ST/A	54		2		ACN=46	Brazil (RJ)	C-78
Scorpaena	*izensis*			48			56				Japan	Y-22
Scorpaena	*miostoma*	*neglecta miostoma*		48			82				Japan	Y-22
Scorpaena	*miostoma*	*neglecta miostoma*		48	6SM + 22ST + 20A	54	76			ACN=48	Japan (Suruga Bay)	M-118
Scorpaena	*notata*		F, M	34	26ST + 8A	34	60	2			Italy (Senigallia)	C-16
Scorpaena	*notata*			34	24ST + 10A	34	58	2	1.1 FD	ACN=48	Spain (Malaga)	T-34
Scorpaena	*notata*	*ustulata*		34	10 M/SM + 24 ST/A	44					Croatia	S-195
Scorpaena	*onaria*	*neglecta neglecta*		48	6SM + 32ST + 10A	54	86			ACN=48	Japan (Suruga Bay)	M-118
Scorpaena	*onaria*	*neglecta neglecta*		47	7SM + 32ST + 8A	54	86				Japan (Suruga Bay)	M-118
Scorpaena	*onaria*	*neglecta neglecta*		48	6M + 14SM + 18ST + 10A	68	86			ACN=48	Japan	N-33
Scorpaena	*onaria*	*neglecta neglecta*		47	7M + 14SM + 18ST + 8A	68	86			ACN=48	Japan	N-33
Scorpaena	*onaria*	*neglecta neglecta*		48			56				Japan	Y-22
Scorpaena	*porcus*		F, M	42	4M + 2SM + 10ST + 26A	48	58	2	(2.8 FCM)	ACN=46	Italy (Senigallia)	C-16, V-86
Scorpaena	*porcus*			42	6M + 10ST + 26A	48	58	2	1.8 FD	ACN=46	Spain (Malaga)	T-34, C-9
Scorpaena	*porcus*		M	42	6M + 10ST + 26A	48	58				Italy (Roma)	C-31
Scorpaena	*porcus*			42	16 M/SM + 26A	58	58				Croatia	S-195
Scorpaena	*scrofa*			46	20ST + 26A	46	66				Italy	T-34
Scorpaenodes	*littoralis*		F	36	32 M/SM + 2ST + 2A	68	70	2			Japan (Chiba)	A-54, T-6
Scorpaenodes	*littoralis*			36	32M + 2ST + 2A	68	70				Japan (Suruga Bay)	M-118
Scorpaenopsis	*cirrosa*	*cirrhosa*		48	4ST + 44A	48	52			ACN=48	Japan (Suruga Bay)	M-118
Scorpaenopsis	*gibbosa*		M	48	10ST + 38A	48	58	2		ACN=48	Indonesia (Java)	C-23
Tetraroginae												
Paracentropogon	*rubripinnis*	*Hypodytes*		48	4 M/SM + 44 ST/A	52			(2.1* FCM)	ACN=48	Japan (Chiba, Kanagawa)	A-54, O-48
Paracentropogon	*rubripinnis*	*Hypodytes*	F	48	2M + 46A	50	50			$X_1X_1X_2X_2$	S. Japan	U-41
Paracentropogon	*rubripinnis*	*Hypodytes*	M	47	3M + 44A	50	50			X_1X_2Y, ACN=48	S. Japan	U-41
Tetraroge	*niger*		M	50							India (Portonovo)	N-13

Table 6.35 Order SCORPAENIFORMES (continued)

A		B	C	D	E	F	G	H	I	J	K	L
Current scientific name of taxon Suborder/family/subfamily/species		Reported in karyotype paper	Sex	2n	Karyotype	NF_1	NF_2	Ag–NORs	Genome size (pg/cell)	Comments	Locality	Reference
Synanceiinae												
Inimicus	*japonicus*			48	10M + 4SM + 2ST + 32A	62	64			ACN=48	Japan	N-33
Suborder Platycephaloidei												
Triglidae												
Chelidonichthys	*lucernus*	*Trigla lucerna*	F, M	48	2M + 2SM + 2ST + 42A	52	54			ACN=48	Italy (Palermo)	V-54
Chelidonichthys	*lucernus*	*Trigla lucerna*	F, M	47	3M + 2SM + 2ST + 40A	52	54			ACN=48	Italy (Palermo)	V-54
Chelidonichthys	*lucernus*	*Trigla lucerna*		48	2M + 4SM + 2ST + 40A	54	56	2		ACN=48	Italy (Senigallia)	C-16
Chelidonichthys	*lucernus*	*Trigla lucerna*		47	3M + 4SM + 2ST + 38A	54	56	2		ACN=48	Italy (Senigallia)	C-16
Chelidonichthys	*lucernus*	*Trigla lucerna*		48							Russia	V-72
Prionotus	*punctatus*			100–102		100–102					Brazil (RJ)	B-86
Trigloporus	*lastoviza*		F, M	48	6M + 4ST + 38A	54	58	2		ACN=48	Italy (Palermo)	C-16
Platycephalidae												
Inegocia	*guttata*			48	20M + 10SM + 6ST + 12A	78	84			ACN=48	Japan (Izu)	I-2
Onigocia	*macrolepis*			48	48A	48	48			ACN=48	Japan (Izu)	I-2
Onigocia	*spinosa*			48	32M + 12SM + 4ST	92	96			ACN=48	Japan (Izu, Tokyo)	I-2
Platycephalus	sp. 2	*indicus*		48	2SM + 6ST + 40A	50	56			ACN=48	Japan (Iwate)	I-2
Platycephalus	*indicus*			48	2M + 8SM + 2ST + 36A	58	60	2	(1.6 FIA)	ACN=48	China (Shandong)	K-96, H-41
Platycephalus	*indicus*			48	4M + 6SM + 2ST + 36A	58	60			ACN=48	China (Shandong)	Z-37
Platycephalus	*indicus*			48	10 M/SM + 38 ST/A	58					India (Portonovo)	N-13
Platycephalus	*indicus*	*Thysanophrys*	F	48	2SM + 10ST + 36A	50	60			ACN=48	India (Orissa)	C-61
Platycephalus	*tuberculatus*			48	4M + 4SM + 6ST + 34A	56	62			ACN=48	India (WB)	N-17
Thysanophrys	*chiltonae*			48	22M + 12SM + 10ST + 4A	82	92			ACN=48	Japan (Okinawa)	I-2
Suborder Anoplopomatoidei												
Anoplopomatidae												
Erilepis	*zonifer*			30	22M + 6SM + 2ST	58	60				N. Pacific	I-21
Erilepis	*zonifer*			30	22M + 6SM + 1ST + 1A	58	59				N. Pacific	I-21
Erilepis	*zonifer*			30	18M + 4SM + 4ST + 4A	52	56				N. Pacific	I-21
Erilepis	*zonifer*			30	18M + 2SM + 4ST + 6A	50	54				N. Pacific	I-21

Table 6.35 Order SCORPAENIFORMES (continued)

A		B	C	D	E	F	G	H	I	J	K	L
Current scientific name of taxon Suborder/family/subfamily/species		Reported in karyotype paper	Sex	2n	Karyotype	NF_1	NF_2	Ag-NORs	Genome size (pg/cell)	Comments	Locality	Reference
Suborder Hexagrammoidei												
Hexagrammidae												
Hexagrammos	*agrammus*	*Agrammus*		48	8M + 26SM + 14ST	82	96			ACN=48	Japan (Hokkaido)	M-51
Hexagrammos	*agrammus*	*Agrammus*	F, M	48	8M + 26SM + 14ST	82	96			ACN=48	China (Liaoning)	Z-15
Hexagrammos	*lagocephalus*			48	2M + 6SM + 28ST + 12A	56	84			ACN=48	Japan (Hokkaido)	M-51
Hexagrammos	*otakii*			48	6M + 12SM + 22ST + 8A	66	88		(1.7* FCM)	ACN=48	Japan (Yamaguchi)	N-36, O-48
Hexagrammos	*otakii*		F, M	48	6M + 8SM + 12ST + 22A	62	74			ACN=48	China (Liaoning)	Z-15
Hexagrammos	*otakii*		F, M	48	6M + 16SM + 20ST + 6A	70	90	2		ACN=48	China (Shandong)	Y-18
Hexagrammos	*otakii*			48	6M + 20SM + 16ST + 6A	74	90				China	W-8
Hexagrammos	*stelleri*		F	48	4M + 12SM + 12ST + 20A	64	76			ACN=48	Japan (Hokkaido)	M-51
Pleurogrammus	*azonus*			48	18M + 8SM + 12ST + 10A	74	86			ACN=48	Japan (Hakodate)	M-51
Suborder Cottoidei												
Cottidae												
Alcichthys	*alcicornis*		F, M	48	2M + 6SM + 40 ST/A	56			1.5 FD	ACN=48	Japan (Iwate)	T-64
Artedius	*fenestralis*		M	48	10M + 10SM + 10ST + 18A	68	78			ACN=48	USA (WA)	I-7
Artedius	*lateralis*		M	48	8M + 12SM + 12ST + 16A	68	80			ACN=48	USA (WA)	I-7
Batrachocottus	*baicalensis*		M	48	6M + 6SM + 36 ST/A	60				ACN=50	Russia (Lake Baikal)	S-167, S-168
Batrachocottus	*multiradiatus*		F, M	48	12M + 8SM + 18ST + 10A	68	86			ACN=48	Russia (Lake Baikal)	S-167, S-168
Clinocottus	*analis*			48*					1.9 BFA		USA (CA)	C-46, H-13
Cottocomephorus	*grewingkii*		F, M	48	10M + 6SM + 32 ST/A	64				ACN=48	Russia (Lake Baikal)	S-168
Cottocomephorus	*inermis*		M	48	6M + 8SM + 34 ST/A	62				ACN=48	Russia (Lake Baikal)	S-168
Cottus	*gobio*		F, M	48	10 M/SM + 38 ST/A	58		4		ACN=48	Italy (Vicenza)	V-62
Cottus	*gobio*		F, M	48	6 M/SM + 42 ST/A	54					Bosnia-Herzegovina	B-24
Cottus	*gobio*			52	6 M/SM + 46A	58	58				Poland	S-107
Cottus	*hangiongensis*		F	48	6 M/SM + 42 ST/A	54				ACN=48	Japan (Hokkaido)	A-1
Cottus	*kazika*		F, M	40	18 M/SM + 22 ST/A	58				ACN=48	Japan (Hokuriku)	A-4
Cottus	*nozawae*		F, M	48	10 M/SM + 38 ST/A	58				ACN=48	Japan (Hokkaido)	A-1
Cottus	*paulus*	*pygmaeus*		48		48					USA (AL)	G-32
Cottus	*poecilopus*			48	8 M/SM + 40A	56	56				Europe	S-107
Cottus	*pollux*		F, M	48	10 M/SM + 38 ST/A	58		2		ACN=48	Japan	A-4
Cottus	*pollux*		F, M	48	4M + 6SM + 16ST + 22A	58	74			ACN=48	Japan (Sanriku)	I-7
Cottus	*reinii*		F	48	12 M/SM + 36 ST/A	60				ACN=48	Japan (Iwate)	A-4

Table 6.35 Order SCORPAENIFORMES (continued)

A		B	C	D	E	F	G	H	I	J	K	L
Current scientific name of taxon Suborder/family/subfamily/species		Reported in karyotype paper	Sex	2n	Karyotype	NF_1	NF_2	Ag-NORs	Genome size (pg/cell)	Comments	Locality	Reference
Gymnocanthus	*intermedius*		F, M	44	2M + 4SM + 38 ST/A	50			1.5 FD	ACN=48	Japan (Iwate)	T-64
Gymnocanthus	*tricuspis*			46	4M + 42A	50	50		(1.5 FIA)		Russia	L-87, H-40
Gymnocanthus	*tricuspis*			46	3M + 1SM + 42A	50	50				Russia	L-87
Icelus	*cataphractus*		M	40	12M + 12SM + 16 ST/A	64			1.5 FD		Japan (Iwate)	T-64
Leocottus	*kesslerii*	*Cottus*	M	48	6 M + 4SM + 38 ST/A	58				ACN=48	Russia (Lake Baikal)	S-167, S-168
Myoxocephalus	*brandti*		F, M	44	2M + 18ST + 24A	46	64			ACN=44	Russia (Amur Bay)	R-103
Myoxocephalus	*jaok*		F, M	24	16M + 4SM + 4A	44	44			ACN=42	Russia (Far East)	R-113
Myoxocephalus	*ochotensis*		F, M	42	2M + 20ST + 20A	44	64			ACN=42	Russia (Sea of Okhotsk)	R-114
Myoxocephalus	*scorpius*			36	8M + 2SM + 2ST + 24A	46	48		1.7 FIA		Russia	V-72, H-40
Myoxocephalus	*scorpius*			37	7M + 2SM + 2ST + 26A	46	48				Russia	V-72
Myoxocephalus	*scorpius*			38	6M + 2SM + 2ST + 28A	46	48				Russia	V-72
Myoxocephalus	*stelleri*			40	6 M/SM + 34 ST/A	46		4		ACN=42	Russia (Far East)	M-67
Ocynectes	*maschalis*		F, M	46	4M + 6SM + 36 ST/A	56				ACN=46	Japan (Kanagawa)	A-72
Oligocottus	*maculosus*		F	48	14M + 14SM + 6ST + 14A	76	82			ACN=48	USA (WA)	I-7
Oligocottus	*snyderi*			48*							USA	C-46
Paracottus	*knerii*		F, M	48	8M + 6SM + 34 ST/A	62				ACN=48	Russia (Lake Baikal)	S-167, S-168
Pseudoblennius	*cottoides*		F	46	4M + 8SM + 34 ST/A	58				ACN=46	Japan (Kanagawa)	A-72
Pseudoblennius	*marmoratus*	Type A		46	4M + 8SM + 34 ST/A	58				ACN=48	Japan (Chiba, Kanagawa)	A-72
Pseudoblennius	*marmoratus*	Type B	F	46	4M + 7SM + 35 ST/A	57				ACN=48	Japan (Chiba, Kanagawa)	A-72
Pseudoblennius	*marmoratus*	Type C	M	46	4M + 6SM + 36 ST/A	56				ACN=48	Japan (Chiba, Kanagawa)	A-72
Pseudoblennius	*percoides*			46	10SM + 36 ST/A	56			1.1 FCM	ACN=48	Korea (Busan)	K-125
Trachidermis	*fasciatus*			40	24 M/SM + 16 ST/A	64				ACN=48	Japan (Fukuoka)	A-4
Trachidermis	*fasciatus*			40		60					China (Shanghai)	C-43
Triglopsis	*quadricornis*	*Tryglopsis*		32	12 M/SM + 20ST	44	64		1.8 FIA		Russia, White Sea	L-87, L-88, H-41
Comephoridae												
Comephorus	*baicalensis*		F	48	8M + 40 ST/A	56				ACN=48	Russia (Lake Baikal)	S-168
Comephorus	*dybowski*			48	2M + 2SM + 44 ST/A	52					Russia (Lake Baikal)	S-168

Table 6.35 Order SCORPAENIFORMES (continued)

A		B	C	D	E	F	G	H	I	J	K	L
Current scientific name of taxon Suborder/family/subfamily/species		Reported in karyotype paper	Sex	2n	Karyotype	NF_1	NF_2	Ag–NORs	Genome size (pg/cell)	Comments	Locality	Reference
Abyssocottidae												
Abyssocottus	*gibbosus*			48	10M + 12SM + 26 ST/A	70					Russia (Lake Baikal)	S–168
Abyssocottus	*korotneffi*		F, M	48	8M + 16SM + 14ST + 10A	72	86			ACN=48	Russia (Lake Baikal)	S–167, S–168
Asprocottus	*herzensteini*			48	10M + 10SM + 28 ST/A	68					Russia (Lake Baikal)	S–168
Asprocottus	*platycephalus*		F, M	48	8M + 10SM + 30 ST/A	66				ACN=48	Russia (Lake Baikal)	S–167, S–168
Cottinella	*boulengeri*		F	48	6M + 12SM + 30 ST/A	66				ACN=48	Russia (Lake Baikal)	S–167, S–168
Cyphocottus	*megalops*	*Limnocottus*	F, M	48	14M + 2SM + 32 ST/A	64				ACN=50	Russia (Lake Baikal)	S–167, S–168
Limnocottus	*bergianus*			48	8M + 12SM + 28 ST/A	68					Russia (Lake Baikal)	S–168
Limnocottus	*griseus*			48	8M + 12SM + 28 ST/A	68					Russia (Lake Baikal)	S–168
Limnocottus	*pallidus*			48	8M + 14SM + 26 ST/A	70					Russia (Lake Baikal)	S–168
Procottus	*major*		F, M	48	10M + 16SM + 22 ST/A	74				ACN=48	Russia (Lake Baikal)	S–167, S–168
Hemitripteridae												
Hemitripterus	*villosus*			46	20M + 16SM + 10A	82	82			ACN=46	China (Liaoning)	M–36
Agonidae												
Agonus	*cataphractus*			48							Russia	V–72
Cyclopteridae												
Cyclopterus	*lumpus*			50	8SM + 12ST + 30A	58	70			ACN=50	Canada (Atlantic)	L–33
Cyclopterus	*lumpus*			50		54					Russia (White Sea)	L–88

Table 6.36 Order PERCIFORMES. Part 1 Percoidei and Elassomatoidei

A		B	C	D	E	F	G	H	I	J	K	L
Current scientific name of taxon Suborder/family/subfamily/species		Reported in karyotype paper	Sex	2n	Karyotype	NF_1	NF_2	Ag-NORs	Genome size (pg/cell)	Comments	Locality	Reference
Suborder Percoidei												
Ambassidae (= Chandidae)												
Chanda	*nama*			48	12ST + 36A	48	60			ACN=48	India	Z-20
Chanda	*nama* ?	*Ambassis nama*	F, M	40	2M + 12SM + 26A	54	54			ACN=46	India (Orissa)	T-53
Chanda	*nama*		F	48	2SM + 46A	50	50			ACN=48	India (WB)	K-34
Parambassis	*siamensis*	*Chanda*		40	4M + 6ST + 30A	44	50			ACN=48	Thailand	D-20
Parambassis	*wolffii*			48	48A	48	48				Thailand	D-28
Pseudambassis	*ranga*	*Chanda*	M	44	4M + 40A	48	48			ACN=46	India (Orissa)	K-41
Pseudambassis	*ranga*	*Chanda*		48	2ST + 46A	48	50			ACN=48	India	Z-20
Apogonidae												
Apogon	*binotatus*			36	14 M/SM + 22 ST/A	50				ACN=46	USA (FL)	R-109
Apogon	*binotatus*			36	26 M/SM + 10 ST/A	62				ACN=44	USA (WI)	R-109
Apogon	*binotatus*			35	14 M/SM + 21 ST/A	49				ACN=45	USA (FL)	R-109
Apogon	*doederleini*			46	2M + 6SM + 38 ST/A	54			2.4 FIA	ACN=46	Japan (Wakayama)	O-41, H-40
Apogon	*endekataenia*		M	46	46 ST/A	46			2.1 FIA		India (Andaman Is.)	R-45, H-41
Apogon	*endekataenia*			46	2M + 4SM + 16ST + 24A	52	68			ACN=46	Japan	M-111
Apogon	*imberbis*			36		56				ACN=46	Spain	A-32
Apogon	*lineatus*			46	2M + 4SM + 2ST + 38A	52	54			ACN=46	Japan	M-111
Apogon	*maculatus*			34	27 M/SM + 7 ST/A	61				ACN=44	Puerto Rico	R-110
Apogon	*moluccensis*		M	46	46 ST/A	46			2.1 FIA		India (Andaman Is.)	R-45, H-40
Apogon	*notatus*		F, M	46	2M + 4SM + 40 ST/A	52				ACN=46	Japan (Wakayama)	O-41
Apogon	*notatus*		F, M	46	2M + 5SM + 39 ST/A	53				ACN=46	Japan (Wakayama)	O-41
Apogon	*notatus*		F, M	46	2M * 4SM + 40 ST/A	52				ACN=46	Japan	O-41, M-111
Apogon	*pseudomaculatus*	*Apogon* (*Apogon*)		36	30 M/SM + 2ST + 4A	66	68			ACN=46	(Puerto Rico)	R-79, R-110
Apogon	*semilineatus*			46	2M + 4SM + 20ST + 20A	52	72			ACN=46	Japan (Suruga Bay)	M-101
Apogon	*semilineatus*			46	2M + 6SM + 38 ST/A	54				ACN=46	Japan (Wakayama)	O-41
Nectamia	*fusca*	*Apogon nubilus*		46	2M + 44 SM/ST/A				2.6 FIA	ACN=46	(Pacific)	R-79, H-41
Phaeoptyx	*pigmentaria*			38	6M + 32 SM/ST/A					ACN=46	(Atlantic)	R-79
Sphaeramia	*orbicularis*	*Apogon*		46	4SM + 42 ST/A	50				ACN=46	(Indo-West Pacific)	O-41

Table 6.36 Order PERCIFORMES. Part 1 Percoidei and Elassomatoidei (continued)

A		B	C	D	E	F	G	H	I	J	K	L
Current scientific name of taxon Suborder/family/subfamily/species		Reported in karyotype paper	Sex	2n	Karyotype	NF_1	NF_2	Ag-NORs	Genome size (pg/cell)	Comments	Locality	Reference
Bramidae												
Brama	*japonica*		F, M	54							N. Pacifuc	Y-3
Carangidae												
Alectis	*ciliaris*			48	48A	48	48			ACN=48	Japan (Izu Peninsula)	M-97
Alepes	*djedaba*	*Selar kalla*	F, M	56	56A	56	56			ACN=56	India (Orissa)	C-64
Alepes	*melanoptera*	*Selar malam*	F, M	48	2SM + 46A	50	50			ACN=48	India (Orissa)	C-64
Atropus	*atropos*	*atropus*	M	48	48A	48	48			ACN=48	India (Orissa)	D-3
Atule	*mate*	*Caranx*		50	14SM + 36A	64				ACN=50	USA (Hawaii)	L-8
Carangoides	*armatus*		M	48	2ST + 46A	48	50			ACN=48	India (Orissa)	D-3
Carangoides	*equula*	*Caranx*		48	2ST + 46A	48	50			ACN=48	Japan (Izu Peninsula)	M-97
Carangoides	*praeustus*	*proeustus*	M	48	10 M/SM + 38A	58	58			ACN=48	India (Orissa)	D-3
Caranx	*latus*			46	46A	46	46	2	1.2 FD		Brazil (RJ)	B-86
Caranx	*sansun*		F	48	2SM + 46A	50	50				India (Orissa)	P-16
Caranx	*sexfasciatus*			48	2ST + 46A	48	50			ACN=48	Japan (Izu Peninsula)	M-97
Chloroscombrus	*chrysurus*			48	48A	48	48		1.6 BFA		Brazil (RJ)	B-86, H-13
Megalaspis	*cordyla*		F	50	2ST + 48A	50	52			ACN=50	India (Orissa)	C-64
Scomberoides	*lysan*	*Chorinemus toloobarah*	M	48	4 M/SM + 44A	52	52			ACN=48	India (Orissa)	D-3
Selene	*setapinnis*			46	2SM + 44A	48	48				Brazil (RJ)	B-86
Selene	*vomer*			48	2ST + 46A	48	50	2		ACN=48	Brazil (SP)	R-108
Seriola	*dumerili*			48	2SM + 2ST + 44A	50	52	2	1.5 FIA	ACN=48	Italy (Sicily)	S-91, H-40
Seriola	*dumerili*			48	2SM + 46A	50	50			ACN=48	Italy (Sicily)	V-52
Seriola	*dumerili*			47	1M + 2SM + 44A	50	50			ACN=48	Italy (Sicily)	V-52
Seriola	*nigrofasciata*		F, M	48	48A	48	48		1.4 FD	ACN=48	India (Orissa)	T-53
Seriola	*quinqueradiata*			48	2SM + 2ST + 44A	50	52		1.7* FCM	ACN=48	Japan (Iwate)	I-1, O-48
Trachinotus	*carolinus*			48	8 M/SM + 40A	56	56	2		ACN=48	Brazil (SP)	R-108
Trachinotus	*falcatus*			48	10 M/SM + 38A	58	58	2	1.7 BFA	ACN=48	Brazil (SP)	R-108, H-13
Trachinotus	*goodei*			48	4 M/SM + 44A	52	52	2		ACN=48	Brazil (SP)	R-108
Trachinotus	*ovatus*		M	48	2M + 4SM + 42 ST/A	54				ACN=48	India (Orissa)	C-64
Trachurus	*japonicus*			48	4M + 14SM + 12ST + 18A	66	78			ACN=48	Japan (Izu Peninsula)	M-97
Trachurus	*mediterraneus ponticus*			48	4M + 6SM + 38 ST/A	58				ACN=48	Black Sea	V-5
Trachurus	*mediterraneus*			48	4M + 4SM + 14ST + 26A	56	70	2		ACN=48	Italy (Ancona)	C-14
Trachurus	*trachurus*		F, M	48	2SM + 46A	50	50	2		ACN=48	Italy (Ancona)	C-14

Table 6.36 Order PERCIFORMES. Part 1 Percoidei and Elassomatoidei (continued)

A		B	C	D	E	F	G	H	I	J	K	L
Current scientific name of taxon Suborder/family/subfamily/species		Reported in karyotype paper	Sex	2n	Karyotype	NF_1	NF_2	Ag-NORs	Genome size (pg/cell)	Comments	Locality	Reference
Centracanthidae												
Spicara	*maena*	*smaris*		46	8M + 4SM + 2ST + 32A	58	60			ACN=48	Black Sea	V-4
Spicara	*maena*	*smaris*		45	9M + 4SM + 2ST + 30A	58	60			ACN=48	Black Sea	V-4
Spicara	*maena*	*smaris*		44	10M + 4SM + 2ST + 28A	58	60			ACN=48	Black Sea	V-4
Spicara	*maena*	*flexuosa*		46	8M + 4SM + 2ST + 32A	58	60			ACN=48	Black Sea	S-7, V-11
Spicara	*maena*	*flexuosa*		45	9M + 4SM + 2ST + 30A	58	60			ACN=48	Black Sea	S-7, V-11
Spicara	*maena*	*flexuosa*		44	10M + 4SM + 2ST + 28A	58	60			ACN=48	Black Sea	S-7, V-11
Spicara	*maena*			48	4M + 2SM + 10ST + 32A	54	64			ACN=48	Black Sea	S-7, V-4, V-11
Centrarchidae												
Acantharchus	*pomotis*		F, M	48	48A	48	48			ACN=48	USA (NC)	R-80
Ambloplites	*rupestris*		F, M	48	48A	48	48		2.3 FIA	ACN=48	USA (NC, WV)	R-80, H-40
Archoplites	*interruptus*		F	48	2ST + 46A	48	50			ACN=48	USA (CA)	B-83
Centrarchus	*macropterus*		F	48	48A	48	48			ACN=48	USA (NC)	R-80
Enneacanthus	*chaetodon*			48	48A	48	48			ACN=48	USA (NC)	R-80
Enneacanthus	*gloriosus*			48	48A	48	48			ACN=48	USA (NC)	R-80
Enneacanthus	*obesus*			48	48A	48	48			ACN=48	USA (NC)	R-80
Lepomis	*auritus*			48	48A	48	48		2.1 FD	ACN=48	USA (NC)	R-80
Lepomis	*cyanellus*			48	48A	48	48		(2.0 FD)	ACN=48	USA (NC)	O-6, R-80
Lepomis	*cyanellus*		F, M	46	2SM + 44A	48	48			ACN=48	USA (WV)	R-80
Lepomis	*cyanellus*		M	48	48A	48	48			ACN=48	USA (CA)	B-15
Lepomis	*cyanellus*		M	47	1M + 46A	48	48			ACN=48	USA (CA)	B-15
Lepomis	*cyanellus*		M	46	2SM + 44A	48	48			ACN=48	USA (CA)	B-15
Lepomis	*gibbosus*		F, M	48	48A	48	48		(1.9 FIA)	ACN=48	USA (NC, WV)	R-80, B-63, H-40
Lepomis	*gibbosus*			46	10M + 36 ST/A	56				ACN=48	(USA)	F-30
Lepomis	*gulosus*	*Chaenobryttus*		48	48A	48	48			ACN=48	USA (NC)	R-80
Lepomis	*humilis*			48					2.0 FD		USA	A-38
Lepomis	*humilis*			46	2SM + 44A	48	48			ACN=48	USA (KY)	R-80
Lepomis	*macrochirus*		F, M	48	48A	48	48		1.9* FCM, 1.9 FD	ACN=48	USA (WV)	R-80, B-63, O-48
Lepomis	*marginatus*			48	48A	48	48		2.2 FD	ACN=48	USA (NC)	R-80
Lepomis	*megalotis*			48	48A	48	48		2.1* FCM, 2.1 FD	ACN=48	USA (TN)	R-80, O-48
Lepomis	*microlophus*		M	48	48A	48	48		2.0 FD	ACN=48	USA (NC, VA)	R-80
Lepomis	*symmetricus*		F, M	40	8 M + 32 ST/A	48				ACN=48	USA (MS)	T-43
Lepomis	*symmetricus*		F, M	41	7M + 34 ST/A	48				ACN=48	USA (MS)	T-43

Table 6.36 Order PERCIFORMES. Part 1 Percoidei and Elassomatoidei (continued)

A		B	C	D	E	F	G	H	I	J	K	L
Current scientific name of taxon Suborder/family/subfamily/species		Reported in karyotype paper	Sex	2n	Karyotype	NF_1	NF_2	Ag-NORs	Genome size (pg/cell)	Comments	Locality	Reference
Lepomis	*symmetricus*		F, M	39	9M + 30 ST/A	48				ACN=48	USA (MS)	T-43
Lepomis	*symmetricus*		M	38	10M + 28 ST/A	48				ACN=48	USA (MS)	T-43
Micropterus	*dolomieui*		M	46	2SM + 44A	48	48		2.0 FIA	ACN=48	USA (WV)	R-80, B-63, H-40
Micropterus	*punctulatus*		F, M	46	2 M/SM + 44 ST/A	48				sex chrom.	USA (TX)	T-40
Micropterus	*salmoides*		F, M	46	2 M/SM + 44A	48	48		2.0* FCM	ACN=48	USA (WV)	T-40, T-73, R-80, O-48
Micropterus	*treculi*		F, M	46	2 M/SM + 44 ST/A	48				sex chrom.	USA (TX)	T-40
Pomoxis	*annularis*			48	48A	48	48		2.1 FCM	ACN=48	USA (NC)	R-80
Pomoxis	*nigromaculatus*		F	48	48A	48	48			ACN=48	USA (NC, WV)	R-80
Centropomidae												
Centropomus	*parallelus*			48	48A	48	48	2			Brazil (RJ)	B-86
Chaetodontidae												
Chaetodon	*auriga*			48	48A	48	48		1.4 FIA	ACN=48	Japan (Okinawa)	A-61, H-40
Chaetodon	*auripes*	*collaris*		48	48A	48	48			ACN=48	Japan (Chiba, Kagoshima)	A-61
Chaetodon	*auripes*			48	48A	48	48		1.5* FCM		Japan	O-48
Chaetodon	*collare*			48	48A	48	48	2		ACN=48	India (Kerala)	N-65
Chaetodon	*lunula*			48	48A	48	48			ACN=48	Japan (Yakushima)	A-61
Chaetodon	*plebeius*	*Megaprotodon*		48	2M + 46A	50	50			ACN=48	Japan (Tanegashima)	A-61
Chaetodon	*sedentarius*			48		48					Brazil (ES)	G-12
Chaetodon	*striatus*		F, M	48	48A	48	48	2		ACN=48	Brazil (RN, BA)	A-7
Chaetodon	*strigangulus*	*Megaprotodon*		48	2SM + 46A	50	50			ACN=48	Japan (Okinawa)	A-61
Chaetodon	*trifasciatus*			48	48A	48	48			ACN=48	Japan (Okinawa)	A-61
Chaetodon	*vagabundus*			48	48A	48	48		1.7 FIA	ACN=48	Japan (Okinawa)	A-61, H-40
Heniochus	*acuminatus*			48	48A	48	48			ACN=48	Japan (Wakayama)	A-77
Echeneidae												
Remora	*remora*		M	42	42 ST/A	42					India (Andaman Is.)	R-45
Gerreidae												
Diapterus	*olithostomus*		F, M	48	48A	48	48	2		ACN=48	Brazil (RN)	M-121
Eucinostomus	*argenteus*		F, M	48	48A	48	48	2		ACN=48	USA (FL), Puerto Rico	R-96
Eucinostomus	*gula*		F, M	48	48A	48	48	2	1.6 BFA	ACN=48	USA (FL)	R-96, H-13
Eucinostomus	*harengulus*		F, M	48	48A	48	48	2		ACN=48	USA (FL)	R-96

Table 6.36 Order PERCIFORMES. Part 1 Percoidei and Elassomatoidei (continued)

A		B	C	D	E	F	G	H	I	J	K	L
Current scientific name of taxon Suborder/family/subfamily/species		Reported in karyotype paper	Sex	2n	Karyotype	NF_1	NF_2	Ag-NORs	Genome size (pg/cell)	Comments	Locality	Reference
Eugerres	*plumieri*		F, M	48	48A	48	48	2		ACN=48	USA (FL)	R-96
Gerres	*filamentosus*	*Gerreis*		48	4M + 44A	52	52				India (WB)	K-46
Gerres	*oblongus*		M	50*					1.4* FCM		India (Andaman Is.)	R-45, O-48
Gerres	*oyena*			48	48 ST/A	48				ACN=48	Japan (Wakayama)	A-77
Haemulidae (= Pomadasyidae)												
Anisotremus	*moricandi*			48	48A	48	48	2		ACN=48	Brazil (RN)	G-12, N-59
Anisotremus	*surinamensis*			48	48A	48	48	2		ACN=48	Brazil (RN)	G-12, N-59
Anisotremus	*virginicus*			48	48A	48	48	2		ACN=48	Brazil (RN, RJ)	G-12, N-59
Conodon	*nobilis*			48	48A	48	48			ACN=48	Brazil (RN)	G-12
Haemulon	*aurolineatum*			48	48A	48	48		(1.5 FD)	ACN=48	Brazil (RN, RJ)	G-12, N-59
Haemulon	*aurolineatum*		F, M	48	48A	48	48	2		ACN=48	Venezuela	N-59
Haemulon	*bonariense*		F, M	48	48A	48	48	2		ACN=48	Venezuela	N-59
Haemulon	*flavolineatum*			48					(1.8 BFA)		Brazil (RN)	G-12, H-13
Haemulon	*flavolineatum*		F, M	48	48A	48	48	2	(1.3 FD)	ACN=48	Venezuela	R-106
Haemulon	*parra*			48	48A	48	48	2		ACN=48	Brazil (RN)	G-12, N-59
Haemulon	*plumierii*			48	48A	48	48	2	(2.0 BFA)	ACN=48	Brazil (RN)	G-12, N-59, H-13
Haemulon	*plumierii*		F, M	48	48A	48	48	2		ACN=48	Venezuela	N-59
Haemulon	*sciurus*			48	48A	48	48		(1.7 BFA)	ACN=48	W. Atlantic	R-41, H-13
Haemulon	*sciurus*			46	2 SM/ST + 44A		48		(1.2 FD)		W. Atlantic	R-41
Haemulon	*striatum*			48	48A	48	48	2		ACN=48	Brazil (RN)	G-12, N-59
Hapalogenys	*analis*	*mucronatus*		48	2M + 8SM + 14ST + 24A	58	72			ACN=48	China	Y-20
Hapalogenys	*nigripinnis*	*nitens*		48	2M + 8SM + 2ST + 36A	58	60			ACN=48	China	Y-20
Orthopristis	*ruber*		F, M	48	2SM + 46 ST/A	50		2		ACN=48	Brazil (RJ)	B-65
Pomadasys	*argenteus*	*hasta*		48	48A	48	48			ACN=48	India (WB)	K-139
Pomadasys	*corvinaeformis*			48	48A	48	48	2		ACN=48	Brazil (RN)	G-12, N-59
Pomadasys	*commersonnii*	*Pristipoma operculare*		48	48A	48	48			ACN=48	India (Bombay)	R-63
Kuhliidae												
Kuhlia	*boninensis*			48	2SM + 46A	50	50			ACN=48	Japan (Okinawa)	A-77
Kuhlia	*mugil*	*taeniura*		48	2SM + 46A	50	50			ACN=48	Japan (Chiba)	A-77

Table 6.36 Order PERCIFORMES. Part 1 Percoidei and Elassomatoidei (continued)

A		B	C	D	E	F	G	H	I	J	K	L
Current scientific name of taxon Suborder/family/subfamily/species		Reported in karyotype paper	Sex	2n	Karyotype	NF_1	NF_2	Ag-NORs	Genome size (pg/cell)	Comments	Locality	Reference
Kyphosidae												
Girellinae												
Girella	*melanichthys*			48	48A	48	48	4		ACN=48	Japan (Yamaguchi)	N-29
Girella	*melanichthys*		F	48	48A	48	48	4		ACN=48	Japan (Wakayama)	U-38
Girella	*punctata*			48	48A	48	48	2		ACN=48	Japan (Wakayama)	U-83
Kyphosinae												
Kyphosus	sp.	*bigibbus*		48	2SM + 2ST + 44A	50	52		1.8* FCM		Japan	O-48
Kyphosus	*cinerascens*			48	2SM + 46A	50	50	2		ACN=48	Japan (Noto Peninsula)	T-21
Kyphosus	*vaigiensis*	*lembus*		48	2SM + 46A	50	50	2		ACN=48	Japan (Noto Peninsula)	T-21
Kyphosus	*bigibbus*	sp.		48	2SM + 46A	50	50	2		ACN=48	Japan (Noto Peninsula)	T-21
Microcanthinae												
Microcanthus	*strigatus*			48	2SM + 46A	50	50			ACN=48	Japan (Chiba)	A-61
Scorpidinae												
Labracoglossa	*argentiventris*		F, M	48	48A	48	48		1.8* FCM	ACN=48	Japan (Shizuoka)	M-103, O-48
Latidae												
Lates	*calcarifer*	*calcerifer*	M	48	2M + 6SM + 2ST + 38A	56	58		1.4 FIA	ACN=48	India (WB)	K-31, H-40
Lates	*calcarifer*	Centropomidae		48						ACN=48	India (Portonovo)	N-13
Psammoperca	*waigiensis*			48	48A	48	48	2		ACN=48	Thailand (Gulf of Siam)	U-39
Leiognathidae												
Gazza	*minuta*		F	48	48A	48	48			ACN=48	India (Orissa)	P-16
Leiognathus	*nuchalis*			48	48 ST/A	48				ACN=48	Japan (Suruga Bay)	A-77
Photopectoralis	*bindus*	*Leiognathus*	M	40	40 ST/A	40			1.3 FIA		India (Goa)	R-45, H-40
Lethrinidae												
Lethrinus	*nebulosus*			48	48A	48	48			ACN=48	China	Z-39
Lethrinus	*xanthochilus*			48	48A	48	48		2.8* FCM		Japan	O-48
Lobotidae												
Lobotus	*surinamensis*		F, M	48	2SM + 46A	50	50			ACN=48	India (Orissa)	T-53

Table 6.36 Order PERCIFORMES. Part 1 Percoidei and Elassomatoidei (continued)

A		B	C	D	E	F	G	H	I	J	K	L
Current scientific name of taxon Suborder/family/subfamily/species		Reported in karyotype paper	Sex	2n	Karyotype	NF_1	NF_2	Ag-NORs	Genome size (pg/cell)	Comments	Locality	Reference
Lutjanidae												
Lutjanus	*alexandrei*		F, M	48	48A	48	48	2		ACN=48	Brazil (RN)	R-111
Lutjanus	*analis*		M	48	48A	48	48	2		ACN=48	Brazil (RN)	R-111
Lutjanus	*analis*			48	48A	48	48	2		ACN=48	Venezuela	N-70
Lutjanus	*argentimaculatus*			48	48A	48	48			ACN=48	China	C-12
Lutjanus	*argentimaculatus*		F, M	48	48A	48	48			ACN=48	India (Orissa)	P-16, K-46
Lutjanus	*bohar*			48	48A	48	48		2.4 FIA	ACN=48	China	C-12, H-40
Lutjanus	*cyanopterus*		M	48	48A	48	48	2		ACN=48	Brazil (RN)	R-111
Lutjanus	*erythropterus*			48	48A	48	48			ACN=48	China	C-12, Y-26
Lutjanus	*griseus*		F, M	48	48A	48	48	2-4		ACN=48	Venezuela	N-70
Lutjanus	*jocu*		F, M	48	48A	48	48	4		ACN=48	Brazil (RN)	R-111
Lutjanus	*johnii*			48	48A	48	48			ACN=48	China (Zhanjiang)	L-81
Lutjanus	*kasmira*		M	48	48A	48	48			ACN=48	India (Orissa)	C-61
Lutjanus	*kasmira*		F, M	48	48A	48	48	2	2.9* FCM	ACN=48	Japan (Okinawa)	U-79
Lutjanus	*quinquelineatus*		F	48	48A	48	48	2	2.9* FCM	$X_1X_1X_2X_2$	Japan (Okinawa)	U-79
Lutjanus	*quinquelineatus*		M	47	1M + 47A	48	48	2	2.9* FCM	X_1X_2Y, ACN=48	Japan (Wakayama, Okinawa)	U-79
Lutjanus	*russelli*			48	48A	48	48	2	2.3 FIA	ACN=48	Thailand (Gulf of Siam)	U-38, H-40
Lutjanus	*sanguineus*		M	48	48A	48	48			ACN=48	India (Goa)	R-45
Lutjanus	*sebae*			48	48A	48	48		1.6 FIA	ACN=48	China (Hainan)	Y-26, H-41
Lutjanus	*synagris*		F, M	48	48A	48	48	2		ACN=48	Brazil (RN)	R-111
Lutjanus	*synagris*	cytotype I	F, M	48	48A	48	48	2		ACN=48	Venezuela	N-70
Lutjanus	*synagris*	cytotype II		47	1M + 46A	48	48	2		ACN=48	Venezuela	N-70
Lutjanus	*vitta*			48	48A	48	48		2.0 FIA	ACN=48	China (Zhanjiang)	L-81, H-41
Ocyurus	*chrysurus*			48	48A	48	48		2.2 FD, 2.6 BFA		Brazil (RN)	G-12, H-13
Ocyurus	*chrysurus*			48	48A	48	48	2		ACN=48	Brazil (RN), Venezuela	R-111, N-76
Rhomboplites	*aurorubens*		F, M	48	2ST + 46A	48	50	2		ACN=48	Venezuela	N-76
Monodactylidae												
Monodactylus	*argenteus*		F, M	48	48A	48	48		(1.8 BFA)	ACN=48	(Asia)	S-129, H-13
Monodactylus	*argenteus*			48							India (Portonovo)	N-13
Monodactylus	*sebae*		M	47	1M + 46A	48	48			XO, ACN=48	W. Africa	S-129
Monodactylus	*sebae*		F	48	48A	48	48			XX, ACN=48	W. Africa	S-129

Table 6.36 Order PERCIFORMES. Part 1 Percoidei and Elassomatoidei (continued)

A		B	C	D	E	F	G	H	I	J	K	L
Current scientific name of taxon Suborder/family/subfamily/species		Reported in karyotype paper	Sex	2n	Karyotype	NF_1	NF_2	Ag-NORs	Genome size (pg/cell)	Comments	Locality	Reference
Moronidae												
Dicentrarchus	*labrax*	Serranidae	F, M	48	48A	48	48	2		XX/XY	Spain (Malaga)	C-11
Dicentrarchus	*labrax*		F, M	48	2ST + 46A	48	50	2		ACN=48	Italy (Tyrrhenian Sea)	V-63, C-31
Dicentrarchus	*labrax*			48	48 ST/A	48		1-2	(1.6 FCM)	ACN=48	Italy	S-90
Dicentrarchus	*labrax*	Serranidae		48	2SM + 46A	50	50	2			USSR (Black Sea)	A-90
Dicentrarchus	*punctatus*		F, M	48	48 ST/A	48		2		ACN=48	Egypt	S-90
Dicentrarchus	*punctatus*		F, M	48	2ST + 46A	48	50	2		ACN=48	Italy (Palermo)	V-63
Lateolabrax	*japonicus*	Serranidae	F, M	48	48A	48	48	2		ACN=48	Japan (Shimane)	K-70, K-82, T-27
Lateolabrax	*japonicus*	Serranidae		48	48A	48	48			ACN=48	China (Shandong)	W-7, Y-20, Y-21
Lateolabrax	*japonicus*	sea bass	F, M	48	48A	48	48		1.7 FCM	ACN=48	Korea (Chungchongnam-do)	P-12
Lateolabrax	*latus*	Serranidae	F, M	48	2ST + 46A	48	50	2		ACN=48	Japan (Shimane)	K-70, K-82, T-27
Lateolabrax	sp.	spotted sea bass	F, M	48	48A	48	48		1.7 FCM	ACN=48	Korea (Chungchongnam-do)	P-12
Morone	*americana*	Serranidae		48	48A	48	48			ACN=48	E. USA	B-64
Morone	*saxatilis*			48	2SM + 8ST + 38A	50	58		(1.9 FIA, 1.8 BFA)	ACN=48	USA (NY)	R-27, H-13, H-41
Morone	*saxatilis*			48	2SM + 6ST + 40A	50	56			ACN=48	USA (NY)	R-27
Mullidae												
Mulloidichthys	*flavolineatus*			48	48A	48	48		1.2* FCM		Japan	O-48
Mullus	*argentinae*			44	2SM + 42A	46	46				Brazil	B-86
Mullus	*barbatus*			44	4 M/SM + 40A	48	48			ACN=48	Monaco	L-5
Mullus	*barbatus*			44	6 M/SM + 38A	50	50			ACN=48	Monaco	L-5
Mullus	*barbatus*		F, M	44	6 M/SM + 16ST + 22A	50	66			ACN=48	Italy (Palermo)	V-66
Mullus	*surmuletus*		F, M	48		50			1.3 FD		Spain (Malaga)	C-9
Mullus	*surmuletus*		F, M	44	8 M/SM + 16ST + 20A	52	68			ACN=48	Italy (Palermo)	V-66
Parupeneus	*spilurus*			44	8M + 8SM + 28 ST/A	60		2		ACN=48	Japan (Chiba)	A-74
Upeneus	*parvus*		F, M	44	8 M/SM + 36 ST/A	52				0-4 B, ACN=48	Brazil (RJ)	P-83
Upeneus	*tragula*		M	50	50 ST/A	50					India (Andaman Is.)	R-45
Nandidae												
Badinae												
Badis	*badis*			48	26 M/SM + 22A	74	74				India (Assam)	K-46
Badis	*badis*		F, M	46	10M + 4SM + 32A	60	60			ACN=48	India (Orissa)	T-53
Badis	*badis*		F, M	46	6M + 40A	52	52			ACN=48	India (Jammu)	S-54

Table 6.36 Order PERCIFORMES. Part 1 Percoidei and Elassomatoidei (continued)

A		B	C	D	E	F	G	H	I	J	K	L
Current scientific name of taxon Suborder/family/subfamily/species		Reported in karyotype paper	Sex	2n	Karyotype	NF_1	NF_2	Ag-NORs	Genome size (pg/cell)	Comments	Locality	Reference
Nandinae												
Nandus	*nandus*		F	48	6M + 28SM + 14A	82	82			ACN=48	India (WB)	N-16
Nandus	*nandus*		F, M	48	4M + 36SM + 6ST + 2A	88	94			ACN=48	India (WB)	M-26
Nandus	*nandus*		F, M	48	4M + 30SM + 6ST + 8A	82	88	4		ACN=48	India (WB)	K-136
Nandus	*nandus*		M	48	4M + 18SM + 14ST + 12A	70	84			ACN=48	India (Jammu)	S-54
Nandus	*oxyrhynchus*			48	14M + 8SM + 6ST + 20A	70	76			ACN=48	Thailand	D-28
Pristolepidinae												
Pristolepis	*marginata*			48	48A	48	48	2		ACN=48	India (Kerala)	N-58
Oplegnathidae												
Oplegnathus	*fasciatus*			48	2SM + 46A	50	50		1.8* FCM	ACN=48	Japan	N-29, M-111, O-48
Oplegnathus	*fasciatus*		F, M	48	2 M/SM + 46A	50	50	2	1.3 FD	ACN=48	Korea	K-53
Oplegnathus	*fasciatus*			48	2M + 2ST + 44A	50	52			ACN=48	(China)	Z-39
Oplegnathus	*punctatus*			48	2SM + 46A	50	50			ACN=48	Japan	M-111
Pempheridae												
Pempheris	*schwenkii*	*xanthoptera*		48	2SM + 46 ST/A	50			1.4* FCM	ACN=48	Japan (Tanegashima)	A-74, O-48
Percichthyidae												
Coreoperca	*herzi*			48	2SM + 46 ST/A	50				ACN=48	Korea (Chungchongnam-do)	U-28
Coreoperca	*herzi*			48							Korea	L-15
Coreoperca	*kawamebari*			48	4SM + 44 ST/A	52			1.8* FCM	ACN=48	Japan (Hyogo)	U-28, O-48
Siniperca	*chuatsi*		F, M	48	24 SM/ST + 24A		72		(1.7* FD)	ACN=48	China	Y-10, C-83
Siniperca	*chuatsi*			48	22ST + 26A	48	70			ACN=48	China (Hubei)	L-53
Siniperca	*kneri*			48	6SM + 14ST + 28A	54	68			ACN=48	China	Y-15
Siniperca	*obscura*			48	4SM + 14ST + 30A	52	66			ACN=48	China	Y-15
Siniperca	*roulei*	*Coreosiniperca*	M	48	2SM + 10ST + 36A	50	60			ACN=48	China	Y-15
Siniperca	*scherzeri*			48	6SM + 14ST + 28A	54	68			ACN=48	China	Y-15
Siniperca	*scherzeri*	black variant	F, M	48	4SM + 44 ST/A	52			1.4 FD	ACN=48	Korea (Han R.)	P-11
Siniperca	*scherzeri*	gold variant	F, M	48	4SM + 44 ST/A	52			1.4 FD	ACN=48	Korea (Han R.)	P-11
Siniperca	*undulata*		F, M	48	2SM + 16ST + 30A	50	66			ACN=48	China	Y-15

Table 6.36 Order PERCIFORMES. Part 1 Percoidei and Elassomatoidei (continued)

A		B	C	D	E	F	G	H	I	J	K	L
Current scientific name of taxon Suborder/family/subfamily/species		Reported in karyotype paper	Sex	2n	Karyotype	NF_1	NF_2	Ag-NORs	Genome size (pg/cell)	Comments	Locality	Reference
Percidae												
Ammocrypta	*vivax*			48	12M + 32SM + 4 ST/A	92				ACN=48	USA (TX)	G-30
Etheostoma	*blennioides blennioides*			48	2M + 46A	50				ACN=48	USA (OH)	R-91
Etheostoma	*blennioides pholidotum*			48	2M + 1SM + 45A	51				ACN=48	USA (OH)	R-91
Etheostoma	*caeruleum*			48	8 M/SM + 40 ST/A	56				ACN=48	Canada (Ontario)	D-1
Etheostoma	*caeruleum*	*camurum caeruleum*		48	4M + 2SM + 42A	54	54			ACN=48	USA (OH)	R-91
Etheostoma	*camurum*			48	2M + 2SM + 44A	52	52			ACN=48	USA (OH)	R-91
Etheostoma	*ditrema*			48		52					N. America	G-32
Etheostoma	*exile*			48	4M/SM + 44 ST/A	52				ACN=48	Canada (Ontario)	D-1
Etheostoma	*flabellare*			48	4M/SM + 44 ST/A	52				ACN=48	Canada (Ontario)	D-1
Etheostoma	*flabellare flabellare*			48	4M + 44A	52	52			ACN=48	USA (OH)	R-91
Etheostoma	*nigrum*			48	6 M/SM + 42 ST/A	54				ACN=48	Canada (Ontario)	D-1
Etheostoma	*nigrum nigrum*			48	2M + 2SM + 44A	52	52			ACN=48	USA (OH)	R-91
Etheostoma	*swaini*			48		52					N. America	G-32
Gymnocephalus	*baloni*		M	48	10M + 32SM + 6ST	90	96			ACN=48	Danube R., Europe	R-11
Gymnocephalus	*cernuus*	*Acerina cernua*	M	48							Sweden	N-40
Gymnocephalus	*cernuus*	*cernua*	F, M	48	4M + 22SM + 16ST + 6A	74	90	2		ACN=48	Germany (Baltic Sea)	K-77
Gymnocephalus	*cernuus*	*Acerina cernua*		48	4M + 22SM + 16ST + 6A	74	90			ACN=48	Hungary	B-84
Gymnocephalus	*cernuus*		F, M	48	2M + 32SM + 8ST + 6A	82	90			ACN=48	Danube R., Europe	R-11
Gymnocephalus	*schraester*		F, M	48	10M + 32SM + 6ST	90	96			ACN=48	Danube R., Europe	R-11
Perca	*flavescens*			48	48 ST/A	48			1.9 FIA, 2.4 BFA	ACN=48	Canada (Ontario)	D-1, H-13, H-40
Perca	*fluviatilis*		F, M	48	2M + 24SM + 12ST + 10A	74	86	2		ACN=48	Germany (Baltic Sea)	K-78
Perca	*fluviatilis*		F, M	48	2M + 28SM + 10ST + 8A	78	88		(2.4 FCM)	ACN=48	Elbe R., Danube R	R-11, V-86
Perca	*fluviatilis*			48	2M + 32SM + 6ST + 8A	82	88			ACN=48	Hungary	B-84
Perca	*fluviatilis*			48	16SM + 20ST + 12A	64	84			ACN=48	Macedonia	K-8
Percarina	*demidoffi*			48	14M + 14SM + 20 ST/A	76				ACN=48	Danube Delta, Europe	S-119
Percina	*caprodes*		F, M	48	8M + 36SM + 4ST	92			(2.1 FIA)	ACN=48	USA (TX)	G-30, G-32, H-40
Percina	*caprodes*			48	1M + 47 ST/A	49				ACN=48	Canada (Ontario)	D-1
Percina	*caprodes*			46	1M + 2SM + 43 ST/A	49				ACN=48	Canada (Ontario)	D-1
Percina	*maculata*			48	5 M/SM + 43 ST/A	53				ACN=48	Canada (Ontario)	D-1
Percina	*nigrofasciata*			48	8M + 38SM + 2ST	94	96			ACN=48	USA (LA)	G-30
Percina	*sciera*		F, M	48	10M + 22SM + 16 ST/A	80				ACN=48	USA (TX)	G-30
Sander	*lucioperca*	*Stizostedion*	M	48	2M + 30SM + 10ST + 6A	80	90	2	(2.3 FCM)	ACN=48	Baltic Sea	G-46, R-11, V-86
Sander	*lucioperca*	*Lucioperca*		48	14M + 14SM + 14ST + 6A	76	90			ACN=48	Hungary	B-84

Table 6.36 Order PERCIFORMES. Part 1 Percoidei and Elassomatoidei (continued)

A		B	C	D	E	F	G	H	I	J	K	L
Current scientific name of taxon Suborder/family/subfamily/species		Reported in karyotype paper	Sex	2n	Karyotype	NF_1	NF_2	Ag-NORs	Genome size (pg/cell)	Comments	Locality	Reference
Sander	*lucioperca*	*Lucioperca*	M	48							Sweden	N-40
Sander	*vitreum vitreum*	*Stizostedion*		48	48 ST/A	48			2.1 FIA	ACN=48	Canada (Ontario)	D-1, H-40
Sander	*volgense*	*Stizostedion*										
Zingel	*streber*		M	48	2M + 30SM + 10ST + 6A	80	90			ACN=48	Danube R., Europe	R-11
Zingel	*zingel*		F, M	48	6M + 20SM + 22 ST/A	74				ACN=48	Danube R., Europe	R-11
Zingel	*zingel*			48	4M + 20SM + 24 ST/A	72				ACN=48	Danube R., Europe	R-11
Zingel	*zingel*		F	48	4M + 16SM + 28 ST/A	68				$X_1X_1X_2X_2$, ACN=48	Czech (Morava R.)	K-122
			M	47	5M + 16SM + 26 ST/A	68				X_1X_2Y, ACN=48	Czech (Morava R.)	K-122
Plesiopidae												
Plesiops	*coeruleolineatus*			48	2M + 46 ST/A	50				ACN=48	Japan (Ogasawara, Okinawa)	A-66
Polycentridae												
Polycentrus	*schomburgkii*			46*							(W. Atlantic)	S-30
Polynemidae												
Eleutheronema	*tetradactylum*		M	48	48A	48	48			ACN=48	India (WB)	K-42
Pomacanthidae												
Centropyge	*aurantonotus*			48	4M + 14SM + 26ST + 4A	66	92	2		ACN=48	Brazil (ES, RJ)	A-9
Centropyge	*bicolor*		F, M	52	2M + 50 ST/A	54		2	1.4 FIA		W. Pacific	T-67, H-41
Centropyge	*bispinosa*		F	48	48 ST/A	48		2		ACN=48	W. Pacific	T-67
Centropyge	*ferrugata*		F	48	48A	48		2		ACN=48	W. Pacific	T-67
Centropyge	*ferrugata*	*ferrugatus*		48	48A	48		2		ACN=48	Philippines Sea	A-9
Centropyge	*heraldi*		M	52	8M + 22SM + 22 ST/A	82		2			W. Pacific	T-67
Centropyge	*loricula*			48	48A	48		2		ACN=48	W. Pacific	T-67
Centropyge	*nox*		F	52	12M + 18SM + 22 ST/A	82		2			W. Pacific	T-67
Centropyge	*tibicen*		F	52	52A	52	52	2			W. Pacific	T-67
Centropyge	*vrolickii*			48	48A	48	48			ACN=48	Japan (Yakushima Is.)	A-61
Holacanthus	*ciliaris*			48	48A	48	48	2		ACN=48	Brazil (CE, PE, BA, ES)	A-7
Holacanthus	*tricolor*			48	48A	48	48	2		ACN=48	Brazil (BA, ES, RJ)	A-7
Pomacanthus	*arcuatus*			48	2ST + 46A	48	50	2		ACN=48	Brazil (CE, AL, BA, RJ)	A-8
Pomacanthus	*paru*			48	2ST + 46A	48	50	2		ACN=48	Brazil (CE, RN, AL, RJ, SP)	A-8
Pomacanthus	*semicirculatus*			48	48A	48	48			ACN=48	Japan (Yakushima Is.)	A-61

Table 6.36 Order PERCIFORMES. Part 1 Percoidei and Elassomatoidei (continued)

A		B	C	D	E	F	G	H	I	J	K	L
Current scientific name of taxon Suborder/family/subfamily/species		Reported in karyotype paper	Sex	2n	Karyotype	NF_1	NF_2	Ag-NORs	Genome size (pg/cell)	Comments	Locality	Reference
Pomatomidae												
Pomatomus	*saltatrix*			48	48A	48			1.6 FIA, 1.9 BFA		Brazil	B-86, H-13, H-40
Priacanthidae												
Priacanthus	*arenatus*		F, M	52	4ST + 48A	52	56	2	2.2 BFA	ACN=48	Brazil (RJ)	G-12, M-121, H-13
Priacanthus	*arenatus*			50	50A	50					Brazil (RJ)	B-86
Sciaenidae												
Aplodinotus	*grunniens*			48	48A	48	48			ACN=48	USA (Mississippi R.)	L-23
Argyrosomus	*amoyensis*	*Nibea miichthioides*		48	48A	48	48			ACN=48	China	W-3
Bairdiella	*chrysoura*			48	48A	48	48		1.5 BFA	ACN=48	Gulf of Mexico	L-23, H-13
Boesemania	*microlepis*			48	48A	48	48			ACN=48	Thailand	D-28
Corvina	*nigra*			46	2M + 44A	48	48			ACN=48	Croatia	S-195
Cynoscion	*acoupa*			48	48A	48	48				Brazil (RJ)	B-86
Cynoscion	*arenarius*			48	2 M/SM + 46 ST/A	50				ACN=48	Gulf of Mexico	F-60
Cynoscion	*nebulosus*			48	2 M/SM + 46 ST/A	50				ACN=48	Gulf of Mexico	F-60
Johnius	*belangerii*			48							India	L-1
Johnius	*belangerii*	*belengeri*		48	48A	48	48			ACN=48	China (Shandong)	W-7
Johnius	*borneensis*	*vogleri*	F, M	48	48A	48	48			ACN=48	India (Orissa)	P-16
Johnius	*carutta*		M	48	48A	48	48			ACN=48	India (Orissa)	P-16
Johnius	*dorsalis*			48	48A	48	48			ACN=48	India	R-125
Johnius	*dussumieri*		M	48	48A	48	48			ACN=48	India (Assam)	K-46, C-108
Kathala	*axillaris*		F, M	48	48A	48	48			ACN=48	India (Orissa)	T-53
Larimichthys	*crocea*	*Pseudosciaena*	F, M	48	48A	48	48			ACN=48	China (Fujian)	Z-32
Larimichthys	*polyactis*	*Pseudosciaena*		48	48A	48	48			ACN=48	China	W-7
Leiostomus	*xanthurus*			48	48A	48	48			ACN=48	Gulf of Mexico	L-23
Menticirrhus	*americanus*			48	48A	48	48	2	1.6 FD	ACN=48	Brazil (RN)	G-51, A-124
Menticirrhus	*littoralis*	*litoralis*		48	48A	48	48	2		ACN=48	Brazil	R-123
Micropogonias	*furnieri*		F, M	48	48A	48	48	2	1.2 FD	ACN=48	Brazil (SP, RJ)	G-50, B-59
Micropogonias	*undulatus*			48	48A	48	48		1.6 BFA	ACN=48	Gulf of Mexico	L-23, H-13
Nibea	*albiflora*			48	48A	48	48			ACN=48	China (Shandong)	W-7, Y-21
Nibea	*mitsukurii*		F	48	48A	48	48	2	1.4* FCM	ACN=48	Japan (Hyogo)	O-46, O-48
Ophioscion	*punctatissimus*			48	48A	48	48	2		ACN=48	Brazil (RN)	A-124
Otolithes	*cuvieri*			48	48A	48	48			ACN=48	India	C-42

Table 6.36 Order PERCIFORMES. Part 1 Percoidei and Elassomatoidei (continued)

A		B	C	D	E	F	G	H	I	J	K	L
Current scientific name of taxon Suborder/family/subfamily/species		Reported in karyotype paper	Sex	2n	Karyotype	NF_1	NF_2	Ag-NORs	Genome size (pg/cell)	Comments	Locality	Reference
Otolithoides	*pama*		M	48	48A	48	48			ACN=48	India (WB)	K-139
Pachyurus	*bonariensis*			48							Argentina	F-20
Paranibea	*semiluctuosa*	*Nibea*		48							India	L-1
Pareques	*acuminatus*			48	48A	48	48	2		ACN=48	Brazil (RN)	A-124
Plagioscion	*montei*			48	2M + 46A	50	50	2			Brazil	P-86
Plagioscion	*squamosissimus*		F, M	48	48A	48	48	2		ACN=48	Brazil (AM)	F-66, P-86
Plagioscion	*ternetzi*			48							Argentina	F-20
Plagioscion	sp.	Cytotype a		48	2M + 46A	50	50	2		ACN=48	Brazil (AM)	F-66
Plagioscion	sp.	Cytotype b		48	48A	48	48	2		ACN=48	Brazil (AM)	F-66
Pogonias	*cromis*			48	48A	48	48		2.0 BFA	ACN=48	Gulf of Mexico	L-23, H-13
Protonibea	*diacanthus*	*Nibea*		48	48A	48	48			ACN=48	China	W-13
Protonibea	*diacanthus*	*Nibea*		48	48A	48	48			ACN=48	India	C-42
Sciaena	*umbra*			48	48A	48	48			ACN=48	Black Sea	V-5
Sciaenops	*ocellatus*			48	48A	48	48			ACN=48	Gulf of Mexico	L-23
Umbrina	*coroides*			46	4SM + 42A	50	50				Brazil (RJ)	B-86
Serranidae												
Epinephelinae												
Alphestes	*afer*			48	48A	48	48	2		ACN=48	Brazil (RN, Bahia)	M-78
Chromileptes	*altivelis*			48	2ST + 46A	48	50	2		ACN=48	(Indo-West Pacific)	T-20
Epinephelus	*adscensionis*			48	48A	48	48	4		ACN=48	Brazil (RN, Bahia)	M-78
Epinephelus	*alexandrinus*			48	48 ST/A	48		2		ACN=48	Spain (Malaga)	M-39
Epinephelus	*alexandrinus*			48	48A	48	48	4		ACN=48	Spain	A-30
Epinephelus	*awoara*			48	48A	48	48			ACN=48	China	Z-23, A-11
Epinephelus	*caninus*			48	48A	48	48	2		ACN=48	Spain (Alboran Sea)	R-89
Epinephelus	*coioides*			48	2SM + 46A	50	50			ACN=48	China (Fujian)	D-13
Epinephelus	*diacanthus*			48	2SM + 46A	50	50			ACN=48	India (Portonovo)	N-13
Epinephelus	*fario*			48	14 M/SM + 34 ST/A	62				ACN=48	China	C-57
Epinephelus	*fasciatomaculosus*			48	48A	48	48	2		ACN=48	China	L-47
Epinephelus	*fasciatus*			48	48A	48	48	2		ACN=48	China	L-47
Epinephelus	*guaza*			48	48 ST/A	48		2		ACN=48	Spain (Malaga)	M-39
Epinephelus	*guaza*			48	48A	48	48			ACN=48	Brazil (RJ)	A-11, A-12
Epinephelus	*guttatus*			48	48A	48	48		2.1 FD, 2.4 BFA	ACN=48	(W. Atlantic)	M-132, H-13
Epinephelus	*malabaricus*		F	48	2SM + 46A	50	50	2		ACN=48	Thailand (Gulf of Siam)	U-39

Table 6.36 Order PERCIFORMES. Part 1 Percoidei and Elassomatoidei (continued)

A	B	C	D	E	F	G	H	I	J	K	L
Current scientific name of taxon Suborder/family/subfamily/species	Reported in karyotype paper	Sex	2n	Karyotype	NF_1	NF_2	Ag–NORs	Genome size (pg/cell)	Comments	Locality	Reference
Epinephelus marginatus			48	48A	48	48			ACN=48	Brazil (RJ)	G–12
Epinephelus merra			48	6ST + 42A	48	54		2.1* FCM, 2.2 FIA	ACN=48	Japan	O–48, H–41
Epinephelus sexfasciatus			48	2SM + 46A	50	50			ACN=48	China	C–56
Epinephelus tauvina		F	48	2SM + 46A	50	50			ACN=48	India (Orissa)	P–16
Epinephelus ongus	*Serranus tumilabris*	F	48	48A	48	48			ACN=48	India (Bombay)	R–63
Mycteroperca acutirostris			48	48A	48	48			ACN=48	Brazil (RJ)	G–12
Mycteroperca rubra			48	48A	48	48			ACN=48	Brazil (RJ)	A–11, A–12
Serraninae											
Centropristes ocyurus			48	28M + 20SM	96	96			ACN=48	W. Atlantic	A–11
Diplectrum eumelum			48	2M + 4SM + 42A	54	54			ACN=48	E. Pacific	A–11
Diplectrum formosum			48	2SM + 46A	50	50	2–4		ACN=48	Brazil (RJ)	A–11, A–12
Diplectrum radiale		F, M	48	48A	48	48	2		ACN=48	Brazil (RJ, SP)	A–11, P–41
Paracentropistis hepatus		F, M	48	48A	48	48	2	1.1 FD	ACN=48	Spain (Malaga)	C–8, C–9, M–39
Serranus cabrilla		F, M	48	48A	48	48	2	1.0 FD	ACN=48	Spain (Malaga)	C–8, C–9, M–39
Serranus cabrilla			48	48A	48	48			ACN=48	Spain (Malaga)	C–11
Serranus cabrilla		F, M	48	48A	48	48	2		ACN=48	Italy (Palermo)	V–68
Serranus flaviventris			48	48A	48	48	2		ACN=48	Brazil (RJ, RN)	A–11, M–78
Serranus scriba			48	48A	48	48	2		ACN=48	Spain (Malaga)	M–39
Serranus scriba		F, M	48	48A	48	48	2		ACN=48	Italy (Palermo)	V–68
Serranus scriba			48	48A	48	48			ACN=48	Croatia	S–86
Serranus scriba			48	48A	48	48			ACN=48	USSR	V–72
Sillaginidae											
Sillago sihama		M	48	48 ST/A	48				ACN=48	India (Andaman Is.)	R–45
Sparidae											
Acanthopagrus latus			48	4SM + 6ST + 38A	52	58			ACN=48	Japan	M–106
Acanthopagrus latus			48	4M + 2SM + 4ST + 38A	54	58			ACN=48	China	Z–23
Acanthopagrus schlegeli		M	48	8 M/SM + 40A	56	56	2	1.2 FD	ACN=48	Korea	K–53
Acanthopagrus schlegeli			48	6SM + 4ST + 38A	54	58			ACN=48	Japan	M–106
Acanthopagrus schlegeli	*macrocephalus*		48	4M + 4SM + 2ST + 38A	56	58	2		ACN=48	China	Y–19
Archosargus probatocephalus			48	1M + 4SM + 43 ST/A	53				ACN=48	Gulf of Mexico	L–6
Archosargus probatocephalus		F, M	48	12 M/SM + 36 ST/A	60				ACN=48	USA (Atlantic)	F–69
Archosargus probatocephalus			48	48A	48		2		ACN=48	Brazil	A–124

Table 6.36 Order PERCIFORMES. Part 1 Percoidei and Elassomatoidei (continued)

A		B	C	D	E	F	G	H	I	J	K	L
Current scientific name of taxon Suborder/family/subfamily/species		Reported in karyotype paper	Sex	2n	Karyotype	NF_1	NF_2	Ag-NORs	Genome size (pg/cell)	Comments	Locality	Reference
Boops	*boops*		F, M	48		54			1.0 FD		Spain (Malaga)	C-9
Dentex	*dentex*			48	2M + 2SM + 44 ST/A	52		2–4		ACN=48	Italy	L-48
Dentex	*hypselosomus*	*tumifrons*		48	4ST + 44A	48	52			ACN=48	Japan	M-106
Diplodus	*annularis*		F, M	48	6 M/SM + 2ST + 40A	54	56			ACN=48	Italy (Tyrrhenian Sea)	C-36
Diplodus	*annularis*			48	2M + 4SM + 42 ST/A	54				ACN=48	Black Sea	V-5
Diplodus	*annularis*		F, M	47	5M + 2SM + 40A	54	54			ACN=48	Italy (Palermo)	V-68
Diplodus	*annularis*		F, M	48	4M + 3SM + 41A	55	55			ACN=48	Italy (Palermo)	V-68
Diplodus	*annularis*		F	48	4M + 2SM + 42A	54	54			ACN=48	Italy (Palermo)	V-68
Diplodus	*annularis*		M	48	4M + 4SM + 40A	56	56			ACN=48	Italy (Palermo)	V-68
Diplodus	*annularis*			48	6 M/SM + 42A	54	54	10		ACN=48	Italy (Palermo)	V-96
Diplodus	*annularis*			47	7 M/SM + 40A	54	54			ACN=48	Italy (Palermo)	V-96
Diplodus	*argenteus*			48	48A	48	48		1.6 FD, 1.9 BFA	ACN=48	Brazil (BA)	G-12, H-13
Diplodus	*bellottii*			46	2M + 6SM + 38 ST/A	54		4		ACN=48	Spain (Malaga)	A-45
Diplodus	*cervinus*			48		54					Spain	A-45
Diplodus	*puntazzo*			48	8 M/SM + 40A	56	56	12		ACN=48	Italy (Sicily)	V-96
Diplodus	*sargus*		F, M	48	6 M/SM + 2ST + 40A	54	56			ACN=48	Italy (Tyrrhenian Sea)	C-36
Diplodus	*sargus*			48	6 M/SM + 42A	54		14		ACN=48	Italy (Sicily)	V-96
Diplodus	*sargus*		F, M	48		52			1.2 FD		Spain (Malaga)	C-9
Diplodus	*vulgaris*			48	6 M/SM + 42 ST/A	54		8		ACN=48	Italy (Palermo)	V-96
Diplodus	*vulgaris*		F, M	48		52			1.1 FD		Spain (Malaga)	C-9
Evynnis	*tumifrons*	*Evinnis japonica*		48	2SM + 46A	50	50			ACN=48	Japan	M-106
Lagodon	*rhomboides*		F, M	48	6 M/SM + 42 ST/A	54				ACN=48	USA (Atlantic)	F-69
Lithognathus	*mormyrus*		F, M	48		52			1.2 FD		Spain (Malaga)	C-9
Lithognathus	*mormyrus*		F, M	48	6M + 16ST + 26A	54	70			ACN=48	Italy (Tyrrhenian Sea)	C-36
Oblada	*melanura*		M	46	6 M/SM + 6ST + 34A	52	58			ACN=48	Italy (Tyrrhenian Sea)	C-36
Pagellus	*acarne*		F, M	48		52			1.3 FD		Spain (Malaga)	C-9
Pagellus	*acarne*		F, M	48	2M + 6ST + 40A	50	56			ACN=48	Italy (Tyrrhenian Sea)	C-36
Pagellus	*bogaraveo*			48	8M + 10SM + 8ST + 22A	66	74	2		ACN=48	Italy (Sicily)	V-61
Pagellus	*erythrinus*		F, M	48	48A	48	48			ACN=48	Italy (Tyrrhenian Sea)	C-36
Pagrus	*auratus*	*Sparus latus*		48	4M + 2SM + 4ST + 38A	54	58		1.3 FIA	ACN=48	China	L-57, H-40
Pagrus	*auriga*			48	48A	48	48	2		ACN=48	Italy (Palermo)	V-105
Pagrus	*caeruleostictus*	*coeruleostictus*		48	2SM + 46A	50	50	2		ACN=48	Italy (Palermo)	V-105
Pagrus	*major*	*Pagrosomus*		48	2ST + 46A	48	50	2		ACN=48	China	Y-19

Table 6.36 Order PERCIFORMES. Part 1 Percoidei and Elassomatoidei (continued)

A		B	C	D	E	F	G	H	I	J	K	L
Current scientific name of taxon Suborder/family/subfamily/species		Reported in karyotype paper	Sex	2n	Karyotype	NF_1	NF_2	Ag-NORs	Genome size (pg/cell)	Comments	Locality	Reference
Pagrus	*major*		F, M	48	48A	48	48	2	1.3 FD	ACN=48	Korea	K-53
Pagrus	*major*	*Chrysophrys*		48	2ST + 46A	48	50		1.8* FCM	ACN=48	Japan (Yamaguchi)	N-29, O-48
Pagrus	*pagrus*			48			50				Brazil (RJ)	G-12
Pagrus	*pagrus*			48	2SM + 46A	50	50	2		ACN=48	Italy (Palermo)	V-105
Rhabdosargus	*sarba*	*Sparus*		48	6SM + 6ST + 34A	54	60		1.5 FIA	ACN=48	Japan	M-106, H-40
Sarpa	*salpa*		F, M	48		58			1.1 FD		Spain (Malaga)	C-9
Sarpa	*salpa*		F, M	48	6M + 10ST + 32A	54	64			ACN=48	Italy (Tyrrhenian Sea)	C-36
Sparus	*aurata*			48		56					Spain	A-32
Sparus	*aurata*		F, M	48	8 M/SM + 10ST + 30A	56	66	2	1.9 FCM	ACN=48	Italy (Tyrrhenian Sea)	C-36, V-105, G-85
Terapontidae (= Teraponidae)												
Rhynchopelates	*oxyrhynchus*	*Rhyncopelates*	F	48	2SM + 46A	50	50	2	1.7* FCM	ACN=48	Japan (Hyogo)	O-48, O-46
Terapon	*jarbua*	*Therapon*	F, M	48	48A	48	48		1.3 FIA	ACN=48	India	S-116, R-45, H-41
Terapon	*puta*	*Therapon*	F, M	48	48A	48	48		1.6 FIA	ACN=48	India	S-116, H-41
Terapon	*theraps*			48	48A	48	48		1.4 FIA	ACN=48	India (WB)	K-46, G-85
Suborder Elassomatoidei												
Elassomatidae												
Elassoma	*zonatum*	Centrarchidae		48	2SM + 46 ST/A	50				ACN=48	USA (NC)	R-80

Table 6.37 Order PERCIFORMES. Part 2 Labroidei and Zoarcoidei

A	B	C	D	E	F	G	H	I	J	K	L
Current scientific name of taxon Suborder/family/subfamily/species	Reported in karyotype paper	Sex	2n	Karyotype	NF_1	NF_2	Ag-NORs	Genome size (pg/cell)	Comments	Locality	Reference
Suborder Labroidei											
Cichlidae											
American Cichlidae											
Acarichthys heckelii		M	48	6 M/SM + 42 ST/A	54				ACN=48	(Amazon R.)	T-41
Aequidens metae		F, M	48	6 M/SM + 16ST + 26A	54	70			ACN=48	(Orinico R.)	T-41
Aequidens plagiozonatus			48	48 ST/A	48		2			(Brazil)	M-149
Aequidens pulcher			48			(82)				(Venezuela, Trinidad)	S-184
Amphilophus citrinellus	*Cichlasoma citrinellum*	F, M	48	8 M/SM + 12ST + 28A	56	68			ACN=48	(C. America, Atlantic side)	T-41
Amphilophus citrinellus	*Cichlasoma citrinella*		48	36 SM/ST + 12A		84			ACN=48	(C. America, Atlantic side)	N-30
Apistogramma agassizii	*agassizi*	F, M	46	24 M/SM + 22 ST/A	70				ACN=46		T-41
Apistogramma agassizii	*agassizi*		40			(78)				(S. America)	S-184
Apistogramma borellii		M	38	22 M/SM + 2ST + 14A	60	62			ACN=46		T-41
Apistogramma borellii			46			(90)				(S. America)	S-184
Apistogramma borellii	*reitzigi*		46			(86)				(S. America)	S-184
Apistogramma cacatuoides			46			(86)				(S. America)	S-184
Apistogramma ortmanni		M	46	24 M/SM + 22 ST/A	70				ACN=48		T-41
Apistogramma ortmanni			38			(64)				(S. America)	S-184
Apistogramma trifasciata			46	16 M/SM + 30 ST/A	62					Argentina	F-20, R-107
Apistogramma sp.			46			(86)				(S. America)	S-184
Archocentrus centrarchus	*Cichlasoma*	F	48	6 M/SM + 14ST + 28A	54	68			ACN=48	(C. America, Pacific side)	T-41
Archocentrus nigrofasciatus	*Cichlasoma*	F, M	48	4 M/SM + 44 ST/A	52				ACN=48	Costa Rica	T-41
Archocentrus septemfasciatus	*Cichlasoma*	F	48	6 M/SM + 14ST + 28A	54	68			ACN=48	Costa Rica (Atlantic side)	T-41
Archocentrus spilurus	*Cichlasoma spilurum*		48	6 M/SM + 42 ST/A	54				ACN=48	Costa Rica	T-39
Archocentrus spilurus	*Cichlasoma cutteri*		48			(68)				(S. America)	S-184
Astronotus ocellatus		F, M	48	12 M/SM + 36 ST/A	60		2		ACN=48	Brazil (MS, AM)	F-9, F-10
Astronotus ocellatus			48			(96)				(S. America)	S-184
Astronotus ocellatus		F	48	6 M/SM + 42 ST/A	54			(2.1* FCM)		(S. America)	T-41, O-48
Astronotus ocellatus		F, M	48	16 M/SM + 32 ST/A	64		2			Brazil (Tiete R., SP)	M-136
Australoheros facetus	*Cichlasoma facetum*	F, M	48	10 M/SM + 38 ST/A	58		2		ACN=48	Brazil (SP)	F-9, F-10
Australoheros facetus	*Cichlasoma facetum*	F, M	48	10SM + 38 ST/A	58		2		ACN=48	Brazil (PR)	V-87
Australoheros facetus	*Cichlasoma facetum*	F, M	48	8SM + 40A	56	56			ACN=48	Uruguay	O-66
Bujurquina vittata			44	22 M/SM + 8 ST/A + 14 MC	66				ACN=46	Argentina	F-20, R-107
Bujurquina vittata	*Aequidens paraguayensis*	F, M	44	26 M/SM +18 ST/A	70				ACN=46	(Paraguay R.)	T-41
Caquetaia kraussii	*Cichlasoma*	F, M	50	6 M/SM + 14ST + 30A	56	70				Colombia	T-41
Chaetobranchopsis australe		F, M	48	48A	48	48	2		ACN=48	Brazil (MS)	F-9, F-10
Cichla monoculus		F, M	48	48A	48	48	2		1-3 B, ACN=48	Brazil (Amazon R.)	F-14, B-85

Table 6.37 Order PERCIFORMES. Part 2 Labroidei and Zoarcoidei (continued)

A		B	C	D	E	F	G	H	I	J	K	L
Current scientific name of taxon Suborder/family/subfamily/species		Reported in karyotype paper	Sex	2n	Karyotype	NF_1	NF_2	Ag-NORs	Genome size (pg/cell)	Comments	Locality	Reference
Cichla	*temensis*		F, M	48	48A	48	48	2		ACN=48	Brazil (Amazon R.)	T-41, B-85
Cichlasoma	*beani*		F, M	48	6 M/SM + 14ST + 28A	54	68			ACN=48	Mexico (Sonora)	T-41
Cichlasoma	*bimaculatum*		M	48	6 M/SM + 14ST + 28A	54	68			ACN=48	(Brazil, Suriname)	T-41
Cichlasoma	*dimerus*			48	8 M/SM + 22ST + 18A	56	78			ACN=48	Argentina	F-20, R-107
Cichlasoma	*dovii*		F, M	48	8 M/SM + 40 ST/A	56				ACN=48	Costa Rica	T-41, S-6
Cichlasoma	*friedrichstalii*			48	6 M/SM + 42 ST/A	54				ACN=48	Costa Rica	S-6
Cichlasoma	*friedrichstalii*			47	6 M/SM + 41 ST/A	53				ACN=48	Costa Rica	S-6
Cichlasoma	*istlanum fusca*	*istlana fusca*	F, M	48	8SM + 40 ST/A	56				ACN=48	Mexico (Michoacán)	U-49
Cichlasoma	*istlanum istlanum*	*istlana istlana*	F	48	8SM + 40 ST/A	56				ACN=48	Mexico (Morelos)	U-49
Cichlasoma	*octofasciatum*		F, M	48	6 M/SM + 14ST + 28A	54	68		2.6 BFA	ACN=48	(Atlantic side of C. America)	T-41, H-13
Cichlasoma	*paranaense*		F, M	48	20SM + 12ST + 16A	68	80	2		ACN=48	Brazil (PR)	M-41
Cichlasoma	*portalegrensis*	*Aequidens*		48			(68)		2.4 BFA		(Brazil)	O-50, S-184, H-13
Cichlasoma	*salvini*		F, M	52	28 M/SM + 24 ST/A	80					Belize	T-41
Cichlasoma	*tetracanthum*			48	6SM + 28ST + 14A	54	82			ACN=48	Cuba	R-6
Cichlasoma	*trimaculatum*		F, M	48	6 M/SM + 14ST + 28A	54	68				(Mexico)	T-41
Cichlasoma	*trimaculatum*		F, M	48	8SM + 40 ST/A	56					Mexico (near Acapulco)	U-48
Cichlasoma	sp.		F, M	48	6 M/SM + 42 ST/A	54				ACN=48	Mexico (Coahuila)	T-41
Cleithracara	*maronii*	*Aequidens maroni*		48			(82)				(Guyana, Suriname)	S-184
Crenicichla	*britskii*			48	8 M/SM + 40 ST/A	56		2		ACN=48	Paraná-Paraguai basin	B-82
Crenicichla	*cincta*			48	8 M/SM + 40 ST/A	56		2		ACN=48	Brazil (AM)	B-82
Crenicichla	*iguassuensis*		F, M	48	2M + 6SM +14ST + 26A	56	70	2		ACN=48	Brazil (PR)	M-159
Crenicichla	*iguassuensis*			48	6 M/SM + 42 ST/A	54		2			Brazil (PR)	B-82
Crenicichla	*inpa*			48	6 M/SM + 42 ST/A	54		2		ACN=48	Brazil (Amazon basin)	B-82
Crenicichla	cf. *johanna*			48	8 M/SM + 40 ST/A	56		2		ACN=48	Brazil (Amazon basin)	B-82
Crenicichla	*lacustris*		M	48	6 M/SM + 42 ST/A	54		2		ACN=48	Brazil (SP.)	F-9, F-10
Crenicichla	*lepidota*		F	48	6 M/SM + 8ST + 34A	54	62			ACN=48	(S. America)	T-41
Crenicichla	*lepidota*		M	48	6 M/SM + 42 ST/A	54		2		ACN=50	Brazil (MS)	F-9, F-10
Crenicichla	*lepidota*			48	6 M/SM + 42 ST/A	54				ACN=48	Argentina	F-20, R-107
Crenicichla	*lepidota*		F, M	48	2M + 4SM + 6ST + 36A	54	60	2		ACN=48	Brazil (PR)	M-41
Crenicichla	*lucius*		F	48							(Amazon R.)	T-41
Crenicichla	*lugubris*			48	8 M/SM + 40 ST/A	56		2		ACN=48	Brazil (Amazon basin)	B-82
Crenicichla	*niederleinii*		F, M	48	2M + 12SM + 4ST + 30A	62	66	4			Brazil (PR)	M-41
Crenicichla	*niederleinii*		M	48	2M + 8SM + 38 ST/A	58		2		ACN=48	Brazil (PR)	L-71
Crenicichla	*niederleinii*			48	6 M/SM + 42 ST/A	54		2		ACN=48	Argentina	F-20, R-107
Crenicichla	*notophthalmus*		F	48	6 M/SM + 8ST + 34A	54	62				(Amazon R.)	T-41
Crenicichla	*reticulata*			48	6 M/SM + 42 ST/A	54		2		1-3 B, ACN=48	Brazil (Amazon R.)	B-82, F-14

Table 6.37 Order PERCIFORMES. Part 2 Labroidei and Zoarcoidei (continued)

A		B	C	D	E	F	G	H	I	J	K	L
Current scientific name of taxon Suborder/family/subfamily/species		Reported in karyotype paper	Sex	2n	Karyotype	NF_1	NF_2	Ag-NORs	Genome size (pg/cell)	Comments	Locality	Reference
Crenicichla	*saxatilis*	*sexatilis*	M	48	4M + 44A	52	52		2.2 BFA	ACN=48	S. America (Atlantic side)	O-66, H-13
Crenicichla	*semifasciata*			48	6 M/SM + 42 ST/A	54					Argentina	F-20
Crenicichla	*semifasciata*	*Batrachops semifasciatus*	F, M	48	6 M/SM + 42 ST/A	54		2		ACN=48	Brazil (MS)	F-9, F-10
Crenicichla	*strigata*		F, M	48	6 M/SM + 42 ST/A	54				ACN=48	(Amazon R.)	T-41
Crenicichla	*vittata*		M	48	6 M/SM + 42 ST/A	54		2		ACN=48	Brazil (MS)	F-9, F-10
Crenicichla	sp.		M	48	2M + 6SM + 40 ST/A	56		2		ACN=48	Brazil (SC)	L-71
Dicrossus	*filamentosus*	*Crenicara filamentosa*	F	46	12 M/SM + 34 ST/A	58				ACN=46	Brazil, Venezuela	T-41
Dicrossus	*maculatus*	*Crenicara*		46			(86)				(Brazil)	S-184
Geophagus	*brasiliensis*		F, M	48	4 M/SM + 44 ST/A	52				ACN=48	(Brazil)	T-41
Geophagus	*brasiliensis*		F, M	48	2 M/SM + 46 ST/A	50		2		ACN=48	Brazil (SP)	F-9, F-10
Geophagus	*brasiliensis*		F, M	48	8SM + 18ST + 22A	56	74	2		ACN=48	Brazil (PR)	M-41
Geophagus	*brasiliensis*		F, M	48	6SM + 42 ST/A	54		2			Brazil (PR)	V-87
Geophagus	*brasiliensis*		F, M	48	4SM + 44 ST/A	52		2		ACN=48	Brazil (PR)	L-89
Geophagus	*surinamensis*		F, M	48	4 M/SM + 44 ST/A	52		2		ACN=48	Brazil (AM)	T-41, F-9, F-10
Gymnogeophagus	*balzanii*		F, M	48	2 M/SM + 46 ST/A	50		2		0-4 B, ACN=48	Brazil (MS)	F-9, F-10
Gymnogeophagus	*balzanii*			48	2 M/SM + 46 ST/A	50		2		ACN=48	Argentina	F-20, R-107
Gymnogeophagus	*gymnogenys*			48	4 M/SM + 44 ST/A	52		2			Brazil	P-87
Gymnogeophagus	*labiatus*			48	4 M/SM + 44 ST/A	52		2			Brazil	P-87
Gymnogeophagus	*lacustris*			48	4 M/SM + 44 ST/A	52					Brazil	P-87
Gymnogeophagus	*rhabdotus*			48	4 M/SM + 44 ST/A	52					Brazil	P-87
Gymnogeophagus	sp.			48	2 M/SM + 46 ST/A	50				ACN=48	Argentina	F-20, R-107
Herichthys	*cyanoguttatus*	*Cichlasoma cyanoguttatum*	F, M	48	6 M/SM + 14ST + 28A	54	68			ACN=48	USA (TX), Mexico	T-41
Herichthys	*labridens*	*Cichlasoma*	F	48	6 M/SM + 14ST + 28A	54	68			ACN=48	Mexico (Atlantic side)	T-41
Heros	*severus*	*Cichlasoma severum*	F, M	48	4 M/SM + 44 ST/A	52				ACN=48	Brazil, Venezuela	T-41
Herotilapia	*multispinosa*		F, M	48	6 M/SM + 42 ST/A	54				ACN=48	C. America	T-41
Hypselecara	*coryphaenoides*	*Cichlasoma*	F, M	48	6 M/SM + 14ST + 28A	54	68			ACN=48	(Amazones)	T-41
Krobia	*itanyi*	*Aequidens*		46			(64)				(Suriname)	S-184
Laetacara	*curviceps*	*Aequidens*		38			(76)				(Amazon R.)	S-184
Laetacara	cf. *dorsigera*			45	3M + 42A	48				0-2 B	Brazil (PR)	C-98
Mesonauta	*festivus*	*Cichlasoma festivum*	F, M	48	8 M/SM + 40 ST/A	56				ACN=48	(S. America)	T-41
Nannacara	*anomala*		F, M	44	18 M/SM + 26 ST/A	62					S. America	T-41
Neetroplus	*nematopus*		F, M	48	8 M/SM + 12ST + 28A	56	68			ACN=48	Costa Rica	T-41
Papiliochromis	*ramirezi*	*Microgeophagus*		48			(64)				(S. America)	S-184
Parachromis	*managuensis*	*Cichlasoma managuense*		48	8 M/SM + 40 ST/A	56				ACN=48	Costa Rica	S-6
Parachromis	*managuensis*	*Cichlasoma managuense*	F, M	48	6 M/SM + 14ST + 28A	54	68			ACN=48	(Costa Rica)	T-41
Pterophyllum	*scalare*		F, M	48	4 M/SM + 44 ST/A	52			(2.0 FCM, 2.4 BFA)	ACN=48	(S. America)	T-41, V-86, H-13

Table 6.37 Order PERCIFORMES. Part 2 Labroidei and Zoarcoidei (continued)

A	B	C	D	E	F	G	H	I	J	K	L
Current scientific name of taxon Suborder/family/subfamily/species	Reported in karyotype paper	Sex	2n	Karyotype	NF_1	NF_2	Ag-NORs	Genome size (pg/cell)	Comments	Locality	Reference
Pterophyllum scalare		F, M	48	12 M/SM + 36 ST/A	60		2			Brazil (PA)	N-2
Satanoperca jurupari			48	6SM + 42 ST/A	54				0–3 B, XY	S. America	M-41, O-86
Satanoperca jurupari	*Geophagus*	F, M	48	4 M/SM + 44 ST/A	52			(2.4 BFA)	ACN=48	(S. America)	T-41, H-13
Satanoperca papaterra		F, M	48	6SM + 6ST + 36A	54	60	2		ACN=48	Brazil (PR.)	M-41
Symphysodon aequifasciatus	*aequifasciata*	F, M	60	58 M/SM + 2 ST/A	118			(2.5* FCM, 2.4 FD)		(Amazon R.)	T-41, O-4, O-48
Symphysodon aequifasciatus	cytotype 1	F, M	60	48 M/SM + 8 ST/A + 4 MC			2		ACN=62	Brazil (Amazon)	M-146
Symphysodon aequifasciatus	cytotype 2	M	60	50 M/SM + 6 ST/A + 4 MC			2		ACN=62	Brazil (Amazon)	M-146
Symphysodon discus	cytotype 1	F, M	60	50 M/SM + 10 ST/A	110		2		ACN=62	Brazil (Amazon)	M-146
Symphysodon discus	cytotype 2	M	60	54 M/SM + 6 ST/A	114		2		ACN=62	Brazil (Amazon)	M-146
Symphysodon haraldi	*aequifasciata axelrodi*		60	26M + 26SM + 8 ST/A	112		2	2.5 FCM		(Amazon R.)	T-24
Symphysodon haraldi		F, M	60	52 M/SM + 4 ST/A + 4 MC			2–5		ACN=62	Brazil (Amazon)	M-146
Thorichthys ellioti	*Cichlasoma*	F, M	48	6 M/SM + 42 ST/A	54					Mexico (Veracruz)	U-48
Thorichthys meeki	*Cichlasoma*	F, M	48	6 M/SM + 14ST + 28A	54	68		2.8 BFA	ACN=48	(S. America)	T-41, H-13
Uaru amphiacanthoides		F, M	46	8 M/SM + 38 ST/A	54				ACN=48	(Amazon R.)	T-41
Asian Cichlidae											
Etroplus maculatus		F, M	46	18 M/SM + 28 ST/A	64				ACN=48?	(Asia)	T-39
Etroplus maculatus		F, M	46							India (Portonovo)	N-13
Etroplus suratensis		F, M	48	48A	48	48			ACN=48	India (Orissa, WB)	R-57, K-46
Etroplus suratensis		F, M	48							India (Portonovo)	N-13
African Cichlidae											
Anomalochromis thomasi	*Pelmatochromis*		48			(58)				(Africa)	S-25
Astatotilapia burtoni		F, M	40	16 M/SM + 24 ST/A	56			1.9 FCM	ACN=48	Africa (Lake Tanganyika)	T-39, T-42, G-85
Aulonocara baenschi			44	6M + 8SM + 30A	58					Malawi	F-68
Aulonocara hueseri			44	4M + 6SM + 34A	54					Malawi	F-68
Aulonocara korneliae			44	4M + 8SM + 32A	56					Malawi	F-68
Aulonocara stuartgranti			44	4M + 6SM + 34A	54					Malawi	F-68
Aulonocara stuartgranti			44	6M + 6SM + 32A	56					Malawi	F-68
Aulonocara sp.			44	6M + 6SM + 32A	56					Malawi	F-68
Chromidotilapia finleyi			48			(68?)				(Africa)	S-184
Chromidotilapia batesii	*Pelmatochromis* (southern form)		44			(80)				E. Cameroon	S-25
Chromidotilapia batesii	*Pelmatochromis* (northern form)		42			(80)				E. Cameroon	S-25
Haplochromis flavijosephi		F	44	10 M/SM + 34 ST/A	54			2.2 FD	ACN=48	Sea of Galilee	K-98
Haplochromis obliquidens		F, M	44	12 M/SM + 32 ST/A	56				0–2 B	(E. Africa)	P-95
Hemichromis bimaculatus			44			(64)				(Africa)	S-184
Hemihaplochromis kirki			44			(76)				(Africa)	S-184
Hemihaplochromis multicolor			44			(78)				(Africa)	S-184

Table 6.37 Order PERCIFORMES. Part 2 Labroidei and Zoarcoidei (continued)

A		B	C	D	E	F	G	H	I	J	K	L
Current scientific name of taxon Suborder/family/subfamily/species		Reported in karyotype paper	Sex	2n	Karyotype	NF_1	NF_2	Ag-NORs	Genome size (pg/cell)	Comments	Locality	Reference
Heterochromis	*multidens*	*Pelmatochromis taeniatus*		42			(74)				(Africa)	S-25
Julidochromis	*ornatus*			44			(58)				(Africa)	S-184
Labeotropheus	*fuelleborni*			44	12 M/SM + 32 ST/A	56					Africa (Lake Malawi)	K-120
Lamprologus	*congoensis*			44			(46)				(Africa)	S-184
Melanochromis	*auratus*		F	46	10 M/SM + 36 ST/A	56				ACN=48	(E. Africa)	T-42
Melanochromis	*auratus*	*Pseudotropheus*		46	10 M/SM + 36 ST/A	56					(E. Africa)	T-39
Metriaclima	*zebra*	*Pseudotropheus*		44	12 M/SM + 32 ST/A	56					Africa (Lake Malawi)	K-120
Neolamprologus	*savoryi*	*Lamprologus savory*		32?			(54?)				(Africa)	S-184
Oreochromis	*andersonii*	*Tilapia*	F, M	44	6SM + 38 ST/A	50				ACN=48	S. Africa	V-35
Oreochromis	*aureus*			44	14M/SM + 30ST	58	88		2.4 FD	ACN=48	Egypt	M-125
Oreochromis	*aureus*	*Sarotherodon*	F, M	44	10SM + 34ST	54	88		2.5 FD	ACN=48	Sea of Galilee	K-98
Oreochromis	*aureus*	*Sarotherodon*	F, M	44	8SM + 2ST + 34A	52	54			ACN=48	(Africa)	T-42
Oreochromis	*aureus*	*Tilapia*		44	44 ST/A	44					(Africa)	T-39
Oreochromis	*karongae*		F, M	38	4M + 22 SM/ST + 12A		64			ACN=48	Africa (Lake Malawi)	H-36
Oreochromis	*lepidurus*	*Tilapia*		44			(86)				(C. Africa)	S-184
Oreochromis	*leucostictus*			44					2.4 BFA		(E. Africa)	H-13
Oreochromis	*macrochir*			44	10SM + 34ST	54	88		1.7 FD	ACN=48	Botswana	M-125
Oreochromis	*macrochir*	*Tilapia*	F, M	44	6SM + 38 ST/A	50				ACN=48	Zaire	V-35
Oreochromis	*mortimeri*			44	18 M/SM + 26A	62	62			ACN=48	Zambezi	H-11
Oreochromis	*mossambicus*			44	6SM + 38ST	50	88		2.0 FD	ACN=48	(Africa)	M-125
Oreochromis	*mossambicus*	*Sarotherodon*	F, M	44	6SM + 38 ST/A	50				ACN=48	(Africa)	T-42
Oreochromis	*mossambicus*	*Tilapia*	M	44	6ST + 38A	44	50			ACN=48	(introduced, India)	P-49
Oreochromis	*mossambicus*	*Tilapia*		44	8SM + 34ST + 2A	52	86		(1.8* FD)		(introduced, China)	C-44, C-83
Oreochromis	*mossambicus*	*Tilapia*		44	12ST + 32A	44	56			ACN=48	(introduced, China)	L-53
Oreochromis	*mossambicus*	*Tilapia*	F, M	44	18ST + 26A	44	62			ACN=48	(introduced, Japan)	F-54
Oreochromis	*niloticus*			44	2M + 18SM + 24ST	64	88		1.9 FD	ACN=48	Egypt	M-125
Oreochromis	*niloticus*	*Tilapia*		44	8SM + 34ST + 2A	52	86		2.2* FD		(introduced, China)	C-44, C-83
Oreochromis	*niloticus*	*Sarotherodon*		44	2M + 16SM + 26 ST/A	62				ACN=48	(introduced, Japan)	A-76
Oreochromis	*spilurus*			44	6SM + 38ST	50	88		1.9 FD	ACN=48	Kenya	M-125
Oreochromis	*urolepis hornorum*			44	8SM + 2ST + 34A	52	54			ACN=48	(introduced, Mexico)	U-47
Pelvicachromis	*pulcher*	*Pelmatochromis*		44			(82)				(Africa)	S-25
Pelvicachromis	*subocellatus*	*Pelmatochromis*		40			(70)				(Africa)	S-25
Pelvicachromis	*taeniatus*	*Pelmatochromis klugei*		40			(70)		2.0 BFA		(Africa)	S-25, H-13
Sarotherodon	*galilaeus*		F, M	44	6SM + 38 ST/A	50		2	2.2 FD	ACN=48	Sea of Galilee	K-98
Sarotherodon	*galilaeus*	*Tilapia galilaea*	F, M	44	6SM + 38 ST/A	50				ACN=48	(Africa)	V-35
Sarotherodon	*galilaeus*	*Tilapia*		44	8SM + 34ST + 2A	52	86				(introduced, China)	C-44

Table 6.37 Order PERCIFORMES. Part 2 Labroidei and Zoarcoidei (continued)

A		B	C	D	E	F	G	H	I	J	K	L
Current scientific name of taxon Suborder/family/subfamily/species		Reported in karyotype paper	Sex	2n	Karyotype	NF_1	NF_2	Ag-NORs	Genome size (pg/cell)	Comments	Locality	Reference
Sarotherodon	*galilaeus*			44	4SM + 40ST	48	88		1.7 FD	ACN=48	Kenya	M-125
Sarotherodon	*melanotheron*			44	4M + 30 SM/ST + 10A		78			ACN=48	Ivory Coast	H-11
Tilapia	*congica*		F, M	44	10 M/SM + 34 ST/A	54				ACN=48	Zaire	V-35
Tilapia	*guineensis*		F, M	44	8 M/SM + 36 ST/A	52				ACN=48	(Guinea)	V-35
Tilapia	*mariae*		F, M	40	4 M/SM + 36 ST/A	44				ACN=48	(Africa)	V-35
Tilapia	*mariae*		F, M	40	8 M/SM + 32 ST/A	48					(Africa)	T-42
Tilapia	*rendalli*		F, M	44	10SM + 8ST + 26A	54	62	4		ACN=48	Brazil (PR)	K-123, M-65
Tilapia	*sparrmanii*		F, M	42	8 M/SM + 34 ST/A	50				ACN=48	Zaire	V-35
Tilapia	*sparrmanii*		F, M	42	10 M/SM + 32 ST/A	52				ACN=48	(Africa)	T-42
Tilapia	*zillii*		F, M	44	10 M/SM + 34 ST/A	54			2.4 FD, (2.4 BFA)	ACN=48	Sea of Galilee	K-98, H-13
Tilapia	*zillii*			44	4M + 18SM + 22ST	66	88			ACN=48	Egypt	M-125
Tristramella	*sacra*		M	44	6 M/SM + 38 ST/A	50			2.6 FD	ACN=48	Sea of Galilee	K-98
Tristramella	*simonis*		F, M	44	6 M/SM + 38 ST/A	50			2.5 FD	ACN=48	Sea of Galilee	K-98
Thysochromis	*ansorgii*	*Pelmatochromis ansorgei*		46			(54)				(Africa)	S-25
Tylochromis	*lateralis*	*Pelmatochromis guentheri*		48			(50)				(Africa)	S-25
Embiotocidae												
Embiotoca	*jacksonii*			48*					2.0 BFA		USA (CA)	C-46, H-13
Micrometrus	*aurora*	*Amphigonopterus*		48*							USA (CA)	C-46
Neoditrema	*ransonneti*			48	6SM + 42 ST/A	54				ACN=48	Japan (Kanagawa)	A-77
Pomacentridae												
Amphiprioninae												
Amphiprion	*clarkii*			48	14M + 16SM + 18 ST/A	78			2.2* FCM, 2.1 FIA	ACN=48	Japan (Tanegashima Is.)	A-65, O-48, H-41
Amphiprion	*clarkii*			48	12M + 26SM + 10 ST/A	86				ACN=48	(W. Pacific)	T-70
Amphiprion	*frenatus*			48	14M + 22SM + 10ST + 2A	84	94			ACN=48	Japan (Okinawa)	A-66
Amphiprion	*frenatus*			48	12M + 26SM + 10 ST/A	86				ACN=48	(W. Pacific)	T-70
Amphiprion	*frenatus*			48	14M + 22SM + 8ST + 4A	84	92	2		ACN=48	Philippines	M-79
Amphiprion	*ocellaris*			48	14M + 22SM + 12 ST/A	84				ACN=48	Japan (Okinawa)	A-66
Chrominae												
Chromis	*chromis*		F, M	48	48A	48	48		2.6* FD	ACN=48	Spain (Malaga)	A-29
Chromis	*chrysura*			48	2M + 46 ST/A	50				ACN=48	(Japan)	O-49
Chromis	*flavicauda*		M	39	9M + 6SM + 24A	54	54	2		ACN=48	Brazil (ES)	M-143
Chromis	*insolata*			46	4M + 6SM + 36A	56	56	2		ACN=48	Brazil (ES)	M-143
Chromis	*insolata*			47	3M + 6SM + 38A	56	56	2		ACN=48	Brazil (ES)	M-143
Chromis	*multilineata*		F	48	48A	48	48	2		ACN=48	Brazil (BA)	M-143

Table 6.37 Order PERCIFORMES. Part 2 Labroidei and Zoarcoidei (continued)

A		B	C	D	E	F	G	H	I	J	K	L
Current scientific name of taxon Suborder/family/subfamily/species		Reported in karyotype paper	Sex	2n	Karyotype	NF_1	NF_2	Ag-NORs	Genome size (pg/cell)	Comments	Locality	Reference
Chromis	*ternatensis*	*caeruleus*		48	48A	48	48			ACN=48	(Pacific)	T-23
Dascyllus	*aruanus*			32	16 M/SM + 16A	48	48	2	(2.1* FCM, 1.7 FIA)	ACN=48	S. Japan	K-15, O-48, H-41
Dascyllus	*aruanus*			31	17 M/SM + 14A	48	48			ACN=48	S. Japan	K-15
Dascyllus	*aruanus*			30	18 M/SM + 12A	48	48			ACN=48	S. Japan	O-36, K-15
Dascyllus	*aruanus*			29	19 M/SM + 10A	48	48			ACN=48	S. Japan	K-15
Dascyllus	*aruanus*			28	20 M/SM + 8A	48	48	2		ACN=48	S. Japan	O-36, K-15
Dascyllus	*melanurus*			48	48A	48	48	2		ACN=48	S. Japan	O-36, K-15
Dascyllus	*reticulatus*			36	12 M/SM + 24A	48	48	2	(1.7 FIA)	ACN=48	S. Japan	K-15, H-40
Dascyllus	*reticulatus*			35	13 M/SM + 22A	48	48			ACN=48	S. Japan	O-36, K-15
Dascyllus	*reticulatus*			34	14 M/SM + 20A	48	48			ACN=48	S. Japan	O-36, K-15
Dascyllus	*trimaculatus*			48	48A	48	48	2	1.8 FIA	ACN=48	S. Japan	O-36, K-15, H-40
Dascyllus	*trimaculatus*			47	1M + 46A	48	48			ACN=48	Japan (Tanegashima Is.)	A-65
Pomacentrinae												
Abudefduf	*notatus*			48	2M + 2ST + 44A	50	52			ACN=48	Japan (Tanegashima Is.)	A-65
Abudefduf	*saxatilis*		F	48	2M + 2SM + 44A	52	52	2	(2.0 BFA)		Brazil (RJ)	A-12, B-59, H-13
Abudefduf	*saxatilis*			48	2M + 2SM + 44A	52	52	2		ACN=48	Brazil (RN)	M-79
Abudefduf	*sexfasciatus*	*coelestinus*		48	2M + 46A	50	50	2	1.4 FIA		(W. Pacific)	T-7, H-40
Abudefduf	*sordidus*			48	2M + 2SM + 2ST + 42A	52	54		1.7 FIA	ACN=48	Japan (Tanegashima Is.)	A-65, H-40
Abudefduf	*vaigiensis*			48	2M + 2SM + 2ST + 42A	52	54			ACN=48	Japan (Chiba)	A-65
Abudefduf	*vaigiensis*			48	2M + 2SM + 44A	52	52	2		ACN=48	(W. Pacific)	T-7
Amblyglyphidodon	*curacao*			48	6M + 22SM + 20 ST/A	76		2	2.2* FCM, 1.8 FIA		(Indo-W. Pacific)	T-6, O-48, O-49, H-40
Cheiloprion	*labiatus*			48	26SM + 22 ST/A	74		2		ACN=48	Japan (Ryukyu)	T-7
Chrysiptera	*cyanea*			42	6M + 16SM + 2ST + 18A	64	66	2	3.4* FCM, 1.6, 2.1 FIA	ACN=48	(W. Pacific)	T-19, O-48, H-41
Chrysiptera	*cyanea*	*Glyphidodontops cyaneus*		48	48 ST/A	48					(Indo-W. Pacific)	O-49
Chrysiptera	*hemicyanea*			48	30SM + 10ST + 8A	78	88	2	2.7* FCM	ACN=48	(W. Pacific)	T-16, T-23, O-48
Chrysiptera	*hemicyanea*	*Glyphidodontops hemicyaneus*		48	32SM + 16 ST/A	80					(Indo-W. Pacific)	O-49
Chrysiptera	*leucopoma*			48	4M + 22SM + 6ST + 16A	74	80	2		ACN=48	(W. Pacific)	T-19
Chrysiptera	*rex*			36	12M + 10SM + 14 ST/A	58		2		ACN=48	(W. Pacific)	T-19
Chrysiptera	*rex*	*Glyphidodontops rex*		48	8M + 22SM + 18 ST/A	78					(Indo-W. Pacific)	O-49
Chrysiptera	*starcki*			48	2M + 10SM + 36A	60	60	2		ACN=48	(W. Pacific)	T-6, T-7
Dischistodus	*prosopotaenia*			48				2	1.5 FIA		(Indo-W. Pacific)	T-6, H-41
Microspathodon	*chrysurus*			48	6M + 10ST + 32A	54	64	2		ACN=48	Brazil (BA)	G-12, M-79
Neoglyphidodon	*melas*			48	8M + 26SM + 2ST + 12A	82	84				Japan (Wakayama)	T-84
Neoglyphidodon	*nigroris*	*Paraglyphidodon*		48	8M + 26SM + 2ST + 12A	82	84	2		ACN=48	(W. Pacific)	T-16
Neoglyphidodon	*oxyodon*	*Paraglyphidodon*		48	2M + 20SM + 12ST + 14A	70	82	2		ACN=48	(W. Pacific)	T-16
Plectroglyphidodon	*lacrymatus*			48				2			(Indo-W. Pacific)	T-6

Table 6.37 Order PERCIFORMES. Part 2 Labroidei and Zoarcoidei (continued)

A		B	C	D	E	F	G	H	I	J	K	L
Current scientific name of taxon Suborder/family/subfamily/species		Reported in karyotype paper	Sex	2n	Karyotype	NF_1	NF_2	Ag-NORs	Genome size (pg/cell)	Comments	Locality	Reference
Plectroglyphidodon	*leucozonus*	*Abudefduf*		48	4ST + 44A	48	52			ACN=48	Japan (Yakushima Is.)	A-65
Pomacentrus	*chrysurus*	*rhodonotus*		48	8M + 22SM + 12ST + 6A	78	90	2	2.1 FIA	ACN=48	(W. Pacific)	T-16, H-40
Pomacentrus	*coelestis*			48	48A	48	48		(1.8 FIA)	ACN=48	Japan (Tanegashima Is.)	A-65, H-40
Pomacentrus	*coelestis*	*Abudefduf*		48	48A	48	48	2		ACN=48	(W. Pacific)	T-7, T-23
Pomacentrus	*moluccensis*			48	10M + 26SM + 10ST + 2A	84	94	2	1.5 FIA	ACN=48	(W. Pacific)	T-16, H-40
Pomacentrus	cf. *nagasakiensis*	*Paraglyphidodon melas*		48	2M + 46A	50	50	2		ACN=48	Japan (Wakayama)	T-7, T-84
Pomacentrus	*philippinus*			48	8M + 24SM + 10ST + 6A	80	90	2		ACN=48	Japan (Yaeyama)	T-16
Pomacentrus	*trilineatus*	*biocellatus*	M	50	8 M/SM + 42 ST/A	58					India (Andaman Is.)	R-45
Pomacentrus	sp.			48	10M + 26SM + 12 ST/A	84					(Pacific)	O-49
Stegastes	*fuscus*		F, M	48	20M + 22SM + 6A	90	90	2	1.5 FD	ACN=48	Brazil (RN)	G-12, G-85, M-145
Stegastes	*insularis*			48	14M + 24SM + 6ST + 4A	86	92	2		ACN=48	India (Kerala)	N-65
Stegastes	*leucostictus*		F, M	48	18M + 22SM + 8A	88	88	2		ACN=48	Brazil (BA)	G-12, M-145
Stegastes	*lividus*	*Eupomacentrus*		48	6M + 24SM + 18 ST/A	78					(Pacific)	O-49
Stegastes	*nigricans*	*Eupomacentrus*		48	2M + 2SM + 44 ST/A	52			1.5 FIA		(Indo-W. Pacific)	O-49, H-40
Stegastes	*pictus*		F, M	48	14M + 28SM + 2ST + 4A	90	92	2		ACN=48	Brazil (BA, CE)	G-12, M-145
Stegastes	*sanctipauli*			48			92				Brazil (SPR)	G-12
Stegastes	*variabilis*		F, M	48	18M + 22SM + 8A	88	88	2		ACN=48	Brazil (RN)	G-12, M-145
Labridae												
Bodianus	*axillaris*			48	8M + 30SM + 10 ST/A	86				ACN=48	S. Japan	O-31
Bodianus	*loxozonus*			48	8M + 26SM + 14 ST/A	82					S. Japan	O-31
Bodianus	*mesothorax*			48	8M + 18SM + 22 ST/A	74					S. Japan	O-31
Bodianus	*rufus*			48	48A	48	48				Brazil	B-86
Bodianus	*rufus*			48			80				Brazil (RN)	G-12
Cheilinus	*bimaculatus*			32	4M + 2SM + 26A	38	38			ACN=34	Japan (Wakayama)	O-27
Cheilinus	*fasciatus*			48	12SM + 36 ST/A	60					(Japan)	O-49
Cheilio	*inermis*			48	12M + 12SM + 24 ST/A	72			1.8* FCM		S. Japan	O-31, O-48
Cheilio	*inermis*			48	4M + 2SM + 42A	54	54			ACN=48	Japan (Yakushima Is.)	A-75
Choerodon	*azurio*			48	6M + 2SM + 40 ST/A	56				ACN=48	Japan (Wakayama)	A-75
Cirrhilabrus	*cyanopleura*			34	10M + 2SM + 22 ST/A	46					S. Japan	O-31
Cirrhilabrus	*temminckii*			34	12M + 22A	46	46			ACN=46	Japan (Wakayama)	O-27
Coris	*aygula*			48	6M + 6SM + 36 ST/A	60					(Japan)	O-49
Coris	*dorsomacula*	*multicolor*		48	6M + 8SM + 34 ST/A	62			1.4 FIA	ACN=48	Japan (Wakayama)	O-27, H-40
Coris	*gaimardi*			48	2M + 10SM + 36 ST/A	60					S. Japan	O-31
Coris	*julis*		F	48	10M + 38A	58	58			ACN=48	Elba, Europe	D-24
Coris	*julis*		M	48	11M + 37A	59	59			ACN=48	Elba, Europe	D-24

Table 6.37 Order PERCIFORMES. Part 2 Labroidei and Zoarcoidei (continued)

A		B	C	D	E	F	G	H	I	J	K	L
Current scientific name of taxon Suborder/family/subfamily/species		Reported in karyotype paper	Sex	2n	Karyotype	NF_1	NF_2	Ag-NORs	Genome size (pg/cell)	Comments	Locality	Reference
Coris	*julis*		F, M	48	4M + 6SM + 38A	58	58		2.4 FD	ACN=48	Italy (Palermo)	V-58, G-85
Coris	*julis*		F, M	48		56			1.4 FD		Spain (Malaga)	C-9
Ctenolabrus	*rupestris*			48	4M + 22SM + 10ST + 12A	74	84			ACN=48	Spain (Mediterranean)	A-31
Epibulus	*insidiator*			48	4M + 8SM + 36 ST/A	60				ACN=48	S. Japan	O-31
Gomphosus	*varius*			48	48 ST/A	48			2.1* FCM		S. Japan	O-31, O-48
Halichoeres	*argus*	*binotopsis*		48	48 ST/A	48				ACN=48	Japan (Yakushima Is.)	A-75
Halichoeres	*hortulanus*	*centriquadrus*		48	48 ST/A	48					S. Japan	O-31
Halichoeres	*melanochir*			48	2M + 46 ST/A	50				ACN=48	Japan (Tanegashima Is.)	A-75
Halichoeres	*melanochir*			48	2SM + 46 ST/A	50				ACN=48	Japan	O-27, O-31
Halichoeres	*melanurus*	*kallochroma*		48	48 ST/A	48			1.7 FIA		S. Japan	O-31, H-40
Halichoeres	*poecilopterus*			48	4M + 2SM + 42 ST/A	54			1.8* FCM	ACN=48	S. Japan	O-31, O-48
Halichoeres	*poecilopterus*			48	4M + 44 ST/A	52				ACN=48	Japan (Chiba)	A-75
Halichoeres	*poecilopterus*		F, M	48	2M + 2SM + 44 ST/A	52				ACN=48	Korea (Cheju Is.)	P-67
Halichoeres	*poeyi*			48			58				Brazil (RN, BA, RJ)	G-12
Halichoeres	*prosopeion*			48	2SM + 46 ST/A	50				ACN=48	Japan (Wakayama)	O-27
Halichoeres	*radiatus*			48	48A	48	48				Brazil (SPR)	G-12
Halichoeres	*tenuispinnis*			48	2SM + 46 ST/A	50			1.5* FCM	ACN=48	Japan (Wakayama)	O-27, M-2
Halichoeres	*tenuispinnis*			48	2SM + 46 ST/A	50				ACN=48	Japan (Chiba)	A-75
Halichoeres	*tenuispinnis*		F, M	48	2M + 46 ST/A	50				ACN=48	Korea (Cheju Is.)	P-67
Halichoeres	*trimaculatus*			48	48 ST/A	48					S. Japan	O-31
Hemigymus	*fasciatus*			48	6M + 6SM + 36 ST/A	60					(Japan)	O-49
Hologymnosus	*annulatus*	*semidiscus*		48	2M + 2SM + 44 ST/A	52					S. Japan	O-31
Labroides	*dimidiatus*			48	48A	48	48		1.8* FCM	ACN=48	Japan (Wakayama)	O-27, O-48
Labrus	*bimaculatus*			48	48A	48	48			ACN=48	Italy (Palermo)	V-53
Labrus	*merula*			48	48A	48	48		2.9 FD	ACN=48	Italy (Palermo)	V-53, G-85
Labrus	*viridis*			48	48A	48	48			ACN=48	Italy (Palermo)	V-53
Novaculichthys	*taeniurus*	*Hemipteronotus*		48	4SM + 44 ST/A	52					(Japan)	O-49
Pseudolabrus	*eoethinus*	*japonicus*		48	2M + 2SM + 44 ST/A	52			2.1* FCM	ACN=48	Japan (Wakayama)	O-27, M-2
Pseudolabrus	*sieboldi*	*japonicus*		42	20M + 8SM + 14 ST/A	70			2.1* FCM	ACN=46	Japan (Kanagawa, Wakayama)	A-75, M-2
Pseudolabrus	*sieboldi*	*japonicus*	F, M	42	4M + 24SM + 14 ST/A	70		2		ACN=46	Korea (Cheju Is.)	P-67
Pteragogus	*aurigarius*	*flagellifera*		44	2M + 10SM + 32 ST/A	56			3.5* FCM	ACN=44	Japan (Izu Peninsula)	A-75, M-2
Pteragogus	*aurigarius?*	*flagellifera*	F, M	42	4M + 24SM + 14 ST/A	70		2		ACN=46	Korea (Cheju Is.)	P-67
Stethojulis	*bandanensis*			48	4M + 44 ST/A	52					S. Japan	O-31
Stethojulis	*interrupta*			48	2M + 46 ST/A	50				ACN=48	Japan (Wakayama)	O-27
Stethojulis	*interrupta*			48	2SM + 46 ST/A	50				ACN=48	Japan (Chiba, Kagoshima)	A-75
Stethojulis	*strigiventer*			48	2M + 46 ST/A	50					S. Japan	O-31

Table 6.37 Order PERCIFORMES. Part 2 Labroidei and Zoarcoidei (continued)

A		B	C	D	E	F	G	H	I	J	K	L
Current scientific name of taxon Suborder/family/subfamily/species		Reported in karyotype paper	Sex	2n	Karyotype	NF_1	NF_2	Ag-NORs	Genome size (pg/cell)	Comments	Locality	Reference
Symphodus	*doderleini*		F, M	48	24M + 6SM + 10ST + 8A	78	88			ACN=48	Italy (Palermo)	C-29
Symphodus	*cinereus*	*griseus*		48	32 M/SM/ST + 16A		80		2.5 FD	ACN=48	USSR	V-8, G-85
Symphodus	*mediterraneus*			48	22M + 20SM + 6A	90	90			ACN=48	Italy (Palermo)	V-53
Symphodus	*mediterraneus*		F, M	46		52			1.1 FD		Spain (Malaga)	C-9
Symphodus	*melops*			46	2M + 42 SM/ST + 2A		90	2		ACN=48	Spain (Malaga)	L-67
Symphodus	*melops*	*Crenilabrus*		46	10M + 36ST	56	92			ACN=48	Italy (Roma)	C-31
Symphodus	*ocellatus*			48	10M + 20SM + 18A	78			2.2 FD	ACN=48	Italy (Palermo)	V-53, G-85
Symphodus	*ocellatus*			48	8M + 18SM + 8ST + 14A	74	82			ACN=48	Spain (Mediterranean)	A-31
Symphodus	*ocellatus*	*Crenilabrus*		48	36 M/SM/ST + 12A		84			ACN=48	USSR	V-8
Symphodus	*roissali*			38	32M + 4SM + 2A	74	74			ACN=48	Italy (Palermo)	V-53
Symphodus	*roissali*			38	36 M/SM + 2A	74	74	2		ACN=48	Spain (Malaga)	L-67
Symphodus	*roissali*	*Crenilabrus quinquemaculatus*		38	36 M/SM/ST + 2A		74			ACN=48	Black Sea	V-8
Symphodus	*roissali*		F, M	46		60			1.0 FD		Spain (Malaga)	C-9
Symphodus	*rostratus*	*scina*		48	40 M/SM/ST + 8A		88		2.0 FD	ACN=48	Black Sea	V-8, G-85
Symphodus	*tinca*	*Crenilabrus*		48	34 M/SM/ST + 14A		82		1.5 FCM	ACN=48	Black Sea	V-8, V-86
Symphodus	*tinca*			48	16M + 14SM + 4ST + 14A	78	82		2.2 FD	ACN=48	Italy (Palermo)	V-53, G-85
Thalassoma	*amblycephala*			48	48 ST/A	48				ACN=48	S. Japan	O-31
Thalassoma	*cupido*			48	48 ST/A	48			2.3* FCM	ACN=48	Japan	O-31, A-75, M-2
Thalassoma	*cupido*		F, M	48	48A	48	48	6		ACN=48	Japan (Wakayama)	U-42
Thalassoma	*lunare*			48	48A	48	48			ACN=48	Japan (Wakayama)	O-27
Thalassoma	*lutescens*			48	48A	48	48			ACN=48	Japan (Wakayama)	O-27
Thalassoma	*lutescens*			48	48 ST/A	48				ACN=48	Japan (Yakushima Is.)	A-75
Thalassoma	*pavo*			48	48A	48	48			ACN=48	Italy (Palermo)	V-53
Thalassoma	*pavo*			48		48			1.6 FD		Spain (Malaga)	C-9
Thalassoma	*quinquevittata*			48	48 ST/A	48					S. Japan	O-31
Xyrichthys	*dea*		F	44	44A	44	44	2		ACN=46	Japan (Wakayama, Tokushima)	U-43
Xyrichthys	*dea*	*Hemipteronotus*		44	44A	44	44				Japan (Wakayama)	O-27
Xyrichthys	*novacula*			48	8SM + 40A	56	56	2		ACN=48	Italy (Palermo)	V-53, V-60
Xyrichthys	*pavo*		F, M	44	44A	44	44	2	1.8* FCM	ACN=46	Japan (Wakayama, Tokushima)	U-43
Xyrichthys	*twistii*		F	22	18 M/SM + 4A	40	40	2	1.7* FCM		Japan (Wakayama)	U-43

Table 6.37 Order PERCIFORMES. Part 2 Labroidei and Zoarcoidei (continued)

A	B	C	D	E	F	G	H	I	J	K	L
Current scientific name of taxon Suborder/family/subfamily/species	Reported in karyotype paper	Sex	2n	Karyotype	NF_1	NF_2	Ag-NORs	Genome size (pg/cell)	Comments	Locality	Reference
Scaridae											
Calotomus japonicus			48	8M + 10SM + 30 ST/A	66				ACN=48	Japan (Chiba)	A-75
Chlorurus sordidus	*Scarus rhoduropterus*		48	10M + 8SM + 30 ST/A	66				ACN=48	Japan (Ogasawara)	A-75
Chlorurus sordidus	*Scarus*		48					4.1* FCM		Japan	O-48
Scarus trispinosus	*coelestinus*		48	6M + 10SM + 24ST + 8A	64	88	2			Brazil (RN)	S-178
Sparisoma axillare		F, M	46	6M + 14SM + 4ST + 22A	66	70	2		ACN=48	Brazil (BA)	S-178
Suborder Zoarcoidei											
Bathymasteridae											
Ronquilus jordani			26	22 M/SM + 4A	48	48			ACN=48	(Canada)	M-61
Zoarcidae											
Lycodichthys dearborni		F, M	48	2M + 2SM + 44 ST/A	52		2	3.2 FD, 2.2 FIA	ACN=48	Ross Sea	M-89, H-41
Pachycara brachycephalum		F, M	48	2M + 4SM + 42 ST/A	54		2	2.9 FD	ACN=48	Ross Sea	M-89
Pachycara brachycephalum		M	72	3M + 6SM + 63A	81	81	3		3X	Ross Sea	M-89
Zoarces elongatus			48	30M + 14SM + 4A	92	92			ACN=48	China (Liaoning)	M-36
Zoarces viviparus		F, M	48	2M + 6SM + 18ST + 22A	56	74	2		ACN=48	Baltic Sea, White Sea	K-80
Stichaeidae											
Alectrias benjamini			48	18M + 18SM + 12A	84	84			ACN=48	China (Liaoning)	M-34
Chirolophis japonicus	*Azuma emmnion*		56	6M + 10SM + 40A	72	72				China (Liaoning)	M-36
Dictyosoma burgeri			48	12M + 18SM + 6ST + 12A	78	84			ACN=48	Japan (Kanagawa)	A-59
Dictyosoma burgeri		M	46	14M + 18SM + 8ST + 6A	78	86			ACN=48	Japan (Yamaguchi)	N-34
Ernogrammus hexagrammus			48	48A	48	48			ACN=48	China (Liaoning)	M-31
Lumpenus fabricii			48	2M + 14SM + 32 ST/A	64					Russia, White Sea	L-87, L-88
Lumpenus fabricii			47	2M + 15SM + 30 ST/A	64					Russia, White Sea	L-87, L-88
Zoarchias microstomus			28	24M + 4A	52	52			ACN=46?	China (Liaoning)	M-36
Pholidae											
Pholis nebulosa	*Enedrias*		26	26M	52	52			ACN=48?	China (Liaoning)	M-33
Pholis picta	*pictus*	M	46							Japan (Hokkaido)	M-14

Table 6.38 Order PERCIFORMES. Part 3 Notothenioidei, Blennioidei, and Callionymoidei

A	B	C	D	E	F	G	H	I	J	K	L
Current scientific name of taxon Suborder/family/species	Reported in karyotype paper	Sex	2n	Karyotype	NF_1	NF_2	Ag-NORs	Genome size (pg/cell)	Comments	Locality	Reference
Suborder Notothenioidei											
Artedidraconidae											
Artedidraco mirus			46	2M + 2SM + 42A	50	50			ACN=46	South Georgia Is.	P-57
Artedidraco orianae		F, M	46	2M + 6SM + 38A	54	54	2		ACN=48	Weddell Sea	O-70
Artedidraco shackletoni		F, M	46	2M + 6SM + 38A	54	54	2		ACN=48	Weddell Sea	O-70
Hitiodraco velifer		F, M	46	6SM + 40 ST/A	52		2			Ross Sea	C-21
Pogonophryne barsukovi		F, M	46	2M + 4SM + 40A	52	52	2		ACN=48	Weddell Sea	O-70
Pogonophryne dolichobranchiata			46	4M + 2SM + 2ST + 38A	52	54			ACN=48	South Orkney Is.	P-57
Pogonophryne marmorata		M	46	2M + 4SM + 40A	52	52	2		ACN=48	Weddell Sea	O-70
Pogonophryne mentella		F, M	46	2M + 4SM + 40A	52	52	2		ACN=48	Weddell Sea	O-70
Pogonophryne scotti		F	46	2M + 4SM + 40A	52	52	2		ACN=48	Weddell Sea	O-70
Pogonophryne scotti		M	46	6SM + 40A	52	52	2	4.1 FD	ACN=48	Ross Sea	M-89
Bathydraconidae											
Bathydraco marri		F	38	4SM + 34A	42	42			XX, ACN=40	Weddell Sea	O-70
Bathydraco marri		M	39	3SM + 36A	42	42			XY_1Y_2, ACN=40	Weddell Sea	O-70
Cygnodraco mawsoni		F	44-46	44-46 ST/A	44-46					Weddell Sea	O-70
Cygnodraco mawsoni		F	48	2M + 4SM + 42 ST/A	54		2	2.8 FD	ACN=48	Ross Sea	M-89, C-20
Gerlachea australis		F	48	2M + 2-4 SM + 42-44 A	52-54	52-54			ACN=48	Weddell Sea	O-70
Parachaenichthys georgianus			48								O-70
Prionodraco evansii			20								O-70
Psilodraco breviceps			48	48A	48	48					O-70
Racovitzia glacialis		F	36	4M + 32A	40	40				Weddell Sea	O-70
Bovichthyidae (= Bovichtidae)											
Bovichtus angustifrons			48	48A	48	48			ACN=48	Tasmania	M-123
Bovichtus diacanthus		F, M	48	48A	48	48			ACN=48	Tristan da Cunha	M-123
Bovichtus variegatus		F, M	48	48A	48	48			ACN=48	New Zealand	M-123
Cottoperca gobio			48-50	48-50 A	48-50	48-50				Chile	P-55
Cottoperca gobio		F	48	48A	48	48	2		ACN=48	Magellan Strait	P-58

Table 6.38 Order PERCIFORMES. Part 3 Notothenioidei, Blennioidei, and Callionymoidei (continued)

A		B	C	D	E	F	G	H	I	J	K	L
Current scientific name of taxon Suborder/family/species		Reported in karyotype paper	Sex	2n	Karyotype	NF_1	NF_2	Ag-NORs	Genome size (pg/cell)	Comments	Locality	Reference
Channichthyidae												
Chaenocephalus	*aceratus*		M	48	4M + 44A	52	52				South Sandwich Is.	M-88
Chaenodraco	*wilsoni*		F	48	4M + 6SM + 38A	58	58			$X_1X_1X_2X_2$, ACN=48	Antarctica	O-68, O-70
Chaenodraco	*wilsoni*		M	47	5M + 6SM + 36A	58	58			X_1X_2Y, ACN=48	Weddell Sea	O-70, M-88
Champsocephalus	*gunnari*			48						ACN=48	Antarctica	D-4
Channichthys	*rhinoceratus*		F, M	48	2M + 4SM + 40A + 2A-sat.	54	54			ACN=48	Antarctica	D-4
Chionobathyscus	*dewitti*		M	47	5SM + 4-6 SM + 38-36 A	56-58	56-58			X_1X_2Y, ACN=48	Weddell Sea	O-70, M-88
Chionodraco	*hamatus*		F	48	2M + 4SM + 42A	54	54	3		$X_1X_1X_2X_2$, ACN=48	Weddell Sea	O-70
Chionodraco	*hamatus*		F	48	2M + 4SM + 42A	54	54		3.7 FD	$X_1X_1X_2X_2$, ACN=48	Ross Sea	M-88
Chionodraco	*hamatus*		M	47	2M + 4SM + 41A	53	53	2		X_1X_2Y, ACN=48	Ross Sea	M-88
Chionodraco	*myersi*		F	48	2M + 6SM + 40A	56	56			$X_1X_1X_2X_2$, ACN=48	Weddell Sea	O-70
Chionodraco	*myersi*		M	47	2M + 6SM + 39A	55	47			X_1X_2Y, ACN=48	Weddell Sea	O-70, M-88
Chionodraco	*rastrospinosus*		M	48	4M + 44A	52	52				South Orkney Is.	M-88
Cryodraco	*atkinsoni*	*antarcticus*	M	48	2M + 4SM + 42A	54	54			ACN=48	Weddell Sea	O-70
Cryodraco	*atkinsoni*	*antarcticus*	M	48	2M + 4SM + 42A	54	54	2	3.9 FD	ACN=48	Ross Sea	M-89
Neopagetopsis	*ionah*		F, M	48	2M + 8SM + 38A	58	58	2		ACN=48	Weddell Sea	O-70
Pagetopsis	*macropterus*		M	47	3M + 12SM + 32A	62	62			ACN=48	Weddell Sea	O-70
Pagetopsis	*macropterus*		F	48	2M + 12SM + 34A	62	62	2	4.4 FD	$X_1X_1X_2X_2$, ACN=48	Ross Sea	M-88
Pagetopsis	*macropterus*		M	47	3M + 12SM + 32A	62	62			X_1X_2Y, ACN=48	Ross Sea	M-88
Pagetopsis	*maculatus*		F	48	2M + 6SM + 40A	56	56				Weddell Sea	O-70
Pseudochaenichthys	*georgianus*			48	4M + 8SM + 36A	60	60				South Georgia	M-88
Eleginopidae												
Eleginops	*maclovinus*		F, M	48	4M + 2SM + 42A	54	54			ACN=48	S. America (subantarctic)	M-124
Harpagiferidae												
Harpagifer	*antarcticus*			48	2M + 4ST + 42A	50	54			ACN=48	South Orkney Is.	P-56
Harpagifer	sp.			48	4M + 2SM + 8ST + 34A	54	62			ACN=48	Macquarie Is.	P-56

Table 6.38 Order PERCIFORMES. Part 3 Notothenioidei, Blennioidei, and Callionymoidei (continued)

A		B	C	D	E	F	G	H	I	J	K	L
Current scientific name of taxon Suborder/family/species		Reported in karyotype paper	Sex	2n	Karyotype	NF_1	NF_2	Ag-NORs	Genome size (pg/cell)	Comments	Locality	Reference
Nototheniidae												
Dissostichus	*eleginoides*			48	2M + 46A	50	50			ACN=48	Antarctica	D-4
Dissostichus	*eleginoides*		F, M	48	2M + 2SM + 44A	52	52	2			Southern Ocean	G-83
Dissostichus	*mawsoni*		F, M	48	2M + 4SM + 42A	54	54	4			Southern Ocean	G-83
Dissostichus	*mawsoni*		F	48	2M + 2SM + 44A	52	52		(2.1 FIA)	ACN=48	Weddell Sea	O-70, H-41
Gobionotothen	*acuta*	*Notothenia*	F	46	6M + 8SM + 32A	60	60			4B	Antarctica (Heard Is.)	O-67
Gobionotothen	*gibberifrons*	*Notothenia*	F, M	46	4M + 2SM + 40A	52	52			ACN=48	Antarctica (Admiralty Bay)	P-26, P-27
Lepidonotothen	*kempi*	*Notothenia*	F, M	48	4M + 44A	52	52				Antarctica (Prydz Bay)	O-67
Lepidonotothen	*squamifrons*	*Notothenia*	F, M	48	4M + 44A	52	52				Antarctica	O-67
Lindbergichthys	*mizops*	*Notothenia*	F, M	48	4M + 44A	52	52				Antarctica	O-67
Notothenia	*angustata*		F, M	26	24 M/SM + 2A	50	50	2		ACN=48	New Zealand	P-36
Notothenia	*coriiceps neglecta*			22	20M + 2SM	44	44			ACN=44	Duville Is.	P-52
Notothenia	*coriiceps neglecta*		F	22	18M + 2SM + 2ST	42	44			ACN=44	Antarctica (Admiralty Bay)	P-27
Notothenia	*coriiceps*		M	22	18M + 4SM	44	44	2		ACN=44	Ross Sea	M-126
Notothenia	*cyanobrancha*		M	48	4M + 44A	52	52			ACN=48	Antarctica	D-4
Notothenia	*rossi marmorata*			24	24M	48	48			ACN=46	South Georgia	P-52
Notothenia	*rossii rossii*		F, M	24	24M	48	48			ACN=46	Antarctica	D-4
Pagothenia	*borchgrevinki*		F	46	4M + 2SM + 40A	52	52	2		$X_1X_1X_2X_2$, ACN=48	Antarctica	M-126
Pagothenia	*borchgrevinki*		M	45	5M + 2SM + 38A	52	52	2		X_1X_2Y, ACN=48	Antarctica	M-126
Paranotothenia	*magellanica*			26	24M + 2A	50	50			ACN=48	Antarctica	D-4
Paranotothenia	*microlepidota*	*Notothenia*		26	22M + 2SM + 2A	50	50			ACN=48	Cambell Is.	P-51
Patagonotothen	*longipes*		F, M	48	2M + 46A	50	50			ACN=48	Magellanic Region	P-51
Patagonotothen	*ramsayi*			46	4M + 42 ST/A	50					Magellanic Region	P-51
Pleuragramma	*antarcticum*		F, M	48	8M + 12SM + 28A	68	68			ACN=48	Weddell Sea	O-70
Pseudotrematomus	*bernacchii*	*Trematomus*		48	2M + 46 ST/A	50					Antarctica (Bransfield Strait)	P-26
Pseudotrematomus	*bernacchii*	*Trematomus*		48	2M + 2SM + 44A	52	52		(2.4 FIA)	ACN=48	Weddell Sea	O-70, H-41
Pseudotrematomus	*bernacchii*	*Trematomus*	F, M	48	2M + 2SM + 44A	52	52	2	3.6 FD	ACN=48	Ross Sea	M-89
Pseudotrematomus	*bernacchii*	*Pagothenia*	F, M	48	2M + 2SM + 44A	52	52	2		ACN=48	Ross Sea	M-126
Pseudotrematomus	*eulepidotus*	*Trematomus*	F, M	24	20M + 4SM	48	48	2		ACN=46	Weddell Sea	O-70
Pseudotrematomus	*eulepidotus*	*Trematomus*	F, M	24	22M + 2SM	48	48	2		ACN=46	Ross Sea	M-126
Pseudotrematomus	*eulepidotus*	*Trematomus*	F, M	24	8M + 14SM + 2A	46	46			ACN=46	Antarctica (Prydz Bay)	O-67, O-70
Pseudotrematomus	*hansoni*	*Trematomus*		48	2M + 4SM + 42 ST/A	54					Antarctica (Bransfield Strait)	P-26
Pseudotrematomus	*hansoni*	*Trematomus*		48	2M + 4SM + 42A	54	54				Weddell Sea	O-70
Pseudotrematomus	*hansoni*	*Pagothenia*	F	46	4M + 2SM + 40A	52	52	2		$X_1X_1X_2X_2$, ACN=48	Ross Sea	M-126

Table 6.38 Order PERCIFORMES. Part 3 Notothenioidei, Blennioidei, and Callionymoidei (continued)

A	B	C	D	E	F	G	H	I	J	K	L
Current scientific name of taxon Suborder/family/species	Reported in karyotype paper	Sex	2n	Karyotype	NF_1	NF_2	Ag-NORs	Genome size (pg/cell)	Comments	Locality	Reference
Pseudotrematomus hansoni	*Pagothenia*	M	45	5M + 2SM + 38A	52	52	2		X_1X_2Y, ACN=48	Ross Sea	M-126
Pseudotrematomus lepidorhinus	*Trematomus*	F, M	48	4M + 44A	52	52				Weddell Sea	O-67, O-70
Pseudotrematomus pennellii	*Trematomus*	M	28	9M + 2SM + 17A	39	39			4B	Weddell Sea	O-70
Pseudotrematomus pennellii	*Trematomus*	F, M	32	12M + 2SM + 18A	46	46	2		ACN=44	Ross Sea	M-126
Pseudotrematomus scotti	*Trematomus*	F	50	4M + 2-4 SM + 42-44 A	56-58	56-58			ACN=50	Weddell Sea	O-70
Trematomus loenbergii		F, M	28	18M + 6SM + 4A	52	52	2		ACN=48	Ross Sea	M-126
Trematomus loenbergii		F	30	16M + 6SM + 8A	52	52	2			Ross Sea	M-126
Trematomus newnesi		F	46	4M + 2SM + 40A	52	52	2		$X_1X_1X_2X_2$, ACN=48	Ross Sea	M-126
Trematomus newnesi		M	45	5M + 2SM + 38A	52	52	2		X_1X_2Y, ACN=48	Ross Sea	M-126
Trematomus newnesi			69						3X	Ross Sea	M-89
Trematomus nicolai		F	58	2M + 6SM + 18ST + 32A	66	84	2		$X_1X_1X_2X_2$, ACN=58	Ross Sea	M-126
Trematomus nicolai		M	57	3M + 6SM + 18ST + 30A	66	84	2		X_1X_2Y, ACN=58	Ross Sea	M-126
Pseudaphritidae											
Pseudaphritis urvillii	Bovichtidae	F	48	4M + 44A	52	52	6		ACN=48	Tasmania	P-58
Suborder Trachinoidei											
Ammodytidae											
Gymnammodytes cicerelus	*cicerellus*	F, M	46	22 M/SM + 24 ST/A	68				ACN=46	Italy (Palermo)	V-57
Pinguipedidae (= Parapercidae)											
Parapercis kamoharai			48	48A	48	48			ACN=48	Japan (Wakayama)	A-80
Parapercis pulchella			42	8M + 34A	50	50	2		ACN=44	Japan (Shizuoka, Yamaguchi)	M-109
Parapercis sexfasciata			26	22M + 2ST + 2A	48	50	2		ACN=48	Japan (Shizuoka, Yamaguchi)	M-109
Parapercis sexfasciata		F	26	22M + 2ST + 2A	48	50			XX, ACN=48	Japan (Hyogo)	O-40
Parapercis sexfasciata		M	26	22M + 2ST + 2A	48	50			XY, ACN=48	Japan (Hyogo)	O-40
Parapercis sexfasciata		F, M	26	22M + 2ST + 2A	48	50		1.9* FCM		Korea (Busan)	P-70
Trachinidae											
Echiichthys vipera		F, M	48	48A	48	48	2		ACN=48	Italy (Ancona)	C-22
Trachinus draco			48	48A	48	48				USSR	V-72
Trachinus draco		F, M	48	48A	48	48			ACN=48	Italy (Palermo)	V-57

Table 6.38 Order PERCIFORMES. Part 3 Notothenioidei, Blennioidei, and Callionymoidei (continued)

A		B	C	D	E	F	G	H	I	J	K	L
Current scientific name of taxon Suborder/family/species		Reported in karyotype paper	Sex	2n	Karyotype	NF_1	NF_2	Ag-NORs	Genome size (pg/cell)	Comments	Locality	Reference
Trichodontidae												
Arctoscopus	*japonicus*		F, M	48	2SM + 46A	50	50	2	1.5 FCM	XY/XX?	Japan	T-66
Uranoscopidae												
Uranoscopus	*scaber*			26	20M + 2SM + 4 ST/A	48					Croatia	S-86
Uranoscopus	*scaber*			30	18M + 12 ST/A	48		2		ACN=48	Italy (Ancona)	C-22
Uranoscopus	*scaber*			28	20M + 8 ST/A	48				ACN=48	Italy (Ancona)	C-22
Uranoscopus	*scaber*			27	21M + 6 ST/A	48				ACN=48	Italy (Ancona)	C-22
Uranoscopus	*scaber*			32	18 M/SM + 14A	50	50				USSR	V-72
Uranoscopus	*scaber*	A-type		30	22M + 4SM + 4A	56	56	2		ACN=48	Italy (Palermo)	V-104
Uranoscopus	*scaber*			31	23M + 4SM + 4A	58	58				Italy (Palermo)	V-104
Uranoscopus	*scaber*	B-type		29	23M + 2SM + 4A	54	54			ACN=48	Italy (Palermo)	V-104
Uranoscopus	*scaber*	C-type		28	24M + 4A	52	52	2		ACN=48	Italy (Palermo)	V-104
Suborder Blennioidei												
Blenniidae												
Aidablennius	*sphynx*		F, M	48	48 ST/A	48		2		ACN=48	Italy (Ancona)	C-19
Aidablennius	*sphynx*	*Blennius sphinx*		48	48A	48	48			ACN=48	Italy	C-33
Aidablennius	*sphynx*	*Blennius sphinx*	F, M	48	4M + 4SM + 40 ST/A	56			1.2 FD	ACN=48	Spain (Malaga)	C-9, C-10
Atrosalarias	*fuscus holomelas*	*fuscus*		48	48A	48	48			ACN=48	Japan (Ishigakijima)	A-55
Blennius	*ocellaris*		F, M	48	2SM + 2ST + 44A	50	52		1.7 FD	ACN=48	Spain (Malaga)	C-9, C-10
Blennius	*ocellaris*			48	2M + 2ST + 44A	50	52			ACN=48	Italy (Palermo)	V-51
Coryphoblennius	*galerita*	*Blennius*	F, M	48	2M + 2SM + 44 ST/A	52		2			Spain (Mediterranean)	G-18
Istiblennius	*enosimae*			48	2SM + 46A	50	50			ACN=48	Japan (Chiba, Kanagawa)	A-55
Istiblennius	*lineatus*		F	48	48 ST/A	48				ACN=48	Japan (Ishigakijima)	A-59
Lipophrys	*adriaticus*	*Lypophris*	F, M	48	48 ST/A	48		2		ACN=48	Italy (Ancona)	C-19
Lipophrys	*canevae*	*Blennius*		48*	8ST + 40A	48	56			ACN=48	Italy	C-33
Lipophrys	*pholis*	*Blennius*	F, M	46	6M + 2SM + 8ST + 30A	54	62	2	1.6 FD	ACN=48	Spain (Malaga)	C-9, C-10, G-18
Omobranchus	*elegans*		F, M	42	10M + 2SM + 6ST + 24A	54	60			ACN=48	Japan (Kanagawa)	A-59
Omobranchus	*punctatus*			44	4M + 40A	48	48			ACN=48	Japan (Wakayama)	A-80
Parablennius	*gattorugine*		F	48	48 ST/A	48		2		ACN=48	Italy (Ancona)	C-19
Parablennius	*gattorugine*	*Blennius*	F, M	48	2M + 2SM + 44A	52	52	2	1.2 FD	ACN=48	Spain (Malaga)	C-9, C-10, G-18
Parablennius	*gattorugine*	*Blennius*	F, M	48	2M + 2SM + 44A	52	52			ACN=48	Italy (Palermo)	V-51
Parablennius	*incognitus*	*Blennius*		48	48A	48	48			ACN=48	Italy	C-33

Table 6.38 Order PERCIFORMES. Part 3 Notothenioidei, Blennioidei, and Callionymoidei (continued)

A	B	C	D	E	F	G	H	I	J	K	L
Current scientific name of taxon Suborder/family/species	Reported in karyotype paper	Sex	2n	Karyotype	NF_1	NF_2	Ag-NORs	Genome size (pg/cell)	Comments	Locality	Reference
Parablennius incognitus	*Blennius ponticus incognitus*	F, M	48	4ST + 44A	48	52	2	1.6 FD	ACN=48	Spain (Malaga)	C-9, C-10, G-18
Parablennius pilicornis		F, M	48	8ST + 40A	48	56				Italy (Palermo)	C-28
Parablennius pilicornis			48	48A	48	48				Brazil (RJ)	B-86
Parablennius ponticus		F, M	48	48 ST/A	48		2		ACN=48	Italy (Ancona)	C-19
Parablennius sanguinolentus	*Blennius*	F, M	48	8ST + 40A	48	56			ACN=48	Italy (Tyrrhenian Sea)	C-31, C-33
Parablennius sanguinolentus	*Blennius*	F, M	48	6ST + 42A	48	54	2	1.8 FD	ACN=48	Spain (Malaga)	C-9, C-10, G-18
Parablennius sanguinolentus	*Blennius*	F, M	48	22ST + 26A	48	70			ACN=48	Black Sea	A-95
Parablennius sanguinolentus	*Blennius*	M	47	1M + 22ST + 24A	48	70			ACN=48	Black Sea	A-95
Parablennius tentacularis	*Blennius*	F	48	48 ST/A	48				XX, ACN=48	Italy (Palermo)	C-24
Parablennius tentacularis	*Blennius*	M	48	1ST + 47 ST/A	48				XY, ACN=48	Italy (Palermo)	C-24
Parablennius tentacularis	*Blennius*	M	47	1SM + 46 ST/A	48				neo Y, ACN=48	Italy (Palermo)	C-24
Parablennius tentacularis		F	48	48 ST/A	48		2		XX, ACN=48	Italy (Ancona)	C-19
Parablennius tentacularis		M	48	1ST + 47 ST/A	48		2		XY, ACN=48	Italy (Ancona)	C-19
Parablennius tentacularis		M	47	1SM + 46 ST/A	48		2		neo Y, ACN=48	Italy (Ancona)	C-19
Parablennius tentacularis	*Blennius*	F	48	48A	48	48			ACN=48	Black Sea	A-95
Parablennius tentacularis	*Blennius*	M	47	1 large A + 46 ST/A	47				ACN=48	Black Sea	A-95
Parablennius tentacularis	*Blennius*	M	47	1 large SM + 46 ST/A	48				ACN=48	Black Sea	A-95
Parablennius tentacularis	*Blennius*		48	48 ST/A	48				ACN=48	USSR	V-72
Parablennius yatabei	*Blennius*	F, M	48	6SM + 12ST + 30A	54	66			ACN=48	Japan (Kanagawa)	A-59
Paralipophrys trigloides	*Blennius*		48	2M + 6SM + 18ST + 22A	56	74			ACN=48	Italy	C-33
Paralipophrys trigloides	*Blennius*	F, M	46	4M + 4SM + 10ST + 28A	54	64	2	2.4 FD	ACN=48	Spain (Malaga)	C-9, C-10, G-18
Paralipophrys trigloides	*Blennius*	F, M	48	2M + 6SM + 40 ST/A	56				ACN=48	Italy (Palermo)	V-51
Petroscirtes breviceps	*Dasson trossulus*	F	40	8M + 32 ST/A	48			1.6* FCM	ACN=44	Japan (Chiba)	A-59, O-48
Salaria fluviatilis	*Blennius*	M	48							Italy (Roma)	C-31, C-33
Salaria pavo	*Blennius*	M	48	8ST + 40A	48	56			ACN=48	Italy (Tyrrhenian Sea)	C-31, C-33
Salaria pavo	*Blennius*	F, M	48	4SM + 12ST + 32A	52	64	2	2.1 FD	ACN=48	Spain (Malaga)	C-9, C-10, G-18
Salarias fasciatus			48	48A	48	48		1.7 FIA	ACN=48	Japan (Okinawa)	A-55, H-40
Salarias luctuosus			48	48 ST/A	48				ACN=48	Japan (Okinawa)	A-59
Scartella cristata	*Blennius cristatus*	F, M	48	2ST + 46A	48	50			ACN=48	Italy (Palermo)	V-51
Scartella cristata		F, M	48	2SM + 46 ST/A	50		2			Brazil (RJ)	B-65
Clinidae											
Clinitrachus argentatus	*Clinethrachus*	F, M	48	48 ST/A	48				ACN=48	Italy (Palermo)	V-51

Table 6.38 Order PERCIFORMES. Part 3 Notothenioidei, Blennioidei, and Callionymoidei (continued)

A		B	C	D	E	F	G	H	I	J	K	L
Current scientific name of taxon Suborder/family/species		Reported in karyotype paper	Sex	2n	Karyotype	NF_1	NF_2	Ag-NORs	Genome size (pg/cell)	Comments	Locality	Reference
Labrisomidae												
Labrisomus	*nuchipinnis*			48			50				Brazil (RN, RJ)	G-12
Suborder Gobiesocoidei												
Gobiesocidae												
Conidens	*laticephalus*			42	12M + 10SM + 20 ST/A	64					Japan (Chiba)	A-68
Diademichthys	*lineatus*			47	3M + 11SM + 33 ST/A	61				sex chrom.?, ACN=48	Japan (Okinawa)	A-68
Lepadichthys	*frenatus*			48	8 M/SM + 40 ST/A	56				ACN=48	Japan (Yakushima)	A-68
Lepadogaster	*candolii*	*candollei*	F, M	46	12M + 18SM + 16ST	76	92	2		XX/XY, ACN=46	Spain (Malaga)	T-36
Lepadogaster	*lepadogaster*		M	46	13M + 17SM + 16 ST/A	76		2		ACN=46	Black Sea	A-89
Suborder Callionymoidei												
Callionymidae												
Eleutherochir	*mirabilis*	*Draculo*		36	36A	36	36			ACN=36	Japan (Hokkaido)	S-16
Repomucenus	*beniteguri*	*Callionymus*	F	38	38ST	38	76			$X_1X_1X_2X_2$, ACN=38	Japan (Yamaguchi, Shizuoka)	M-105
Repomucenus	*beniteguri*	*Callionymus*	M	37	1M + 36ST	38	74	2		X_1X_2Y, ACN=38	Japan (Yamaguchi, Shizuoka)	M-105
Repomucenus	*huguenini*	*Callionymus doryssus*	M	32	2M + 30A	34	34	2		ACN=34	Japan (Yamaguchi, Shizuoka)	M-110
Repomucenus	*ornatipinnis*	*Callionymus*	F	38	38ST	38	76			$X_1X_1X_2X_2$, ACN=38	Japan (Yamaguchi)	M-105
Repomucenus	*ornatipinnis*	*Callionymus*	M	37	1M + 36ST	38	74	2		X_1X_2Y, ACN=38	Japan (Yamaguchi)	M-105
Repomucenus	*richardsonii*	*Callionymus punctatus*	F, M	38	38A	38	38	2		ACN=38	Japan (Yamaguchi, Shizuoka)	M-110
Repomucenus	*richardsonii*	*Callionymus punctatus*	M	38	36 SM/ST + 2A		74		1.6* FCM	ACN=38	Japan (Hyogo)	O-46, O-48

Table 6.39 Order PERCIFORMES. Part 4 Gobioidei, Kurtoidei, Acanthuroidei, and Scombroidei

A		B	C	D	E	F	G	H	I	J	K	L
Current scientific name of taxon Suborder/family/subfamily/species		Reported in karyotype paper	Sex	2n	Karyotype	NF_1	NF_2	Ag-NORs	Genome size (pg/cell)	Comments	Locality	Reference
Suborder Gobioidei												
Odontobutidae												
Micropercops	*swinhonis*	*Hypseleotris*		44	44A	44	44				China (Wuhan)	D-18
Odontobutis	*obscura*		F, M	44	44 ST/A	44			(2.4* FCM)	ACN=44	Japan (Oita, Kagoshima)	A-60, O-48
Odontobutis	*obscura*			44	44A	44	44			ACN=44	Japan (Yamaguchi)	N-31
Odontobutis	*obscura*			44	44A	44	44				Korea	L-15
Odontobutis	*obscura*			44	44A	44	44				China (Hubei)	D-18
Odontobutis	*obscura*		F, M	44	4SM + 40A	48	48		(2.8* FD)	ACN=44	China (Wuhan)	G-67, Y-15, C-83
Odontobutis	*platycephalus*		F, M	44	44 ST/A	44				ACN=44	Korea (Kwanchon)	L-79, L-14
Perccottus	*glehni*		F, M	44	44 ST/A	44			2.2 FCM	ACN=44	Russia (Moscow)	K-104, V-86
Perccottus	*glehni*			44	1SM + 43 ST/A	45				ACN=44	Russia (Moscow)	K-104
Eleotridae												
Butinae												
Bostrychus	*sinensis*	*Bostrichthys*		48	4 M/SM + 44 ST/A	52				ACN=48	Japan (Okinawa)	A-60
Bostrychus	*sinensis*	*Bostrichthys*		48	4SM + 2ST + 42A	52	52			ACN=48	China (Zhejiang)	F-8
Ophiocara	*porocephala*			48	48 ST/A	48				ACN=48	Thailand	A-73
Oxyeleotris	*lineolatus*			46	2SM + 8ST + 36A	48	56				(Australia)	C-90
Oxyeleotris	*marmorata*			46	2M + 2SM + 42 ST/A	50			(2.5 FIA)	ACN=46	Thailand (Khon Kaen)	A-73, H-40
Oxyeleotris	*marmorata*			46	2M + 2SM + 42A	50	50				Thailand	D-19
Oxyeleotris	*marmorata*			46	2M + 2ST + 42A	48	50				China (Guangdong)	Z-30
Oxyeleotris	*urophthalmoides*			46	6M + 6SM + 8ST + 26A	58	66				Thailand	D-28
Eleotrinae												
Dormitator	*latifrons*			46	12M + 22SM + 10ST + 2A	80	90			ACN=46	Mexico (Pacific coast)	U-45, U-46
Dormitator	*maculatus*		F, M	46	12M + 22SM + 10ST + 2A	80	90			ACN=46	Mexico (Veracruz)	U-46, M-18
Dormitator	*maculatus*		F, M	46	40 M/SM + 6 ST/A	86		2		ACN=46	Brazil (RN)	M-80
Dormitator	*maculatus*		F	46	14M + 28SM + 2ST + 2A	88	90	2		XX, ACN=46	Brazil (SP)	O-76
Dormitator	*maculatus*		M	46	13M + 28SM + 3ST + 2A	87	90	2		XY, ACN=46	Brazil (SP)	O-76
Eleotris	*acanthopoma*		M	46	46 ST/A	46				ACN=46	Japan (Okinawa)	A-57
Eleotris	*oxycephala*		F, M	46	46A	46	46			ACN=46	China (Guangdong)	G-67, Y-15
Eleotris	*picta*			52	52A	52	52				Mexico	U-82
Eleotris	*pisonis*			46	2M + 44A	48					Mexico	M-80
Eleotris	*pisonis*		F, M	46	46A	46	46	2		ACN=46	Brazil (RN)	M-80
Gobiomorus	*dormitor*		F, M	48	2M + 4SM + 42A	54	54			ACN=48	Mexico (Veracruz)	M-18
Hypseleotris	*cyprinoides*			48	48A	48	48			ACN=48	Japan	S-122
Mogurnda	*mogurnda*		F, M	46	6SM + 40 ST/A	52				ACN=46	Australia	A-60

Table 6.39 Order PERCIFORMES. Part 4 Gobioidei, Kurtoidei, Acanthuroidei, and Scombroidei (continued)

A		B	C	D	E	F	G	H	I	J	K	L
Current scientific name of taxon Suborder/family/subfamily/species		Reported in karyotype paper	Sex	2n	Karyotype	NF_1	NF_2	Ag-NORs	Genome size (pg/cell)	Comments	Locality	Reference
Ptereleotridae												
Parioglossus	*raoi*		F, M	46	46A	46	46			ACN=46	Fiji	W-22
Gobiidae												
Gobiinae												
Acentrogobius	*pflaumi*			50	48 M/SM/ST + 2A		98			ACN=50	Japan (Shimonoseki)	N-31
Acentrogobius	*pflaumi*		F, M	50	36 M/SM + 14 ST/A	86				ACN=50	Korea (Kunsan)	L-14
Amblygobius	*phalaena*	*albimaculatus*		44	2M + 42 ST/A	46			2.0 FIA	ACN=44	Japan (Okinawa)	A-60, H-40
Aphia	*minuta*		F, M	44	44A	44	44			ACN=44	Italy (Ortona)	C-18
Aphia	*minuta*		M	43	1ST + 42A	43	44			ACN=44	Italy (Ortona)	C-18
Aphia	*minuta*		F, M	42	1M + 1ST + 40A	43	44			ACN=44	Italy (Ortona)	C-18
Aphia	*minuta*		M	42	2M + 40A	44	44			ACN=44	Italy (Ortona)	C-18
Aphia	*minuta*		F, M	41	2M + 1ST + 38A	43	44			ACN=44	Italy (Ortona)	C-18
Bathygobius	*fuscus*			48	48A	48	48			ACN=48	Japan (Chiba, Kanagawa)	A-52, A-62
Bathygobius	*fuscus?*			44	44A	44	44				Thailand	D-19
Bathygobius	*soporator*			46			48				Brazil (RN)	G-12
Benthophilus	*leobergius*	*stellatus leobergius*		44	2SM + 2ST + 40A	46	48				Russia (Volga R.)	G-63
Benthophilus	*stellatus*	*stellatus stellatus*		44	2ST + 42A	44	46			ACN=46	Russia (Volga R.)	G-63
Benthophilus	*leobergius*	*stellatus stellatus*		44	1SM + 2ST + 41A	45	47				Russia (Volga R.)	G-63
Caspiosoma	*caspium*			48	4SM + 30ST + 14A	52	82				Russia (Don R.)	G-63
Favonigobius	*gymnauchen*		F, M	48	48 M/SM	96	96			ACN=50	Korea (Kunsan)	L-14
Glossogobius	*giuris*		F, M	46	46A	46	46			ACN=46	India (Orissa)	R-57, M-24
Glossogobius	*olivaceus*	*fasciapunctatus*		46	16SM + 6ST + 24A	62	68			ACN=46	China (Zhejiang)	F-8
Gobiodon	*citrinus*		F	44	2M + 42 ST/A	46			2.1 FIA	ACN=46	Japan (Okinawa)	A-57, H-41
Gobiodon	*citrinus*		M	43	1M + 42 ST/A	44				ACN=44	Japan (Okinawa)	A-57
Gobiodon	*quinquestrigatus*			44	44 ST/A	44			2.2 FIA	ACN=44	Japan (Okinawa)	A-73, H-40
Gobiodon	*rivulatus rivulatus*			44	44 ST/A	44				ACN=44	Australia (Heron Is.)	A-73
Gobiopsis	*macrostoma*	*Gobiophis*		46	10M + 4SM + 32A	60	60				India (WB)	K-46
Gobiosoma	*macrodon*		F, M	38	38A	38	38				Venezuela	A-13
Gobiosoma	*zebrellus*	*zebrella*	F, M	38	38A	38	38				Venezuela	A-13
Gobiosoma	*zebrellus*	*zebrella*	F	37	1M + 36A	38	38				Venezuela	A-13
Gobius	*bucchichi*		M	40	4M + 2SM + 34A	46	46		0.9 FD		Spain (Málaga)	T-32, C-9
Gobius	*bucchichi*		F, M	44	2SM + 42A	46	46			ACN=44	Spain (Málaga)	T-31
Gobius	*cobitis*		F, M	46	46A	46	46		1.3 FD	ACN=46	Spain (Málaga)	T-32, C-9
Gobius	*cobitis*			46	46A	46	46			ACN=46	Italy (Tyrrhenian Sea)	C-31

Table 6.39 Order PERCIFORMES. Part 4 Gobioidei, Kurtoidei, Acanthuroidei, and Scombroidei (continued)

A		B	C	D	E	F	G	H	I	J	K	L
Current scientific name of taxon Suborder/family/subfamily/species		Reported in karyotype paper	Sex	2n	Karyotype	NF_1	NF_2	Ag-NORs	Genome size (pg/cell)	Comments	Locality	Reference
Gobius	*cobitis*			46	46A	46	46			ACN=46	Black Sea	G-63
Gobius	*couchi*		M	46	2SM + 44A	48	48				Europe	M-116
Gobius	*cruentatus*		F, M	46	2ST + 44A	46	48			ACN=46	Spain (Málaga)	T-31
Gobius	*fallax*		M	38	8 M/SM + 30A	46	46			ACN=46	Spain (Málaga)	T-37
Gobius	*fallax*		M	39	7 M/SM + 32A	46	46			ACN=46	Spain (Málaga)	T-37
Gobius	*fallax*		F, M	40	6 M/SM + 34A	46	46	1-4	0.9 FD	ACN=46	Spain (Málaga)	T-30, T-32, T-37
Gobius	*fallax*			40	7 M/SM + 33A	47	47			ACN=46	Spain (Málaga)	T-37
Gobius	*fallax*		F, M	41	5 M/SM + 36A	46	46			ACN=46	Spain (Málaga)	T-37
Gobius	*fallax*		F, M	42	4 M/SM + 38A	46	46			ACN=46	Spain (Málaga)	T-37
Gobius	*fallax*		M	43	3 M/SM + 40A	46	46			ACN=46	Spain (Málaga)	T-37
Gobius	*niger*			46							UK	M-116
Gobius	*niger*		M	50	1M + 12SM + 37A	63	63		2.0 FD	XY, ACN=46	Spain (Málaga)	T-32, C-9
Gobius	*niger*		F	48	4M + 4SM + 10ST + 30A	56	66	2		ACN=50	Norway	K-131
Gobius	*niger*		F	50	4M + 4SM + 10ST + 32A	58	68			ACN=52	Norway	K-131
Gobius	*niger*		F	51	3M + 4SM + 10ST + 34A	58	68				Norway	K-131
Gobius	*niger jozo*			48	2M + 6SM + 8ST + 32A	56	64				Italy (Rome)	C-31
Gobius	*niger jozo*		M	50*							Italy (Venice)	C-76
Gobius	*niger jozo*	A-type	F, M	52	8 M/SM + 10ST + 28A + 6MC	60	70			ACN=46	Italy (Palermo)	V-59
Gobius	*niger jozo*	B-type	F, M	51	9 M/SM + 10ST + 26A + 6MC	60	70			ACN=46	Italy (Palermo)	V-59
Gobius	*niger jozo*	C-type	F, M	50	10 M/SM + 10ST + 30A	60	70			ACN=46	Italy (Palermo)	V-59
Gobius	*niger jozo*	D-type	M	49	11 M/SM + 10ST + 28A	60	70			ACN=46	Italy (Palermo)	V-59
Gobius	*niger jozo*			52	4SM + 48 ST/A	56					Black Sea	V-22
Gobius	*niger jozo*			52	1M + 4SM + 47 ST/A	57					Black Sea	V-22
Gobius	*ophiocephalus*			45	1M + 44A	46	46			ACN=46	Black Sea	V-72
Gobius	*ophiocephalus*			46	46A	46	46			ACN=46	Black Sea	V-72
Gobius	*paganellus*			45	1M + 44 ST/A	46		4			Spain (Mediterranean)	G-76
Gobius	*paganellus*			45	3M + 42 ST/A	48		4			Spain (Mediterranean)	G-76
Gobius	*paganellus*		F, M	45	2 M/SM + 43 ST/A	47		4		ACN=46	Mediterranean	G-76, V-49
Gobius	*paganellus*		M	46	46 ST/A	46				ACN=46	Spain (Málaga)	T-33
Gobius	*paganellus*		F	46	1 M/SM + 45 ST/A	47				ACN=46	Italy (Palermo)	V-49
Gobius	*paganellus*		F, M	46	2 M/SM + 44 ST/A	48				ACN=48	Mediterranean	V-49, T-33
Gobius	*paganellus*		F	46	2M + 44A	48	48		0.8 FD	XX, ACN=46	Spain (Málaga)	T-32, A-44, C-9
Gobius	*paganellus*		M	46	1M + 45A	47	47			XY, ACN=46	Spain (Málaga)	T-32
Gobius	*paganellus*		F, M	47	1 M/SM + 46 ST/A	48				ACN=47	Italy (Palermo)	V-49
Gobius	*paganellus*		F, M	47	1 M/SM + 46 ST/A	48				ACN=48	Mediterranean	V-49, T-33
Gobius	*paganellus*		M	47	47 ST/A	47				ACN=47	Spain (Málaga)	T-33

Table 6.39 Order PERCIFORMES. Part 4 Gobioidei, Kurtoidei, Acanthuroidei, and Scombroidei (continued)

A	B	C	D	E	F	G	H	I	J	K	L
Current scientific name of taxon Suborder/family/subfamily/species	Reported in karyotype paper	Sex	2n	Karyotype	NF_1	NF_2	Ag–NORs	Genome size (pg/cell)	Comments	Locality	Reference
Gobius paganellus		F	48	48A	48	48			ACN=48	Italy (Palermo)	V–49
Gobius paganellus		F, M	48	48 ST/A	48				ACN=48	Mediterranean	V–49, T–33
Gobius paganellus		F	48	1SM + 47A	49	49			ACN=48	Italy (Palermo)	V–49
Gobius paganellus		M	48	1M/SM + 47A	49	49				Spain (Málaga)	T–33
Gobiusculus flavescens		F, M	46	6 M/SM + 40A	52	52	2		ACN=46	Norway	K–131
Mesogobius batrachocephalus			30	16 M/SM + 14A	46	46			ACN=46	Black Sea	G–64
Mesogobius batrachocephalus	*Gobius*		30	16M + 14A	46	46				Black Sea	I–18
Neogobius cephalargoides			46	46A	46	46				Black Sea	E–9
Neogobius constructor	*platygobius construtor*		42	4 M/SM + 38A	46	46			ACN=46	Georgia (Tbilisi Reservoir)	V–18, V–23, V–103
Neogobius constructor	*platygobius construtor*		44	2M + 42A	46	46			ACN=46	Black Sea basin	V–18, V–23, V–103
Neogobius constructor	*platygobius construtor*		44	2M + 2SM + 40A	48	48			ACN=46	Black Sea basin	V–18, V–23, V–103
Neogobius cyrius	*platygobius construtor*		37	9 M/SM + 2ST + 26A	46	48			ACN=46	Georgia (Kura R.)	V–18, V–23, V–103
Neogobius cyrius	*platygobius construtor*		38	8 M/SM + 30A	46	46			ACN=46	Georgia (Tbilisi Reservoir)	V–18, V–23, V–103
Neogobius cyrius	*platygobius construtor*		40	6 M/SM + 34A	46	46			ACN=46	Georgia (Kura R.)	V–18, V–23, V–103
Neogobius cyrius	*platygobius construtor*		41	5 M/SM + 1ST + 35A	46	47			ACN=46	Georgia (Kura R.)	V–23, V–103
Neogobius eurycephalus		F, M	32	12M + 2SM + 18A	46	46	2		ACN=46	Danube Delta system	E–9
Neogobius eurycephalus		F, M	31	13M + 2SM + 16A	46	46	2		ACN=46	Danube Delta system	E–9
Neogobius eurycephalus		M	30	14M + 2SM + 14A	46	46	2		ACN=46	Danube Delta system	E–9
Neogobius fluviatilis fluviatilis			46	46A	46	46			ACN=46	Russia (Don R.)	G–63
Neogobius fluviatilis pallasi			46	46A	46	46			ACN=46	Russia (Don R.)	G–63
Neogobius gorlap		F, M	46	46 ST/A	46					Caspian basin	V–21
Neogobius gymnotrachelus			46	46A	46	46			ACN=46	Black Sea basin	G–64
Neogobius gymnotrachelus			46	2 SM/ST + 44A		48			ACN=46	Black Sea basin	G–64
Neogobius gymnotrachelus			46	1M + 1SM + 44A	48	48			ACN=46	Black Sea basin	G–64
Neogobius kessleri		F	30	14M + 2SM + 14 ST/A	46					Black Sea basin	V–21
Neogobius kessleri		M	29	15M + 2SM + 12 ST/A	46					Black Sea basin	V–21
Neogobius kessleri kessleri			30	14M + 2SM + 14A	46	46			ACN=46	Black Sea basin	G–64
Neogobius kessleri kessleri			29	17 M/SM + 12A	46	46			ACN=45	Black Sea basin	G–64
Neogobius melanostomus	*Gobius*		46	46A	46	46		(2.5 FIA)		Black Sea basin	I–18, H–41
Neogobius melanostomus affinis			46	46A	46	46				Black Sea basin	V–72
Neogobius melanostomus	*cephalarges*		46	46A	46	46			ACN=46	Azov Sea basin	V–23
Neogobius rhodioni	*platyrostris*		46	46A	46	46			ACN=46	Black Sea basin	V–23, V–103
Padogobius bonelli	*martensii*	F	46	3M + 2SM + 3ST + 38A	51	54			ACN=47	Italy (Liguria)	C–31
Pomatoschistus lozanoi		F, M	37	3M + 12SM + 10ST + 12A	52	62				Dutch North Sea	W–21
Pomatoschistus microps		F, M	46	30SM + 16ST	76	92			ACN=46	UK (Devon)	W–23

Table 6.39 Order PERCIFORMES. Part 4 Gobioidei, Kurtoidei, Acanthuroidei, and Scombroidei (continued)

A		B	C	D	E	F	G	H	I	J	K	L
Current scientific name of taxon Suborder/family/subfamily/species		Reported in karyotype paper	Sex	2n	Karyotype	NF_1	NF_2	Ag-NORs	Genome size (pg/cell)	Comments	Locality	Reference
Pomatoschistus	*microps*		F, M	46	4M + 16SM + 20ST + 6A	66	86			ACN=46	Germany (Baltic Sea)	K-74
Pomatoschistus	*minutus*		F, M	46	4M + 16SM + 16ST + 10A	66	82			ACN=46	Germany (Baltic Sea)	K-74
Pomatoschistus	*minutus*		F, M	46	18SM + 18ST + 10A	64	82	2		ACN=46	Norway	K-131
Pomatoschistus	*minutus*		F, M	46	6SM + 24ST + 16A	52	76			ACN=46	UK (Plymouth)	W-21
Pomatoschistus	*norvegicus*		F, M	32	10M + 10SM + 8ST + 4A	52	60				UK (Plymouth)	W-21
Pomatoschistus	*pictus*		F, M	46	22 M/SM + 12ST + 12A	68	80	2		ACN=46	Norway	K-131
Proterorhinus	*marmoratus*		F, M	46	46A	46	46			XX/XY	Slovakia	R-8
Proterorhinus	*marmoratus*			46	46A	46	46				Russia (Don R.)	G-63, V-72
Valenciennea	*muralis*	*Eleotris*		46	46A	46	46			ACN=46	India (WB)	K-139
Valenciennea	*strigata*	*Eleotriodes strigatus*		44	2M + 42 ST/A	46				ACN=44	Japan (Okinawa)	A-57
Yongeichthys	*criniger*	*Ctenogobius*	F, M	50	34 M/SM + 6ST + 10A	84	90			ACN=50	Japan (Okinawa)	A-57
Zosterisessor	*ophiocephalus*		F, M	46	2M + 44 ST/A	48		2		ACN=46	Italy (Ancona)	C-15
Zosterisessor	*ophiocephalus*	*Gobius*		46	46A	46	46				Black Sea	V-72
Gobionellinae												
Acanthogobius	*elongata?*	sp.	F, M	42	8 M/SM + 34 ST/A	50				ACN=46	Korea (Kunsan)	L-14
Acanthogobius	*flavimanus*		M	44	36ST + 8A	44	80			ACN=44	Japan (Kanagawa)	A-52, A-62
Acanthogobius	*flavimanus*			44	10ST + 34A	44	54			ACN=44	Japan (Yamaguchi)	N-31
Acanthogobius	*flavimanus*		F, M	44	44A	44	44				Korea (Dadaepo)	L-14
Acanthogobius	*flavimanus*			44	44A	44	44			ACN=44	China (Shandong)	W-7
Acanthogobius	*hasta*		F, M	44	6 M/SM + 38 ST/A	50				ACN=44	Korea (Kunsan)	L-14
Acanthogobius	*hasta*		F, M	44	2M + 42 ST/A	46				ACN=44	Korea (Gyehwato)	L-79
Acanthogobius	*hasta*	*Synechogobius*		44	2M + 42 ST/A	46				ACN=44	Japan (Ariake Sea)	A-62
Acanthogobius	*hasta*	*Synechogobius*		44	2M + 42 ST/A	46				ACN=44	China	Y-21
Acanthogobius	*lactipes*		F, M	40	40A	40	40			ACN=46	Korea (Pi-in)	L-14
Acanthogobius	*lactipes*	*Aboma*	F, M	40	40 ST/A	40		2		ACN=44	Japan (Lake Kasumigaura)	A-52, A-57
Awaous	*flavus*	*strigatus*		46						X_1X_2Y	(Brazil)	O-86
Awaous	*grammepomus*		F, M	46	46A	46	46			ACN=46	India (WB)	K-42
Awaous	*tajasica*			46	46A	46	46				Brazil	S-196
Brachygobius	*nunus*			48							(S. Asia)	P-44
Chaenogobius	*annularis*	*Chasmichthys dolichognathus*		44	44 ST/A	44				ACN=44	Japan (Kanagawa)	A-52, A-62
Chaenogobius	*annularis*	*Chasmichthys dolichognathus*	F, M	44	44A	44	44			ACN=44	Korea (Kunsan)	L-79, L-14
Chaenogobius	*gulosus*	*Chasmichthys*		44	44 ST/A	44				ACN=44	Japan (Chiba, Kanagawa)	A-62
Chaenogobius	*gulosus*	*Chasmichthys*		44	16ST + 28A	44	60			ACN=44	Japan (Yamaguchi)	N-31
Chaenogobius	*gulosus*	*Chasmichthys*	F, M	44	8 M/SM + 36 ST/A	52				ACN=44	Korea (Woolsan)	L-14
Ctenogobius	*shufeldti*	*Gobionellus*	F	48	48A	48	48			$X_1X_1X_2X_2$, ACN=48	USA (LA)	P-24

Table 6.39 Order PERCIFORMES. Part 4 Gobioidei, Kurtoidei, Acanthuroidei, and Scombroidei (continued)

A		B	C	D	E	F	G	H	I	J	K	L
Current scientific name of taxon Suborder/family/subfamily/species		Reported in karyotype paper	Sex	2n	Karyotype	NF_1	NF_2	Ag-NORs	Genome size (pg/cell)	Comments	Locality	Reference
Ctenogobius	*shufeldti*	*Gobionellus*	M	47	1M + 46A	48	48			X_1X_2Y, ACN=48	USA (LA)	P-24
Gillichthys	*mirabilis*		F, M	44	12ST + 32A	44	56			ACN=44	USA (CA)	C-53
Gillichthys	*seta*		F	44	6M + 14ST + 24A	50	64			ACN=44	Gulf of California	C-53
Gobionellus	*microdon*			56	4M + 6SM + 46 ST/A	66					Mexico	U-82
Gymnogobius	*breunigii*	*Chaenogobius*		44	36 M/SM/ST + 8A		80			ACN=44	Japan (Yamaguchi)	N-31
Gymnogobius	*breunigii*	*Chaenogobius*	F, M	42	22 M/SM + 20 ST/A	64				ACN=42	Korea (Samchuk)	L-14
Gymnogobius	*heptacanthus*	*Chaenogobius*	M	44	44 ST/A	44					Korea (Dolsanto)	L-79
Gymnogobius	*isaza*	*Chaenogobius*		44	12SM + 32ST	56	88			ACN=44	Japan (Lake Biwa)	A-62
Gymnogobius	*castaneus*	*Rhodoniichthys laevis*	F, M	42	14 M/SM + 28ST	56	84				Japan (Tokyo)	A-60
Gymnogobius	*castaneus*	*Chaenogobius urotaenia*		42	14SM + 28ST	56	84			ACN=42	Japan (Hokkaido)	Y-5
Gymnogobius	*mororanus*	*Chaenogobius*	F, M	42	12 M/SM + 30 ST/A	54				ACN=42	Korea (Kunsan)	L-14
Gymnogobius	*urotaenia*	*Chaenogobius annularis*		44	36 M/SM + 8ST	80	88		(2.4* FCM)	ACN=44	Japan (Kanto district)	A-60, O-48
Gymnogobius	*urotaenia*	*Chaenogobius annularis*		44	18SM + 26ST	62	88			ACN=44	Japan (Lake Biwa)	A-62
Gymnogobius	*urotaenia*	*Chaenogobius annularis*		44	38 SM/ST + 6A		82			ACN=44	Japan (Yamaguchi)	N-31
Gymnogobius	*urotaenia*	*Chaenogobius annularis*	F, M	44	40 M/SM + 4 ST/A	84				ACN=44	Korea (Kum R.)	L-14
Gymnogobius	*urotaenia*	*Chaenogobius annularis*		44	20M + 22SM + 2A	86	86			ACN=44	China (Liaoning)	M-35
Luciogobius	*grandis*			44						ACN=44	Japan (Izu Peninsula)	A-123
Luciogobius	*guttatus*		F	44	12ST + 32A	44	56			ACN=44	Japan (Izu Peninsula)	M-102
Luciogobius	*guttatus*			44							Japan (Kanagawa)	A-52
Luciogobius	*guttatus*			44	14M + 14SM + 16A	72	72			ACN=44	China (Liaoning)	M-32
Mugilogobius	*abei*		F, M	46	46 ST/A	46				ACN=46	Korea (Kunsan)	L-14
Mugilogobius	*abei*			46		46					Japan (Okayama)	A-52
Pterogobius	*elapoides*			44	14SM + 30ST	58	88			ACN=44	Japan (Chiba, Kanagawa)	A-52, A-62
Pterogobius	*zonoleucus*			44	14SM + 30 ST/A	58				ACN=44	Japan (Kanagawa)	A-62
Quietula	*guaymasiae*		F, M	42	6M + 4SM + 32A	52	52			ACN=42	Gulf of California	C-77
Quietula	*y-cauda*		F	42	42A	42	42			ACN=42	Gulf of California	C-77
Rhinogobius	sp. CB	*brunneus*, Shimayoshinobori		44	44 ST/A	44				ACN=44	Japan (Chiba)	A-62
Rhinogobius	sp. LD	*brunneus*, Ooyoshinobori	F, M	44	44 ST/A	44				ACN=44	Japan (Itoh)	A-62
Rhinogobius	sp. DA	*brunneus*, Kuroyoshinobori	F, M	44	44 ST/A	44				ACN=44	Japan (Tokyo)	A-62
Rhinogobius	*brunneus*		F, M	44	44A	44	44			ACN=44	W. Japan	N-1, N-31
Rhinogobius	*brunneus*		F, M	44	44A	44	44			ACN=44	Korea (Pongdong)	L-14
Rhinogobius	*flumineus*	*Tukugobius*	F, M	44	44A	44	44		3.0* FCM	ACN=44	W. Japan	N-1, O-48
Rhinogobius	*flumineus*			44	44 ST/A	44				ACN=44	Japan	A-52
Rhinogobius	*giurinus*			44	44A	44	44			ACN=44	Japan (Yamaguchi)	N-31

Table 6.39 Order PERCIFORMES. Part 4 Gobioidei, Kurtoidei, Acanthuroidei, and Scombroidei (continued)

A		B	C	D	E	F	G	H	I	J	K	L
Current scientific name of taxon Suborder/family/subfamily/species		Reported in karyotype paper	Sex	2n	Karyotype	NF_1	NF_2	Ag-NORs	Genome size (pg/cell)	Comments	Locality	Reference
Rhinogobius	*giurinus*	*Ctenogobius*		44	44A	44	44			ACN=44	China (Guangdong)	G-67, Y-15
Rhinogobius	*giurinus*			44	44 ST/A	44				ACN=44	China (Kunming)	L-38
Rhinogobius	*giurinus*?	*Ctenogobius*		44	8M + 12SM + 24A	64		2–4		ACN=44	China (Anhui)	Z-38
Rhinogobius	*shennongensis*	*Ctenogobius*		44	44A	44	44				China (Wuhan)	D-18
Stigmatogobius	*sadanundio*	*Gobius*		48					2.8 BFA			H-13
Stigmatogobius	*sadanundio*			46	2SM + 44A	48	48				India (WB)	K-46
Tridentiger	*brevispinis*	*obscurus brevispinis*	F, M	44	10 M/SM + 34ST	54	88			ACN=44	Japan (Lake Kasumigaura)	A-52, A-60
Tridentiger	*nudicervicus*			44		62					Korea	L-80
Tridentiger	*obscurus*		F, M	44	12 M/SM + 32 ST/A	56				ACN=44	Korea (Samchuk)	L-14
Tridentiger	*obscurus*	*obscurus obscurus*		44	10 M/SM + 34ST	54	88			ACN=44	Japan (Chiba)	A-60
Tridentiger	*obscurus*	*obscurus obscurus*		44	26 M/SM/ST + 18A		70			ACN=44	Japan (Yamaguchi)	N-31
Tridentiger	*trigonocephalus*		F, M	44	16 M/SM + 28 ST/A	60		2		ACN=44	Japan (Kanagawa)	A-52, A-60
Tridentiger	*trigonocephalus*			44	28 M/SM/ST + 16A		72			ACN=44	Japan (Yamaguchi)	N-31
Tridentiger	*trigonocephalus*		F, M	44	20 M/SM + 24 ST/A	64				ACN=44	Korea (Gyeyato)	L-79
Tridentiger	*trigonocephalus*			44	20M + 12SM + 12A	76	76			ACN=44	China (Liaoning)	M-32
Tridentiger	*trigonocephalus*			44	10M + 28SM + 2ST + 4A	82	84			ACN=44	China (Zhejiang)	F-8
Tridentiger	sp.		F, M	44	18 M/SM + 26 ST/A	62				ACN=44	Korea (Kunsan)	L-14
Sicydiinae												
Sicyopterus	*japonicus*			44	10 M + 10SM + 24 ST/A	64				ACN=46	Japan (Chiba)	A-73
Oxudercinae												
Periophthalmini												
Periophthalmodon	*schlosseri*			46	46A	46	46		1.9 FIA	ACN=46	Malaysia	M-52, H-40
Periophthalmus	*chrysospilos*			56							Malaysia	M-52
Periophthalmus	*modestus*	*cantonensis*		46	18M + 12SM + 16 ST/A	76				ACN=46	Japan (Tokyo Bay)	A-52, A-62
Periophthalmus	*modestus*			46	32 M/SM + 14 ST/A	78				ACN=46	Japan	M-52
Periophthalmus	*modestus*	*cantonensis*		46	34 M/SM/ST + 12A		80			ACN=46	Japan (Ariake Sea)	N-31
Periophthalmus	*modestus*	*cantonensis*	F, M	46	16 M/SM + 30A	62	62			ACN=46	Korea (Kohwato)	L-79
Apocryptei												
Apocryptes	*bato*		M	46	24M + 10SM + 12A	80	80			ACN=46	India (WB)	N-15
Apocryptodon	*madurensis*			48							India	V-33
Apocryptodon	*punctatus*			42	24 M/SM + 18 ST/A	66				ACN=44	Japan (Ariake Sea)	M-52
Parapocryptes	*serperaster*	*Apocryptes*		46	8M + 16SM + 4ST + 18A	70	74			ACN=46	India (WB)	K-139
Pseudapocryptes	*borneensis*			48							India	V-33
Pseudapocryptes	*lanceolatus*	*Apocryptes*	F	38	14M + 22SM + 2ST	74	76			ACN=46	India (WB)	N-15

Table 6.39 Order PERCIFORMES. Part 4 Gobioidei, Kurtoidei, Acanthuroidei, and Scombroidei (continued)

A		B	C	D	E	F	G	H	I	J	K	L
Current scientific name of taxon Suborder/family/subfamily/species		Reported in karyotype paper	Sex	2n	Karyotype	NF_1	NF_2	Ag-NORs	Genome size (pg/cell)	Comments	Locality	Reference
Boleophthalmi												
Boleophthalmus	*boddarti*			46	46A	46	46			ACN=46	Malaysia	M-52
Boleophthalmus	*boddarti*	*boddaerti*	F, M	46	46A	46	46			ACN=46	India	S-177
Boleophthalmus	*boddarti*	*Gobius striatus*		46							India	V-33
Boleophthalmus	*dussumieri*	*dentatus*	F, M	46	14M + 18SM + 14A	78	78			ACN=46	India (WB)	K-39
Boleophthalmus	*dussumieri*			46							India	V-33, K-102
Boleophthalmus	*dussumieri*			46	4M + 8ST + 34A	50	58			ACN=46	Malaysia	M-52
Boleophthalmus	*pectinirostris*			46	46 ST/A	46				ACN=46	Japan (Ariake Sea)	A-62
Boleophthalmus	*pectinirostris*			46	46A	46	46			ACN=46	Japan (Ariake Sea)	N-31, M-52
Boleophthalmus	*pectinirostris*			46	2ST + 44A	46	48			ACN=46	China (Zhejiang)	F-8
Scartelaos	*cantoris*	*Boleophthalmus glaucus*	F, M	46	12M + 20SM + 2ST + 12A	78	80			ACN=46	India (WB)	M-24
Scartelaos	*histophorus*			48	8 M/SM + 40 ST/A	56					Malaysia	M-52
Amblyopinae												
Odontamblyopus	*rubicundus*			46	4M + 16SM + 26 ST/A	66				ACN=46	Japan (Ariake Sea)	A-62
Odontamblyopus	*rubicundus?*	*rubicundus*		38	20M + 18SM	76	76				China	M-36
Odontamblyopus	*rubicundus*		F, M	46	4M + 24SM + 8ST + 10A	74	82			ACN=46	India (WB)	K-42
Odontamblyopus	*rubicundus*	*Gobioides*	M	46	2M + 26SM + 10ST + 8A	74	84			ACN=46	India (WB)	M-24
Trypauchen	*vagina*		M	46	12M + 6SM + 10ST + 18A	64	74			ACN=46	India (WB)	K-29
Suborder Kurtoidei												
Kurtidae												
Kurtus	*gulliveri*		F, M	44	2SM + 4A + 38 non-identified					ACN=46	Australia (N. Territory)	E-10
Suborder Acanthroidei												
Acanthuridae												
Acanthurus	*bahianus*			36	16 non-A + 20A		52				Brazil (RN, BA)	G-12
Acanthurus	*chirurgus*			34	16 non-A + 18A		50		1.4 FD		Brazil (RN)	G-12, G-85, M-133
Acanthurus	*coeruleus*			48	4 non-A + 44A		52				Brazil (RN, BA)	G-12
Acanthurus	*triostegus*			48	48A	48	48		1.6* FCM	ACN=48	Japan (Yakushima)	A-65, O-48
Ctenochaetus	*striatus*			48	48A	48	48		1.7* FCM		Japan	O-48
Prionurus	*scalprum*	*microlepidotus*		48	48A	48	48			ACN=48	Japan (Chiba)	A-65
Ephippidae												
Platax	*teira*	*orbicularis*		48	48A	48	48			ACN=48	Japan (Yakushima)	A-66

Table 6.39 Order PERCIFORMES. Part 4 Gobioidei, Kurtoidei, Acanthuroidei, and Scombroidei (continued)

A		B	C	D	E	F	G	H	I	J	K	L
Current scientific name of taxon Suborder/family/subfamily/species		Reported in karyotype paper	Sex	2n	Karyotype	NF_1	NF_2	Ag-NORs	Genome size (pg/cell)	Comments	Locality	Reference
Scatophagidae												
Scatophagus	*argus*		F	48	2SM + 46A	50	50		(1.5 BFA)	ACN=48	India (Orissa)	C-61, H-13
Scatophagus	*argus*			48							India (Portonovo)	N-13
Scatophagus	*argus*		M	48	1M + 47A	49	49			XY, ACN=48	India	K-26
Scatophagus	*argus*		F	48	48A	48	48			XX, ACN=48	India	K-26
Scatophagus	*argus*			48	48A	48	48	2		ACN=48	(W. Pacific)	S-130
Scatophagus	*argus*			48	48A	48	48				China	Z-39
Selenotoca	*multifasciata*		M, F	48	48A	48	48		1.4 FIA	ACN=48	(W. Pacific)	S-130, H-41
Siganidae												
Siganus	*fuscescens*			48	2ST + 46A	48	50		1.2 FIA	ACN=48	Japan (Shimane)	K-68, H-41
Siganus	*javus*		F	48	48A	48	48			ACN=48	India (Orissa)	C-61
Siganus	*spinus*			42	6M + 36A	48	48		1.1* FCM, 1.2 FIA		Japan	O-48, H-40
Suborder Scombroidei												
Sphyraenidae												
Sphyraena	*tome*			48	48A	48	48				Brazil (RJ)	B-86
Scombridae												
Auxis	*rochei*			48		50			1.9 FD	ACN=48	Japan	I-9
Auxis	*thazard*			48	48A	48	48		1.8 FD, 1.7 FIA	ACN=48	Japan (Sanriku)	I-9, H-40
Katsuwonus	*pelamis*		M	48	48A	48	48		(1.6 FIA, 2.0 BFA)	ACN=48	Japan (Sanriku)	I-9, H-13, H-40
Katsuwonus	*pelamis*			48	48A	48	48			ACN=48	N. Pacific	R-38, S-95, S-96
Scomber	*australasicus*	*tapaeinocephalus*		48	2ST + 46A	48	50		1.7 FIA	ACN=48	Japan (Iwate)	I-1, H-40
Scomber	*japonicus*		F, M	48	2ST + 46A	48	50	2		ACN=48	Japan	M-99, K-50
Thunnus	*alalunga*			48	2M + 2SM + 2ST + 42A	52	54		1.8 FIA	ACN=48	N. Pacific	R-38, S-96, H-41
Thunnus	*albacares*			48	2M + 2SM + 2ST + 42A	52	54		1.8 FIA	ACN=48	N. Pacific	R-38, S-96, H-40
Thunnus	*orientalis*		M	48	4M + 6SM + 4ST + 34A	58	62			ACN=48	India (WB)	K-136
Thunnus	*thynnus*		F	48	2M + 2ST + 44A	50	52		1.7 FD, (1.6 FIA)		Japan (Sanriku)	I-1, I-9, H-40
Thunnus	*thynnus*		F	48	2M + 4ST + 42A	50	54		1.7 FD	ACN=48	Japan (Sanriku)	I-9

Table 6.40 Order PERCIFORMES. Part 5 Anabantoidei, Channoidei, and Caproidei

A	B	C	D	E	F	G	H	I	J	K	L
Current scientific name of taxon Suborder/family/subfamily/species	Reported in karyotype paper	Sex	2n	Karyotype	NF_1	NF_2	Ag-NORs	Genome size (pg/cell)	Comments	Locality	Reference
Suborder Anabantoidei											
Anabantidae											
Anabantinae											
Anabas cobojius	*oligolepis*		46						ACN=46	India (Andhra Pradesh)	D-25
Anabas testudineus			48							India (Andhra Pradesh)	D-25
Anabas testudineus			46	2M + 6SM + 6ST + 32A	54	60	2		ACN=46	India (Manipur)	K-138
Anabas testudineus		F, M	46	2SM + 2ST + 42A	48	50	2		ACN=46	India	K-44
Anabas testudineus			46	2SM + 44 ST/A	48		2		ACN=46	India (Delhi)	R-68
Anabas testudineus		F, M	46	4ST + 42A	46	50			ACN=46	India (WB)	M-22
Anabas testudineus		F	46	4 M/SM + 42A	50	50			ACN=46	(Asia)	A-3
Anabas testudineus			46							(Bangladesh, Dhaka)	T-68
Ctenopominae											
Ctenopoma acutirostre			48	48A	48	48			ACN=48	(Africa)	K-110
Ctenopoma muriei			48	48A	48	48			ACN=48	W. Ethiopia	K-110
Ctenopoma ocellatum			48	48 ST/A	48				ACN=48	(Africa)	K-110
Ctenopoma petherici			48	48A	48	48			ACN=48	W. Ethiopia	K-110
Ctenopoma sp.			48	48A	48	48			ACN=48	(Africa)	K-110
Microctenopoma ansorgii	*ansorgei*		46	2M + 44A	48	48				(Africa)	K-110
Microctenopoma congicum			46	2M + 44A	48	48			ACN=48	(Africa)	K-110
Microctenopoma pekkolai			48	2M + 2SM + 44 ST/A	52					W. Ethiopia	K-110
Helostomatidae											
Helostoma temminkii	*temmincki*	F, M	48	48A	48	48		1.8 BFA	ACN=48	(S. E. Asia)	A-3, H-13
Osphronemidae (= Belontiidae)											
Luciocephalinae											
Ctenops nobilis		F, M	44	8M + 8SM + 28A	60	60			ACN=44	India	R-77, R-78
Luciocephalus pulcher			20	20A	20	20				Thailand	D-28
Sphaerichthys osphromenoides		F, M	16	14 M/SM + 2A	30	30			ACN=42	(Asia)	C-89
Sphaerichthys osphromenoides			16	10M + 4SM + 2A	30	30				(India)	K-134
Trichogaster chuna	*Colisa*	F, M	46	10M + 8SM + 28A	64	64		(1.3 FCM)	ACN=46	(India)	R-77, V-86
Trichogaster chuna	*Colisa sota*		46	20M + 8SM + 6ST + 12A	74	80				India	L-1
Trichogaster chuna	*Colisa chuna* var. *sota*		46	28M + 12SM + 6A	86	86				India (Assam)	K-46, C-108
Trichogaster fasciatus	*Colisa fasciatus*		48	16M + 16SM + 6ST + 10A	80	86	2		ACN=48	India (Manipur)	K-138
Trichogaster fasciatus		F	48	15M + 12 SM/ST + 21A		75			ZW, ACN=48	India (Haryana)	R-47
Trichogaster fasciatus		M	48	14M + 12 SM/ST + 22A		74			ZZ, ACN=48	India (Haryana)	R-47

Table 6.40 Order PERCIFORMES. Part 5 Anabantoidei, Channoidei, and Caproidei (continued)

A	B	C	D	E	F	G	H	I	J	K	L
Current scientific name of taxon Suborder/family/subfamily/species	Reported in karyotype paper	Sex	2n	Karyotype	NF_1	NF_2	Ag-NORs	Genome size (pg/cell)	Comments	Locality	Reference
Trichogaster fasciatus	*Colisa fasciata*	F	48	17M + 16SM + 15A	81	81			ZW, ACN=48	India (Haryana)	R-54, R-77, R-78, S-57
Trichogaster fasciatus	*Colisa fasciata*	M	48	16M + 16SM + 16A	80	80			ZZ, ACN=48	India (Haryana)	R-54, R-77, R-78, S-57
Trichogaster fasciatus	*Colisa fasciata*	F	48	15M + 16SM + 4ST + 13A	79	83	6		ZW, ACN=48	India (Manipur)	S-175
Trichogaster fasciatus	*Colisa fasciata*	M	48	16M + 16SM + 4ST + 12A	80	84			ZZ, ACN=48	India (Manipur)	S-175
Trichogaster fasciatus		F, M	48	8M + 20SM + 12ST + 8A	76	88				India (WB)	M-26
Trichogaster fasciatus		F, M	46	18M + 12SM + 16A	76	76			ACN=46	India (Orissa)	T-50
Trichogaster labiosus	*Colisa*		48	22M + 12SM + 4ST + 12A	82	86				India	L-1
Trichogaster labiosus		M	48	12M + 6SM + 12ST + 18A	66	78				India (WB)	M-26
Trichogaster labiosus	*Colisa*	F, M	48	22M + 16SM + 10A	86	86			ACN=48	(India)	R-77
Trichogaster lalius	*Colisa*	F	45	14M + 12 SM/ST + 19A		71			ZO	India (Haryana)	R-50
Trichogaster lalius	*Colisa*	M	46	14M + 12 SM/ST + 20A		72			ZZ	India (Haryana)	R-50
Trichogaster lalius	*Colisa*	F	46	14M + 6SM + 26A	66	66			XX, ACN=46	India (Haryana)	R-77, R-78
Trichogaster lalius	*Colisa*	M	45	14 M + 6SM + 25A	65	65			XO	India (Haryana)	R-77, R-78
Trichogaster lalius	*larius*	F, M	46	14M + 10SM + 12ST + 10A	70	82			ACN=46	India (WB)	K-34
Trichogaster lalius	*Colisa lalia*		46	24 M/SM + 22 ST/A	70			(1.2 FCM, 1.2 BFA)	ACN=46	(Asia)	A-3, V-101, H-39
Trichopodus cantoris	*Trichogaster pectoralis*		46	46A	46	46				(SE Asia)	K-114
Trichopodus leeri	*Trichogaster*	F, M	46	46A	46	46		(1.4 FCM)	ACN=46	(India)	R-77, V-86
Trichopodus leeri	*Trichogaster*	M	46	46A	46	46				(Asia)	A-3
Trichopodus microlepis	*Trichogaster*		46	46A	46	46					K-134
Trichopodus microlepis	*Trichogaster*		48	48A	48	48		1.8* FCM		(Asia)	O-48
Trichopodus sumatranus	*Trichogaster*		48		48					(Asia)	C-89
Trichopodus trichopterus	*Trichogaster*	M	46	46A	46	46		(1.2 FCM)	ACN=46	(Asia)	A-3, V-86
Trichopodus t. sumatranus	*Colisa*	F, M	46	46A	46	46			ACN=46	(India)	R-77
Trichopodus t. trichopterus	*Colisa*	F, M	46	46A	46	46			ACN=46	(India)	R-77
Macropodusinae											
Betta splendens			42					1.3 FIA, 1.3 BFA		(Thailand)	H-13, H-41
Macropodus chinensis			48	6M + 2SM + 40A	56	56				Korea	L-15
Macropodus chinensis			46	4M + 4SM + 38A	54	54		2.8* FD		China (Wuhan)	D-18, C-83
Macropodus ocellatus		F, M	46	8M + 8SM + 14ST + 16A	62	76			ACN=48	(Asia)	K-97
Macropodus opercularis		F, M	46	8M + 8SM + 16ST + 14A	62	78		(1.1 FCM)		(Asia)	K-97, V-101
Macropodus opercularis		F, M	46	4M + 10SM + 12ST + 20A	60	72		(1.2 BFA)	ACN=46	China (Guilin)	Y-15, H-13
Macropodus opercularis			46	12 M/SM + 34 ST/A	58				ACN=46	(Asia)	A-3
Macropodus spechti	*concolor*	F, M	46	10M + 2SM + 22ST + 12A	58	80			ACN=46	(Asia)	K-97
Osphronemus goramy			48	2SM + 46A	50	50			ACN=48	(S. E. Asia)	K-97
Parosphromenus sumatranus	*Colisa*		46	46A	46	46			ACN=46	(India)	R-77

Table 6.40 Order PERCIFORMES. Part 5 Anabantoidei, Channoidei, and Caproidei (continued)

A		B	C	D	E	F	G	H	I	J	K	L
Current scientific name of taxon Suborder/family/subfamily/species		Reported in karyotype paper	Sex	2n	Karyotype	NF_1	NF_2	Ag-NORs	Genome size (pg/cell)	Comments	Locality	Reference
Suborder Channoidei												
Channidae												
Channa	*argus*		F, M	48	4SM + 44 ST/A	52			1.7* FD	ACN=48	China (Wuhan)	L-31, Y-15, C-83
Channa	*argus*			48	2M + 4SM + 42ST	54	96	4		ACN=48	China (Shashi)	Z-21
Channa	*argus*	*Ophiocephalus*		48	20ST + 28A	48	68				China (Hubei)	L-53
Channa	*argus*	*Ophiocephalus*	F, M	48	4SM + 22ST + 22A	52	74			ACN=48	China (Beijing)	Z-31
Channa	*asiatica*		F, M	44	4M + 8SM + 32 ST/A	56			1.8* FD	ACN=46	China (Shaoguan)	L-31, Y-15, C-83
Channa	*asiatica*		F, M	46	2M + 8SM + 36 ST/A	56				ACN=46	China (Guangzhou)	L-31
Channa	*asiatica*			44	2M + 8SM + 28ST + 6A	54	82	4			China (Shashi)	Z-21
Channa	*barca*			38	6M + 6SM + 4ST + 22A	50	54			ACN=42	India (Assam, Meghalaya)	D-9
Channa	*gachua*		F, M	78	12M + 12SM + 54A	102	102				India (WB)	M-21
Channa	*gachua*		F, M	78	10M + 16SM + 52 ST/A	104			2.0 FD		India	B-2
Channa	*maculata*		F, M	42	4M + 2SM + 36 ST/A	48				ACN=46	China (Guangdong)	L-31, Y-15
Channa	*maculata*	*Ophiocephalus*	F, M	42							China	C-85
Channa	*marulius*			44	4M + 40A	48	48			ACN=44	India (Simlipal Hills)	K-41
Channa	*orientalis?*	*orientalis*		42	2M + 2SM + 38A	46	46			ACN=44	India (Assam)	K-41
Channa	*orientalis*			76	2M + 6SM + 68A	84	84				India (Assam, Meghalaya)	D-9
Channa	*orientalis*			78	34M + 2SM + 42 ST/A	114				ACN=78	Bangladesh (Dhaka)	R-102
Channa	*punctata*	*punctatus*	F, M	32	20M + 12SM	64	64	2			India (Haryana, U.P.)	R-70, R-73, S-2
Channa	*punctata*	*punctatus*	F, M	32	24M + 8SM	64	64		1.3 FD	ACN=42	India	B-2
Channa	*punctata*	*punctatus*	F, M	32	18M + 12SM + 2ST	62	64				India (WB)	M-21
Channa	*punctata*	*punctatus*	F, M	32	10M + 18SM + 4A	60	60				India (Haryana)	R-46, R-53
Channa	*punctata*	*punctatus*	F, M	32	16M + 14SM + 2A	62	62				India (Haryana)	R-75
Channa	*punctata*	var. A		34	16M + 14SM + 4A	64	64			ACN=42	India (Assam, Meghalaya)	D-9
Channa	*punctata*	var. B		32	16M + 16SM	64	64			ACN=42	India (Assam, Meghalaya)	D-9
Channa	*punctata*			32	24M + 2SM + 2ST + 4A	58	60				Bangladesh (Dhaka)	R-102
Channa	*stewartii*			66	12M + 6SM + 6ST + 42A	84	90				India (Assam, Meghalaya)	D-9
Channa	*stewartii*		M	104	2M + 102A	106	106				India (Nagaland)	R-65
Channa	*striata*	*striatus*	F, M	40	8M + 2SM + 2ST + 28A	50	52		1.5 FD	ACN=46	India	B-2
Channa	*striata*	*striatus*		40	8M + 6ST + 26A	48	54			ACN=46	India (Assam, Meghalaya)	D-9
Channa	*striata*	*striatus*	F, M	40	8M + 2SM + 30A	50	50				India (WB)	M-21
Parachanna	*obscura*	*Ophiocephalus obscurus*		42					2.0 BFA		(Africa)	H-13
Suborder Caproidei												
Caproidae												
Capros	*aper*			46	2M + 2SM + 8ST + 34A	50	58	2		ACN=48	France (Gulf of Lion)	V-67
Capros	*aper*			44	4M + 2SM + 8ST + 30A	50	58	2		ACN=48	France (Gulf of Lion)	V-67
Capros	*aper*			42	6M + 2SM + 8ST + 26A	50	58	2		ACN=48	France (Gulf of Lion)	V-67

Table 6.41 Order PLEURONECTIFORMES

A		B	C	D	E	F	G	H	I	J	K	L
Current scientific name of taxon Suborder/family/species		Reported in karyotype paper	Sex	2n	Karyotype	NF_1	NF_2	Ag-NORs	Genome size (pg/cell)	Comments	Locality	Reference
Suborder Pleuronectoidei												
Scophthalmidae												
Psetta	*maeotica*	*Rhombus maeoticus*		40	20M + 20 SM/A					ACN=44	Black Sea	I-19
Psetta	*maeotica*	*Rhombus maeoticus*		48						ACN=48	Black Sea	I-19
Psetta	*maeotica*	*Rhombus maeoticus*		44							Black Sea	V-72
Psetta	*maxima*	*Scophthalmus maximus*	F, M	44	4M + 12ST + 28A	48	60	2	(1.7 FIA)	ACN=44	Spain	B-53, P-63, H-41
Scophthalmus	*rhombus*			44	4M + 2SM + 38 ST/A	50		2		ACN=44	Spain	P-63
Paralichthyidae												
Citharichthys	*spilopterus*		F, M	28	14M + 6SM + 2ST + 6A	48	50			ACN=46	USA (LA)	L-17
Citharichthys	*spilopterus*		F, M	26	18 M/SM + 8 ST/A	44				ACN=38	Brazil (SP)	A-120
Etropus	*crossotus*		F, M	38	38 M/SM/ST		76		(2.0 BFA)	ACN=44	USA (LA)	L-17, H-13
Etropus	*crossotus*		F, M	38	26 M/SM + 12 ST/A	64				ACN=44	Brazil (SP)	A-120
Hippoglossina	*macrops*			48	48A	48	48			ACN=48	Chile (Coquimbo bay)	W-36
Paralichthys	*adpersus*			46	2M + 44 ST/A	48				ACN=48	Chile (Coquimbo bay)	W-36
Paralichthys	*dentatus*			48	48A	48	48				(W.N. Atlantic)	X-3
Paralichthys	*lethostigma*		M	48	20ST + 28A	48	68			ACN=48	USA (LA)	L-17
Paralichthys	*microps*			46	2M + 44 ST/A	48				ACN=48	Chile (Coquimbo bay)	W-36
Paralichthys	*olivaceus*			48	48A	48	48	2	(1.4* FCM)	ACN=48	Japan	K-48, O-48
Paralichthys	*olivaceus*		F	46	46A	46	46	2		ACN=46	Japan (Yamaguchi)	S-5
Paralichthys	*olivaceus*			48	48A	48	48				China	Y-20, X-3, Z-37
Paralichthys	*orbignyanus*			46	2M + 44 ST/A	48				ACN=46	Brazil (SC)	A-120
Paralichthys	*patagonicus*		F, M	46	46A	46	46			ACN=46	Brazil (SP)	A-120
Pseudorhombus	*arsius*		M	46	46 ST/A	46			1.0 FIA	ACN=46	India (Orissa)	P-17, H-40
Pseudorhombus	*cinnamoneus*			48	48A	48	48				China	Y-20
Pseudorhombus	*triocellatus*		F	48	48A	48	48			ACN=48	India (Orissa)	P-17
Xystreurys	*liolepis*		M	48	48A	48	48		1.6 FD, 1.5 BFA	ACN=48	(E. Pacific)	O-4, H-13
Pleuronectidae												
Cleisthenes	*herzensteini*	*pinetorum herzensteini*		44	4 SM/ST + 40A		48			ACN=48	Japan (Hokkaido)	F-53
Glyptocephalus	*stelleri*		M	46							Japan (Hokkaido)	Y-1
Hippoglossus	*hippoglossus*			48	48 ST/A	48			1.5 FIA		UK	B-71, H-41
Kareius	*bicoloratus*			48	48A	48	48	2	(1.3* FCM)	ACN=48	Japan (Hokkaido)	F-53, S-142, O-48
Kareius	*bicoloratus*		F	48	48A	48	48			ACN=48	China (Yellow Sea)	Z-15, Z-37, Y-20
Limanda	*ferruginea*	*Pleuronectes*		48	48A	48	48		1.5 FIA		East coast of Canada	A-120, H-41

Table 6.41 Order PLEURONECTIFORMES (continued)

A		B	C	D	E	F	G	H	I	J	K	L
Current scientific name of taxon Suborder/family/species		Reported in karyotype paper	Sex	2n	Karyotype	NF_1	NF_2	Ag-NORs	Genome size (pg/cell)	Comments	Locality	Reference
Limanda	*limanda*		F, M	46	46 ST/A	46				ACN=46	Europe (Irish Sea)	D-10
Limanda	*limanda*			46		46					Russia (White Sea)	L-88
Liopsetta	*glacialis*			48	48A	48	48				Russia (White Sea)	L-87, L-88
Microstomus	*achne*			48	48A	48	48			ACN=48	Japan (Hokkaido)	F-53
Microstomus	*achne*			48	48A	48	48				China	Y-20
Platichthys	*flesus*	*Pleuronectes*		48	48A	48	48				Russia (White Sea)	L-87
Platichthys	*flesus*			48	48A	48	48				Germany (Baltic Sea)	K-132
Platichthys	*flesus*			48	48A	48	48	2		ACN=48	Spain	P-63
Platichthys	*stellatus*			48	48A	48	48	2		ACN=48	Japan (Hokkaido)	F-53, S-142
Pleuronectes	*herzensteini*	*Limanda*	F, M	48	48A	48	48				Japan (Hokkaido)	F-53
Pleuronectes	*obscurus*			48	48A	48	48		1.2 FCM	ACN=48	Korea (Busan)	K-125
Pleuronectes	*platessa*			48	48A	48	48				(UK)	B-6
Pleuronectes	*platessa*			48	48A	48	48			ACN=48	UK	F-4
Pleuronectes	*platessa*			47	1M + 47A	48	48			ACN=48	UK	F-4
Pleuronectes	*punctatissimus*	*Limanda*	M	48	48A	48	48				Japan (Hokkaido)	F-53
Pleuronectes	*putnami*	*Liopsetta*		48							WN Atlantic	B-71
Pleuronectes	*schrenki*	*Limanda*		48	48A	48	48				Japan (Hokkaido)	F-53
Pleuronectes	*schrenki*			48	48A	48	48	2		ACN=48	Japan (Hokkaido)	S-142
Pleuronectes	*yokohamae*	*Limanda*		48	48A	48	48		1.3* FCM		Japan (Hokkaido)	F-53, O-48
Pleuronectes	*yokohamae*	*Pseudopleuronectes*		48	48A	48	48				China (Shandong)	Y-20
Pleuronectes	*yokohamae*			48	48A	48	48			ACN=48	Korea (Busan)	P-70
Pleuronichthys	*cornutus*			48	14M + 34A	62	62		1.1* FCM		Japan	O-48
Pleuronichthys	*cornutus*			48	12M + 2SM + 34A	62	62				China (Shandong)	Y-20
Pleuronichthys	*verticalis*		M	48	48A	48	48		1.3 FD	ACN=48	(E. Pacific)	O-4
Pseudopleuronectes	*americanus*			48					1.5 FD, 1.4 BFA		WN Atlantic	B-71, H-13
Verasper	*moseri*		F, M	46	2 SM/ST + 44A		48			ACN=46	Japan (Hokkaido)	F-53
Bothidae												
Bothus	*ocellatus*		F	32	18 M/SM + 14 ST/A	50				ACN=48	Brazil (SP)	A-120
Bothus	*podas*		F	38	12M + 2SM + 24ST	52		2		XX	Italy (Palermo)	V-100
Bothus	*podas*		M	38	12M + 2SM + 24ST	52		2		XY	Italy (Palermo)	V-100
Psettina	*tosana*			44	4M + 2SM + 6ST + 32A	50	56		1.1* FCM		Japan	O-48

Table 6.42 Order LOPHIIFORMES

A		B	C	D	E	F	G	H	I	J	K	L
Current scientific name of taxon Family/species		Reported in karyotype paper	Sex	2n	Karyotype	NF_1	NF_2	Ag-NORs	Genome size (pg/cell)	Comments	Locality	Reference
Antennariidae												
Antennarius	*nummifer*			48	16 M/SM + 32 ST/A	64				ACN=48	Japan (Chiba)	A-69
Histrio	*histrio*			46	46 ST/A	46				ACN=46	Japan (Kanagawa)	A-69
Lophiidae												
Lophius	*piscatorius*		F, M	46	14 M/SM + 32 ST/A	60		2		ACN=46	Italy (Palermo)	V-57, V-65

Table 6.43 Order TETRAODONTIFORMES

A		B	C	D	E	F	G	H	I	J	K	L
Current scientific name of taxon Suborder/family/species		Reported in karyotype paper	Sex	2n	Karyotype	NF_1	NF_2	Ag-NORs	Genome size (pg/cell)	Comments	Locality	Reference
Suborder Balistoidei												
Triacanthidae												
Triacanthus	*biaculeatus*			48	2M + 4ST + 42A	50	54				Japan	I-21
Triacanthus	*biaculeatus*	*brevirostris*	M	48	1M + 47A	49	49			ACN=48	India (Orissa)	C-62
Triacanthus	*biaculeatus*	*Tricanthus brevirostris*	F, M	48	48A	48	48			ACN=48	India (Orissa)	R-57
Pseudotriacanthus	*strigilifer*	*Triacanthus*	M	48	48A	48	48				India (Goa)	R-45
Balistidae												
Balistapus	*undulatus*		F	42	42A	42	42	2	1.5* FCM, 1.3 FIA	ACN=46	Japan (Okinawa)	T-9, O-48, H-41
Balistes	*capriscus*	*carolinensis*	F, M	44	44A	44	44	2	(1.1 FD)	ACN=46	Italy (Palermo)	V-57, V-65, M-133
Balistes	*capriscus*	*carolinensis*	F, M	44	44A	44	44	2		ACN=46	Spain (Malaga)	T-38
Balistes	*vetula*			44	44A	44	44	2		ACN=46	Brazil (Bahia)	S-110
Balistoides	*conspicillus*		F	44	44A	44	44	2		ACN=46	Japan (Okinawa)	T-9
Balistoides	*viridescens*			44	2M + 2SM + 40 ST/A	48		2	1.4* FCM	ACN=46	(Indo-West Pacific)	T-10, O-48
Melichthys	*niger*			40	40A	40	40	2	1.4 FCM	ACN=42	Brazil (Saint Pauls Rocks)	S-110, B-75
Melichthys	*vidua*		M	40	40 ST/A	40		2		ACN=42	(Indo-West Pacific)	K-71
Odonus	*niger*			42	42 ST/A	42		2		ACN=44	(Indo-West Pacific)	K-71
Pseudobalistes	*flavimarginatus*			44	2M + 42 ST/A	46				ACN=46	Japan (Yaku and Tanega Is)	A-64
Rhinecanthus	*aculeatus*			44	44A	44	44	2	1.3 FIA	ACN=46	Japan (Okinawa)	A-64, K-71, T-9, H-40
Rhinecanthus	*echarpe*			44	44 ST/A	44		2		ACN=46	(Indo-West Pacific)	K-71, T-9
Rhinecanthus	*verrucosus*			44	44A	44	44	2		ACN=46	Japan (Tanegashima)	A-64, K-71, T-9
Sufflamen	*chrysopterus*	*Hemibalistes*		46	46A	46	46		1.2 FIA	ACN=46	Japan (Yakushima)	A-64, H-40
Sufflamen	*fraenatus*		F	46	46A	46	46	2	1.3 FIA	ACN=46	Japan (Okinawa)	T-9, H-41

Table 6.42 Order LOPHIIFORMES

A		B	C	D	E	F	G	H	I	J	K	L
Current scientific name of taxon Family/species		Reported in karyotype paper	Sex	2n	Karyotype	NF_1	NF_2	Ag-NORs	Genome size (pg/cell)	Comments	Locality	Reference
Antennariidae												
Antennarius	*nummifer*			48	16 M/SM + 32 ST/A	64				ACN=48	Japan (Chiba)	A-69
Histrio	*histrio*			46	46 ST/A	46				ACN=46	Japan (Kanagawa)	A-69
Lophiidae												
Lophius	*piscatorius*		F, M	46	14 M/SM + 32 ST/A	60		2		ACN=46	Italy (Palermo)	V-57, V-65

Table 6.43 Order TETRAODONTIFORMES

A		B	C	D	E	F	G	H	I	J	K	L
Current scientific name of taxon Suborder/family/species		Reported in karyotype paper	Sex	2n	Karyotype	NF_1	NF_2	Ag-NORs	Genome size (pg/cell)	Comments	Locality	Reference
Suborder Balistoidei												
Triacanthidae												
Triacanthus	*biaculeatus*			48	2M + 4ST + 42A	50	54				Japan	I-21
Triacanthus	*biaculeatus*	*brevirostris*	M	48	1M + 47A	49	49			ACN=48	India (Orissa)	C-62
Triacanthus	*biaculeatus*	*Tricanthus brevirostris*	F, M	48	48A	48	48			ACN=48	India (Orissa)	R-57
Pseudotriacanthus	*strigilifer*	*Triacanthus*	M	48	48A	48	48				India (Goa)	R-45
Balistidae												
Balistapus	*undulatus*		F	42	42A	42	42	2	1.5* FCM, 1.3 FIA	ACN=46	Japan (Okinawa)	T-9, O-48, H-41
Balistes	*capriscus*	*carolinensis*	F, M	44	44A	44	44	2	(1.1 FD)	ACN=46	Italy (Palermo)	V-57, V-65, M-133
Balistes	*capriscus*	*carolinensis*	F, M	44	44A	44	44	2		ACN=46	Spain (Malaga)	T-38
Balistes	*vetula*			44	44A	44	44	2		ACN=46	Brazil (Bahia)	S-110
Balistoides	*conspicillus*		F	44	44A	44	44	2		ACN=46	Japan (Okinawa)	T-9
Balistoides	*viridescens*			44	2M + 2SM + 40 ST/A	48		2	1.4* FCM	ACN=46	(Indo-West Pacific)	T-10, O-48
Melichthys	*niger*			40	40A	40	40	2	1.4 FCM	ACN=42	Brazil (Saint Pauls Rocks)	S-110, B-75
Melichthys	*vidua*		M	40	40 ST/A	40		2		ACN=42	(Indo-West Pacific)	K-71
Odonus	*niger*			42	42 ST/A	42		2		ACN=44	(Indo-West Pacific)	K-71
Pseudobalistes	*flavimarginatus*			44	2M + 42 ST/A	46				ACN=46	Japan (Yaku and Tanega Is)	A-64
Rhinecanthus	*aculeatus*			44	44A	44	44	2	1.3 FIA	ACN=46	Japan (Okinawa)	A-64, K-71, T-9, H-40
Rhinecanthus	*echarpe*			44	44 ST/A	44		2		ACN=46	(Indo-West Pacific)	K-71, T-9
Rhinecanthus	*verrucosus*			44	44A	44	44	2		ACN=46	Japan (Tanegashima)	A-64, K-71, T-9
Sufflamen	*chrysopterus*	*Hemibalistes*		46	46A	46	46		1.2 FIA	ACN=46	Japan (Yakushima)	A-64, H-40
Sufflamen	*fraenatus*		F	46	46A	46	46	2	1.3 FIA	ACN=46	Japan (Okinawa)	T-9, H-41

Table 6.43 Order TETRAODONTIFORMES (continued)

A		B	C	D	E	F	G	H	I	J	K	L
Current scientific name of taxon Suborder/family/species		Reported in karyotype paper	Sex	2n	Karyotype	NF_1	NF_2	Ag-NORs	Genome size (pg/cell)	Comments	Locality	Reference
Monacanthidae												
Cantherhines	*macrocerus*			40	40A	40	40				Brazil (RJ)	B-86
Cantherhines	*pardalis*	*Amanses sandwichiensis*		40	40 ST/A	40				ACN=40	Japan (Tanegashima)	A-64
Meuschenia	*scaber*	*Parika*	F, M	40	40A	40	40	2		ACN=40	New Zealand	M-112
Oxymonacanthus	*longirostris*			36	36A	36	36			ACN=36	Japan (Okinawa)	A-64
Paramonacanthus	*japonicus*	*Stephanolepis*	F	34	34A	34	34			ACN=36	Japan (W. Izu Peninsula)	M-98
Paramonacanthus	*japonicus*	*oblongus*	M	34	34A	34	34				Japan (W. Izu Peninsula)	M-98
Rudarius	*ercodes*			36	36A	36	36			ACN=36	Japan (Izu Peninsula)	A-64
Stephanolepis	*cirrhifer*			34	34A	34	34		(1.2* FCM)	ACN=34	Japan (W. Izu Peninsula)	M-98, O-48
Stephanolepis	*cirrhifer*		F	34	34A	34	34			$X_1X_1X_2X_2$, ACN=34	Japan (Izu Peninsula)	M-100
Stephanolepis	*cirrhifer*		M	33	1M + 32A	34	34			X_1X_2Y, ACN=34	Japan (Izu Peninsula)	M-100
Stephanolepis	*cirrhifer*			33	1M + 32A	34	34				China (Shandong)	W-6
Stephanolepis	*hispidus*		F	34	34A	34	34	2	1.2 FCM, 1.4 BFA	$X_1X_1X_2X_2$, ACN=34	Brazil (Bahia, RJ)	S-181, B-75, H-13
Stephanolepis	*hispidus*		M	33	1M + 32A	34	34	2		X_1X_2Y, ACN=34	Brazil (Bahia, RJ)	S-181
Thamnaconus	*modestus*	*Navodon*		40	40 ST/A	40			1.1* FCM	ACN=40	Japan (W. Izu Peninsula)	M-98, O-48
Thamnaconus	*septentrionalis*	*Navodon*	F, M	40	40A	40	40			ACN=40	China (Yellow Sea)	Z-15, W-6, Y-21
Ostraciidae												
Lactoria	*cornuta*			48	2M + 2SM + 44ST	52	96				Japan	I-23, O-72
Lactoria	*diaphana*			48	4M + 44A	52	52				Japan	I-23
Lactoria	*diaphana*			36	10M + 2SM + 24 ST/A	48				ACN=44	Japan (Wakayama)	A-79
Lactoria	*fornasini*			34	12M + 6SM + 16 ST/A	52					Japan	I-23
Ostracion	*cubicus*	*tuberculatus*		50	4SM + 46 ST/A	54				ACN=48	Japan (Ishigaki and Yaku Is)	A-64
Ostracion	*immaculatus*			50	4SM + 46 ST/A	54				ACN=48	Japan (Wakayama)	A-79
Suborder Tetraodontoidei												
Tetraodontidae												
Arothron	*hispidus*			42					0.8* FCM, 1.0 FIA		India (Portonovo)	N-13, O-48, G-85
Arothron	*immaculatus*		F, M	42	12M + 14SM + 16 ST/A	68				ACN=42	India (Orissa)	C-62
Arothron	*manilensis*	*Tetraodon immaculatus*		42	14M + 16SM + 12ST	72	84		1.0 FIA	ACN=42	Japan (Okinawa)	A-64, G-85, H-41
Arothron	*meleagris*			38					0.8* FCM		Japan	O-48
Arothron	*nigropunctatus*	*Tetraodon*		38	14M + 20SM + 4ST	72	76			ACN=40	Japan (Okinawa)	A-64
Arothron	*reticularis*		M	42	12M + 14SM + 16 ST/A	68				ACN=44	India (Orissa)	C-62
Canthigaster	*coronata*			28	6M + 2SM + 20 ST/A	36					Japan (Wakayama)	A-79
Canthigaster	*rivulata*			34	4M + 6SM + 10ST + 14A	44	54		0.7* FCM	ACN=34	Japan (Chiba)	A-64, M-2

Table 6.43 Order TETRAODONTIFORMES (continued)

A		B	C	D	E	F	G	H	I	J	K	L
Current scientific name of taxon Suborder/family/species		Reported in karyotype paper	Sex	2n	Karyotype	NF_1	NF_2	Ag–NORs	Genome size (pg/cell)	Comments	Locality	Reference
Chelonodon	*patoca*			40	14M + 16SM + 10 ST/A	70				ACN=42	Japan (Okinawa)	A-64
Lagocephalus	*inermis*	*Tetraodon*		44	2M + 42A	46					India	L-1
Lagocephalus	*laevigatus*			46							Brazil	S-110
Lagocephalus	*lunaris*	*Gastrophysus*	M	44	10M + 14SM + 20 ST/A	68			0.9 FIA		India (Orissa)	C-62, G-85, H-41
Sphoeroides	*greeleyi*		F, M	46	24 M/SM + 22 ST/A	70		2		ACN=46	Brazil (RJ)	B-56
Sphoeroides	*spengleri*		F, M	46	18 M/SM + 28 ST/A	64				ACN=46	Brazil (RJ)	B-56
Sphoeroides	*spengleri*		F, M	46	20 M/SM + 26 ST/A	66				0–2 B, ACN=46	Brazil (SP)	A-126
Sphoeroides	*testudineus*			46	18M + 4SM + 6ST + 18A	68	74	2		ACN=46	Brazil (RN)	S-110
Sphoeroides	*tyleri*		F, M	46	14 M/SM + 32 ST/A	60				ACN=46	Brazil (RJ)	B-57
Takifugu	*chrysops*	*Sphoeroides*		44	6M + 14SM + 24 ST/A	64				ACN=46	Japan (Izu Peninsula)	A-64
Takifugu	*niphobles*	*Fugu*		44	20 M/SM + 24 ST/A	64			(0.8* FCM)	ACN=46	Japan (Kanagawa)	A-54, O-48
Takifugu	*niphobles*		F	44	4M + 16SM + 24 ST/A	64				ACN=46	Japan (Oomura Bay)	M-134
Takifugu	*pardalis*	*Fugu*		44							Japan (Wakayama, Shizuoka)	A-79
Takifugu	*pardalis*			44	6M + 16SM + 22 ST/A	66				ACN=46	Japan (Oomura Bay)	M-134
Takifugu	*poecilonotus*		F	44	12M + 10SM + 22 ST/A	66				ACN=46	Japan (Tachibana Bay)	M-134
Takifugu	*poecilonotus*	*Fugu*		44							Japan (Wakayama)	A-79
Takifugu	*pseudommus*			44	12M + 8SM + 24A	64	64				China (Yellow Sea)	Z-37
Takifugu	*rubripes*			44	10M + 12SM + 22 ST/A	66				ACN=46	Japan	M-134
Takifugu	*rubripes*	*Fugu*		44	20 M/SM + 24 ST/A	64					China	Y-20, G-85
Takifugu	*rubripes*	*Fugu*		44	12M + 6SM + 26A	62					China (Shandong)	W-6
Takifugu	*vermicularis*	*radiatus*	F	44	8M + 14SM + 22 ST/A	66					Japan (Tachibana Bay)	M-134
Takifugu	*xanthopterus*		M	44	8M + 14SM + 22 ST/A	66				ACN=46	Japan (Tachibana Bay)	M-134
Tetraodon	*cutcutia*		F, M	42	16M + 12SM + 4ST + 10A	70	74		0.8 FCM	ACN=44	India (WB)	K-42, V-86
Tetraodon	*fluviatilis*			42	8M + 14SM + 2ST + 18A	64	66		(0.8 FCM, 0.8 BFA)	ACN=42	India (WB)	B-3, B-75, H-13
Tetraodon	*fluviatilis*		F, M	42	2M + 4SM + 2ST + 34A	48	50	4	(0.8* FCM)		(S. Asia)	M-19, O-48
Tetraodon	*leopardus*	*Arothron*	F, M	40	14M + 14SM + 12 ST/A	68				ACN=42	India (Orissa)	C-62
Tetraodon	*nigroviridis*			42	20 M/SM + 22ST	62	84		0.7 FCM, 1.0 FIA		(S. Asia)	F-24, J-21, H-41
Tetraodon	*palembangensis*			42*					1.0 BFA		(SE Asia)	H-13
Diodontidae												
Chilomycterus	*antennatus*			52	6M + 46 ST/A	58		2			Brazil (RJ)	S-110
Chilomycterus	*spinosus*		F, M	52	16 M/SM + 36 ST/A	68					Brazil (RJ)	B-57
Diodon	*holocanthus*			46	20 M/SM + 26 ST/A	66			1.6 FCM		Brazil	S-110, B-75
Diodon	*liturosus*	*bleekeri*		46	6M + 6SM + 34 ST/A	58			1.7* FCM	ACN=46	Japan (Okinawa)	A-64, O-48
Molidae												
Mola	*mola*		M	46	46A	46	46	2	(1.7, 1.9 FCM)	ACN=46	Japan (Chiba)	N-7, B-75

Table 7 Class SARCOPTERYGII (OSTEICHTHYES)

Table 7.1 Order COELACANTHIFORMES

A		B	C	D	E	F	G	H	I	J	K	L
Current scientific name of taxon Family/species		Reported in karyotype paper	Sex	2n	Karyotype	NF_1	NF_2	Ag-NORs	Genome size (pg/cell)	Comments	Locality	Reference
Latimeriidae												
Latimeria	*chalumnae*			48	10M + 4ST + 18A + 16 MC	58	62		7.2 FD		Comoro	B-41, C-95

Table 7.2 Order CERATODONTIFORMES

A		B	C	D	E	F	G	H	I	J	K	L
Current scientific name of taxon Suborder/family/species		Reported in karyotype paper	Sex	2n	Karyotype	NF_1	NF_2	Ag-NORs	Genome size (pg/cell)	Comments	Locality	Reference
Suborder Ceratodontoidei												
Ceratodontidae												
Neoceratodus	*forsteri*	*Epiceratodus*		32-38					(149.7* FD)		Australia	W-25, P-72
Neoceratodus	*forsteri*			54	6M + 2SM + 26 ST/A + 20 MC	62			105.5 FCM, 109.2 FIA		Australia (Sydney)	R-101, G-85
Suborder Lepidosirenoidei												
Lepidosirenidae												
Lepidosiren	*paradoxa*		F	38	38 M/SM	76	76		(161.1 FCM, 225.6* FD)		(S. America)	O-4, V-101, P-72
Lepidosiren	*paradoxa*			38	38 M/SM/ST		76				Brazil (Manaus)	O-50
Lepidosiren	*paradoxa*			38							Argentina	F-20
Protopteridae												
Protopterus	*aethiopicus congicus*			34					80.2, 265.7* FD		Zaire	V-38, P-72
Protopterus	*annectens*			34	14M + 10SM + 4ST + 6A	58	62				(Africa)	S-128
Protopterus	*annectens*		F, M	34	16M + 6 SM + 12A	56	56	2	125.2 FCM		Nigeria	M-115
Protopterus	*annectens annectens*			34	16M + 6SM + 12A	56	56		80.9 FD		Senegal	V-38
Protopterus	*dolloi*?	*dolloi*							130.8 FCM		(Africa)	V-101
Protopterus	*dolloi*			68	36 M/SM + 32 ST/A	104			163.2 FD	4X	Zaire	V-38

References

(Papers marked with an asterisk were not seen directly.)

[A]

A- 1 Abe, S. (1972) Notes on the chromosomes of two species of fresh-water cottid fishes. Chrom. Inform. Serv., 13: 25–27.

A- 2 Abe, S. and Muramoto, J. (1974) Differential staining of chromosomes of two salmonoid species, *Salvelinus leucomaenis* (Pallas) and *Salvelinus malma* (Walbaum). Proc. Japan Acad., 50: 507–511.

A- 3 Abe, S. (1975) Karyotypes of 6 species of anabantoid fishes. Chrom. Inform. Serv., 19: 5–7.

A- 4 Abe, S. (1976) A cytotaxonomical study in some freshwater cottid fishes (Cottidae, Pisces). Cytologia, 41: 323–329.

A- 5 Abucarma, M. and Martins-Santos, I.C. (2001) Karyotype and B chromosome of *Rhamdia* species (Pisces, Pimelodidae) endemic in the River Iguaçu basin. Cytologia, 66: 299–306.

A- 6 Abuin, M., Martinez, P. and Sánchez, L. (1996) Localization of the repetitive telomeric sequence (TTAGGG)n in four salmonid species. Genome, 39: 1035–1038.

A- 7 Affonso, P.R.A.M., Guedes, W., Pauls, E. and Galetti Jr., P.M. (2001) Cytogenetic analysis of coral reef fishes from Brazil (families Pomacanthidae and Chaetodontidae). Cytologia, 66: 379–384.

A- 8 Affonso, P.R.A.M., Guedes, W., Pauls, E. and Galetti Jr., P.M. (2002) Close karyotypical relationship between two species of marine angelfishes from South Atlantic: *Pomacanthus arcuatus* and *P. paru* (Perciformes, Pomacanthidae). Caryologia, 55: 323–329.

A- 9 Affonso, P.R.A.M. and Galetti Jr., P.M. (2005) Chromosomal diversification of reef fishes from genus *Centropyge* (Perciformes, Pomacanthidae). Genetica, 123: 227–233.

A- 10 Agnèse, J.F., Oberdorff, T. and Ozouf-Costaz, C. (1990) Karyotypic study of some species of family Mochokidae (Pisces, Siluriformes): evidence of female heterogamety. J. Fish Biol., 37: 375–381.

A- 11 Aguilar, C.T. and Galetti Jr., P.M. (1997) Chromosomal studies in South Atlantic serranids (Pisces, Perciformes). Cytobios, 89: 105–114.

A- 12 Aguilar, C.T., Oliveira-Corrêa, M.M. and Galetti Jr., P.M. (1998) Chromosome associations by centromeric heterochromatin in marine fishes. Chrom. Sci., 2: 73–76.

A- 13 Ahmed, M. (1974) A chromosome study of two species of *Gobiosoma* from Venezuela (Gobiidae: Teleostei). Bol. Inst. Oceanogr. Univ. Oriente, 13: 11–16.

A- 14 Alberdi, A.J. and Fenocchio, A.S. (1997) Karyotypes of five Tetragonopterinae species (Pisces, Characidae) from Argentina. Cytologia, 62: 171–176.

A- 15 Albert, J.S., Fernandes-Matioli, F.M.C. and Almeida-Toledo, L.F. (1999) New species of *Gymnotus* (Gymnotiformes, Teleostei) from southeastern Brazil: toward the deconstruction of *Gymnotus carapo*. Copeia, 1999: 410–421.

A- 16 Allendorf, F.W. and Thorgaard, G.H. (1984) Tetraploidy and the evolution of salmonid fishes. In: Turner, B.J. ed. Evolutionary genetics of fishes. Plenum Press, New York, pp. 1–53.

A- 17 Almeida Marques, M.B., Moreira-Filho, O., Garcia, C. and Margarido, V.P. (2008) Cytogenetic analyses of two endemic fish species from the São Francisco River basin: *Conorhynchus conirostris* and *Lophiosilurus alexandri* (Siluriformes). Genet. Mol. Biol., 31 (suppl.): 215–221.

A- 18 Almeida Toledo, L.F., Foresti, F. and Almeida Toledo, S. (1981) Constitutive heterochromatin and nucleolus organizer region in the knifefish, *Apteronotus albifrons* (Pisces, Apteronotidae). Experientia, 37: 953–954.

A- 19 Almeida Toledo, L.F., Foresti, F. and Toledo Filho, S.A. (1985) Spontaneous triploidy and NOR activity in *Eigenmannia* sp. (Pisces, Sternopygidae) from the Amazon basin. Genetica, 66: 85–88.

A- 20 Almeida-Toledo, L.F., Foresti, F., Trajano, E. and Toledo Filho, S.A. (1992) Cytogenetic analysis of the Brazilian blind catfish *Pimelodella kronei* and its presumed ancestor, *P. transitoria*. Caryologia, 45: 255–262.

A- 21 Almeida Toledo, L.F., Foresti, F., Zambelli Daniel, M.F. and Toledo-Filho, S.A. (1993) Nucleolar chromosome variants in *Sternopygus macrurus* (Pisces, Sternopygidae) from three Brazilian river basins. Caryologia, 46: 53–61.

A- 22 Almeida-Toledo, L.F., Bigoni, A.P., Bernardino, G., Foresti, F. and Toledo-Filho, S.A. (1996) Karyotype and NOR conservatism with heterochromatin reorganization in Neotropical bryconids. Caryologia, 49: 35–43.

A- 23 Al-Sabti, K., Kurelec, B. and Fijan, N. (1983) Cytological studies of some Cyprinidae in Croatia. Veterinar. Arhiv., 53(2): 65–70.

A- 24 Al-Sabti, K., Kurelec, B. and Fijan, N. (1983) Spontaneous triploidy and tetraploidy in the common carp (*Cyprinus carpio* L.). Veterinar. Arhiv., 53(5): 217–223.

A- 25 Al-Sabti, K., Fijan, N. and Kurelec, B. (1983) A simple and fast technique for the chromosome preparation in the fish. Veterinar. Arhiv., 53(6): 283–290.

A- 26 Al-Sabti, K. (1985) Chromosomal studies by blood leukocyte culture technique on three salmonids from Yugoslavian waters. J. Fish Biol., 26: 5–12.

A- 27 Al-Sabti, K. (1986) Karyotypes of *Cyprinus carpio* and *Leuciscus cephalus*. Cytobios, 47: 19–25.

A- 28 Al-Sabti, K. (1987) Cytogenetic studies on five species of Pisces from Yugoslavia. Cytobios, 49: 175–188.

A- 29 Alvarez, M.C., Cano, J. and Thode, G. (1980) DNA content and chromosome complement of *Chromis chromis* (Pomacentridae, Perciformes). Caryologia, 33: 267–274.

A- 30 Alvarez, M.C., Thode, G. and Cano, J. (1983) Somatic karyotypes of two Mediterranean teleost species: *Phycis phycis* (Gadidae) and *Epinephelus alexandrinus* (Serranidae). Cytobios, 38: 91–95.

A- 31 Alvarez, M.C., Garcia, E. and Thode, G. (1986) Contribution to the karyoevolutive study of the Labridae (Perciformes). The karyotypes of *Ctenolabrus rupestris* and *Symphodus ocellatus*. Caryologia, 39: 353–357.

A- 32 Alvarez, M.C., Otis, J., Amores, A. and Guise, K. (1991) Short-term cell culture technique for obtaining chromosomes in marine and freshwater fish. J. Fish Biol., 39: 817–824.

A- 33 Alves, A.L. and Martins-Santos, I.C. (2002) Cytogenetic studies in two populations of *Astyanax scabripinnis* with 2n = 48 chromosomes (Teleostei, Characidae). Cytologia, 67: 117–122.

A- 34 Alves, A.L., Oliveira, C. and Foresti, F. (2003) Karyotype variability in eight species of the subfamilies Loricariinae and Ancistrinae (Teleostei, Siluriformes, Loricariidae). Caryologia, 56: 57–63.

A- 35 Alves, A.L., Oliveira, C. and Foresti, F. (2005) Comparative cytogenetic analysis of eleven species of subfamilies Neoplecostominae and Hypostominae (Siluriformes: Loricariidae). Genetica, 124: 127–136.

A- 36 Amemiya, C.T., Bickham, J.W. and Gold, J.R. (1984) A cell culture technique for chromosome preparation in cyprinid fishes. Copeia, 1984: 232–235.

A- 37 Amemiya, C.T., Kelsch, S.W., Hendricks, F.S. and Gold, J.R. (1986) The karyotype of the Mexican blindcat, *Prietella phreatophila carranza* (Ictaluridae). Copeia, 1986: 1024–1028.

A- 38 Amemiya, C.T. and Gold, J.R. (1986) Chromomycin A_3 stains nucleolus organizer regions of fish chromosomes. Copeia, 1986: 226–231.

A- 39 Amemiya, C.T. and Gold, J.R. (1987) Karyology of 12 species of North American Cyprinidae (minnows) from the southern United States. Cytologia, 52: 715–719.

A- 40 Amemiya, C.T. and Gold, J.R. (1988) Chromosomal NORs as taxonomic and systematic characters in North American cyprinid fishes. Genetica, 76: 81–90.

A- 41 Amemiya, C.T. and Gold, J.R. (1990) Chromosomal NOR phenotypes of seven species of North American Cyprinidae, with comments on cytosystematic relationships of the *Notropis volucellus* species-group, *Opsopoeodus emiliae*, and the genus *Pteronotropis*. Copeia, 1990: 68–78.

A- 42 Amemiya, C.T., Powers, P.K. and Gold, J.R. (1992) Chromosomal evolution in North American cyprinids. In: Mayden, R. ed. Systematics, historical ecology, & North American freshwater fishes. Stanford University Press, Stanford, pp. 515–533.

A- 43 Amores, A., Giles, V., Thode, G. and Alvarez, M.C. (1990) Adaptive character of a Robertsonian fusion in chromosomes of the fish *Gobius paganellus* (Pisces, Perciformes). Heredity, 65: 151–155.

A- 44 Amores, A., Giles, V., Thode, G. and Alvarez, M.C. (1990) A tandem fusion in the fish *Gobius paganellus* (Gobiidae, Perciformes), a karyotypically polymorphic species. Genome, 33: 57–59.

A- 45 Amores, A., Martinez, G., Reina, J. and Alvarez, M.C. (1993) Karyotype, C-banding, and Ag-NOR analysis in *Diplodus bellottii* (Sparidae, Perciformes). Intra-individual polymorphism involving heterochromatic regions. Genome, 36: 672–675.

A- 46 Amores, A., Bejar, J. and Alvarez, M.C. (1995) Replication, C and Ag-NOR chromosome-banding in two anguilliform fish species. Mar. Biol., 123: 845–849.

A- 47 Andreata, A.A., Almeida-Toledo, L.F., Oliveira, C. and Toledo Filho, S.A. (1992) Chromosome studies in Hypoptopomatinae (Pisces, Siluriformes, Loricariidae): I. XX/XY sex chromosome heteromorphism in *Pseudotocinclus tietensis*. Cytologia, 57: 369–372.

A- 48 Andreata, A.A., Almeida-Toledo, L.F., Oliveira, C. and Toledo Filho, S.A. (1993) Chromosome studies in Hypoptopomatinae (Pisces, Siluriformes, Loricariidae). II. ZZ/ZW sex-chromosome system, B chromosomes, and constitutive heterochromatin differentiation in *Microlepidogaster leucofrenatus*. Cytogenet. Cell Genet., 63: 215–220.

A- 49 Andreata, A.A., Almeida-Toledo, L.F., Oliveira, C. and Toledo Filho, S.A. (1994) Cytogenetic studies on the subfamily Hypoptopomatinae (Pisces, Siluriformes, Loricariidae). III. Analysis of seven species. Caryologia, 47: 27–37.

A- 50 Anjum, R., Jankun, M., Kohlman, K. and Kersten, P. (1997) Silver and chromomycin A_3 staining of nucleolus organizer regions in the chromosomes of ornamental (Koi) common carp, *Cyprinus carpio*. Cytobios, 90: 73–79.

A- 51 Anjum, R. and Jankun, M. (1998) NOR-bearing chromosomal associations revealed through silver and sequential chromomycin A_3 staining in the mirror carp, *Cyprinus carpio*. Caryologia, 51: 167–171.

A- 52 Arai, R. and Kobayasi, H. (1973) A chromosome study on thirteen species of Japanese gobiid fishes. Japan. J. Ichthyol., 20: 1–6.

A- 53 Arai, R. (1973) Preliminary notes on chromosomes of the medaka, *Oryzias latipes*. Bull. Natn. Sci. Mus. Tokyo, 16: 173–176.

A- 54 Arai, R. and Katsuyama, I. (1973) Notes on the chromosomes of three species of shore-fishes. Bull. Natn. Sci. Mus. Tokyo, 16: 405–408.

A- 55 Arai, R. and Shiotsuki, K. (1973) A chromosome study on three species of the tribe Salariini from Japan (Pisces, Blenniidae). Bull. Natn. Sci. Mus. Tokyo, 16: 581–584.

A- 56 Arai, R. and Hirano, H. (1974) First record of the clariid catfish, *Clarias fuscus*, from Japan. Japan. J. Ichthyol., 21: 53–60.

A- 57 Arai, R. and Sawada, Y. (1974) Chromosomes of Japanese gobioid fishes (I). Bull. Natn. Sci. Mus. Tokyo, 17: 97–102.

A- 58 Arai, R. and Katsuyama, I. (1974) A chromosome study on four species of Japanese catfishes (Pisces, Siluriformes). Bull. Natn. Sci. Mus. Tokyo, 17: 187–192.

A- 59 Arai, R. and Shiotsuki, K. (1974) Chromosomes of six species of Japanese blennioid fishes. Bull. Natn. Sci. Mus. Tokyo, 17: 261–268.

A- 60 Arai, R., Katsuyama, I. and Sawada, Y. (1974) Chromosomes of Japanese gobioid fishes (II). Bull. Natn. Sci. Mus. Tokyo, 17: 269–274.

A- 61 Arai, R. and Inoue, M. (1975) Chromosomes of nine species of Chaetodontidae and one species of Scorpidae from Japan. Bull. Natn. Sci. Mus. Tokyo, (A), 1: 217–224.

A- 62 Arai, R. and Sawada, Y. (1975) Chromosomes of Japanese gobioid fishes (III). Bull. Natn. Sci. Mus. Tokyo, (A), 1: 225–232.

A- 63 Arai, R., Nagaiwa, K. and Sawada, Y. (1976) Chromosomes of *Chanos chanos* (Gonorynchiformes, Chanidae). Japan. J. Ichthyol., 22: 241–242.

A- 64 Arai, R. and Nagaiwa, K. (1976) Chromosomes of tetraodontiform fishes from Japan. Bull. Natn. Sci. Mus. Tokyo, (A), 2: 59–72.

A- 65 Arai, R. and Inoue, M. (1976) Chromosomes of seven species of Pomacentridae and two species of Acanthuridae from Japan. Bull. Natn. Sci. Mus. Tokyo, (A), 2: 73–78.

A- 66 Arai, R., Inoue, M. and Ida, H. (1976) Chromosomes of four species of coral fishes from Japan. Bull. Natn. Sci. Mus. Tokyo, (A), 2: 137–142.

A- 67 Arai, R. and Nagaiwa, K. (1976) Chromosomes of two species of beryciform fishes from Japan. Bull. Natn. Sci. Mus. Tokyo, (A), 2: 199–203.

A- 68 Arai, R. and Nagaiwa, K. (1977) Chromosomes of three species of clingfishes from Japan. Bull. Natn. Sci. Mus. Tokyo, (A), 3: 117–120.

A- 69 Arai, R. and Kawai, M. (1977) Chromosomes of two species of antennariid fishes from Japan. Bull. Natn. Sci. Mus. Tokyo, (A), 3: 121–124.

A- 70 Arai, R. and Fujiki, A. (1977) Chromosomes of three races of goldfish, Kuro-demekin, Sanshiki-demekin and Ranchu. Bull. Natn. Sci. Mus. Tokyo, (A), 3: 187–192.

A- 71 Arai, R. and Fujiki, A. (1978) Chromosomes of two species of atherinoid fishes. Bull. Natn. Sci. Mus. Tokyo, (A), 4: 147–150.

A- 72 Arai, R. and Fujiki, A. (1978) Chromosomes of three species of cottid fishes from Japan. Bull. Natn. Sci. Mus. Tokyo, (A), 4: 233–239.

A- 73 Arai, R. and Fujiki, A. (1979) Chromosomes of Japanese gobioid fishes (IV). Bull. Natn. Sci. Mus. Tokyo, (A), 5: 153–159.

A- 74 Arai, R. and Koike, A. (1979) Chromosomes of a sweeper and a goatfish (Teleostei, Percoidei) from Japan. Bull. Natn. Sci. Mus. Tokyo, (A), 5: 219–223.

A- 75 Arai, R. and Koike, A. (1980) Chromosomes of labroid fishes from Japan. Bull. Natn. Sci. Mus. Tokyo, (A), 6: 119–135.

A- 76 Arai, R. and Koike, A. (1980) A karyotype study on two species of freshwater fishes transplanted into Japan. Bull. Natn. Sci. Mus. Tokyo, (A), 6: 275–278.

A- 77 Arai, R. and Yamamoto, T. (1981) Chromosomes of six species of percoid fishes from Japan. Bull. Natn. Sci. Mus. Tokyo, (A), 7: 87–100.

A- 78 Arai, R. (1982) A chromosome study on two cyprinid fishes, *Acrossocheilus labiatus* and *Pseudorasbora pumila pumila*, with notes on Eurasian cyprinids and their karyotypes. Bull. Natn. Sci. Mus. Tokyo, (A), 8: 131–152.

A- 79 Arai, R. (1983) Karyological and osteological approach to phylogenetic systematics of tetraodontiform fishes. Bull. Natn. Sci. Mus. Tokyo, (A), 9: 175–210.

A- 80 Arai, R. (1984) Karyotypes of a mugiloidid, *Parapercis kamoharai*, and a blenniid, *Omobranchus punctatus* (Pisces, Perciformes). Bull. Natn. Sci. Mus. Tokyo, (A), 10: 209–213.

A- 81 Arai, R. and Tokoro, T. (1986) Karyotypes of two types of a Chinese cyprinid fish, *Tanichthys albonubes*. Bull. Natn. Sci. Mus. Tokyo, (A), 12: 37–43.

A- 82 Arai, R., Suzuki, A. and Akai, Y. (1988) The karyotype and DNA value of a cypriniform algae eater, *Gyrinocheilus aymonieri*. Japan. J. Ichthyol., 34: 515–517.

A- 83 Arai, R., Suzuki, A. and Akai, Y. (1988) A karyotype of a Chinese bitterling, *Paracheilognathus himantegus* (Cypriniformes, Cyprinidae). Bull. Natn. Sci. Mus. Tokyo, (A), 14: 43–46.

A- 84 Arai, R. and Magtoon, W. (1991) Karyotypes of four cyprinid fishes from Thailand. Bull. Natn. Sci. Mus. Tokyo, (A), 17: 183–188.

A- 85 Arai, R., Akai, Y. and Suzuki, N. (1992) The karyotype of a Chinese bitterling, *Acheilognathus tonkinensis* (Pisces, Cyprinidae). Bull. Natn. Sci. Mus. Tokyo, (A), 18: 73–77.

A- 86 Arai, R., Jeon, S.R. and Ueda, T. (2001) *Rhodeus pseudosericeus* sp. nov., a new bitterling from South Korea (Cyprinidae, Acheilognathinae). Ichthyol. Res., 48: 275–282.

A- 87 Arcement, R.J. and Rachlin, J.W. (1976) A study of the karyotype of a population of banded killifish (*Fundulus diaphanus*) from the Hudson River. J. Fish Biol., 8: 119–125.

A- 88 Arefjev, V.A. (1983) Polykaryographic analysis of sheap sturgeon, *Acipenser nudiventris* (Acipenseridae). J. Ichthyol., 23(2): 26–35.

A- 89 Arefjev, V.A. (1989) Male karyotype of Black sea clingfish *Lepadogaster lepadogaster* (Bonnaterre) (Gobiesociformes, Pisces). Nucleus, 32: 131–134.

A- 90 Arefjev, V.A. (1989) Cytogenetic analysis and nucleolar organization of the sea bass *Dicentrarchus labrax*. J. Ichthyol., 29(8): 1–12.

A- 91 Arefjev, V.A. (1989) Chromosome sets of four characid fish species (Teleostei, Characidae). Zool. Zhurnal, 68: 82–91. (In Russian with English summary)

A- 92 Arefjev, V.A. (1990) Karyotypic diversity of characid families (Pisces, Characidae). Caryologia, 43: 291–304.

A- 93 Arefjev, V.A. (1990) Problems of karyotypic variability in the family Characidae (Pisces, Characiformes) with the description of somatic karyotypes for six species of tetras. Caryologia, 43: 305–319.

A- 94 Arefjev, V.A. and Nikolaev, A.I. (1991) Cytological analysis of the reciprocal hybrids between low- and high-chromosome acipenserids, the great sturgeon, *Huso huso* (L.), and the Russian sturgeon, *Acipenser gueldenstaedtii* Brandt. Cytologia, 56: 495–502.

A- 95 Arefjev, V.A. (1991) Different rate of intrapopulation chromosome polymorphism in two species of Black sea blennies (Blenniidae, Pisces). Genetica, 83: 181–187.

A- 96 Arkhipchuk, V.V., Androshchuk, A.F. and Berdyshev, G.D. (1986) Utilization of variation index of chromosomes and their morphology in karyosystematics in several taxa of subclass Teleostei. Doklady Acad. Sci. Ukrain. SSR, Ser. B, 1986(1): 66–68.

A- 97 Arkhipchuk, V.V. (1999) Nucleolar variations during the ontogenesis of diploid and tetraploid cyprinid species. J. Fish Biol., 54: 513–524.

A- 98 Artoni, R.F. and Bertollo, L.A.C. (1996) Cytogenetic studies on Hypostominae (Pisces, Siluriformes, Loricariidae). Considerations on karyotype evolution in the genus *Hypostomus*. Caryologia, 49: 81–90.

A- 99 Artoni, R.F., Venere, P.C. and Bertollo, L.A.C. (1998) A heteromorphic ZZ/ZW sex chromosome system in fish, genus *Hypostomus* (Loricariidae). Cytologia, 63: 421–425.

A- 100 Asahida, T., Ida, H. and Inoue, S. (1987) Karyotypes of three rays in the order Myliobatiformes. Japan. J. Ichthyol., 33: 426–430.

A- 101 Asahida, T., Ida, H. and Inoue, T. (1988) Karyotypes and cellular DNA contents of two sharks in the family Scyliorhinidae. Japan. J. Ichthyol., 35: 215–219.

A- 102 Asahida, T. and Ida, H. (1989) Karyological notes on four sharks in the order Carcharhiniformes. Japan. J. Ichthyol., 36: 275–280.

A- 103 Asahida, T. and Ida, H. (1990) Karyotypes of two rays, *Torpedo tokionis* and *Dasyatis matsubarai*, and their systematic relationships. Japan. J. Ichthyol., 37: 71–75.

A- 104 Asahida, T., Ida, H., Terashima, H. and Chang, H.Y. (1993) The karyotype and cellular DNA content of a ray, *Mobula japonica*. Japan. J. Ichthyol., 40: 317–322.

A- 105 Asahida, T., Ida, H. and Hayashizaki, K. (1995) Karyotypes and cellular DNA contents of some sharks in the order Carcharhiniformes. Japan. J. Ichthyol., 42: 21–26.

A- 106 Atkin, N.B. and Ohno, S. (1967) DNA values of four primitive chordates. Chromosoma, 23: 10–13.

A- 107 Avise, J.C. and Gold, J.R. (1977) Chromosomal divergence and speciation in two families of North American fishes. Evolution, 31: 1–13.

A- 108 Andreata, A.A., Oliveira, C. and Foresti, F. (2006) Karyological characterization of four Neotropical fish species of the genus *Hisonotus* (Teleostei, Loricariidae, Hypoptopomatinae) from distinct Brazilian river basins. Genet. Mol. Biol., 29: 62–66.

A- 109 Artoni, R.F. and Bertollo, L.A.C. (2001) Trends in the karyotype evolution of Loricariidae fish (Siluriformes). Hereditas, 134: 201–210.

A- 110 Asahida, T. and Ida, H. (1995) Karyotype and cellular DNA content of a guitarfish, *Rhinobatos schlegelii*. La Kromosomo, II, 79/80: 2725–2730.

A- 111 Alves, M.J., Coelho, M.M. and Collares-Pereira, M.J. (2001) Evolution in action through hybridisation and polyploidy in an Iberian freshwater fish: a genetic review. Genetica, 111: 375–385.

A- 112 Alves, A.L., Oliveira, C., Nirchio, M., Granado, A. and Foresti, F. (2006) Karyotypic relationships among the tribes of Hypostominae (Siluriformes: Loricariidae) with description of XO sex chromosome system in a Neotropical fish species. Genetica, 128: 1–9.

A- 113 Almeida-Toledo, L.F., Daniel-Silva, M.F.Z., Lopes, C.E. and Toledo-Filho, S.A. (2000) Sex chromosome evolution in fish. II. Second occurrence of an X_1X_2Y sex chromosome system in Gymnotiformes. Chrom. Res., 8: 335–340.

A- 114 Almeida-Toledo, L.F., Foresti, F., Péquignot, E.V. and Daniel-Silva, M.F.Z. (2001) XX: XY sex chromosome system with X heterochromatinization: an early stage of sex chromosome differentiation in the Neotropic electric eel *Eigenmannia virescens*. Cytogenet. Cell Genet., 95: 73–78.

A- 115 Almeida-Toledo, L.F., Daniel-Silva, M.F.Z., Moyses, C.B., Fonteles, S.B.A., Lopes, C.E., Akama, A. and Foresti, F. (2002) Chromosome evolution in fish: sex chromosome variability in *Eigenmannia virescens* (Gymnotiformes: Sternopygidae). Cytogenet. Genome Res., 99: 164–169.

A- 116 Almeida-Toledo, L.F., Foresti, F., Daniel, M.F.Z. and Toledo-Filho, S.A. (2000) Sex chromosome evolution in fish: the formation of the neo-Y chromosome in *Eigenmannia* (Gymnotiformes). Chromosoma, 109: 197–200.

A- 117 Artoni, R.F., Falcão, J.D.N., Moreira-Filho, O. and Bertollo, L.A.C. (2001) An uncommon condition for a sex chromosome system in Characidae fish. Distribution and differentiation of the ZZ/ZW system in *Triportheus*. Chrom. Res., 9: 449–456.

A- 118 Arefjev, V.A. and Karnauchov, G.I. (1989) Species-specificity of electrophoretic patterns of haemoglobin and uniformity of karyotypes in fishes, genus *Abramis* (Pisces: Cyprinidae). Biochem. Syst. Ecol., 17: 479–488.

A- 119 Artoni, R.F., Vicari, M.R., Endler, A.L., Cavallaro, Z.I., Jesus, C.M., Almeida, M.C., Moreira-Filho, O. and Bertollo, L.A.C. (2006) Banding pattern of A and B chromosomes of *Prochilodus lineatus* (Characiformes, Prochilodontidae), with comments on B chromosomes evolution. Genetica, 127: 277–284.

A- 120 Azevedo, M.F.C., Oliveira, C., Pardo, B.G., Martinez, P. and Foresti, F. (2007) Cytogenetic characterization of six species of flatfishes with comments to karyotype differentiation patterns in Pleuronectiformes (Teleostei). J. Fish Biol., 70 (suppl. A): 1–15.

A- 121 Azevedo, M.F.C., Oliveira, C., Pardo, B.G., Martinez, P. and Foresti, F. (2005) Chromosome banding and 18S rDNA *in situ* hybridization analysis of seven species of the family Achiridae (Teleostei: Pleuronectiformes). Genetica, 125: 125–132.

A- 122 Adriano Moreira, P.W., Bertollo, L.A.C. and Moreira-Filho, O. (2007) Comparative cytogenetics between three Characidae fish species from the Sao Francisco River basin. Caryologia, 60: 64–68.

A- 123 Arai, R. (1981) Fishes of *Luciogobius* and *Inu* (Gobiidae) from the Izu Peninsula, Central Japan. Mem. Natn. Sci. Mus. Tokyo, 14: 151–166. (In Japanese with English summary)

A- 124 Accioly, I.V. and Molina, W.F. (2008) Cytogenetic studies in Brazilian marine Sciaenidae and Sparidae (Perciformes). Genet. Mol. Res., 7: 358–370.

A- 125 Almeida-Toledo, L.F., Daniel-Silva, M.F.Z., Moysés, C.B. and Foresti, F. (2007) Chromosome variability in Gymnotiformes (Teleostei: Ostariophysi). In: Pisano, E., Ozouf-Costaz, C., Foresti, F. and Kapoor, B.G. eds. Fish cytogenetics. Science Publishers, Enfield, pp. 17–39.

A- 126 Alves, A.L., Porto-Foresti, F., Oliveira, C. and Foresti, F. (2008) Supernumerary chromosomes in the pufferfish *Sphoeroides spengleri*—First occurrence in marine teleostean Tetraodontiformes fish. Genet. Mol. Biol., 31 (suppl.): 243–245.

A- 127 Arratia, G. and Campos, H. (1997) Karyotypes of diplomystid and certain loricarioid catfishes. International symposium on phylogeny and classification of neotropical fishes and neotropical ichthyological association 9th annual meeting. Porto Alegre, Brazil. Program and Abstracts, pp. 53–54.

A- 128 Adriano, M.P.W., Bertollo, L.A.C. and Moreira-Filho, O. (2006) Karyotypic characterization of *Myleus micans* (Lütken, 1875) (Pisces, Characidae, Serrasalminae). Caryologia, 59: 125–130.

A- 129 Almeida-Toledo, L.F., Foresti, F., Toledo, S.A., Bernardino, G., Ferrari, W. and Alcantara, R.C.G. (1987) Cytogenetic studies of *Colossoma mitrei*, *Colossoma macropomum* and their interspecific hybrid. Proc. World symp. On selection, hybridization, and genetic engineering in aquaculture, Vol. I, pp. 189–195.

A- 130 Almeida Toledo, L.F., Foresti, F. and Toledo Filho, S.A. (1984) Complex sex chromosome system in *Eigenmannia* sp. (Pisces, Gymnotiformes). Genetica, 64: 165–169.

A- 131 Almeida-Toledo, L.F., Stocker, A.J., Foresti, F. and Toledo-Filho, S.A. (1996) Fluorescence *in situ* hybridization with rDNA probes on chromosomes of two nucleolus organizer region phenotypes of a species of *Eigenmannia* (Pisces, Gymnotoidei, Sternopygidae). Chrom. Res., 4: 301–305.

A- 132 Artoni, R.F. and Bertollo, L.A.C. (1999) Nature and distribution of constitutive heterochromatin in fishes, genus *Hypostomus* (Loricariidae). Genetica, 106: 209–214.

A- 133 Artoni, R.F., Molina, W.F., Bertollo, L.A.C. and Galetti Jr., P.M. (1999) Heterochromatin analysis in the fish species *Liposarcus anisitsi* (Siluriformes) and *Leporinus elongatus* (Characiformes). Genet. Mol. Biol., 22: 39–44.

[B]

B- 1 Bachmann, K., Goin, O.B. and Goin, C.J. (1972) The nuclear DNA of *Polypterus palmas*. Copeia, 1972: 363–365.

B- 2 Banerjee, S.K., Misra, K.K., Banerjee, S. and Ray-Chaudhuri, S.P. (1988) Chromosome numbers, genome sizes, cell volumes and evolution of snake-head fish (family Channidae). J. Fish Biol., 33: 781–789.

B- 3 Barat, A. and Khuda-Bukhsh, A.R. (1984) Karyomorphology of a sea-frog *Tetraodon fluviatilis* (Tetraodontidae, Pisces). Curr. Sci., 53: 1108–1109.

B- 4 Barat, A. and Khuda-Bukhsh, A.R. (1986) Karyomorphometrical studies in two species of fishes, *Lepidocephalichthys guntea* (fam.: Cobitidae) and *Mystus corsula* (fam.: Bagridae). Perspect. Cytol. Genet., 5: 115–118.

B- 5 Barat, A., Sahoo, P.K., Nagpure, N.S. and Ponniah, A.G. (1997) Variation in NOR pattern in different populations of *Schizothorax richardsonii* (Cyprinidae: Pisces). Cytobios, 91: 181–185.

B- 6 Barker, C.J. (1972) A method for the display of chromosomes of plaice, *Pleuronectes platessa*, and other marine fishes. Copeia, 1972: 365–368.

B- 7 Baršiene, J.V. (1977) The karyotype of dwarf males of *Salmo salar* L. from the Neva population. Tsitologiya, 19: 906–913. (In Russian with English summary)

B- 8 Baršiene, J.V. (1977) Variability of chromosome sets in cells of various organs and tissues of the Atlantic salmon *Salmo salar* L. Tsitologiya, 19: 791–797. (In Russian with English summary)

B- 9 Baršiene, J.V. (1980) Mechanism of ontogenic variability of chromosome sets in Atlantic salmon (*Salmo salar* L.). In: Kirpichnikov, V.S. ed. Karyological variability, mutagenesis and gynogenesis in fishes. Inst. Cytol. Acad. Sci. USSR, Leningrad, pp. 3-9. (In Russian with English summary)

B- 10 Baruffaldi, A. and Cucchi, C. (1988) Karyological analysis in *Ictalurus nebulosus marmoratus* and *Ictalurus* species (Siluriformes: Ictaluridae). Cytobios, 54: 191–194.

B- 11 Beamish, R.J. and Tsuyuki, H. (1971) A biochemical and cytological study of the longnose sucker (*Catostomus catostomus*) and large and dwarf forms of the white sucker (*Catostomus commersoni*). J. Fish. Res. Board Can., 28: 1745–1748.

B- 12 Beamish, R.J., Merrilees, M.J. and Crossman, E.J. (1971) Karyotypes and DNA values for members of the suborder Esocoidei (Osteichthyes: Salmoniformes). Chromosoma, 34: 436–447.

B- 13 Beamish, R.J. and Miller, R.R. (1977) Cytotaxonomic study of gila trout, *Salmo gilae*. J. Fish. Res. Board Can., 34: 1041–1045.

B- 14 Beamish, R.J. and Uyeno, T. (1978) Karyotype of *Hiodon tergisus* and DNA values of *Hiodon tergisus* and *Hiodon alosoides*. Chrom. Inform. Serv., 24: 5–7.

B- 15 Beçak, W., Beçak, M.L. and Ohno, S. (1966) Intraindividual chromosomal polymorphism in green sunfish (*Lepomis cyanellus*) as evidence of somatic segregation. Cytogenetics, 5: 313–320.

B- 16 Berberović, Lj., Hadziselimovic, R. and Sofradžija, A. (1969) Chromosome number of the species *Paraphoxinus alepidotus* (Heckel). Bull. Sci. Yougoslavie, Sec. A, 14: 386.

B- 17 Berberović, Lj., Hadziselimovic, R. and Sofradžija, A. (1970) Comparative review of the basic data on the chromosome complements of *Chondrostoma phoxinus* Heckel and *Chondrostoma kneri* Heckel. Ichthyologia, 2(1): 25–30. (In Serbo-Croatian with English summary)

B- 18 Berberović, Lj., Curic, M., Hadziselimovic, R. and Sofradžija, A. (1970) Chromosome complement of *Salmothymus obtusirostris oxyrhynchus* (Steindachner). Genetika, Acta Biol. Iugoslavica, 2: 55–63.

B- 19 Berberović, Lj. and Sofradžija, A. (1972) A review of the caryological data regarding freshwater fishes of Yugoslavia. Ichthyologia, 4(3): 1–21. (In Serbo-Croatian with English summary)

B- 20 Berberović, Lj., Hadziselimovic, R., Pavlovic, B. and Sofradžija, A. (1973) Chromosome set of the species *Aulopyge hugeli* Heckel 1841. Bull. Sci. Yougoslavie, Sec. A, 18: 10–11.

B- 21 Berberović, Lj. and Sofradžija, A. (1974) Basic data about diploid chromosome complement of *Phoxinus phoxinus* (Cyprinidae, Pisces). Bull. Sci. Yougoslavie, Sec. A, 19: 142–143.

B- 22 Berberović, Lj., Sofradžija, A. and Obradovic, S. (1975) Chromosome complement of *Ictalurus nebulosus* (Le Sueur), Ictaluridae, Pisces. Bull. Sci. Yougoslavie, Sec. A, 20: 149–150.

B- 23 Berberović, Lj., Sofradžija, A. and Vagner, Lj. (1978) The chromosomes of kinch (*Tinca tinca*). Ichthyologia, 10(1): 9–15. (In Serbo-Croatian with English summary)

B- 24 Berberović, Lj. and Sofradžija, A. (1974) The chromosomes of the species *Cottus gobio* L. (Cottidae, Pisces). God. Biol. Inst. Univ. Sarajevu, 27: 5–15. (In Serbo-Croatian with English summary)

B- 25 Berdyshev, G.D. and Arkhipchuk, V.V. (1985) Method of chromosome preparation from the tissue of phytophagous fishes. Hydrobiol. J., 21(1): 61–66. (In Russian with English summary)

B- 26 Berdyshev, G.D. and Arkhipchuk, V.V. (1986) Characteristics of karyotypes and corresponding evolutionary transformations in various taxa of Pisces. Progress Modern Biol., 1986(1): 30–42. (In Russian)

B- 27 Bermejo, J.V.D. and Millán, M.M. (1988) Chromosomes of *Valencia lozanoi* (Pisces, Cyprinodontidae). Cytobios, 54: 149–152.

B- 28 Berrebi, P. and Ráb, P. (1998) The '*Barbus*' *intermedius* species flock in Lake Tana (Ethiopia): III-Cytogenetic and molecular genetic data. Ital. J. Zool., 15 (suppl), 15–20.

B- 29 Bertollo, L.A.C., Takahashi, C.S. and Moreira Filho, O. (1978) Cytotaxonomic considerations on *Hoplias lacerdae* (Pisces, Erythrinidae). Brazil. J. Genet., 1: 103–120.

B- 30 Bertollo, L.A.C., Takahashi, C.S. and Moreira Filho, O. (1979) Karyotypic studies of two allopatric populations of the genus *Hoplias* (Pisces, Erythrinidae). Brazil. J. Genet., 2: 17–37.

B- 31 Bertollo, L.A.C., Takahashi, C.S. and Moreira Filho, O. (1983) Multiple sex chromosomes in the genus *Hoplias* (Pisces: Erythrinidae). Cytologia, 48: 1–12.

B- 32 Bertollo, L.A.C., Moreira-Filho, O. and Galetti Jr., P.M. (1986) Cytogenetics and taxonomy: considerations based on chromosome studies of freshwater fish. J. Fish Biol., 28: 153–159.

B- 33 Bertollo, L.A.C. (1996) The nucleolar organizer regions of Erythrinidae fish. An uncommon situation in the genus *Hoplias*. Cytologia, 61: 75–81.

B- 34 Bertollo, L.A.C., Fontes, M.S., Fenocchio, A.S. and Cano, J. (1997) The X_1X_2Y sex chromosome system in the fish *Hoplias malabaricus*. I. G-, C- and chromosome replication banding. Chrom. Res., 5: 493–499.

B- 35 Bertollo, L.A.C., Oliveira, C., Molina, W.F., Margarido, V.P., Fontes, M.S., Pastori, M.C., Falcão, J. N. and Fenocchio, A.S. (2004) Chromosome evolution in the erythrinid fish, *Erythrinus erythrinus* (Teleostei: Characiformes). Heredity, 93: 228–233.

B- 36 Bianco, P.G., Aprea, G., Balletto, E., Capriglione, T., Fulgione, D. and Odierna, G. (2004) The karyology of the cyprinid genera *Scardinius* and *Rutilus* in southern Europe. Ichthyol. Res., 51: 274–278.

B- 37 Birstein, V.J. and Vasiliev, V.P. (1987) Tetraploid-octoploid relationships and karyological evolution in the order Acipenseriformes (Pisces). Karyotypes, nucleoli, and nucleolus-organizer regions in four acipenserid species. Genetica, 72: 3–12.

B- 38 Black, D.A. and Howell, W.M. (1978) A distinctive chromosomal race of the cyprinodontid fish, *Fundulus notatus*, from the Upper Tombigbee River system of Alabama and Mississippi. Copeia, 1978: 280–288.

B- 39 Black, D.A. and Howell, W.M. (1979) The North American mosquitofish, *Gambusia affinis*: a unique case in sex chromosome evolution. Copeia, 1979: 509–513.

B- 40 Blaxhall, P.C. (1983) Chromosome karyotyping of fish using conventional and G-banding methods. J. Fish Biol., 22: 417–424.

B- 41 Bogart, J.P., Balon, E.K. and Bruton, M.N. (1994) The chromosomes of the living coelacanth and their remarkable similarity to those of one of the most ancient frogs. J. Heredity, 85: 322–325.

B- 42 Bolla, S. (1987) Cytogenetic studies in Atlantic salmon and rainbow trout embryos. Hereditas, 106: 11–17.

B- 43 Booke, H.E. (1968) Cytotaxonomic studies of the coregonine fishes of the Great Lakes, USA: DNA and karyotype analysis. J. Fish. Res. Board Can., 25: 1667–1687.

B- 44 Booke, H.E. (1974) A cytotaxonomic study of the round whitefishes, genus *Prosopium*. Copeia, 1974: 115–119.

B- 45 Booke, H.E. (1975) Cytotaxonomy of the salmonid fish *Stenodus leucichthys*. J. Fish. Res. Board Can., 32: 295–296.

B- 46 Borin, L.A. and Martins-Santos, I.C. (2002) Cytogenetic aspects in species of the genus *Pimelodus* (Pisces, Siluriformes, Pimelodidae) of the River Paraná basin. Cytologia, 67: 199–204.

B- 47 Borin, L.A. and Martins-Santos, I.C. (2004) Study on karyotype and occurrence of B chromosomes in two endemic species of the genus *Pimelodus* (Siluriformes, Pimelodidae) from the river Iguaçu. Hereditas, 140: 201–209.

B- 48 Boroń, A. (1992) Karyotype study of diploid and triploid *Cobitis taenia* (Pisces, Cobitididae) from Vistula river basin. Cytobios, 72: 201–206.

B- 49 Boroń, A. (1994) Use of erythrocyte measurements to detect natural triploids of spined loach *Cobitis taenia* (L.). Cytobios, 78: 197–202.

B- 50 Boroń, A. (1994) Karyotypes of diploid and triploid silver crucian carp *Carassius auratus gibelio* (Bloch). Cytobios, 80: 117–124.

B- 51 Boroń, A. (1995) Chromosome banding studies of *Noemacheilus barbatulus* (Linnaeus, 1758) from Poland. Caryologia, 48: 239–246.

B- 52 Boroń, A. (1995) Chromosome banding studies of spined loach *Cobitis taenia* (L.). Cytobios, 81: 97–102.

B- 53 Bouza, C., Sanchez, L. and Martinez, P. (1994) Karyotypic characterization of turbot (*Scophthalmus maximus*) with conventional, fluorochrome and restriction endonuclease-banding techniques. Mar. Biol., 120: 609–613.

B- 54 Bozhko, S.I., Horváth, Á. and Mészáros, B. (1976) Karyological examinations for four species of Cyprinidae from Hungary. Acta Biol. Debrecina, 13: 237–254.

B- 55 Brum, M.J.I., Galetti Jr., P.M., Oliveira Corrêa, M.M. and Aguilar, C.T. (1992) Multiple sex chromosomes in South Atlantic fish, *Brevoortia aurea*, Clupeidae. Brazil. J. Genet., 15: 547–553.

B- 56 Brum, M.J.I., Oliveira, C.C. and Galetti Jr., P.M. (1995) Cytogenetic studies of two puffer species (*Sphoeroides*, Tetraodontidae) from Rio de Janeiro Coast, Brazil. Cytologia, 60: 369–374.

B- 57 Brum, M.J.I. (2000) Cytogenetic studies in tetraodontiforms *Sphoeroides tyleri* (Tetraodontidae) and *Chilomycterus spinosus* (Diodontidae) from Rio de Janeiro Coast, Brazil. Chrom. Sci., 4: 103–105.

B- 58 Brum, M.J.I., Affonso, P.R.A.M., Mota, L.C.G., Pauls, E. and Netto, M.R.C.B. (2001) Cytogenetic characterization of *Porichthys porosissimus* (Valenciennes, 1857) (Batrachoididae, Batrachoidiformes) from the Rio de Janeiro Coast, Brazil. Chrom. Sci., 5: 15–18.

B- 59 Brum, M.J.I., Corrêa, M.M.O. and Aguilar, C.T. (2001) Karyotype analysis in South Atlantic perciforms *Micropogonias furnieri* and *Abudefduf saxatilis*. Chrom. Sci., 5: 83–89.

B- 60 Busack, C.A., Thorgaard, G.H., Bannon, M.P. and Gall, G.A.E. (1980) An electrophoretic, karyotypic and meristic characterization of the Eagle Lake trout, *Salmo gairdneri aquilarum*. Copeia, 1980: 418–424.

B- 61 Bertollo, L.A.C., Born, G.G., Dergam, J.A., Fenocchio, A.S. and Moreira-Filho, O. (2000) A biodiversity approach in the neotropical Erythrinidae fish, *Hoplias malabaricus*. Karyotypic survey, geographic distribution of cytotypes and cytotaxonomic considerations. Chrom. Res., 8: 603–613.

B- 62 Boroń, A. (2001) Comparative chromosomal studies on two minnow fish, *Phoxinus phoxinus* (Linnaeus, 1758) and *Eupallasella perenurus* (Pallas, 1814); an associated cytogenetic-taxonomic considerations. Genetica, 111: 387–395.

B- 63 Benirschke, K. and Hsu, T.C. (eds) (1971) Chromosome atlas: Fish, amphibians, reptiles and birds. Volume 1. Springer-Verlag, Berlin.

B- 64 Benirschke, K. and Hsu, T.C. (eds) (1973) Chromosome atlas: Fish, amphibians, reptiles and birds. Volume 2. Springer-Verlag, Berlin.

B- 65 Brum, M.J.I., Corrêa, M.M.O., Oliveira, C.C. and Galetti Jr., P.M. (1995) Cytogenetic studies on the Perciformes *Orthopristis ruber* (Haemulidae) and *Scartella cristata* (Blenniidae). Caryologia, 48: 309–318.

B- 66 Boron, A., Porycka, K., Ito, D., Abe, S. and Kirtiklis, L. (2009) Comparative molecular cytogenetic analysis of three *Leuciscus* species (Pisces, Cyprinidae) using chromosome banding and FISH with rDNA. Genetica, 135: 199–207.

B- 67 Borin, L.A., Martins-Santos, I.C. and Oliveira, C. (2002) A natural triploid in *Trichomycterus davisi* (Siluriformes, Trichomycteridae): mitotic and meiotic characterization by chromosome banding and synaptonemal complex analyses. Genetica, 115: 253–258.

B- 68 Birstein, V.J., Poletaev, A.I. and Goncharov, B.F. (1993) DNA content in Eurasian sturgeon species determined by flow cytometry. Cytometry, 14: 377–383.

B- 69 Blacklidge, K.H. and Bidwell, C.A. (1993) Three ploidy levels indicated by genome quantification in Acipenseriformes of North America. J. Heredity, 84: 427–430.

B- 70 Biavati, S.H. and Maistro, E.L. (2007) Cytogenetical analyses in three *Astyanax scabripinnis* populations (Pisces, Characidae) from Minas Gerais State, Brazil. Cytologia, 72: 357–361.

B- 71 Brown, N.P., Bromage, N.R. and Penman, D.J. (1997) The karyotype of the Atlantic halibut, *Hippoglossus hippoglossus* (Linnaeus). Aquacult. Res., 28: 489–491.

B- 72 Bohlen, J., Völker, M., Rábová, M. and Ráb, P. (2008) Note on the banded karyotype of the enigmatic South Asian loach *Vaillantella maassi* (Cypriniformes: Vaillantellidae). Ichthyol. Res., 55: 82–84.

B- 73 Barat, A. and Sahoo, P.K. (2007) Karyotype analysis of *Channa punctatus* (Pisces) using restriction endonucleases. Cytologia, 72: 471–473.

B- 74 Brassesco, M.S., Pastori, M.C., Roncati, H.A. and Fenocchio, A.S. (2004) Comparative cytogenetic studies of Curimatidae (Pisces, Characiformes) from the middle Paraná River (Argentina). Genet. Mol. Res., 3: 293–301.

B- 75 Brainerd, E.L., Slutz, S.S., Hall, E.K. and Phillis, R.W. (2001) Patterns of genome size evolution in tetraodontiform fishes. Evolution, 55: 2363–2368.

B- 76 Born, G.G. and Bertollo, L.A.C. (2000) An XX/XY sex chromosome system in a fish species, *Hoplias malabaricus*, with a polymorphic NOR-bearing X chromosome. Chrom. Res., 8: 111–118.

B- 77 Born, G.G. and Bertollo, L.A.C. (2001) Comparative cytogenetics among allopatric populations of the fish, *Hoplias malabaricus*. Cytotypes with 2n = 42 chromosomes. Genetica, 110: 1–9.

B- 78 Bertollo, L.A.C. (2007) Chromosome evolution in the neotropical Erythrinidae fish family: an overview. In: Pisano, E., Ozouf-Costaz, C., Foresti, F. and Kapoor, B.G. eds. Fish cytogenetics. Science Publisher, Enfield, pp. 195–211.

B- 79 Borin, L.A. and Martins-Santos, I.C. (2000) Intra-individual numerical chromosomal polymorphism in *Trichomycterus davisi* (Siluriformes, Trichomycteridae) from the Iguaçu River basin in Brazil. Genet. Mol. Biol., 23: 605–607.

B- 80 Borin, L.A. and Martins-Santos, I.C. (1999) Karyotype characterization of three species of the genus *Trichomycterus* (Teleostei, Siluriformes) from Iguaçu river basin. Genetica, 106: 215–221.

B- 81 Bellafronte, E., Margarido, V.P. and Moreira-Filho, O. (2005) Cytotaxonomy of *Parodon nasus* and *Parodon tortuosus* (Pisces, Characiformes). A case of synonymy confirmed by cytogenetic analyses. Genet. Mol. Biol., 28: 710–716.

B- 82 Benzaquem, D.C., Feldberg, E., Porto, J.I.R., Gross, M.C. and Zuanon, J.A.S. (2008) Cytotaxonomy and karyoevolution of the genus *Crenicichla* (Perciformes, Cichlidae). Genet. Mol. Biol., 31 (suppl.): 250–255.

B- 83 Busack, C.A. and Thorgaard, G.H. (1980) Karyotype of the Sacramento perch, *Archoplites interruptus*. Calif. Fish Game, 66: 189–191.

B- 84 Bozhko, S.I., Meszaros, B. and Tóth, E. (1978) The karyological studies on some species of Percidae. Ichthyologia, 10: 17–28.

B- 85 Brinn, M.N.A., Porto, J.I.R. and Feldberg, E. (2004) Karyological evidence for interspecific hybridization between *Cichla monoculus* and *C. temensis* (Perciformes, Cichlidae) in the Amazon. Hereditas, 141: 252–257.

B- 86 Brum, M.J.I. (1996) Cytogenetic studies of Brazilian marine fish. Brazil. J. Genet., 19: 421–427.

[C]

C- 1 Chiarelli, A.B. and Capanna, E. (eds) (1973) Cytotaxonomy and vertebrate evolution. Academic Press, London.

C- 2 Camacho, J.P.M., Sharbel, T.F. and Beukeboom, L.W. (2000) B-chromosome evolution. Phil. Trans. R. Soc. London, (B), 355: 163–178.

C- 3 Camilo, F.M. and Moreira Filho, O. (2005) Karyotypic description of *Corumbataia cuestae* (Pisces, Loricariidae, Hypoptopomatinae). Cytologia, 70: 47–51.

C- 4 Campos, H.H. and Hubbs, C. (1971) Cytomorphology of six species of gambusiine fishes. Copeia, 1971: 566–569.

C- 5 Campos, H.H. (1972) Karyology of three galaxiid fishes *Galaxias maculatus*, *G. platei*, and *Brachygalaxias bullocki*. Copeia, 1972: 368–370.

C- 6 Campos, H.H. and Hubbs, C. (1973) Taxonomic implications of the karyotype of *Opsopoeodus emiliae*. Copeia, 1973: 161–163.

C- 7 Campos, H.H., Arratia, G. and Cuevas, C. (1997) Karyotypes of the most primitive catfishes (Teleostei: Siluriformes: Diplomystidae). J. Zool. Syst. Evol. Res., 35: 113–119.

C- 8 Cano, J., Thode, G. and Sánchez, M.L. (1982) Estudio cariológico en dos especies de Serránidos del Mediterráneo (Peces: Perciformes). Doñana, Acta Vertebrata, 9: 5–12.

C- 9 Cano, J., Thode, G. and Alvarez, M.C. (1982) Karyoevolutive considerations in 29 Mediterranean teleost fishes. Vie Milieu, 32: 21–24.

C- 10 Cano, J., Alvarez, M.C., Thode, G. and Muñoz, E. (1982) Phylogenetic interpretation of chromosomal and nuclear-DNA-content data in the genus *Blennius* (Blenniidae: Perciformes). Genetica, 58: 11–16.

C- 11 Cano, J., Pretel, A., Melendez, S., Garcia, F., Caputo, V., Fenocchio, A.S. and Bertollo, L.A.C. (1996) Determination of early stages of sex chromosome differentiation in the sea bass *Dicentrarchus labrax* L. (Pisces: Perciformes). Cytobios, 87: 45–59.

C- 12 Cao, F.J., Li, C.L. and Liu, C.W. (2002) Studies on the karyotypes of *Lutjanus erythopterus*, *Lutjanus argentimaculatus* and *Lutjanus bohar*. Marine Sciences (Beijing), 26 (11): 43–46. (In Chinese with English abstract)

C- 13 Capanna, E. and Cataudella, S. (1973) The chromosomes of *Calamoichthys calabaricus* (Pisces, Polypteriformes). Experientia, 29: 491–492.

C- 14 Caputo, V., Marchegiani, F. and Olmo, E. (1996) Karyotype differentiation between two species of carangid fishes, genus *Trachurus* (Perciformes: Carangidae). Mar. Biol., 127: 193–199.

C- 15 Caputo, V., Vitturi, R., Odierna, G., Cano, J., Olmo, E. and Colomba, M.S. (1996) Characterization of mitotic chromosomes in the gobiid fish *Zosterisessor ophiocephalus* (Pallas, 1811) (Perciformes, Gobiidae). Biol. Zentralbl., 115: 328–336.

C- 16 Caputo, V., Sorice, M., Vitturi, R., Magistrelli, R. and Olmo, E. (1998) Cytogenetic studies in some species of Scorpaeniformes (Teleostei: Percomorpha). Chrom. Res., 6: 255–262.

C- 17 Caputo, V. (1998) Nucleolar organizer (NOR) location and cytotaxonomic implications in six species of gobiid fishes (Perciformes, Gobiidae). Ital. J. Zool., 65: 93–99.

C- 18 Caputo, V., Caniglia, M.L. and Machella, N. (1999) The chromosomal complement of *Aphia minuta*, a paedomorphic goby. J. Fish Biol., 55: 455–458.

C- 19 Caputo, V., Machella, N., Nisi-Cerioni, P. and Olmo, E. (2001) Cytogenetics of nine species of Mediterranean blennies and additional evidence for an unusual multiple sex-chromosome system in *Parablennius tentacularis* (Perciformes, Blenniidae). Chrom. Res., 9: 3–12.

C- 20 Caputo, V., Nisi Cerioni, P., Splendiani, A., Capriglione, T., Odierna, G. and Olmo, E. (2002) Chromosomal studies on ten species of notothenioid fishes (Notothenioidei: Bathydraconidae, Channichthyidae, Nototheniidae). Cytogenet. Genome Res., 98: 285–290.

C- 21 Caputo, V., Splendiani, A., Nisi Cerioni, P. and Olmo, E. (2003) The chromosomal complement of the artedidraconid fish *Histiodraco velifer* (Perciformes: Notothenioidei) from Terra Nova Bay, Ross Sea. Cytogenet. Genome Res., 101: 29–32.

C- 22 Caputo, V., Colomba, M., Nisi Cerioni, P., Vitturi, R., Giovannotti, M. and Olmo, E. (2003) Chromosome banding and molecular cytogenetic study of two Mediterranean trachinoid fish species (Teleostei: Trachinidae, Uranoscopidae). Cytogenet. Genome Res., 103: 139–143.

C- 23 Caputo, V., Nisi Cerioni, P., Caniglia, M.L. and Giovannotti, M. (2003) Chromosomal studies of five tropical scorpaeniform fishes (Teleostei, Scorpaenidae). Ital. J. Zool., 70: 201–204.

C- 24 Carbone, P., Vitturi, R., Catalano, E. and Macaluso, M. (1987) Chromosome sex determination and Y-autosome fusion in *Blennius tentacularis* Brunnich, 1765 (Pisces, Blenniidae). J. Fish Biol., 31: 597–602.

C- 25 Carvalho, M.L., Oliveira, C. and Foresti, F. (2001) Cytogenetic analysis of three species of the families Characidae and Curimatidae (Teleostei, Characiformes) from the Acre River. Chrom. Sci., 5: 91–96.

C- 26 Carvalho, M.L., Oliveira, C. and Foresti, F. (2002) Cytogenetic analysis of five species of the subfamily Tetragonopterinae (Teleostei, Characiformes, Characidae). Caryologia, 55: 181–188.

C- 27 Capistano, T.G., Castro, A.L.B.P. and Julio-Junior, H.F. (2008) Chromosome divergence and NOR polymorphism in *Bryconamericus* aff. *iheringii* (Teleostei, Characidae) in the hydrographic systems of the Paranapanema and Ivai rivers, Paraná, Brazil. Genet. Mol. Biol., 31 (suppl.): 203–207.

C- 28 Catalano, E., Vitturi, R., Zava, B. and Macaluso, M. (1985) Ritrovamento di *Parablennius pilicornis* (Cuvier, 1829) nelle acque italiane e suo cariotipo. Atti Soc. Ital. Sci. nat. Museo civ. Stor. nat. Milano, 126: 155–164.

C- 29 Catalano, E., Vitturi, R. and Valvo, M.L. (1988) Osservazioni morfologiche e cariologiche su *Symphodus doderleini* Jordan, 1891 del golfo di Palermo. Atti Soc. Ital. Sci. nat. Museo civ. Stor. nat. Milano, 129: 261–271.

C- 30 Cataudella, S. and Capanna, E. (1973) Chromosome complements of three species of Mugilidae (Pisces, Perciformes). Experientia, 29: 489–491.

C- 31 Cataudella, S., Civitelli, M.V. and Capanna, E. (1973) The chromosomes of some Mediterranean teleosts: Scorpaenidae, Serranidae, Labridae, Blenniidae, Gobiidae (Pisces-Scorpaeniformes, Perciformes). Boll. Zool., 40: 385–389.
C- 32 Cataudella, S., Civitelli, M.V. and Capanna, E. (1974) Chromosome complements of the Mediterranean mullets (Pisces Perciformes). Caryologia, 27: 93–105.
C- 33 Cataudella, S. and Civitelli, M.V. (1975) Cytotaxonomical consideration of the genus *Blennius* (Pisces-Perciformes). Experientia, 31: 167–169.
C- 34 Cataudella, S., Sola, L., Accame Muratori, R. and Capanna, E. (1977) The chromosomes of 11 species of Cyprinidae and one Cobitidae from Italy, with some remarks on the problem of polyploidy in the Cypriniformes. Genetica, 47: 161–171.
C- 35 Cataudella, S. and Sola, L. (1977) Sex chromosomes of the mosquitofish (*Gambusia affinis*): an interesting problem for American ichthyologists. Copeia, 1977: 382–384.
C- 36 Cataudella, S., Perin Riz, P. and Sola, L. (1980) A chromosome study of eight Mediterranean species of Sparidae (Pisces, Perciformes). Genetica, 54: 155–159.
C- 37 Cavallaro, Z.I., Bertollo, L.A.C., Perfectti, F. and Camacho, J.P.M. (2000) Frequency increase and mitotic stabilization of a B chromosome in the fish *Prochilodus lineatus*. Chrom. Res., 8: 627–634.
C- 38 Cavender, T.M. and Kimura, S. (1989) Cytotaxonomy and interrelationships of Pacific basin *Salvelinus*. Physiol. Ecol. Japan, Spec. vol. 1: 49–68.
C- 39 Centofante, L., Bertollo, L.A.C., Justi, Á.J. and Moreira-Filho, O. (2003) Correlation of chromosomal and morphologic characters in two *Astyanax* species (Teleostei: Characidae). Ichthyol. Explor. Freshwaters, 14: 361–368.
C- 40 Centofante, L., Bertollo, L.A.C., Miyazawa, C.S. and Moreira-Filho, O. (2003) Chromosomal differentiation among allopatric populations of *Hyphessobrycon anisitsi* (Pisces, Tetragonopterinae). Cytologia, 68: 283–288.
C- 41 Cestari, M.M. and Galetti Jr., P.M. (1992) Chromosome studies of *Serrasalmus spilopleura* (Characidae, Serrasalminae) from the Paraná-Paraguay rivers: evolutionary and cytotaxonomic considerations. Copeia, 1992: 108–112.
C- 42 Chakraborty, S.K. and Kagwade, P.V. (1989) Somatic chromosomes of two marine teleosts. In: Das, P. and Jhingran, A.G. eds. Fish Genetics in India. Today & Tomorrow's Printers and Publishers, New Delhi, pp. 63–68.
C- 43 Chen, J.H., Zhao, Z.A. and Li, K.B. (1984) Karyotype of *Trachidermus fasciatus* Heckel. Zool. Res., 5: 103–104. (In Chinese)
C- 44 Chen, M.R. and Chen, H.X. (1983) A comparative study of karyotypes in three *Tilapia* fishes. Acta Genet. Sinica, 10: 56–62. (In Chinese with English abstract)
C- 45 Chen, T.R. and Ebeling, A.W. (1966) Probable male heterogamety in the deep-sea fish *Bathylagus wesethi* (Teleostei: Bathylagidae). Chromosoma, 18: 88–96.
C- 46 Chen, T.R. (1967) Comparative karyology of selected deep-sea and shallow-water teleost fishes. Ph. D. Thesis, Yale University.
C- 47 Chen, T.R. and Ebeling, A.W. (1968) Karyological evidence of female heterogamety in the mosquitofish, *Gambusia affinis*. Copeia, 1968: 70–75.
C- 48 Chen, T.R. (1969) Karyological heterogamety of deep-sea fishes. Postilla, 130: 1–29.
C- 49 Chen, T.R. and Ruddle, F.H. (1970) A chromosome study of four species and a hybrid of the killifish genus *Fundulus* (Cyprinodontidae). Chromosoma, 29: 255–267.
C- 50 Chen, T.R. and Reisman, H.M. (1970) A comparative chromosome study of the North American species of sticklebacks (Teleostei: Gasterosteidae). Cytogenetics, 9: 321–332.
C- 51 Chen, T.R. (1970) Fish chromosome preparation: air-dried displays of cultured ovarian cells in two killifishes (*Fundulus*). J. Fish. Res. Board Can., 27: 158–161.
C- 52 Chen, T.R. (1971) A comparative chromosome study of twenty killifish species of the genus *Fundulus* (Teleostei: Cyprinodontidae). Chromosoma, 32: 436–453.

C- 53 Chen, T.R. and Ebeling, A.W. (1971) Chromosomes of the goby fishes in the genus *Gillichthys*. Copeia, 1971: 171–174.

C- 54 Chen, T.R. and Ebeling, A.W. (1974) Cytotaxonomy of Californian myctophoid fishes. Copeia, 1974: 839–848.

C- 55 Chen, T.R. and Ebeling, A.W. (1975) Karyotypes from short- and long-term cultures of hybrid-killifish and platyfish tissues. Copeia, 1975: 178–181.

C- 56 Chen, Y.H., Rong, S.B., Liu, S.Q., Zhang, H.Q. and Pei, M.H. (1990) The karyotype of *Epinephelus sexfasciatus*. J. Zhanjiang Fish. Coll., 10(1): 43–44, 51. (In Chinese with English abstract)

C- 57 Chen, Y.H., Rong, S.B. and Liu, S.Q. (1990) The karyotype of *Epinephelus fario*. Fujian Fish., 1990(1): 23–25. (In Chinese)

C- 58 Chen, Y.R., Uwa, H. and Chu, X.L. (1989) Taxonomy and distribution of the genus *Oryzias* in Yunnan, China. Acta Zootax. Sinica, 14: 239–246. (In Chinese with English summary)

C- 59 Chenuil, A., Galtier, N. and Berrebi, P. (1999) A test of the hypothesis of an autopolyploid vs. allopolyploid origin for a tetraploid lineage: application to the genus *Barbus* (Cyprinidae). Heredity, 82: 373–380.

C- 60 Chiarelli, B., Ferrantelli, O. and Cucchi, C. (1969) The caryotype of some teleostean fish obtained by tissue culture *in vitro*. Experientia, 25: 426–427.

C- 61 Choudhury, R.C., Prasad, R. and Das, C.C. (1979) Chromosomes of six species of marine fishes. Caryologia, 32: 15–21.

C- 62 Choudhury, R.C., Prasad, R. and Das, C.C. (1982) Karyological studies in five tetraodontiform fishes from the Indian Ocean. Copeia, 1982: 728–732.

C- 63 Choudhury, R.C., Prasad, R. and Das, C.C. (1993) Chromosomes of four Indian marine catfishes (Bagridae, Ariidae: Siluriformes) with a heteromorphic pair in male *Mystus gulio*. Caryologia, 46: 233–243.

C- 64 Choudhury, R.C., Prasad, R. and Das, C.C. (1993) Karyomorphology of four carangids from the Indian Ocean (Carangidae: Perciformes). La Kromosomo, II, 69: 2329–2335.

C- 65 Clark, B. and Mathis, P. (1982) Karyotypes of middle Tennessee bullheads: *Ictalurus melas* and *Ictalurus natalis* (Cypriniformes: Ictaluridae). Copeia, 1982: 457–460.

C- 66 Colihueque, N., Corrales, O. and Parraguez, M. (2006) Karyotype and nuclear DNA content of *Trichomycterus areolatus* (Siluriformes, Trichomycteridae). Genet. Mol. Biol., 29: 278–282.

C- 67 Colihueque, N., Iturra, P., Diaz, N., Veloso, A. and Estay, F. (1992) Karyological analysis and identification of heterochromosomes in experimental gynogenetic offspring of rainbow trout (*Oncorhynchus mykiss*, Walbaum). Brazil. J. Genet., 15: 535–546.

C- 68 Collares-Pereira, M.J. (1983) Cytotaxonomic studies in Iberian cyprinids. I. Karyology of *Chondrostoma lusitanicum* Collares-Pereira, 1980. Cytologia, 48: 753–760.

C- 69 Collares-Pereira, M.J. (1985) Cytotaxonomic studies in Iberian cyprinids. II. Karyology of *Anaecypris hispanica* (Steindachner, 1866), *Chondrostoma lemmingi* (Steindachner, 1866), *Rutilus arcasi* (Steindachner, 1866) and *R. macrolepidotus* (Steindachner, 1866). Cytologia, 50: 879–890.

C- 70 Collares-Pereira, M.J. and Coelho, M.M. (1989) Polyploidy versus diploidy: a new model for the karyological evolution of Cyprinidae. Arquivos Mus. Bocage, n.s., 1: 375–383.

C- 71 Collares-Pereira, M.J. (1989) Hybridization in European cyprinids: evolutionary potential of unisexual populations. In: Dawley, R.M. and Bogart, J.P. eds. Evolution and ecology of unisexual vertebrates. Mus. Bull. 466. New York State Mus., New York, pp. 281–288.

C- 72 Collares-Pereira, M.J. and Madeira, J.M. (1990) Cytotaxonomic studies in Iberian cyprinids. III. Karyology of Portuguese populations of *Barbus* Cuvier, 1817, with some reconsiderations on the karyological evolution of Cyprinidae. Caryologia, 43: 17–26.

C- 73 Collares-Pereira, M.J. and Madeira, J.M. (1995) Spontaneous triploidy in the stone loach *Noemacheilus barbatulus* (Balitoridae). Copeia, 1995: 483–484.

C- 74 Collares-Pereira, M.J. (1994) The karyology of barbins and the possible plesiomorphic condition of polyploidy in Cyprinidae. Bull. Fr. Pêche Piscic., 334: 191–199.

C- 75 Collares-Pereira, M.J. and Costa, L.M. (1999) Intraspecific and interspecific genome size variation in Iberian Cyprinidae and the problem of diploidy and polyploidy, with review of genome sizes within the family. Folia Zool., 48: 61–76.

C- 76 Colombera, D. and Rasotto, M. (1982) Chromosome studies in males of *Gobius niger jozo* (Padoa) and *Gobius paganellus* (Linneo) (Gobiidae, Osteichthyes). Caryologia, 35: 257–260.

C- 77 Cook, P.C. (1978) Karyotypic analysis of the gobiid fish genus *Quietula* Jordan and Evermann. J. Fish Biol., 12: 173–179.

C- 78 Corrêa, M.M.O. and Galetti Jr., P.M. (1997) Chromosomal diversity in Scorpaenidae (Teleostei, Scorpaeniformes): cytogenetic studies in *Scorpaena brasiliensis* and *Scorpaena isthmensis* from the coast of Rio de Janeiro, Brazil. Cytologia, 62: 397–404.

C- 79 Cristina, A., Malacrida, C.P., Dias, A.L. and Giuliano-Caetano, L. (2003) Natural triploidy in *Astyanax* aff. *scabripinnis* (Pisces, Characidae) of the Tibagi river bay–PR. Cytologia, 68: 267–270.

C- 80 Crossman, E.J. and Ráb, P. (2001) Chromosomal NOR phenotype and C-banded karyotype of Olympic mudminnow, *Novumbra hubbsi* (Euteleostei: Umbridae). Copeia, 2001: 860–865.

C- 81 Cuellar, O. and Uyeno, T. (1972) Triploidy in rainbow trout. Cytogenetics, 11: 508–515.

C- 82 Cuevas, C.C., Campos, H.H. and Busse, K. (1999) Cytotaxonomic studies on Chilean galaxiid fishes. The karyotypes, C-bands, Ag-NORs and hybrids of *Brachygalaxias gothei* and *B. bullocki* (Osteichthyes: Galaxiidae). Cytologia, 64: 379–385.

C- 83 Cui, J.X., Ren, X.H. and Yu, Q.X. (1991) Nuclear DNA content variation in fishes. Cytologia, 56: 425–429.

C- 84 Cui, Y.L. and Zan, R.G. (1982) Studies on gynogenesis of the high-backed form of *Carassius auratus* in Kunming Lake. Zool. Res., 3 (suppl.): 237–242. (In Chinese with English summary)

C- 85 Cytogenetic Laboratory of Wuhan University (1983) A preliminary report on the chromosome number of 52 species of fishes of the Pearl River system in Guangdong Province. J. Wuhan Univ. (Nat. Sci.), 1983(3): 58, 123–124. (In Chinese)

C- 86 Cytogenetic Laboratory of Wuhan University (1981) A preliminary report of fish chromosomes of 7 families and 23 species. J. Wuhan Univ. (Nat. Sci.), 1981(3): 14. (In Chinese)

C- 87 Cestari, M.M. and Galetti Jr., P.M. (1992) Chromosome evolution in the genus *Serrasalmus* and cytotaxonomic considerations about Serrasalminae (Characidae, Pisces). Brazil. J. Genet., 15: 555–567.

C- 88 Collares-Pereira, M.J. and Ráb, P. (1999) NOR polymorphism in the Iberian species *Chondrostoma lusitanicum* (Pisces: Cyprinidae)—Re-examination by FISH. Genetica, 105: 301–303.

C- 89 Calton, M.S. and Denton, T.E. (1974) Chromosomes of the chocolate gourami: a cytogenetic anomaly. Science, 185: 618–619.

C- 90 Chen, Y.L., Liu, Y.H., Chen, K.C. and Zhu, X.P. (2006) The morphological character and karyotype of *Oxyeleotris lineolatus*. J. Fish. China, 30: 562–565. (In Chinese with English abstract)

C- 91 Centofante, L., Bertollo, L.A.C., Buckup, P.A. and Moreira-Filho, O. (2003) Chromosomal divergence and maintenance of sympatric *Characidium* fish species (Crenuchidae, Characidiinae). Hereditas, 138: 213–218.

C- 92 Cataudella, S., Sola, L. and Capanna, E. (1978) Remarks on the karyotype of the Polypteriformes. The chromosomes of *Polypterus delhezi, P. endlicheri congicus* and *P. palmas*. Experientia, 34: 999–1000.

C- 93 Carvalho, M.L., Oliveira, C. and Foresti, F. (1998) Nuclear DNA content of thirty species of neotropical fishes. Genet. Mol. Biol., 21: 47–54.

C- 94 Carvalho, M.L., Oliveira, C., Navarrete, M.C., Froehlich, O. and Foresti, F. (2002) Nuclear DNA content determination in Characiformes fish (Teleostei, Ostariophysi) from the Neotropical region. Genet. Mol. Biol., 25: 49–55.

C- 95 Cimino, M.C. and Bahr, G.F. (1974) The nuclear DNA content and chromatin ultrastructure of the coelacanth *Latimeria chalumnae*. Exp. Cell Res., 88: 263–272.

C- 96 Calgaro, M.R., Fenocchio, A.S., Pastori, M.C. and Roncati, H.A. (2004) Karyology of *Apareiodon affinis* from Paraná river (Argentina). I. Chromosome polymorphism. Cytologia, 69: 475–479.

C- 97 Carvalho, R.A., Giuliano-Caetano, L. and Dias, A.L. (2004) Cytogenetic analysis of A- and B-chromosomes of *Iheringichthys labrosus* (Pisces, Pimelodidae) from the Tibagi river, Paraná, Brazil. Cytologia, 69: 381–385.

C- 98 Carvalho, R.A., Martins-Santos, I.C. and Dias, A.L. (2008) B chromosomes: an update about their occurrence in freshwater neotropical fishes (Teleostei). J. Fish Biol., 72: 1907–1932.

C- 99 Cau, A., Salvadori, S., Deiana, A.M., Bella, J.L. and Mezzanotte, R. (1988) The characterization of *Muraena helena* L. mitotic chromosomes: karyotype, C-banding, nucleolar organizer regions, and *in situ* digestion with restriction endonucleases. Cytogenet. Cell Genet., 47: 223–226.

C- 100 Chang, H.Y., Sang, T.K., Jan, K.Y. and Chen, C.T. (1995) Cellular DNA contents and cell volumes of batoids. Copeia, 1995: 571–576.

C- 101 Carvalho, M.L., Oliveira, C. and Foresti, F. (2002) Description of a ZZ/ZW sex chromosome system in *Thoracocharax* cf. *stellatus* (Teleostei, Characiformes, Gasteropelecidae). Genet. Mol. Biol., 25: 299–303.

C- 102 Collares-Pereira, M.J. (1985) The "*Rutilus alburnoides* (Steindachner, 1866) complex" (Pisces, Cyprinidae). II. First data on the karyology of a well-established diploid-triploid group. Arquivos Mus. Bocage, (Ser. A), 3: 69–89.

C- 103 Centofante, L., Porto, J.I.R. and Feldberg, E. (2002) Chromosomal polymorphism in *Serrasalmus spilopleura* Kner, 1858 (Characidae, Serrasalminae) from Central Amazon basin. Caryologia, 55: 37–45.

C- 104 Carvalho, R.A. and Dias, A.L. (2005) Cytogenetic characterization of B chromosomes in two populations of *Iheringichthys labrosus* (Pisces, Pimelodidae) from the Capivara Reservoir (Paraná, Brazil). Caryologia, 58: 269–273.

C- 105 Carvalho, R.A. and Dias, A.L. (2007) Interindividual size heteromorphism of NOR and chromosomal location of 5S rRNA genes in *Iheringichthys labrosus*. Brazil. Archiv. Biol. Technol., 50: 141–146.

C- 106 Costa, G.W.W.F. and Molina, W.F. (2009) Karyoevolution of the toadfish *Thalassophryne nattereri* (Batrachoidiformes: Batrachoididae). Genet. Mol. Res., 8: 1099–1106.

C- 107 Cereali, S.S., Pomini, E., Rosa, R., Zawadzki, C.H., Froehlich, O. and Giuliano-Caetano, L. (2008) Karyotype description of two species of *Hypostomus* (Siluriformes, Loricariidae) of the Planalto da Bodoquena, Brazil. Genet. Mol. Res., 7: 583–591.

C- 108 Chanda, T. (1998) *Aspidoparia morar, Chagunius chagunio, Crossocheilus latius latius, Mystus bleekeri, Mystus menoda menoda, Ailia coila, Clupisoma garua, Pseudeutropius atherinoides, Glyptothorax trilineatus, Laguvia rebeiroi, Nangra punctata, Johnius dussumieri, Colisa sota.* In: Ponniah, A.G. and John, G. eds. Fish chromosome atlas. National Bureau of Fish Genetic Resources, Lucknow, India, pp. 22–23, 48–49, 58–59, 158–159, 164–165, 190–191, 192–193, 196–197, 208–213, 244–245, 286–287.

C- 109 Cimino, M.C. (1974) The nuclear DNA content of diploid and triploid *Poeciliopsis* and other poeciliid fishes with reference to the evolution of unisexual forms. Chromosoma, 47: 297–307.

C- 110 Cioffi, M.B., Martins, C. and Bertollo, L.A.C. (2010) Chromosome spreading of associated transposable elements and ribosomal DNA in the fish *Erythrinus erythrinus*. Implications for genome change and karyoevolution in fish. BMC Evol. Biol. 2010, 10: 271 doi: 10.1186/1471-2148-10-271.

[D]

D- 1 Danzmann, R.G. (1979) The karyology of eight species of fish belonging to the family Percidae. Can. J. Zool., 57: 2055–2060.

D- 2 Das, R.K. and Kar, R.N. (1977) Somatic chromosome analysis of a siluroid fish, *Rita chrysea* Day. Caryologia, 30: 247–253.

D- 3 Das, N.K., Prasad, R. and Das, C.C. (1980) Chromosomes of four species of marine fishes (Fam: Carangidae). Ind. Biologist, 12: 21–25.

D- 4 Doussau de Bazignan, M. and Ozouf-Costaz, C. (1985) Une technique rapide d'analyse chromosomique appliquée a sept espèces de poissons Antarctiques. Cybium, 9: 57–74.

D- 5 Deiana, A.M., Salvadori, S. and Cau, A. (1990) The characterization of somatic chromosomes of *Gymnothorax unicolor* (Delaroche, 1809) by C-banding and NOR staining (Osteichthyes, Anguilliformes). Genetica, 81: 17–20.

D- 6 Denton, T.E. and Howell, W.M. (1969) A technique for obtaining chromosomes from the scale epithelium of teleost fishes. Copeia, 1969: 392–393.

D- 7 Denton, T.E. (1973) Fish chromosome methodology. Charles C. Thomas Publisher, Springfield, USA.

D- 8 Denton, T.E. and Howell, W.M. (1973) Chromosomes of the African polypterid fishes, *Polypterus palmas* and *Calamoichthys calabaricus* (Pisces: Brachiopterygii). Experientia, 29: 122–124.

D- 9 Dhar, N.J. and Chatterjee, K. (1984) Chromosomal evolution in Indian murrels (Channiformes: Channidae). Caryologia, 37: 359–371.

D- 10 Di, S.J. and Knowles, J.F. (1994) The karyotype of the dab (*Limanda limanda* L.). J. Fish Biol., 44: 354–356.

D- 11 Dias, A.L. and Foresti, F. (1993) Cytogenetic studies on fishes of the family Pimelodidae (Siluroidei). Brazil. J. Genet., 16: 585–600.

D- 12 Dias, A.L., Foresti, F. and Oliveira, C. (1998) Synapsis in supernumerary chromosomes of *Prochilodus lineatus* (Teleostei: Prochilodontidae). Caryologia, 51: 105–113.

D- 13 Ding, S.X., Wang, S.F., Wang D.X. and Wang J. (2004) Analysis of the karyotype of *Epinephelus coioides*. J. Xiamen Univ. (Nat. Sci.), 43: 426–428. (In Chinese with English abstract)

D- 14 Dingerkus, G. and Howell, W.M. (1976) Karyotypic analysis and evidence of tetraploidy in the North American paddlefish, *Polyodon spathula*. Science, 194: 842–844.

D- 15 Dingerkus, G. (1979) Chordate cytogenetic studies: an analysis of their phylogenetic implications with particular reference to fishes and the living coelacanth. Occ. Pap. Calif. Acad. Sci., 134: 111–127.

D- 16 Diniz, D. and Bertollo, L.A.C. (2003) Karyotypic studies on *Hoplerythrinus unitaeniatus* (Pisces, Erythrinidae) populations. A biodiversity analysis. Caryologia, 56: 303–313.

D- 17 Donahue, W.H. (1974) A karyotypic study of three species of Rajiformes (Chondrichthyes, Pisces). Can. J. Genet. Cytol., 16: 203–211.

D- 18 Dong, Y.K. (1984) Karyotypes of four freshwater fishes (Perciformes). Freshwater Fish., 4: 23–24. (In Chinese)

D- 19 Donsakul, T. and Magtoon, W. (1997) A chromosome study of two fishes, *Oxyeleotris marmoratus* and *Bathygobius fuscus*, from Thailand. Proceedings of the 23th Congress on Science and Technology of Thailand (Biology Section), pp. 430–431. (In Thai)

D- 20 Donsakul, T. (1998) A chromosome study of four teleostean fishes, *Microphis boaja*, *Xenentodon cancila*, *Acanthophthalmus kuhlii* and *Chanda siamensis*, from Thailand. Proceedings of the 36th Kasetsart University Annual Conference, pp. 1–11. (In Thai)

D- 21 Donsakul, T. and Magtoon, W. (1999) A chromosome study of two species of catfishes, *Pangasius gigas* and *P. sanitwongsei* from Thailand. Proceedings of the 37th Kasetsart University Annual Conference, pp. 221–226. (In Thai)

D- 22 Dorofeyeva, Ye.A. and Rukhkyan, R.G. (1982) Divergence of the Sevan trout, *Salmo ischchan*, in the light of karyological and morphological data. J. Ichthyol., 22(1): 23–36.

D- 23 Doucette Jr., A.J. and Fitzsimons, J.M. (1988) Karyology of elopiform and clupeiform fishes. Copeia, 1988: 124–130.

D- 24 Duchac, B., Huber, F., Müller, Hj. and Senn, D. (1982) Mating behavior and cytogenetical aspects of sex-inversion in the fish *Coris julis* L. (Labridae, Teleostei). Experientia, 38: 809–811.

D- 25 Dutt, S. and Ramaseshaiah, M. (1980) Chromosome number in *Anabas testudineus* (Bloch, 1795) and *A. oligolepis* Bleeker, 1855 (Osteichthyes: Anabantidae). Matsya, 6: 71–74.

D- 26 Doucette Jr., A.J. and Fitzsimons, J.M. (1982) Karyology of the ladyfish *Elops saurus*. Japan. J. Ichthyol., 29: 223–226.

D- 27 Durand, J.D., Tsigenopoulos, C.S., Unlu, E. and Berrebi, P. (2002) Phylogeny and biogeography of the family Cyprinidae in the Middle East inferred from cytochrome *b* DNA—Evolutionary significance of this region. Mol. Phylogenet. Evol., 22: 91–100.

D- 28 Donsakul, T., Rangsiruji, A. and Magtoon, W. (2005, personal communication)

D- 29 Domingues, M.S., Vicari, M.R., Abilhoa, V., Wamser, J.P., Cestari, M.M., Bertollo, L.A.C., de Almeida, M.C. and Artoni, R.F. (2007) Cytogenetic and comparative morphology of two allopatric populations of *Astyanax altiparanae* Garutti & Britski, 2000 (Teleostei: Characidae) from upper rio Paraná basin. Neotrop. Ichthyol., 5: 37–44.

D- 30 Diniz, D., Moreira-Filho, O. and Bertollo, L.A.C. (2008) Molecular cytogenetics and characterization of a ZZ/ZW sex chromosome system in *Triportheus nematurus* (Characiformes, Characidae). Genetica, 133: 85–91.

D- 31 Dantas, E.S.O., Vicari, M.R., Souza, I.L., Moreira-Filho, O., Bertollo, L.A.C. and Artoni, R.F. (2007) Cytotaxonomy and karyotype evolution in *Moenkhausia* Eigenmann, 1903 (Teleostei, Characidae). Nucleus, 50: 509–522.

D- 32 Della-Rosa, V.A., Bertollo, L.A.C., Ferrari, I., Takahashi, C.S., Moreira-Filho, O. and Foresti, F. (1980) Estudios citogenéticos em peixes da regiao Amazônia. II. Ordem Siluriformes. Ciênc. Cult., 32: 735. (In Portuguese)

D- 33 Das, J.K. and Khuda-Bukhsh, A.R. (2007) Preponderance of GC-rich sites in silver-stained nucleolus organizing regions of *Rita rita* (Hamilton) and *Mystus gulio* (Hamilton) (Bagridae, Pisces), as revealed by chromomycin A_3-staining technique and scanning electron microscopic studies. Genet. Mol. Res., 6: 284–291.

D- 34 Das, J.K. and Khuda-Bukhsh, A.R. (2007) GC-rich heterochromatin in silver stained nucleolar organizer regions (NORs) fluoresces with chromomycin A_3 (CMA_3) staining in three species of teleostean fishes (Pisces). Ind. J. Exp. Biol., 45: 413–418.

[E]

E- 1 Ebeling, A.W. and Chen, T.R. (1970) Heterogamety in teleostean fishes. Trans. Amer. Fish. Soc., 99: 131–138.

E- 2 Ebeling, A.W., Atkin, N.B. and Setzer, P.Y. (1971) Genome sizes of teleostean fishes: increases in some deep-sea species. Amer. Natur., 105: 549–561.

E- 3 Elder Jr., J.F., Turner, B.J., Thomerson, J.E. and Taphorn, D.C. (1991) Chromosomal divergence and heterogameity in two annual killifishes of the genus *Pterolebias*. Genome, 34: 674–676.

E- 4 Elder Jr., J.F., Turner, B.J., Thomerson, J.E. and Taphorn, D.C. (1993) Karyotypes of nine Venezuelan annual killifishes (Cyprinodontidae), with comments on karyotype differentiation in annual killifishes. Ichthyol. Explor. Freshwaters, 4: 261–268.

E- 5 Ewulonu, U.K., Haas, R. and Turner, B.J. (1985) A multiple sex chromosome system in the annual killifish, *Nothobranchius guentheri*. Copeia, 1985: 503–508.

E- 6 Ebeling, A.W. and Setzer, P.Y. (1971) Cytological confirmation of female homogamety in the deep-sea fish *Bathylagus milleri*. Copeia, 1971: 560–562.

E- 7 Esmaeili, H.R. and Piravar, Z. (2006) Karyotype of Persian chub, *Petroleuciscus persidis* (Coad, 1981) (Actinopterygii: Cyprinidae) from southern Iran. Turk. J. Zool., 30: 137–139.

E- 8 Elvira, B. (1997) Threatened fishes of the world: *Iberocypris palaciosi* Doadrio, 1980 (Cyprinidae). Environ. Biol. Fishes, 50: 104.

E- 9 Ene, A.-C. (2003) Chromosomal polymorphism in the goby *Neogobius eurycephalus* (Perciformes: Gobiidae). Mar. Biol., 142: 583–588.

E- 10 Ezaz, T., Berra, T.M. and Marshall Graves, J.A. (2006) Karyotype of the Australian nurseryfish, *Kurtus gulliveri* (Kurtidae: Perciformes). Chrom. Sci., 9: 85–88.

E- 11 Eler, E.S., Dergam, J.A., Vênere, P.C., Paiva, L.C., Miranda, G.A. and Oliveira, A.A. (2007) The karyotypes of the thorny catfishes *Wertheimeria maculata* Steindachner, 1877 and *Hassar wilderi* Kindle, 1895 (Siluriformes: Doradidae) and their relevance in doradid chromosomal evolution. Genetica, 130: 99–103.

E- 12 Esmaeili, H.R., Piravar, Z. and Shiva, A.H. (2007) Karyological analysis of two endemic tooth-carps, *Aphanius persicus* and *Aphanius sophiae* (Pisces: Cyprinodontidae), from Southwest Iran. Turk. J. Zool., 31: 69–74.

E- 13 Eschmeyer, W.N. (2009) Catalogue of fishes. http://www.research.calacademy.org/redirect?url=http://researcharchive.calacademy.org/research/ichthyology/catalog/fishcatmain.asp

[F]

F- 1 Fajkus, J., Sykorova, E. and Leitch, A.R. (2005) Telomeres in evolution and evolution of telomeres. Chrom. Res., 13: 469–479.

F- 2 Falcão, J.N. and Bertollo, L.A.C. (1985) Chromosome characterization in Acestrorhynchinae and Cynopotaminae (Pisces, Characidae). J. Fish Biol., 27: 603–610.

F- 3 Fan, Z. and Fox, D.P. (1990) A new method for fish chromosome preparation. J. Fish Biol., 37: 553–561.

F- 4 Fan, Z. and Fox, D.P. (1991) Robertsonian polymorphism in plaice, *Pleuronectes platessa* L., and cod, *Gadus morhua* L. (Pisces Pleuronectiformes and Gadiformes). J. Fish Biol., 38: 635–640.

F- 5 Fan, Z.T., Yin, H.B., Song, S.X. and Pan, F. (1995) DNA content from thirteen species of cultured freshwater fishes in China. J. Fish. China, 19: 322–326. (In Chinese with English abstract)

F- 6 Fauaz, G., Vicente, V.E. and Moreira-Filho, O. (1994) Natural triploidy and B chromosomes in the neotropical fish genus *Astyanax* (Characidae). Brazil. J. Genet., 17: 157–163.

F- 7 Fazoli, L.C., Silva, V.A.B., Portela-Castro, A.L.B. and Jūlio Jūnior, H.F. (2003) Chromosome characterization of *Astyanax* sp B (Characidae, Tetragonopterinae), an endemic species of the Iguaçu river, Paraná, Brazil. Cytologia, 68: 389–394.

F- 8 Fei, Z.Q. and Tao, R.Q. (1987) Studies on the karyotypes of four species in gobioids. J. Zhejiang College Fish., 1987: 127–131. (In Chinese with English abstract)

F- 9 Feldberg, E. and Bertollo, L.A.C. (1985) Nucleolar organizing regions in some species of neotropical cichlid fish (Pisces, Perciformes). Caryologia, 38: 319–324.

F- 10 Feldberg, E. and Bertollo, L.A.C. (1985) Karyotypes of 10 species of neotropical cichlids (Pisces, Perciformes). Caryologia, 38: 257–268.

F- 11 Feldberg, E., Bertollo, L.A.C., Toledo, L.F.A., Foresti, F., Moreira Filho, O. and Santos, A.F. (1987) Biological aspects of Amazonian fishes. IX. Cytogenetic studies in two species of the genus *Semaprochilodus* (Pisces, Prochilodontidae). Genome, 29: 1–4.

F- 12 Feldberg, E., Porto, J.I.R. and Bertollo, L.A.C. (1992) Karyotype evolution in Curimatidae (Teleostei, Characiformes) of the Amazon region. I. Studies on the genera *Curimata*, *Psectrogaster*, *Steindachnerina* and *Curimatella*. Brazil. J. Genet., 15: 369–383.

F- 13 Feldberg, E., Porto, J.I.R., Nakayama, C.M. and Bertollo, L.A.C. (1993) Karyotype evolution in Curimatidae (Teleostei, Characiformes) from the Amazon region. II. Centric fissions in the genus *Potamorhina*. Genome, 36: 372–376.

F- 14 Feldberg, E., Porto, J.I.R., Alves-Brinn, M.N., Mendonça, M.N.C. and Benzaquem, D.C. (2004) B chromosomes in Amazonian cichlid species. Cytogenet. Genome Res., 106: 195–198.

F- 15 Fenocchio, A.S. and Bertollo, L.A.C. (1990) Supernumerary chromosomes in a *Rhamdia hilarii* population (Pisces, Pimelodidae). Genetica, 81: 193–198.

F- 16 Fenocchio, A.S., Jorge, L.C., Vênere, P.C. and Bertollo, L.A.C. (1993) Karyotypic characterization and nucleolus organizer regions in three species of Doradidae (Pisces, Siluriformes). Brazil. J. Genet., 16: 1097–1101.

F- 17 Fenocchio, A.S. and Bertollo, L.A.C. (1992) Karyotype similarities among Pimelodidae (Pisces, Siluriformes) from the Brazilian Amazon region. Cytobios, 69: 41–46.

F- 18 Fenocchio, A.S. and Bertollo, L.A.C. (1992) Karyotype, C-bands and NORs of the neotropical siluriform fish *Ageneiosus brevifilis* and *Ageneiosus atronases* (Ageneiosidae). Cytobios, 72: 19–22.

F- 19 Fenocchio, A.S., Bertollo, L.A.C., Takahashi, C.S., Dias, A.L. and Swarça, A.C. (2003) Cytogenetic studies and correlate considerations on Rhamdiinae relationships (Pisces, Siluroidei, Pimelodidae). Cytologia, 68: 363–368.

F- 20 Fenocchio, A.S., Pastori, M.C., Roncati, H.A., Moreira Filho, O. and Bertollo, L.A.C. (2003) A cytogenetic survey of the fish fauna from Argentina. Caryologia, 56: 197–204.

F- 21 Fernandes, C.A. and Martins-Santos, I.C. (2003) Cytogenetic characterization of two populations of *Astyanax scabripinnis* (Pisces, Characiformes) of the Ivai basin PR Brazil. Cytologia, 68: 289–293.

F- 22 Fernandes, C.A. and Martins-Santos, I.C. (2005) Sympatric occurrence of three cytotypes and four morphological types of B chromosomes of *Astyanax scabripinnis* (Pisces, Characiformes) in the River Ivaí Basin, state of Paraná, Brazil. Genetica, 124: 301–306.

F- 23 Ferris, S.D. (1984) Tetraploidy and the evolution of the catostomid fishes. In: Turner, B.J. ed. Evolutionary genetics of fishes. Plenum press, New York, pp. 55–93.

F- 24 Fischer, C., Ozouf-Costaz, C., Crollius, H.R., Dasilva, C., Jaillon, O., Bouneau, L., Bonillo, C., Weissenbach, J. and Bernot, A. (2000) Karyotype and chromosome location of characteristic tandem repeats in the pufferfish *Tetraodon nigroviridis*. Cytogenet. Cell Genet., 88: 50–55.

F- 25 Fisher, P.B. and Rachlin, J.W. (1972) Karyotypic analysis of 2 sympatric species of fish, *Fundulus heteroclitus* L. and *F. majalis*. J. Fish Biol., 4: 67–71.

F- 26 Fitzsimons, J.M. and Doucette Jr., A.J. (1981) Karyology of the shads *Dorosoma cepedianum* and *D. petenense* (Osteichthyes: Clupeiformes). Copeia, 1981: 908–911.

F- 27 Fitzsimons, J.M., LeGrande, W.H. and Korth, J.W. (1988) Karyology of the marine catfish *Bagre marinus* (Ariidae) with an analysis of chromosome numbers among siluriform fishes. Japan. J. Ichthyol., 35: 189–193.

F- 28 Flajšhans, M. and Ráb, P. (1990) Chromosome study of *Oncorhynchus mykiss kamloops*. Aquaculture, 89: 1–8.

F- 29 Fonseca, Y.M., Oliveira, C., Foresti, F. and Maistro, E.L. (2003) First cytogenetic description of the species *Rhamdella microcephala* (Pisces, Hepapteridae). Cytologia, 68: 31–34.

F- 30 Fontana, F., Chiarelli, B. and Rossi, A.C. (1970) Il cariotipo di alcune specie di Cyprinidae, Centrarchidae, Characidae studiate mediante colture « *in vitro* ». Caryologia, 23: 549–564.

F- 31 Fontana, F., Chiarelli, B. and Rossi, A. (1970) Some new data on the number of chromosomes of teleost fish obtained by means of tissue culture *in vitro*. Experientia, 26: 1021.

F- 32 Fontana, F., Tagliavini, J., Congiu, L., Lanfredi, M., Chicca, M., Laurente, C. and Rossi, R. (1998) Karyotypic characterization of the great sturgeon, *Huso huso*, by multiple staining techniques and fluorescent *in situ* hybridization. Mar. Biol., 132: 495–501.

F- 33 Fontana, F., Lanfredi, M., Chicca, M., Aiello, V. and Rossi, R. (1998) Localization of the repetitive sequence (TTAGGG)n in four sturgeon species. Chrom. Res., 6: 303–306.

F- 34 Foresti, F., Almeida Toledo, L.F. and Toledo, S.A. (1984) Chromosome studies in *Gymnotus carapo* and *Gymnotus* sp. (Pisces, Gymnotidae). Caryologia, 37: 141–146.

F- 35 Foresti, F., Almeida-Toledo, L.F. and Toledo, S.A. (1989) Supernumerary chromosome system, C-banding pattern characterization and multiple nucleolus organizer regions in *Moenkhausia sanctaefilomenae* (Pisces, Characidae). Genetica, 79: 107–114.

F- 36 Foresti, F., Almeida Toledo, L.F. and Pathak, S. (1992) "*In vitro*" occurrence of premature chromosomes condensation in kidney cells of *Eigenmannia virescens* (Pisces, Sternopygidae). Cytologia, 57: 377–381.

F- 37 Foresti, F., Ozouf-Costaz, C. and Ráb, P. (1992) Volume dependent effect of hypotonic treatment on good spreading of fish chromosomes. Cybium, 16: 209–212.

F- 38 Formacion, M.J. and Uwa, H. (1985) Cytogenetic studies on the origin and species differentiation of the Philippine medaka, *Oryzias luzonensis*. J. Fish Biol., 27: 285–291.

F- 39 Freitas, P.D., Navarrete, M.C., Miyazawa, C.S. and Galetti Jr., P.M. (1998) Chromosome characterization of a neotropical fish *Poptella paraguayensis* from Paraguay river basin (Stethaprioninae, Characidae). Cytologia, 63: 73–77.

F- 40 Frolov, S.V. (1981) Studies of Kamchatka trout (*Salmo mykiss* Walbaum) karyotype by differential staining method. Biol. Morya, 1981(6): 73–75. (In Russian with English abstract)

F- 41 Frolov, S.V. (1986) The additional chromosome system in the karyotype of *Coregonus nasus*. Tsitologiya, 28: 215–219. (In Russian with English abstract)

F- 42 Frolov, S.V. (1986) Polymorphism and mosaicism for additional chromosomes in *Coregonus sardinella*. Tsitologiya, 28: 740–744. (In Russian with English abstract)

F- 43 Frolov, S.V. (1989) Differentiation of sex chromosomes in the Salmonidae. I. Karyotype and sex chromosomes in *Parasalmo mykiss*. Tsitologiya, 31: 1391–1394. (In Russian with English abstract)

F- 44 Frolov, S.V. (1990) Differentiation of sex chromosomes in the Salmonidae. II. Karyotype and sex chromosomes in *Oncorhynchus nerka*. Tsitologiya, 32: 509–514. (In Russian with English abstract)

F- 45 Frolov, S.V. (1990) Differentiation of sex chromosomes in the Salmonidae. III. Multiple sex chromosomes in *Coregonus sardinella*. Tsitologiya, 32: 659–663. (In Russian with English abstract)

F- 46 Frolov, S.V. (1991) Karyotype of *Stenodus leucichthys nelma* from Anadyr' River. Chrom. Inform. Serv., 52: 11–12.

F- 47 Frolov, S.V. and Miller, I.N. (1991) Karyotypes of *Salvelinus malma* and *S. leucomaenis* from North Primorye. Chrom. Inform. Serv., 52: 12–14.

F- 48 Frolov, S.V. (1992) Karyotype and chromosomal variability in smallmouth char, *Salvelinus elgyticus*, from Lake Elgygytgyn. J. Ichthyol., 32(1): 61–66.

F- 49 Frolov, S.V. (1993) Unique karyotype and constitutive heterochromatin in the endemic salmonine fish *Salvethymus svetovidovi*. J. Ichthyol., 33(6): 42–49.

F- 50 Frolov, S.V., Sakai, H., Ida, H., Yamazaki, Y. and Takahashi, H. (1999) Karyotype of Siberian taimen *Hucho taimen* from Amur River Basin. Chrom. Sci., 3: 33–35.

F- 51 Fujioka, Y. (1973) A comparative study of the chromosomes in Japanese freshwater fishes. I. A study of the somatic chromosome of common catfish *Parasilurus asotus* (Linné) and fork-tailed bullhead *Pelteobagrus nudiceps* (Sauvage). Bull. Fac. Educ. Yamaguchi Univ., 23: 191–195. (In Japanese with English summary)

F- 52 Fujiwara, A., Nishida-Umehara, C., Sakamoto, T., Okamoto, N., Nakayama, I. and Abe, S. (2001) Improved fish lymphocyte culture for chromosome preparation. Genetica, 111: 77–89.

F- 53 Fukuoka, H. and Niiyama, H. (1970) Notes on the somatic chromosomes of ten species of pleuronectid fishes. Chrom. Inform. Serv., 11: 18–19.

F- 54 Fukuoka, H. and Muramoto, J. (1975) Somatic and meiotic chromosomes of *Tilapia mossambica* Peters. Chrom. Inform. Serv., 18: 4–6.

F- 55 Furlong, R.F. and Holland, P.W.H. (2004) Polyploidy in vertebrate ancestry: Ohno and beyond. Biol. J. Linn. Soc., 82: 425–430.

F- 56 Falcão, J.N., Moreira-Filho, O. and Bertollo, L.A.C. (1984) An additional chromosome in two fish species. Brazil. J. Genet., 7: 109–118.

F- 57 Foresti, F., Oliveira, C. and Tien, O.S. (1992) Cytogenetic studies in fishes of the genus *Synbranchus* (Pisces, Synbranchiformes, Synbranchidae). Naturalia, Sao Paulo, 17: 129–138.

F- 58 Fontana, F. and Colombo, G. (1974) The chromosomes of Italian sturgeons. Experientia, 30: 739–742.

F- 59 Fitzsimons, J.M. (1972) A revision of two genera of goodeid fishes (Cyprinodontiformes, Osteichthyes) from the Mexican Plateau. Copeia, 1972: 728–756.

F- 60 Fitzsimons, J.M., Rogers, J.S. and Cashner, R.C. (1985) Karyologic and electrophoretic studies of the genus *Cynoscion* (Sciaenidae, Perciformes) from the northern gulf of Mexico. Japan. J. Ichthyol., 31: 444–448.

F- 61 Fontana, F., Zane, L., Pepe, A. and Congiu, L. (2007) Polyploidy in Acipenseriformes: cytogenetic and molecular approaches. In: Pisano, E., Ozouf-Costaz, C., Foresti, F. and Kapoor, B.G. eds. Fish cytogenetics. Science Publishers, Enfield, pp. 385–403.

F- 62 Fontana, F., Bruch, R.M., Binkowski, F.P., Lanfredi, M., Chicca, M., Beltrami, N. and Congiu, L. (2004) Karyotype characterization of the lake sturgeon, *Acipenser fulvescens* (Rafinesque 1817) by chromosome banding and fluorescent *in situ* hybridization. Genome, 47: 742–746.

F- 63 Fontana, F., Lanfredi, M., Kirschbaum, F., Garrido-Ramos, M.A., Robles, F., Forlani, A. and Congiu, L. (2008) Comparison of karyotypes of *Acipenser oxyrinchus* and *A. sturio* by chromosome banding and fluorescent *in situ* hybridization. Genetica, 132: 281–286.

F- 64 Fenerich, P.C., Foresti, F. and Oliveira, C. (2004) Nuclear DNA content in 20 species of Siluriformes (Teleostei: Ostariophysi) from the Neotropical region. Genet. Mol. Biol., 27: 360–354.

F- 65 Ferro, D.A.M., Moreira-Filfo, O. and Bertollo, L.A.C. (2003) B chromosome polymorphism in the fish, *Astyanax scabripinnis*. Genetica, 119: 147–153.

F- 66 Feldberg, E., Porto, J.I.R., Santos, E.B.P. and Valentim, F.C.S. (1999) Cytogenetic studies of two freshwater sciaenids of the genus *Plagioscion* (Perciformes, Sciaenidae) from the central Amazon. Genet. Mol. Biol., 22: 351–356.

*F- 67 Feichtinger, W. (1988) Cytogenetic investigations in livebearing toothcarps (Pisces, Poeciliidae). II. Simp. Citogenet. Evol. e Aplic. de Peixes Neotropicais, Sao Carlos, SP, p. 38 (abstract). (After K-114)

*F- 68 Foerster, W. and Schartl, M. (1987) Karyotype and isozyme patterns of five species of *Aulonocara* Regan, 1922. Cour. Forsch. Inst. Senckenberg, 94: 55–62. (After K-114)

F- 69 Fitzsimons, J.M. and Parker, R.M. (1985) Karyology of the sparid fishes *Lagodon rhomboides* and *Archosargus probatocephalus* (Osteichthyes, Perciformes) from coastal Louisiana, North Carolina, and Florida. Proc. Louisiana Acad. Sci., 48: 86–92.

F- 70 Fontana, F. (1994) Chromosomal nucleolar organizer regions in four sturgeon species as markers of karyotype evolution in Acipenseriformes (Pisces). Genome, 37: 888–892.

F- 71 Fernandes-Matioli, F.M.C., Marchetto, M.C.N., Almeida-Toledo, L.F. and Toledo-Filho, S.A. (1998) High intraspecific karyological conservation in four species of *Gymnotus* (Pisces: Gymnotiformes) from Southeastern Brazilian basins. Caryologia, 51: 221–234.

F- 72 Fernandes-Matioli, F.M.C. and Almeida-Toledo, L.F. (2001) A molecular phylogenetic analysis in *Gymnotus* species (Pisces: Gymnotiformes), with inferences on chromosome evolution. Caryologia, 54: 23–30.

[G]

G- 1 Gajardo, G.M. (1992) Karyotypes of *Basilichthys microlepidotus* and *B. australis* (Pisces: Atherinidae). Copeia, 1992: 256–258.

G- 2 Galetti Jr., P.M., Foresti, F., Bertollo, L.A.C. and Moreira Filho, O. (1981) Karyotypic similarity in three genera (*Leporinus*, *Leporellus* and *Schizodon*) of the family Anostomidae (Pisces-Teleostei). Brazil. J. Genet., 4: 11–15.

G- 3 Galetti Jr., P.M., Foresti, F., Bertollo, L.A.C. and Moreira Filho, O. (1981) Heteromorphic sex chromosomes in three species of the genus *Leporinus* (Pisces, Anostomidae). Cytogenet. Cell Genet., 29: 138–142.

G- 4 Galetti Jr., P.M., Foresti, F., Bertollo, L.A.C. and Moreira Filho, O. (1984) Characterization of eight species of Anostomidae (Cypriniformes) fish on the basis of the nucleolar organizing region. Caryologia, 37: 401–406.

G- 5 Galetti Jr., P.M. and Foresti, F. (1986) Evolution of the ZZ/ZW system in *Leporinus* (Pisces, Anostomidae). Cytogenet. Cell Genet., 43: 43–46.

G- 6 Galetti Jr., P.M. (1991) Chromosome complement of induced hybrids of the fish *Prochilodus marggravii* and *Prochilodus affinis* (Prochilodontidae, Characiformes). Brazil. J. Genet., 14: 203–207.

G- 7 Galetti Jr., P.M. and Rasch, E.M. (1993) Chromosome studies in *Poecilia latipunctata* with NOR polymorphism as shown by silver nitrate and chromomycin A_3 (Teleostei: Poeciliidae). Ichthyol. Explor. Freshwaters, 4: 269–277.

G- 8 Galetti Jr., P.M. and Rasch, E.M. (1993) NOR variability in diploid and triploid forms of the Amazon molly *Poecilia formosa* as shown by silver nitrate and chromomycin A_3 staining. Brazil. J. Genet., 16: 927–938.

G- 9 Galetti Jr., P.M., Bertollo, L.A.C. and Moreira-Filho, O. (1994) Trends in chromosome evolution of neotropical characiform fishes. Caryologia, 47: 289–297.

G- 10 Galetti Jr., P.M., Lima, N.R.W. and Venere, P.C. (1995) A monophyletic ZW sex chromosome system in *Leporinus* (Anostomidae, Characiformes). Cytologia, 60: 375–382.

G- 11 Galetti Jr., P.M., Aguilar, C.T. and Molina, W.F. (2000) An overview of marine fish cytogenetics. Hydrobiologia, 420: 55–62.

G- 12 Galetti Jr., P.M., Molina, W.F., Affonso, P.R.A.M. and Aguilar, C.T. (2006) Assessing genetic diversity of Brazilian reef fishes by chromosomal and DNA markers. Genetica, 126: 161–177.

G- 13 Garcia, C., Moreira-Filho, O., Bertollo, L.A.C. and Centofante, L. (2003) B chromosomes and natural triploidy in *Rhamdia* sp. (Pisces, Siluriformes, Heptapteridae). Cytologia, 68: 403–411.

G- 14 Garcia, G., Scvortzoff, E., Máspoli, M.C. and Vaz-Ferreira, R. (1993) Analysis of karyotypic evolution in natural populations of *Cynolebias* (Pisces: Cyprinodontiformes, Rivulidae) using banding techniques. Cytologia, 58: 85–94.

G- 15 Garcia, G., Scvortzoff, E. and Hernández, A. (1995) Karyotypic heterogeneity in South American annual killifishes of the genus *Cynolebias* (Pisces, Cyprinodontiformes, Rivulidae). Cytologia, 60: 103–110.

G- 16 Garcia, G., Lalanne, A.I., Aguirre, G. and Cappetta, M. (2001) Chromosome evolution in the annual killifish genus *Cynolebias* and mitochondrial phylogenetic analysis. Chrom. Res., 9: 437–448.

G- 17 Garcia, G. (2006) Multiple simultaneous speciation in killifishes of the *Cynolebias adloffi* species complex (Cyprinodontiformes, Rivulidae) from phylogeography and chromosome data. J. Zool. Syst. Evol. Res., 44: 75–87.

G- 18 Garcia, E., Alvarez, M.C. and Thode, G. (1987) Chromosome relationships in the genus *Blennius* (Blenniidae Perciformes) C-banding patterns suggest two karyoevolutional pathways. Genetica, 72: 27–36.

G- 19 Giles, L. and Bertollo, L.A.C. (1988) Karyotype variability in *Hoplerythrinus unitaeniatus* (Characiformes, Erythrinidae). I. Chromosome polymorphism in the Rio Negro population (Manaus, State of Amazonas). Brazil. J. Genet., 11: 299–306.

G- 20 Giuliano-Caetano, L. and Bertollo, L.A.C. (1990) Karyotypic variability in *Hoplerythrinus unitaeniatus* (Pisces, Characiformes, Erythrinidae). II. Occurrence of natural triploidy. Brazil. J. Genet., 13: 231–237.

G- 21 Giuliano-Caetano, L., Jorge, L.C., Moreira-Filho, O. and Bertollo, L.A.C. (2001) Comparative cytogenetic studies on *Hoplerythrinus unitaeniatus* populations (Pisces, Erythrinidae). Caryologia, 66: 39–43.

G- 22 Gregory, T.R. (2005) Genome size evolution in animals. In: Gregory, T.R. ed. The evolution of the genome. Elsevier, Amsterdam, pp. 3–87.

G- 23 Gjedrem, T., Eggum, Å. and Refstie, T. (1977) Chromosomes of some salmonids and salmonid hybrids. Aquaculture, 11: 335–348.

G- 24 Gold, J.R. (1974) A fast and easy method for chromosome karyotyping in adult teleosts. Progressive Fish-Culturist, 36: 169–171.

G- 25 Gold, J.R. and Gall, G.A.E. (1975) Chromosome cytology and polymorphism in the California High Sierra golden trout (*Salmo aguabonita*). Can. J. Genet. Cytol., 17: 41–53.

G- 26 Gold, J.R. (1977) Systematics of western North American trout (*Salmo*), with notes on the redband trout of Sheepheaven Creek, California. Can. J. Zool., 55: 1858–1873.

G- 27 Gold, J.R. and Avise, J.C. (1977) Cytogenetic studies in North American minnows (Cyprinidae). I. Karyology of nine California genera. Copeia, 1977: 541–549.

G- 28 Gold, J.R., Womac, W.D., Deal, F.H. and Barlow Jr., J.A. (1978) Gross karyotypic change and evolution in North American cyprinid fishes. Genet. Res. Camb., 32: 37–46.

G- 29 Gold, J.R. (1979) Cytogenetics. In: Hoar, W.S., Randall, D.J. and Brett, J.R. eds. Fish physiology. Vol. VIII. Academic Press, New York, pp. 353–405.

G- 30 Gold, J.R., Janak, B.J. and Barlow Jr., J.A. (1979) Karyology of four North American percids (Perciformes: Percidae). Can. J. Genet. Cytol., 21: 187–191.

G- 31 Gold, J.R., Whitlock, C.W., Karel, W.J. and Barlow Jr., J.A. (1979) Cytogenetic studies in North American minnows (Cyprinidae). VI. Karyotypes of thirteen species in the genus *Notropis*. Cytologia, 44: 457–466.

G- 32 Gold, J.R., Karel, W.J. and Strand, M.R. (1979) Chromosome formulae of North American fishes. Texas A & M Univ. System, Texas.

G- 33 Gold, J.R., Womac, W.D., Deal, F.H. and Barlow Jr., J.A. (1981) Cytogenetic studies in North American minnows (Cyprinidae). VII. Karyotypes of 13 species from the southern United States. Cytologia, 46: 105–115.

G- 34 Gold, J.R. and Ellison, J.R. (1982) Silver staining for nucleolar organizing regions of vertebrate chromosomes. Stain Technol., 58: 51–55.

G- 35 Gold, J.R. (1984) Silver-staining and heteromorphism of chromosomal nucleolus organizer regions in North American cyprinid fishes. Copeia, 1984: 133–139.

G- 36 Gold, J.R. and Amemiya, C.T. (1986) Cytogenetic studies in North American minnows (Cyprinidae). XII. Patterns of chromosomal nucleolus organizer region variation among 14 species. Can. J. Zool., 64: 1869–1877.

G- 37 Gold, J.R., Amemiya, C.T. and Ellison, J.R. (1986) Chromosomal heterochromatin differentiation in North American cyprinid fishes. Cytologia, 51: 557–566.

G- 38 Gold, J.R. and Amemiya, C.T. (1987) Genome size variation in North American minnows (Cyprinidae). II. Variation among 20 species. Genome, 29: 481–489.

G- 39 Gold, J.R., Jenkin, J.D. and Powers, P.K. (1990) Cytogenetic studies in North American minnows (Cyprinidae). XVIII. Chromosomal NOR variation among eight species. Cytologia, 55: 483–492.

G- 40 Gold, J.R., Li, Y.C., Shipley, N.S. and Powers, P.K. (1990) Improved methods for working with fish chromosomes with a review of metaphase chromosome banding. J. Fish Biol., 37: 563–575.

G- 41 Gold, J.R. and Li, Y.C. (1991) Trypsin G-banding of North American cyprinid chromosomes: phylogenetic considerations, implications for fish chromosome structure, and chromosomal polymorphism. Cytologia, 56: 199–208.

G- 42 Gold, J.R., Li, Y.C., Schmidt, T.R. and Tave, D. (1991) Nucleolar dominance in interspecific hybrids of cyprinid fishes. Cytobios, 65: 139–147.

G- 43 Gold, J.R., Zoch, P.K. and Amemiya, C.T. (1988) Cytogenetic studies in North American minnows (Cyprinidae). XIV Chromosomal NOR phenotypes of eight species from the genus *Notropis*. Cytobios, 54: 137–147.

G- 44 Gold, J.R. and Li, Y.C. (1994) Chromosomal NOR karyotypes and genome size variation among squawfishes of the genus *Ptychocheilus* (Teleostei: Cyprinidae). Copeia, 1994: 60–65.

G- 45 Gold, J.R. and Li, Y.C. (1994) Cytosystematic evidence that the genus *Richardsonius* belongs in the Western clade of phoxinin cyprinids. Copeia, 1994: 815–818.

G- 46 Goldammer, T. and Klinkhardt, M.B. (1992) Karyologische Studien an verschiedenen Süsswasserfischarten aus brackigen Küstengewässern der südwestlichen Ostsee. V. Der Zander (*Stizostedion lucioperca* (Linnaeus, 1758)). Zool. Anz., 228: 129–139.

G- 47 Golubtsov, A.S. and Krysanov, E.Yu. (1993) Karyological study of some cyprinid species from Ethiopia. The ploidy differences between large and small *Barbus* of Africa. J. Fish Biol., 42: 445–455.

G- 48 Gold, J.R. and Price, H.J. (1985) Genome size variation among North American minnows (Cyprinidae). I. Distribution of the variation in five species. Heredity, 54: 297–305.

G- 49 Golubtsov, A.S. and Krysanov, E.Yu. (2003) Karyological comparison of the small African barbs *Barbus paludinosus* (Cyprinidae) with developed and reduced spine in the dorsal fin. J. Ichthyol., 43: 245–252.

G- 50 Gomes, V., Vazzoler, A.E.A.M. and Ngan, P.V. (1983) Estudos cariotipicos de peixes da familia Sciaenidae (Teleostei, Perciformes) da região de Cananéia, SP, Brasil. 1. Sobre o cariotipo de *Micropogonias furnieri* (Desmarest, 1823). Bolm Inst. Oceanogr. S Paulo, 32: 137–142. (In Portuguese with English synopsis)

G- 51 Gomes, V., Vazzoler, A.E.A.M. and Ngan, P.V. (1983) Estudos cariotipicos de peixes da familia Sciaenidae (Teleostei, Perciformes) da região de Cananéia, SP, Brasil. 2. Sobre o cariotipo de *Menticirrhus americanus* (Linnaeus, 1758). Bolm Inst. Oceanogr. S Paulo, 32: 187–191. (In Portuguese with English synopsis)

G- 52 Gomes, V., Ngan, P.V. and Passos, M.J.A.C.R. (1990) The karyotype of a marine catfish, *Bagre bagre*, from Brazil. Japan. J. Ichthyol., 37: 321–323.

G- 53 Gonzalez, J.A., Bueno, M.L. and Forero, J.E. (1992) Caracterizacion cromosomica de dos especies icticas nativas; Guapucha, (*Grundulus bogotensis*) y Capitan, (*Eremophilus mutisii*), de la sabana de Bogota. Acta Biol. Colombiana, 7/8: 45–54. (In Spanish with English summary)

G- 54 Gornung, E., Gabrielli, I. and Sola, L. (1998) Localization of the (TTAGGG)n telomeric sequence in zebrafish chromosomes. Genome, 41: 136–138.

G- 55 Gorshkov, S.A. and Gorshkova, G.V. (1981) Analysis of relations between species of Pacific salmons from the genera *Oncorhynchus* and *Salmo*. Zool. Zhurnal, 60: 84–96. (In Russian with English summary)

G- 56 Gorshkova, G.V. (1978) Some peculiarities of karyotypes of the Pacific salmon. Tsitologiya, 20: 1431–1435. (In Russian with English summary)

G- 57 Gorshkova, G.V. (1980) Karyology and chromosomal polymorphism of Pacific salmon. In: Kirpichnikov, V.S. ed. Karyological variability, mutagenesis and gynogenesis in fishes. Inst. Cytol. Acad. Sci. USSR, Leningrad, pp. 29–33. (In Russian with English summary)

G- 58 Gorshkova, G.V. and Gorshkov, S.A. (1985) Karyotypes of Kamchatkan trout of the genus *Salmo*. J. Ichthyol., 25(4): 115–123.

G- 59 Gorshkova, G.V., Gorshkov, S.A. and Golani, D. (2002) Karyotypes of *Barbus canis* and *Capoeta damascina* (Pisces, Cyprinidae) from the Middle East. Ital. J. Zool., 69: 191–194.

G- 60 Greenfield, D.W., Abdel-Hameed, F., Deckert, G.D. and Flinn, R.R. (1973) Hybridization between *Chrosomus erythrogaster* and *Notropis cornutus* (Pisces: Cyprinidae). Copeia, 1973: 54–60.

G- 61 Greenfield, D.W. and Greenfield, T. (1972) Introgressive hybridization between *Gila orcutti* and *Hesperoleucus symmetricus* (Pisces: Cyprinidae) in the Cuyama River basin, California: I. Meristics, morphometrics and breeding. Copeia, 1972: 849–859.

G- 62 Gregory, P.E., Howard-Peebles, P.N., Ellender, R.D. and Martin, B.J. (1980) C-banding of chromosomes from three established marine fish cell lines. Copeia, 1980: 545–547.

G- 63 Grigoryan, K.A. and Vasiliev, V.P. (1993) Karyotypes of five species of goby (Gobiidae) from the basins of the Black and Caspian seas. J. Ichthyol., 33(1): 137–143.

G- 64 Grigoryan, K.A. and Vasiliev, V.P. (1993) Comparative analysis of karyotypes of three species of gobies (Gobiidae) from the Black Sea Basin in the connection with their taxonomic relationships. Vopr. Ichthyol., 33(3): 449–452. (In Russian)

G- 65 Grimm, H. (1972) Cytologische Untersuchungen an westafrikanischen Zahnkarpfen der Gattungen *Aphyosemion* Myers, 1924 und *Roloffia* Stenholt Clausen, 1966 (Pisces, Cyprinodontidae). Mitt. Hamburg. Zool. Mus. Inst., 68: 195–205.

G- 66 Guégan, J.F., Rab, P., Machordom, A. and Doadrio, I. (1995) New evidence of hexaploidy in 'large' African *Barbus* with some considerations of the origin of hexaploidy. J. Fish Biol., 47: 192–198.

G- 67 Gui, J.F., Li, Y.C., Li, K. and Zhou, T. (1984) A comparative study of the karyotypes of three species of gobioid fishes. Zool. Res., 5(3) suppl.: 67–68. (In Chinese)

G- 68 Gui, J.F., Li, Y.C., Li, K., Hong, Y.H. and Zhou, T. (1985) Studies on the karyotypes of Chinese cyprinid fishes. VI. Karyotypes of three tetraploid species in Barbinae and one tetraploid species in Cyprininae. Acta Genet. Sinica, 12: 302–308. (In Chinese with English abstract)

G- 69 Gui, J.F., Li, Y.C., Li, K. and Zhou, T. (1986) Studies on the karyotypes of Chinese cyprinid fishes. (VIII) Karyotypic analyses of fifteen species of Barbinae with a consideration for their phyletic evolution. Trans. Chinese Ichthyol. Soc., 5: 119–127. (In Chinese with English abstract)

G- 70 Gui, J.F. and Zhou, T. (1986) Further improvement of the BrDU-Hoechst-Giemsa method and studies of BrDU-replication band karyotypes of *Mylopharyngodon piceus* and *Ctenopharyngodon idellus*. Acta Hydrobiol. Sinica, 10: 157–161. (In Chinese with English abstract)

G- 71 Gul, S., Colak, A. and Sezgin, I. (2000) Karyotype analysis in *Chalcalburnus mossulensis* (Heckel, 1843). Turk. J. Biol., 24: 657–662. (In Turkish with English abstract)

G- 72 Guo, X.H., Liu, S.J., Zhang, C. and Liu, Y. (2004) Comparative and evolutionary analysis of the cytochrome *b* sequences in cyprinids with different ploidy levels derived from crosses. Genetica, 121: 295–301.

G- 73 Gyldenholm, A.O. and Scheel, J.J. (1971) Chromosome numbers of fishes. I. J. Fish Biol., 3: 479–486.

G– 74 Garcia Morina, F. and Uribe-Alcocer, M. (1989) Análisis cromosómico del bagre marino *Arius felis* (Ariidae: Siluriformes) de la región de la laguna de Términos, Camp. An. Inst. Cienc. Mar Limnol. Univ. Nal. Autón. México, 16: 69–74.

G– 75 Gornung, E., Cordisco, C.A., Rossi, A.R., De Innocentiis, S., Crosetti, D. and Sola, L. (2001) Chromosomal evolution in Mugilidae: karyotype characterization of *Liza saliens* and comparative localization of major and minor ribosomal genes in the six Mediterranean mullets. Mar. Biol., 139: 55–60.

G– 76 Giles, V., Thode, G. and Alvarez, M.C. (1985) A new Robertsonian fusion in the multiple chromosome polymorphism of a Mediterranean population of *Gobius paganellus* (Gobiidae, Perciformes). Heredity, 55: 255–260.

G– 77 Gregory, T.R. and Hebert, P.D.N. (1999) The modulation of DNA content: proximate causes and ultimate consequences. Genome Res., 9: 317–324.

G– 78 Gromicho, M., Coutanceau, J., Ozouf-Costaz, C. and Collares-Pereira, M.J. (2006) Contrast between extensive variation of 28S rDNA and stability of 5S rDNA and telomeric repeats in the diploid-polyploid *Squalius alburnoides* complex and in its maternal ancestor *Squalius pyrenaicus* (Teleostei, Cyprinidae). Chrom. Res., 14: 297–306.

G– 79 Gaffaroglu, M., Yuksel, E. and Ráb, P. (2006) Note on the karyotype and NOR phenotype of leuciscine fish *Acanthobrama marmid* (Osteichthyes, Cyprinidae). Biol. Bratislava, 61: 207–209.

G– 80 Gul, S., Nur, G. and Baysal, A. (2006) Karyotype analysis of *Alburnus filippii*. Ind. Vet. J., 83: 102–103.

G– 81 Garcia, C. and Moreira Filho, O. (2005) Cytogenetical analyses in three fish species of the genus *Pimelodus* (Siluriformes: Pimelodidae) from rio São Francisco: considerations about the karyotypical evolution in the genus. Neotrop. Ichthyol., 3: 285–290.

G– 82 Gold, J.R., Amemiya, C.T., Karel, W.J. and Iida, N. (1988) The karyotype and genome structure of the pirate perch *Aphredoderus sayanus* (Aphredoderidae: Teleostei). Experientia, 44: 68–70.

G– 83 Ghigliotti, L., Mazzei, F., Ozouf-Costaz, C., Bonillo, C., Williams, R., Cheng, C.C. and Pisano, E. (2007) The two giant sister species of the Southern Ocean, *Dissostichus eleginoides* and *Dissostichus mawsoni*, differ in karyotype and chromosomal pattern of ribosomal RNA genes. Polar Biol., 30: 625–634.

G– 84 Guégan, J.F. and Morand, S. (1996) Polyploid hosts: strange attractors for parasites? Oikos, 77: 366–370.

G– 85 Gregory, T.R. (2008) Animal genome size database. http://www.genomesize.com/

G– 86 Gold, J.R., Li, Y.C., Birkner, M.C. and Jenkin, J.D. (1992) Chromosomal NOR karyotypes and genome sizes in *Dionda* (Osteichthyes: Cyprinidae) from Texas and New Mexico. Southwest. Nat., 37: 217–222.

G– 87 Gold, J.R., Ragland, C.J. and Schliesing, L.J. (1990) Genome size variation and evolution in North American cyprinid fishes. Genet. Sel. Evol., 22: 11–29.

G– 88 Goddard, K.A., Megwinoff, O., Wessner, L.L. and Giaimo, F. (1998) Confirmation of gynogenesis in *Phoxinus eos-neogaeus* (Pisces: Cyprinidae). J. Hered., 89: 151–157.

G– 89 Ghigliotti, L., Mazzei, F., Ozouf-Costaz, C., Christiansen, J.S., Fevolden, S. and Pisano, E. (2008) First cytogenetic characterization of the sub-arctic marine fish *Mallotus villosus* (Müller, 1776), Osmeriformes, Osmeridae. Genet. Mol. Biol., 31 (suppl.): 180–187.

G– 90 Galetti Jr., P.M., Da Silva, E.B. and Cerminaro, R.T. (1985) A multiple NOR system in the fish *Serrasalmus spilopleura* (Serrasalminae, Characidae). Brazil. J. Genet., 8: 479–484.

G– 91 Galetti Jr., P.M. and Foresti, F. (1987) Two new cases of ZZ/ZW heterogamety in *Leporinus* (Anostomidae, Characiformes) and their relationships in the phylogeny of the group. Brazil. J. Genet., 10: 135–140.

*G– 92 Garcia, G., Hernandez, A. and Scvortzoff, E. (1994) Estudios citogeneticos en 5 nuevas especies del genero *Cynolebias* Steindachner, 1876 (Cyprinodontiformes, Rivulidae). V.

Simp. Citogenet. Evol. e Aplic. de Peixes Neotropicais, Botucatu, SP, p. 46 (abstract). (After K-114)

*G- 93 Grimm, H. (1974) Zum Problem der Unterscheidung der Gattungen *Roloffia* und *Aphyosemion*. Aquarien-Terrarien, 2: 50–53. (After K-114)

G- 94 Galetti Jr., P.M., Mestriner, C.A., Monaco, P.J. and Rasch, E.M. (1995) Post-zygotic modifications and intra- and inter-individual nucleolar organizing region variations in fish: report of a case involving *Leporinus friderici*. Chrom. Res., 3: 285–290.

G- 95 Gravena, W., Teribele, R., Giuliano-Caetano, L. and Dias, A.L. (2007) Occurrence of B chromosomes in *Cyphocharax modestus* (Fernández-Yépez, 1948) and *Steindachneria insculpta* (Fernández-Yépez, 1948) (Characiformes, Curimatidae) from the Tibagi River basin (Paraná State, Brazil). Brazil. J. Biol., 67 (special number): 631–637.

G- 96 Giuliano-Caetano, L. and Bertollo, L.A.C. (1988) Karyotype variability in *Hoplerythrinus unitaeniatus* (Characiformes, Erythrinidae). I. Chromosome polymorphism in the Rio Negro population (Manaus, state of Amazonas). Brazil. J. Genet., 11: 299–306.

G- 97 Gold, J.R., Avise, J.C. and Gall, G.A.E. (1977) Chromosome cytology in the cutthroat trout series, *Salmo clarki* (Salmonidae). Cytologia, 42: 377–382.

[H]

H- 1 Haas, R. (1979) Intergeneric hybridyzation in a sympatric pair of Mexican cyprinodontid fishes. Copeia, 1979: 149–152.

H- 2 Hafez, R., Labat, R. and Quillier, R. (1978) Etude cytogénétique chez quelques espèces de cyprinidés de la région Midi-Pyrénées. Bull. Soc. Hist. Nat. Toulouse, 114: 122–159.

H- 3 Hafez, R., Labat, R. and Quillier, R. (1978) Aneuploidie observée chez des populations de gardons (*Rutilus rutilus* L.) et d'ablettes (*Alburnus alburnus* L.) de la région Midi-Pyrénées. Bull. Soc. Hist. Nat. Toulouse, 114: 85–92.

H- 4 Hafez, R., Labat, R. and Quillier, R. (1978) Teneurs nucléaires en A.D.N. et relations évolutives dans la famille des cyprinides (Teleostei). Bull. Soc. Hist. Nat. Toulouse, 114: 71–84.

H- 5 Hafez, R., Labat, R. and Quillier, R. (1981) Recherches sur les chromosomes surnuméraires de l'ablette (*Alburnus alburnus* L.). Cybium, 3rd ser., 5: 81–87.

H- 6 Hartley, S.E. and Horne, M.T. (1984) Chromosome relationships in the genus *Salmo*. Chromosoma, 90: 229–237.

H- 7 Hartley, S.E. and Horne, M.T. (1984) Chromosome polymorphism and constitutive heterochromatin in the Atlantic salmon, *Salmo salar*. Chromosoma, 89: 377–380.

H- 8 Hartley, S.E. (1987) The chromosomes of salmonid fishes. Biol. Rev., 62: 197–214.

H- 9 Hartley, S.E. (1989) Chromosomes and constitutive heterochromatin distribution in Arctic charr, *Salvelinus alpinus* (L.) (Pisces: Salmonidae). Genetica, 79: 161–166.

H- 10 Hartley, S.E. (1988) Cytogenetic studies of Atlantic salmon, *Salmo salar* L., in Scotland. J. Fish Biol., 33: 735–740.

H- 11 Harvey, S.C., Powell, S.F., Kennedy, D.D., McAndrew, B.J. and Penman, D.J. (2002) Karyotype analysis of *Oreochromis mortimeri* (Trewavas) and *Sarotherodon melanotheron* (Rüppell). Aquacult. Res., 33: 339–342.

H- 12 Hellmer, A., Voiculescu, I. and Schempp, W. (1991) Replication banding studies in two cyprinid fishes. Chromosoma, 100: 524–531.

H- 13 Hinegardner, R. and Rosen, D.E. (1972) Cellular DNA content and the evolution of teleostean fishes. Amer. Natur., 106: 621–644.

H- 14 Hirata, J. and Urushido, T. (2000) Karyotypes and DNA contents in Osteoglossiformes fishes. Sci. Rep. Res. Inst. Evol. Biol., 9: 83–90. (In Japanese with English abstract)

H- 15 Hitotsumachi, S. and Sasaki, M. (1967) A note on the chromosomes of the loach (*Misgurnus anguillicaudatus*). Chrom. Inform. Serv., 8: 19–20.

H- 16 Hitotsumachi, S., Sasaki, M. and Ojima, Y. (1969) A comparative karyotype study in several species of Japanese loaches (Pisces, Cobitidae). Japan. J. Genet., 44: 157–161.

H- 17 Hollenbeck, P.J. and Chrisman, C.L. (1981) Kidney preparations for chromosomal analyses of Ictaluridae. Copeia, 1981: 216–217.

H- 18 Hong, Y.H. and Zhou, T. (1983) Karyotypic study of two species of Chinese catfishes (Siluridae). J. Wuhan Univ. (Nat. Sci.), 1983(3): 106–108, 128. (In Chinese with English abstract)

H- 19 Hong, Y.H., Zhou, M. and Zhou, T. (1983) Studies on the karyotypes of Chinese cyprinid fishes. III. Comparative analysis of the chromosomes of seven species of acheilognathid fishes. J. Wuhan Univ. (Nat. Sci.), 1983(2): 96–102, 116. (In Chinese with English abstract)

H- 20 Hong, Y.H. and Zhou, T. (1984) Karyotypes of nine species of Chinese catfishes (Bagridae). Zool. Res., 5(3) (suppl.): 21–28. (In Chinese with English abstract)

H- 21 Hong, Y.H. and Zhou, T. (1984) Karyotype and its ZZ-ZO type sex chromosome of *Coilia brachygnathus*. Hereditas (Beijing), 6(4): 12–14. (In Chinese)

H- 22 Hong, Y.H., Li, Y.C., Li, K., Gui, J.F. and Zhou, T. (1984) Studies on the karyotypes of Chinese cyprinid fishes. IV. Comparative analysis of the karyotypes of 11 species of gobionid fishes with special considerations of their phylogenetic relationship. Acta Zool. Sinica, 30: 343–351. (In Chinese with English abstract)

H- 23 Hong, Y.H. and Zhou, T. (1985) Studies on the karyotype and C-banding patterns in *Acheilognathus gracilis* with a discussion on the evolution of acheilognathid fishes. Acta Genet. Sinica, 12: 143–148. (In Chinese with English abstract)

H- 24 Hong, Y.H. and Zhou, T. (1985) Chromosome banding in fishes. I. An improved BrdU-Hoechst-Giemsa method for revealing DNA-replication bands in fish chromosomes. Acta Genet. Sinica, 12: 67–71. (In Chinese with English abstract)

H- 25 Howell, W.M. and Denton, T.E. (1969) Chromosome of ammocoetes of the Ohio brook lampley, *Lampetra aepyptera*. Copeia, 1969: 393–395.

H- 26 Howell, W.M. and Bloom, S.E. (1973) Sex-associated differential fluorescence of mud-minnow chromosomes and spermatozoa. Nature, 245: 261–263.

H- 27 Howell, W.M. and Black, D.A. (1979) Location of the nucleolus organizer regions on the sex chromosomes of the banded killifish, *Fundulus diaphanus*. Copeia, 1979: 544–546.

H- 28 Hartley, S.E. (1990) Variation in cellular DNA content in Arctic charr, *Salvelinus alpinus* (L.). J. Fish Biol., 37: 189–190.

H- 29 Howell, W.M. (1972) Somatic chromosomes of the black ghost knifefish, *Apteronotus albifrons* (Pisces: Apteronotidae). Copeia, 1972: 191–193.

H- 30 Howell, W.M. and Villa, J. (1976) Chromosomal homogeneity in two sympatric cyprinid fishes of the genus *Rhinichthys*. Copeia, 1976: 112–116.

H- 31 Howell, W.M. and Black, D.A. (1980) Controlled silver-staining of nucleolus organizer regions with a protective colloidal developer: a 1-step method. Experientia, 36: 1014–1015.

H- 32 Hong, Y.H. and Zhou, T. (1984) Karyotype of half-beak fish. Zool. Res., 5(3) suppl.: 63–64. (In Chinese)

H- 33 Huber, J.H. and Scheel, J.J. (1981) Revue systématique de la superespèce *Aphyosemion elegans*. Description de *A. chauchei* et *A. schioetzi* n. sp. Rev. fr. Aquariol., 8(2): 33–42.

H- 34 Hu, G.H., Sun, G.H., Zhang, Y.B., Gu, Q., Li, G., Tang, H. and Fan, L.H. (2005) Karyotype analyses of northern sheatfish (*Silurus soldatovi*). J. Dalian Fish. Univ., 20: 318–321. (In Chinese with English abstract)

H- 35 Howell, W.M. and Duckett, C.R. (1971) Somatic chromosomes of the lamprey, *Ichthyomyzon gagei* (Agnatha: Petromyzonidae). Experientia, 27: 222–223.

H- 36 Harvey, S.C., Campos-Ramos, R., Kennedy, D.D., Ezaz, M.T., Bromage, N.R., Griffin, D.K. and Penman, D.J. (2002) Karyotype evolution in Tilapia: mitotic and meiotic chromosome analysis of *Oreochromis karongae* and *O. niloticus* × *O. karongae* hybrids. Genetica, 115: 169–177.

H- 37 Hinegardner, R. (1976) The cellular DNA content of sharks, rays and some other fishes. Comp. Biochem. Physiol., 55B: 367–370.

H- 38 Hattori, R.S., Daniel-Silva, M.F.Z. and Almeida-Toledo, L.F. (2007) Karyotype characterization and gene mapping of 5S and 18S rDNA in three species of *Oligosarcus* (Teleostei: Characidae). Caryologia, 60: 372–378.

H- 39 Hinegardner, R. (1968) Evolution of cellular DNA content in teleost fishes. Amer. Natur., 102: 517–523.

H- 40 Hardie, D.C. and Hebert, P.D.N. (2004) Genome-size evolution in fishes. Can. J. Fish. Aquat. Sci., 61: 1636–1646, table D1.

H- 41 Hardie, D.C. and Hebert, P.D.N. (2003) The nucleotypic effects of cellular DNA content in cartilaginous and ray-finned fishes. Genome, 46: 683–706.

H- 42 Hashimoto, D.T., Laudicina, A., Bortolozzi, J., Foresti, F. and Porto-Foresti, F. (2009) Chromosomal features of nucleolar dominance in hybrids between the Neotropical fish *Leporinus macrocephalus* and *Leporinus elongatus* (Characiformes, Anostomidae). Genetica, 137: 135–140.

*H-43 Hernandez, A. and Garcia, G. (1994) Estudio citogenetico en *Cynolebias cheradophilus* (Cyprinodontiformes, Rivulidae). V. Simp. Citogenet. Evol. E Aplic. De Peixes Neotropicais, Botucatu, SP, p. 22 (abstract). (After K-114)

*H-44 Haimovici, S. and Ciuca, L. (1973) Observations concemant les chromosomes chez *Eudontomyzon mariae* (Berg, 1931) (Cyclostomata, Petromyzonidae). Ann. Sti. Univ. Jasi. Sec. 2, 19: 345–348. (After K-114)

*H-45 Hudson, R.G. (1976) A comparison of karyotypes and erythrocyte DNA quantities in several species of catfish (Siluriformes) with phylogenetic implications. Unpublished Ph D. Dissertation, North Carolina State University, Raleigh. (After K-114)

[I]

I- 1 Ida, H., Murofushi, M., Fujiwara, S. and Fujino, K. (1978) Preparation of fish chromosomes by *in vitro* colchicine treatment. Japan. J. Ichthyol., 24: 281–284.

I- 2 Ida, H. and Yunokawa, K. (1980) Karyotypic variation found among five species of the family Platycephalidae. Japan. J. Ichthyol., 27: 122–128.

I- 3 Ida, H. (1981) Karyotypes of elasmobranchs. Mar. Sci. Monthly, 13(1): 52–58. (In Japanese)

I- 4 Ida, H., Iwasawa, T. and Kamitori, M. (1982) Karyotypes in eight species of *Sebastes* from Japan. Japan. J. Ichthyol., 29: 162–168.

I- 5 Ida, H., Sato, I. and Miyawaki, N. (1985) Karyotypes of two species in the order Torpediniformes. Japan. J. Ichthyol., 32: 107–111.

I- 6 Ida, H., Asahida, T., Yano, K. and Tanaka, S. (1986) Karyotypes of two sharks, *Chlamydoselachus anguineus* and *Heterodontus japonicus*, and their systematic implications. In: Uyeno, T., Arai, R., Taniuchi, T. and Matsuura, K. eds. Indo-Pacific fish biology. Ichthyological Society of Japan, Tokyo, pp. 158–163.

I- 7 Ida, H., Terashima, H. and Fujimi, T. (1989) Karyotypes in four species of the family Cottidae. Japan. J. Ichthyol., 36: 135–140.

I- 8 Ida, H., Oka, N. and Hayashizaki, K. (1991) Karyotypes and cellular DNA contents of three species of the subfamily Clupeinae. Japan. J. Ichthyol., 38: 289–294.

I- 9 Ida, H., Oka, N., Terashima, H. and Hayashizaki, K. (1993) Karyotypes and cellular DNA contents of three species of the family Scombridae from Japan. Bull. Japan. Soc. Sci. Fish., 59: 1319–1323.

I- 10 Ilyinskikh, N.N. (1980) Variability of the normal chromosome set in homoiothermal and poikilothermal vertebrates. Zool. Zhurnal, 59: 1494–1499. (In Russian with English summary)

I- 11 Inokuchi, T., Abe, S., Yamaha, E. and Yamazaki, F. (1993) A preliminary study on BrdU replication banding in the chromosomes of chum salmon, *Oncorhynchus keta*. Chrom. Inform. Serv., 54: 30–32.

I- 12 Ishii, K. and Yabu, H. (1981) Chromosomes in two species of rock fish. Chrom. Inform. Serv., 31: 22–24.

I- 13 Ishii, K. and Yabu, H. (1983) Chromosomes of the longnose eel, *Synaphobranchus kaupi* Johnson (Anguilliformes, Pisces). Chrom. Inform. Serv., 35: 11–13.

I- 14 Ishii, K. and Yabu, H. (1985) Chromosomes in three species of Gadidae (Pisces). Bull. Japan. Soc. Sci. Fish., 51: 25–28.

I- 15 Ishikawa, T., Takai, A. and Ueno, K. (2001) Chromosome aberrations and survival in normal and dechorionated embryos of *Misgurnus anguillicaudatus* (Cobitidae, Pisces) exposed to mitomycin C. Cytologia, 66: 215–219.

I- 16 Itahashi, M. and Kawase, H. (1975) Karyotype analysis in five species belonging to Cyprinodontidae, Teleostomi. Bull. Aichi Univ. Educ. (Nat. Sci.), 24: 153–160.

I- 17 Itoh, Y. and Niiyama, H. (1972) Comparative chromosome studies of two cyprinid fishes, Ugui, *Tribolodon hakonensis* (Günther) and Ezo-ugui, *T. ezoe* Okada et Ikeda. Bull. Fac. Fish. Hokkaido Univ., 23(2): 73–76. (In Japanese with English abstract)

I- 18 Ivanov, V.N. (1975) Chromosomes of the Black Sea Gobiidae-*Gobius melanostomus* (Pallas) and *Gobius batrachocephalus* (Pallas). Tsitol. Genet., 9: 551–552. (In Russian with English summary)

I- 19 Ivanov, V.N. (1969) The chromosomes of the Black Sea flatfish *Rhombus maeoticus* Pallas. Dokl. Biol. Sci., 187: 507–509.

I- 20 Ivanov, V.N. (1969) Chromosomes in the embryonic cells of the Black Sea anchovy. Tsitol. Genet., 3: 285–287. (In Russian with English summary)

I- 21 Ida, H. (1988) A new method for karyological analysis of large fish. In: Uyeno, T. and Okiyama, M. eds. Ichthyology currents 1988. Asakura Shoten, Tokyo, pp. 202–217. (In Japanese)

I- 22 Iliadou, K. and Rackham, B.D. (1990) The chromosomes of a catfish *Parasilurus aristotelis* from Greece. Japan. J. Ichthyol., 37: 144–148.

I- 23 Iwami, T., Ogawa, A. and Matsuura, K. (1992) Karyotypes in three species of the boxfish genus *Lactoria*. Advance abstracts for 25th Annual meeting, 1992, the Ichthyological Society of Japan, p. 50. (In Japanese)

[J]

J- 1 Jamieson, A. (1974) Genetic 'tags' for marine fish stocks. In: Harden Jones, F.R. ed. Sea fisheries research. Elek Science, London, pp. 91–99.

J- 2 Jankun, M., Kucharczyk, D., Woznicki, P. and Kotik, A. (1997) Chromosome study of *Abramis brama* (L.) from Lake Kortowskie, Poland. Cytobios, 90: 175–179.

J- 3 Jankun, M., Rab, P. and Vuorinen, J. (1991) A karyotype study of vendace, *Coregonus albula* (Pisces, Coregoninae). Hereditas, 115: 291–294.

J- 4 Jankun, M., Ocalewicz, K., Pardo, B.G., Martinez, P., Woznicki, P. and Sanchez, L. (2003) Localization of 5S rRNA loci in three coregonid species (Salmonidae). Genetica, 119: 183–186.

J- 5 Jankun, M., Ocalewicz, K., Pardo, B.G., Martinez, P., Woznicki, P. and Sanchez, L. (2003) Chromosomal characteristics of rDNA in European grayling *Thymallus thymallus* (Salmonidae). Genetica, 119: 219–224.

J- 6 Jenkin, J.D., Li, Y.C. and Gold, J.R. (1992) Cytogenetic studies in North American minnows (Cyprinidae). XXVI. Chromosomal NOR phenotypes of 21 species from the Western United States. Cytologia, 57: 443–453.

J- 7 Jesus, C.M. and Moreira-Filho, O. (2000) Karyotypes of three species of *Parodon* (Teleostei: Parodontidae). Ichthyol. Explor. Freshwaters, 11: 75–80.

J- 8 Jesus, C.M. and Moreira-Filho, O. (2000) Cytogenetic studies in some *Apareiodon* species (Pisces, Parodontidae). Cytologia, 65: 397–402.

J- 9 John, G., Barat, A. and Lakra, W.S. (1992) Localization of nucleolar organizer regions in fish species, *Catla catla* and *Mystus vittatus*. Nucleus, 35: 179–181.

J- 10 John, G., Barat, A. and Lakra, W.S. (1993) Localization of nucleolar organizer regions in *Labeo* (Cyprinidae). La Kromosomo, II, 70: 2381–2384.

J- 11 Joswiak, G.R., Starnes, W.C. and Moore, W.S. (1980) Karyotypes of three species of the genus *Phoxinus* (Pisces: Cyprinidae). Copeia, 1980: 913–916.

J- 12 Jurajda, P., Ráb, P., Šlechta, V. and Šlechtová, V. (1996) Occurrence and identification of Kessler's gudgeon (*Gobio kessleri*) in the River Bečva (Czech Republic). Folia Zool., 45: 253–257.

J- 13 Jordão, L.C., Oliveira, C., Foresti, F. and Godinho, H.M. (1992) Cytogenetic characterization of the mullet, *Mugil platanus* (Pisces, Mugilidae). Bol. Inst. Pesca, 19: 63–66. (In Portuguese with English abstract)

J- 14 Johnson, C.R., White, R.W.G. and Bick, Y.A.E. (1981) Cytotaxonomy of seven species of *Galaxias* (Pisces: Galaxiidae) in Tasmania. Genetica, 56: 17–21.

J- 15 Johnson, O.W., Utter, F.M. and Rabinovitch, P.S. (1987) Interspecies differences in salmonid cellular DNA identified by flow cytometry. Copeia, 1987: 1001–1009.

J- 16 Jesus, C.M., Galetti Jr., P.M., Valentini, S.R. and Moreira-Filho, O. (2003) Molecular characterization and chromosomal localization of two families of satellite DNA in *Prochilodus lineatus* (Pisces, Prochilodontidae), a species with B chromosomes. Genetica, 118: 25–32.

J- 17 Jara-Seguel, P., Valdebenito, I., Palma-Rojas, C. and Rebolledo, C. (2008) Nuclear DNA content in *Galaxias maculatus* (Teleostei: Osmeriformes: Galaxiidae). Lat. Am. J. Aquat. Res., 36: 87–91.

J- 18 Jara-Seguel, P., Valdebenito, I. and Palma-Rojas, C. (2008) Nuclear DNA content in *Galaxias platei* (Steindachner, 1898) (Teleostei: Osmeriformes: Galaxiidae). Gayana, 72(2): 196–201. (In Spanish with English abstract)

*J- 19 Jesus, C.F.G., Lister, T.J., Grisolia, C.K. and Silva, A.J. (1988) Estudos citogenéticos preliminares de *Metynnis mola* e *Mylossoma orbignyanum* (Pisces, Characidae) do Pantanal Mato-Grossense. II. Simp. Citogenet. Evol. e Aplic. de Peixes Neotropicais, Sao Carlos, SP, p. 23 (abstract). (After K-114)

*J- 20 Julio, H.F., Garcia, R.M.G. and Venere, P.C. (1988) Estudos cariotipicos em 2 espécies da familia Doradidae (Pisces: Siluriformes). II. Simp. Citogenet. Evol. E Aplic. De Peixes Neotropicais, Sao Carlos, SP, p. 39 (abstract). (After K-114)

J- 21 Jaillon, O., Aury, J.M., Brunet, F., Petit, J.L., Stange-Thomann, N. et al. (2004) Genome duplication in the teleost fish *Tetraodon nigroviridis* reveals the early vertebrate protokaryotype. Nature, 431: 946–957.

[K]

K- 1 Kaidanova, T.I. (1980) A comparative karyological analysis of some salmon and whitefish species. In: Kirpichnikov, V.S. ed. Karyological variability, mutagenesis and gynogenesis in fishes. Inst. Cytol. Acad. Sci. USSR, Leningrad, pp. 16–23. (In Russian with English summary)

K- 2 Kalat, M., Mayr, B., Ráb, P. and Schleger, W. (1988) Heterochromatin studies and localization of NORs in European grayling (*Thymallus thymallus*, Pisces, Thymallidae). Caryologia, 41: 245–249.

K- 3 Kang, Y.S. and Park, E.H. (1973) Somatic chromosomes of the Manchurian trout, *Brachymystax lenok* (Salmonidae). Chrom. Inform. Serv., 15: 10–11.

K- 4 Kang, Y.S. and Park, E.H. (1973) Studies on the karyotypes and comparative DNA values in several Korean cyprinid fishes. Korean J. Zool., 16: 97–108.

K- 5 Kang, Y.S. and Park, E.H. (1974) Studies on the chromosomes and relative DNA values in four species of the family Cyprinidae of Korean fresh-waters. Chrom. Inform. Serv., 16: 24–26.

K- 6 Kang, Y.S. and Park, E.H. (1975) Leucocyte culture of the eel without autologous serum. Japan. J. Genet., 50: 159–161.

K- 7 Kantek, D.L.Z., Fenocchio, A.S. and Cestari, M.M. (2003) Cytogenetic characterization and NOR polymorphism in *Astyanax* sp. C (Tetragonopterinae, Characidae) from the First Plateau of the Iguaçu River (Paraná State, Brazil). Cytologia, 68: 19–24.

K- 8 Karakousis, Y., Vareli, K., Kotsinas, A. and Triantaphyllidis, C. (1993) Genetic characteristics of a perch (*Perca fluviatilis*) population from Lake Volvi (Macedonia, Greece). Zool. Anz., 230(3/4): 131–136.

K- 9 Karbe, L. (1961) Cytologische Untersuchung der Sterilitätserscheinungen bei anatolischen Zahnkarpfen, ein Beitrag zum Speziationsproblem. Mitt. Hamburg. Zool. Mus. Inst., 59: 73–104.

K- 10 Kasama, M., Kobayasi, H. and Jeon, S.R. (1984) Chromosomes of *Coreoleuciscus splendidus* compared with those of *Pungtungia herzi.* Japan Women's Univ. J., 31: 109–111.

K- 11 Kasama, M. and Kobayasi, H. (1987) Hybridization experiments in Cyprinida (I). Cross between *Gnathopogon elongatus elongatus* and *Pseudorasbora parva.* Japan Women's Univ. J., 34: 107–112.

K- 12 Kasama, M. and Kobayasi, H. (1988) Hybridization experiments in Cyprinida (III). Cross between *Gnathopogon elongatus elongatus* and *Sarcocheilichthys biwaensis.* Japan Women's Univ. J., 35: 119–124.

K- 13 Kasama, M. and Kobayasi, H. (1989) Hybridization experiment between female crucian carp and male grass carp. Bull. Japan. Soc. Sci. Fish., 55: 1001–1006.

K- 14 Kasama, M. and Kobayasi, H. (1989) Hybridization experiments in Cyprinida (IV). Cross between *Gnathopogon elongatus elongatus* and *Squalidus gracilis.* Japan Women's Univ. J., 36: 101–106.

K- 15 Kashiwagi, E., Takai, A. and Ojima, Y. (2005) Chromosomal distribution of constitutive heterochromatin and nucleolus organizer regions in four *Dascyllus* fishes (Pomacentridae, Perciformes). Cytologia, 70: 345–349.

K- 16 Kato, F. and Ojima, Y. (1991) Karyotypes of four species of *Corydoras* catfishes (Siluriformes, Callichthyidae). Chrom. Inform. Serv., 52: 26–29.

K- 17 Kato, F. and Ojima, Y. (1991) A new report of the karyotype in a catfish *Corydoras aeneus* Gill. Chrom. Inform. Serv., 51: 29–30.

K- 18 Kaur, D. and Srivastava, M.D.L. (1965) The structure and behaviour of chromosomes in five fresh-water teleosts. Caryologia, 18: 181–191.

K- 19 Kavalco, K.F. and Moreira-Filho, O. (2003) Cytogenetical analyses in four species of the genus *Astyanax* (Pisces, Characidae) from Paraiba do Sul river basin. Caryologia, 56: 453–461.

K- 20 Kavalco, K.F., Pazza, R., Bertollo, L.A.C. and Moreira-Filho, O. (2005) Karyotypic diversity and evolution of Loricariidae (Pisces, Siluriformes). Heredity, 94: 180–186.

K- 21 Kavalco, K.F., Pazza, R., Bertollo, L.A.C. and Moreira-Filho, O. (2005) Molecular cytogenetics of *Oligosarcus hepsetus* (Teleostei, Characiformes) from two Brazilian locations. Genetica, 124: 85–91.

K- 22 Kawamura, K. and Hosoya, K. (1995) Karyotypic variation in Japanese rosy bitterling, *Rhodeus ocellatus kurumeus* (Jordan and Thompson). Bull. Natl. Res. Inst. Aquaculture, 24: 17–21.

K- 23 Kaydanova, T.I. (1975) A study of the karyotypes of the brown trout, *Salmo trutta* morpha *fario,* of the Ropsha population. J. Ichthyol., 15: 1007–1011.

K- 24 Kelly, R.K., Miller, H.R., Nielsen, O. and Clayton, J.W. (1980) Fish cell culture: characteristics of a continuous fibroblastic cell line from walleye (*Stizostedion vitreum vitreum*). Can. J. Fish. Aquat. Sci., 37: 1070–1075.

K- 25 Kendall, C., Valentino, S., Bodine, A.B. and Luer, C.A. (1994) Triploidy in a nurse shark, *Ginglymostoma cirratum*. Copeia, 1994: 825–827.

K- 26 Khuda-Bukhsh, A.R. and Manna, G.K. (1974) Somatic chromosomes in seven species of teleostean fishes. Chrom. Inform. Serv., 17: 5–6.

K- 27 Khuda-Bukhsh, A.R. (1975) Somatic chromosomes of an exotic fish *Puntius japonicus*. J. Cytol. Genet. (India) Congr. Suppl., 1975: 118–120.

K- 28 Khuda-Bukhsh, A.R. and Manna, G.K. (1976) Chromosome analyses in two species of Indian mullets (Pisces). Ind. Biologist, 8: 23–26.

K- 29 Khuda-Bukhsh, A.R. (1978) Somatic chromosomes of an estuarine fish, *Trypauchen vagina* (fam.: Gobiidae) from Sagar Island, West Bengal, India. Curr. Sci., 47: 137–138.

K- 30 Khuda-Bukhsh, A.R. (1979) Karyology of two species of hillstream fishes, *Barilius bendelesis* and *Rasbora daniconius* (fam.: Cyprinidae). Curr. Sci., 48: 793–794.

K- 31 Khuda-Bukhsh, A.R. (1979) Chromosomes in three species of fishes, *Aplocheilus panchax* (Cyprinodontidae), *Lates calcerifer* (Percidae) and *Gadusia chapra* (Clupeidae). Caryologia, 32: 161–169.

K- 32 Khuda-Bukhsh, A.R., Gupta, S.K. and Goswami, S. (1980) Karyotypic studies in *Garra lamta* and *Mystus cavassius* (Pisces). Proc. Ind. Acad. Sci., Sec. B, 89: 557–562.

K- 33 Khuda-Bukhsh, A.R. (1980) A high number of chromosomes in the hillstream cyprinid, *Tor putitora* (Pisces). Experientia, 36: 173–174.

K- 34 Khuda-Bukhsh, A.R., Chattopadhyay, G.K. and Das, S.K. (1981) Karyomorphology of two species of freshwater fishes *Chanda* (*Ambassis*) *nama* (fam: Centropomidae) and *Trichogaster lalius* (fam. Osphronemidae). Perspect. Cytol. Genet., 3: 141–144.

K- 35 Khuda-Bukhsh, A.R. and Nayak, K. (1981) Karyotypic studies in two species of *Barilius* (Cyprinidae: Pisces). Chrom. Inform. Serv., 30: 15–17.

K- 36 Khuda-Bukhsh, A.R. (1982) Karyomorphology of two species of *Tor* (Pisces; Cyprinidae) with a high number of chromosomes. Experientia, 38: 82–83.

K- 37 Khuda-Bukhsh, A.R. and Nayak, K. (1982) Karyomorphological studies in two species of hillstream fishes from Kashmir, India: occurrence of a high number of chromosomes. Chrom. Inform. Serv., 33: 12–14.

K- 38 Khuda-Bukhsh, A.R. (1984) Karyotype studies in two species of hill stream fishes, *Labeo diplostomus* and *Garra gotyla gotyla* (Pisces, Cyprinidae). Perspect. Cytol. Genet., 4: 421–424.

K- 39 Khuda-Bukush, A.R. and Barat, A. (1984) Somatic and germinal chromosomes of a gobiid, *Boleophthalmus dentatus* (Pisces). Life Sci. Adv., 3: 146–148.

K- 40 Khuda-Bukhsh, A.R. (1985) Cytological indications of polyploidy in some Himalayan hillstream fishes. Zool. Orientalis, 2: 93–99.

K- 41 Khuda-Bukhsh, A.R., Chanda, T. and Barat, A. (1986) Karyomorphology and evolution in some Indian hillstream fishes with particular reference to polyploidy in some species. In: Uyeno, T., Arai, R., Taniuchi, T. and Matsuura, K. eds. Indo-Pacific fish biology. Ichthyological Society of Japan, Tokyo, pp. 886–898.

K- 42 Khuda-Bukhsh, A.R. and Barat, A. (1987) Chromosomes in fifteen species of Indian teleosts (Pisces). Caryologia, 40: 131–144.

K- 43 Khuda-Bukhsh, A.R. and Chanda, T. (1989) Somatic chromosomes of three species of hillstream fishes from Assam. In: Das, P. and Jhingran, A.G. eds. Fish genetics in India. Today & Tomorrow's Printers and Publishers, New Delhi, pp. 69–73.

K- 44 Khuda-Bukhsh, A.R. and Tiwary, S. (1991) Localization of nucleolus organizer regions (NORs) in the metaphase chromosomes of the Indian climbing perch, *Anabas testudineus* (Pisces). Proc. Zool. Soc. Calcutta, 44: 101–105.

K- 45 Khuda-Bukhsh, A.R., Rahman, A. and Chanda, T. (1992) A study of chromosomes in three species of hillstream fish belonging to the genus *Barilius* (Cyprinidae: Pisces). Proc. Nat. Acad. Sci. India, 62(B): 199–203.

K- 46 Khuda-Bukhsh, A.R., Rahman, A., Chanda, T., Nayak, K. and Khuda-Bukhsh, A. (1995) Diploid numbers and chromosome formulae of some 29 species of Indian teleosts (Pisces). Chrom. Inform. Serv., 58: 38–39.

K- 47 Kikuma, Y., Inafuku, J., Kubota, S. and Kohno, S.-I. (2000) Banding karyotype and 5S ribosomal DNA loci in the Japanese bitterling, *Rhodeus ocellatus* (Cyprinidae). Chrom. Sci., 3: 101–103.

K- 48 Kikuno, T., Ojima, Y. and Yamashita, N. (1986) Chromosomes of flounder, *Paralichthys olivaceus*. Proc. Japan Acad., Ser. B, 62: 194–196.

K- 49 Kikuno, T. and Ojima, Y. (1987) A karyotypic study of a guitarfish, *Rhinobatus hynnicephalus* Richardson (Pisces, Rajiformes). La Kromosomo, II, 47/48: 1538–1544.

K- 50 Kikuno, T. and Ojima, Y. (1987) Karyotypic study of two species in Atheriniformes and Perciformes (Teleostei, Pisces). Proc. Japan Acad., Ser. B, 63: 205–207.

K- 51 Kiliç Demirok, N. and Ünlü, E. (2001) Karyotypes of cyprinid fish *Capoeta trutta* and *Capoeta capoeta umbla* (Cyprinidae) from the Tigris River. Turk. J. Zool., 25: 389–393.

K- 52 Kim, D.S., Park, E.H. and Kim, J.S. (1982) Karyotypes of nine species of the Korean catfishes (Teleostomi: Siluriformes). Korean J. Genet., 4: 57–68.

K- 53 Kim, D.S., Jeon, I.G. and Lee, J.K. (1989) Karyotypes, DNA values and nuclear sizes of several scups (Teleostomi: Perciformes). Korean J. Ichthyol., 1: 35–41. (In Korean with English abstract)

K- 54 Kim, D.S., Nam, Y.K., Noh, J.K., Park, C.H. and Chapman, F.A. (2005) Karyotype of North American shortnose sturgeon *Acipenser brevirostrum* with the highest chromosome number in the Acipenseriformes. Ichthyol. Res., 52: 94–97.

K- 55 Kim, I.S., Lee, G.Y. and Yang, S.Y. (1985) Systematic study of the subfamily Leuciscinae (Cyprinidae) from Korea. Bull. Korean Fish. Soc., 18: 381–400. (In Korean with English abstract)

K- 56 Kim, I.S. and Lee, J.H. (1986) A chromosomal study on the genus *Cobitis* (Pisces: Cobitidae) in the southern part of Korea. Bull. Korean Fish. Soc., 19: 257–264. (In Korean with English abstract)

K- 57 Kim, I.S. and Moon, K.C. (1987) The karyotype of a ricefish, *Oryzias latipes* from southern Korea. Korean J. Zool., 30: 379–386. (In Korean with English abstract)

K- 58 Kim, I.S., Lee, E.H. and Son, Y.M. (1988) Morphological variation and geographic distribution of two species of Nemacheiline loaches (Pisces, Cobitidae) from Korea. Korean J. Zool., 31: 283–294. (In Korean with English abstract)

K- 59 Kim, I.S. and Shim, J.H. (1991) Karyotypes of the genus *Pseudopungtungia* (Pisces, Cyprinidae). Korean J. Zool., 34: 54–58.

K- 60 Kimizuka, Y., Kobayasi, H. and Mizuno, N. (1982) Geographic distributions and karyotypes of *Cobitis takatsuensis* and *Niwaella delicata* (Cobitididae). Japan. J. Ichthyol., 29: 305–310.

K- 61 Kimizuka, Y. and Kobayasi, H. (1983) Geographic distributions of karyological races of *Cobitis biwae* (Cobitididae). Japan. J. Ichthyol., 30: 308–312.

K- 62 Kingston, D.I. (1978) *Skiffia francesae*, a new species of goodeid fish from Western Mexico. Copeia, 1978: 503–508.

K- 63 Kirby, R.F., Thompson, K.W. and Hubbs, C. (1977) Karyotypic similarities between the Mexican and blind tetras. Copeia, 1977: 578–580.

K- 64 Kirpichnikov, V.S. (ed) (1980) Karyological variability, mutagenesis and gynogenesis in fishes. Inst. Cytol. Acad. Sci. USSR, Leningrad. (In Russian with English summary)

K- 65 Kitada, J. and Tagawa, M. (1972) On the chromosomes of the rice-field eel (*Fluta alba* = *Monopterus albus*). La Kromosomo, 88/89: 2804–2807. (In Japanese with English summary)

K- 66 Kitada, J. and Tagawa, M. (1973) On the chromosomes of two species of Cyclostomata. La Kromosomo, 91: 2913–2916. (In Japanese with English summary)

K- 67 Kitada, J. and Tagawa, M. (1975) Somatic chromosomes of three species of Cyclostomata. Chrom. Inform. Serv., 18: 10–12.

K- 68 Kitada, J., Tatewaki, R. and Tagawa, M. (1979) The karyotype of the common aigo, *Siganus fuscescens*. Chrom. Inform. Serv., 27: 12–13.

K- 69 Kitada, J., Tatewaki, R. and Tagawa, M. (1980) Chromosomes of the pond smelt, *Hypomesus transpacificus nipponensis* M. Chrom. Inform. Serv., 28: 8–9.

K- 70 Kitada, J., Tatewaki, R. and Tagawa, M. (1981) Karyotype analysis in two species of *Lateolabrax* (Pisces: Perciformes: Serranidae). Chrom. Inform. Serv., 30: 9–11.

K- 71 Kitayama, E. and Ojima, Y. (1984) A preliminary report on the phylogeny of five balistid fish in terms of several chromosome banding techniques in combination with a statistical analysis. Proc. Japan Acad., Ser. B, 60: 58–61.

K- 72 Kligerman, A.D. and Bloom, S.E. (1977) Distribution of F-bodies, heterochromatin, and nucleolar organizers in the genome of the central mudminnow, *Umbra limi*. Cytogenet. Cell Genet., 18: 182–196.

K- 73 Kligerman, A.D. and Bloom, S.E. (1977) Rapid chromosome preparations from solid tissues of fishes. J. Fish. Res. Board Can., 34: 266–269.

K- 74 Klinkhardt, M.B. (1989) Investigations into karyology and fecundity of *Pomatoschistus microps* (Krøyer) and *P. minutus* (Pallas) (Teleostei, Gobiidae) living in an inner coastal water of the South-West Baltic Sea. Zool. Anz., 222: 177–190.

K- 75 Klinkhardt, M.B. and Buuk, B. (1990) Karyological studies in several species of freshwater fishes from brackish coastal waters of Southwestern Baltic. III. The three-spined stickleback (*Gasterosteus aculeatus* Linnaeus, 1758) and the nine-spined stickleback (*Pungitius pungitius* (Linnaeus, 1758)). Zool. Anz., 225: 341–352.

K- 76 Klinkhardt, M.B. and Buuk, B. (1990) Karyological studies in several species of freshwater fishes from brackish coastal waters of Southwestern Baltic. II. The roach (*Rutilus rutilus* (Linnaeus, 1758)). Zool. Anz., 224: 359–368.

K- 77 Klinkhardt, M.B. (1990) Karyological studies in several species of freshwater fishes from brackish coastal waters of Southwestern Baltic. I. The ruffe (*Gymnocephalus cernua* (Linnaeus, 1758)). Zool. Anz., 224: 156–164.

K- 78 Klinkhardt, M.B. and Buuk, B. (1991) Karyological studies in several species of freshwater fishes from brackish coastal waters of Southwestern Baltic. IV. The perch (*Perca fluviatilis* Linnaeus, 1758). Zool. Anz., 227: 38–43.

K- 79 Klinkhardt, M.B. (1991) A brief comparison of methods for preparing fish chromosomes: an overview. Cytobios, 67: 193–208.

K- 80 Klinkhardt, M.B. (1991) Comparison of the karyology (C-banding, Ag-NOR) of eelpout (*Zoarces viviparus*) from the White Sea and the Baltic Sea. Cytobios, 68: 29–36.

K- 81 Klinkhardt, M.B. (1993) Cytogenetics of the herring (*Clupea harengus* L.). 1 Karyotypes of North European stocks. Cytobios, 75: 113–128.

K- 82 Klinkhardt, M.B. (1993) Cytogenetics of the herring (*Clupea harengus* L.). 2 Karyotypes of White Sea herring groups. Cytobios, 75: 149–156.

K- 83 Klinkhardt, M.B. (1994) Karyotypic divergence between species of Gadidae (Pisces, Gadiformes). Cytobios, 77: 207–214.

K- 84 Kobayasi, H. (1971) A cytological study on gynogenesis of the triploid ginbuna (*Carassius auratus langsdorfii*). Zool. Mag. Japan, 80: 316–322. (In Japanese with English abstract)

K- 85 Kobayasi, H., Ochi, H. and Takeuchi, N. (1973) Chromosome studies of the silver crucian carps (*Carassius auratus gibelio*) from the valley of the Amur river, and their progenies. Japan Women's Univ. J., 20: 83–88. (In Japanese with English summary)

K- 86 Kobayasi, H., Ochi, H. and Takeuchi, N. (1973) Chromosome studies in the genus *Carassius*: comparison of *C. auratus grandoculis*, *C. auratus buergeri*, and *C. auratus langsdorfii*. Japan. J. Ichthyol., 20: 7–12. (In Japanese with English abstract)

K- 87 Kobayasi, H. and Nakamura, S. (1974) Comparative studies of the erythrocytes between the diploid and the triploid funa, *Carassius auratus*. Japan Women's Univ. J., 21: 155–163. (In Japanese)

K- 88 Kobayasi, H. and Kimura, Y. (1975) On the chromosomes of leucocytes from Japanese and European eels (*Anguilla*). Japan Women's Univ. J., 22: 169–174. (In Japanese with English summary)

K- 89 Kobayasi, H. (1975) Some tissue culture methods for chromosome studies of freshwater fishes. Japan Women's Univ. J., 22: 159–167. (In Japanese with English summary)

K- 90 Kobayasi, H. and Nakano, K. (1976) Morphological and cytological studies in backcross hybrids of F_1 fishes between the ginbuna (*Carassius auratus langsdorfii*) female and the carp (*Cyprinus carpio*) male. Japan Women's Univ. J., 23: 113–122. (In Japanese with English summary)

K- 91 Kobayasi, H. (1976) Comparative study of karyotypes in the small and large races of spinous loaches (*Cobitis biwae*). Zool. Mag. Japan, 85: 84–87. (In Japanese with English abstract)

K- 92 Kobayasi, H., Nakano, K. and Nakamura, M. (1977) On the hybrids, 4n ginbuna (*Carassius auratus langsdorfii*) × kinbuna (*C. auratus* subsp.), and their chromosomes. Bull. Japan. Soc. Sci. Fish., 43: 31–37.

K- 93 Kobayasi, H. (1977) A chromosome study on *Anguilla marmorata* Quoy et Gaimard. Japan Women's Univ. J., 24: 135–137. (In Japanese with English summary)

K- 94 Kobayasi, H. (1982) The distribution of polyploid funa (*Carassius auratus*) in and around Japan. Japan Women's Univ. J., 29: 145–161. (In Japanese with English summary)

K- 95 Kobayasi, H. (1985) The distribution of polyploid funa (*Carassius auratus*) in and around Japan and their origin. Mar. Sci. Monthly, 17(2): 75–81. (In Japanese)

K- 96 Kong, X.Y., Yu, Z.I., Xie, Z.Y. and Xu, W.W. (1994) Studies on karyotype and Ag-NOR banding of *Platycephalus indicus* (LINNAEUS). J. Ocean Univ. Qingdao, 24(3): 344–348. (In Chinese with English abstract)

K- 97 Koref-Santibanez, S., Paepke, H. and Tjio, H. (1991) Karyotypes of the species of the genus *Macropodus* Lac. (Pisces, Anabantoidei). Zool. Anz., 227: 271–278.

K- 98 Kornfield, I., Ritte, U., Richler, C. and Wahrman, J. (1979) Biochemical and cytological differentiation among cichlid fishes of the sea of Galilee. Evolution, 33: 1–14.

K- 99 Kornfield, I. (1981) Distribution of constitutive heterochromatin and the evolution of sex chromosomes in *Fundulus*. Copeia, 1981: 916–918.

K- 100 Korth, J.W. and Fitzsimons, J.M. (1987) Karyology of three species of eastern North American atherinid fishes. Copeia, 1987: 505–509.

K- 101 Krishna, G. and Lakra, W.S. (1994) Karyotype of a marine catfish, *Arius caelatus* (Ariidae, Siluriformes). Chrom. Inform. Serv., 56: 10–12.

K- 102 Krishnaja, A.P. and Rege, M.S. (1980) Some observations on the chromosomes of certain teleosts using a simple method. Ind. J. Exp. Biol., 18: 268–270.

K- 103 Krishnaja, A.P. and Rege, M.S. (1983) A cytogenetic study on the *Gambusia affinis* population from India. Cytologia, 48: 47–49.

K- 104 Krysanov, E.Yu. and Elovenko, V.N. (1981) Structural peculiarities of the karyotype of the Amur "sleeper" *Perccottus glehni* (Eleotridae) in different parts of its range. J. Ichthyol., 21(5): 124–127.

K- 105 Kawahara, R., Miya, M., Mabuchi, K., Lavoué, S., Inoue, J.G., Satoh, T.P., Kawaguchi, A. and Nishida, M. (2008) Interrelationships of the 11 gasterosteiform families (sticklebacks, pipefishes, and their relatives): a new perspective based on whole mitogenome sequences from 75 higher teleosts. Mol. Phylogenet. Evol., 46: 224–236.

K- 106 Krysanov, E.Yu. (1992) Aneuploidy in postnatal ontogenesis of fishes. Acta Zool. Fennica, 191: 177–182.

K- 107 Krysanov, E.Yu. and Golubtsov, A.S. (1993) Karyotypes of three *Garra* species from Ethiopia. J. Fish Biol., 42: 465–467.

K- 108 Krysanov, E.Yu. and Golubtsov, A.S. (1996) Karyotypes of some Ethiopian *Barbus* and *Varicorhinus* from the Nile basin including Lake Tana morphotypes. Folia Zool., 45 (suppl. 1): 67–75.

K- 109 Krysanov, E.Yu. (1999) Karyotypes of *Varicorhinus capoeta* and *Barbus goktschaicus* (Cypriniformes) from Lake Sevan, Armenia. J. Ichthyol., 39(2): 187–189.

K- 110 Krysanov, E.Yu. and Golubtsov, A.S. (2001) Karyotypes of eight species of the African climbing gouramies of the genera *Ctenopoma* and *Microctenopoma* (Anabantidae, Perciformes) with comments on their phylogenetic relationships. J. Ichthyol., 41: 698–702.

K- 111 Kushwaha, B., Srivastava, S.K., Nagpure, N.S., Ogale, S.N. and Ponniah, A.G. (2001) Cytogenetic studies in two species of mahseer, *Tor khudree* and *Tor mussullah* (Cyprinidae, Pisces) from India. Chrom. Sci., 5: 47–50.

K- 112 Kusunoki, T., Arai, K. and Suzuki, R. (1994) Production of viable gynogens without chromosome duplication in the spinous loach *Cobitis biwae*. Aquaculture, 119: 11–23.

*K-113 Kossowski, C., Bracamonte, N.O. and Velazko, J.Q. (1983) Cariotipo del hibrido de *Calossoma macropomus* (Hembra)(Cuvier) 1818 × *Mylossoma duriventris* (Macho)(Cuvier) 1818 y sus progenitores (Pisces, Cypriniformes, Characidae). Acta Cient. Venezolana, 34: 173–175. (After K-114)

K- 114 Klinkhardt, M.B., Tesche, M. and Greven, H. (1995) Database of fish chromosomes. Westarp Wissenschaften, Magdeburg.

K- 115 Kobayasi, H., Kawashima, Y. and Takeuchi, N. (1970) Comparative chromosome studies in the genus *Carassius*, especially with a finding of polyploidy in the ginbuna (*C. auratus langsdorfii*). Japan. J. Ichthyol., 17: 153–160. (In Japanese with English abstract)

K- 116 Kim, I.S. and Lee, J.H. (1990) Diploid-triploid hybrid complex of the spined loach *Cobitis sinensis* and *C. longicorpus* (Pisces: Cobitidae). Korean J. Ichthyol., 2: 203–210. (In Korean with English abstract)

K- 117 Kobayasi, H. (1980) Distribution in Japan of polyploid funa (*Carassius auratus*) and *Cobitis*. Animals and Nature, 10(13): 14–18. (In Japanese)

K- 118 Kim, D.S. and Park, I.S. (1990) Chromosomal polymorphism of Coho salmon, *Oncorhynchus kisutch*. Korean J. Ichthyol., 2: 211–216.

K- 119 Kalous, L., Doadrio, I., Rábová, M. and Ráb, P. (2008) Note on the karyotype and NOR phenotype of the cyprinid fish *Parachondrostoma arrigonis*. Cybium, 32: 211–215.

K- 120 Kornfield, I. (1984) Descriptive genetics of cichlid fishes. In: Turner, B.J. ed. Evolutionary genetics of fishes. Plenum Press, New York, pp. 591–616.

K- 121 Kavalco, K.F., Brandão, K.O., Pazza, R. and Almeida-Toledo, L.F. (2009) *Astyanax hastatus* Myers, 1928 (Teleostei, Characidae): a new species complex within the genus *Astyanax*? Genet. Mol. Biol., 32: 477–483.

K- 122 Karel, H., Vetešnik, L., Lusk, S., Mendel, J. and Papoušek, I. (2007) The $X_1X_1X_2X_2/X_1X_2Y$ multiple sex chromosome system in the *Zingel zingel* (Pisces: Perciformes) from the Morava river (Czech Republic). Caryologia, 60: 222–225.

K- 123 Kakagawa, S.M.H., Portela-Castro, M.A.L.B. and Martins-Santos, I.C. (2007) A study of nucleolus organizing regions by Ag-staining, *in situ* hybridization and chromomycin A_3 techniques in *Tilapia rendalli* (Cichlidae, Perciformes) of the River Iguaçu basin, Paraná, Brazil. Cytologia, 72: 373–378.

K- 124 Kobayasi, H. and Kawashima, Y. (1972) On the chromosomes of all-female population in the ginbuna, *Carassius auratus langsdorfii*. Japan Women's Univ. J., 19: 259–263.

K- 125 Kim, E.M., An, H.S. and Park, I.S. (2007) Cytogenetic study of *Pleuronectes obscurus*, *Konosirus punctatus* and *Pseudoblennius percoides*. J. Fish. Sci. Technol., 10: 24–29.

K- 126 Kavalco, K.F., Pazza, R. and Almeida-Toledo, L.F. (2009) *Astyanax bockmanni* Vari and Castro, 2007: an ambiguous karyotype in the *Astyanax* genus. Genetica, 136: 135–139.

K- 127 Kim, S.Y., Park, J.Y. and Kim, I.S. (1999) Chromosomes of spined loach, *Iksookimia yongdokensis* (Pisces: Cobitidae) from Korea. Korean J. Ichthyol., 11: 172–176. (In Korean with English abstract)

K- 128 Kim, I.S. and Kim, E.J. (2008) Karyotype of dwarf loach, *Kichulchoia brevifasciata* (Pisces: Cobitidae) from Korea. Korean J. Ichthyol., 20: 61–65.

K- 129 Kim, I.S., Park, J.Y. and Nalbant, T.T. (1999) The Far-East species of the genus *Cobitis* with the description of three new taxa (Pisces: Ostariophysi: Cobitidae). Travaux Mus. Natl. Hist. Nat. "Grigore Antipa", 41: 373–391.

K- 130 Kottelat, M. and Freyhof, J. (2007) Handbook of European freshwater fishes. Publications Kottelat, Cornol, Switzerland.

K- 131 Klinkhardt, M. (1992) Chromosome structures of four Norwegian gobies (Gobiidae, Teleostei) and a hypothetical model of their karyo-evolution. Chromatin, 1: 169–183.

*K–132 Klinkhardt, M. (1993) Zur Zytogenetik von *Platichthys flesus* und *Limanda limanda* (Pleuronectidae, Teleostei) und Karyoevolution der Pleuronectiformes. Z. Fischkunde, 2: 65–75. (After K-114)

*K–133 Knezevic, B., Kavaric, M. and Fontana, F. (1976) Chromosome morphology of *Pachychilon pictum* (Heckel et Kner) (Cyprinidae, Pisces) from Shadar Lake. Poljoprivreda i Sumarstvo, 22: 85–89. (After K-114)

*K–134 Koref-Santibanez, S. and Paepke, H.J. (1994) Karyotypes of the Trichogasterinae Liem (Teleostei, Anabantoidei). Abstr. VIII Congr. Soc. Europ. Ichthyol., Oviedo, p. 55. (After K-114)

K- 135 Koehler, M.R., Dehm, D., Guttenbach, M., Nanda, I., Haaf, T., Molina, W.F., Galetti Jr., P.M. and Schmid, M. (1997) Cytogenetics of the genus *Leporinus* (Pisces, Anostomidae). 1. Karyotype analysis, heterochromatin distribution and sex chromosomes. Chrom. Res., 5: 12–22.

K- 136 Khuda-Bukhsh, A.R. and Das, J.K. (2007) Cytogenetic analyses in eight species of teleostean fishes (Pisces): karyotypes, multiple Ag-NORs, sex chromosomes. Res. Rev. BioSciences (India), 1: 47–52.

K- 137 Kumar, R., Kushwaha, B., Nagpure, N.S., Behera, B.K., Srivastava, S.K. and Lakra, W.S. (2009) Physical mapping of rRNA gene in endangered fish *Osteobrama belangeri* (Valenciennes, 1844) (Family: Cyprinidae). Ind. J. Exp. Biol., 47: 597–601.

K- 138 Kushwaha, B., Kumar, R., Nagpure, N.S., Behera, B.K. and Lakra, W.S. (2008) Chromosome banding in two freshwater gouramies from north eastern India. Nucleus, 51: 61–72.

K- 139 Khuda-Bukhsh, A.R. and Nayak, K. (1990) Karyotypic studies in six species of brackish water fishes from India. La Kromosomo, II, 58: 1955–1960.

K- 140 Kojima, N.F., Kojima, K.K., Kobayakawa, S., Higashide, N., Hamanaka, C., Nitta, A., Koeda, I., Yamaguchi, T., Shichiri, M., Kohno, S. and Kubota, S. (2010) Whole chromosome elimination and chromosome terminus elimination both contribute to somatic differentiation in Taiwanese hagfish *Paramyxine sheni*. Chrom. Res., 18: 383–400.

K- 141 Kohn, M., Högel, J., Vogel, W., Minich, P., Kehrer-Sawatzki, H., Graves, J.A.M. and Hameister, H. (2006) Reconstruction of a 450-My-old ancestral vertebrate protokaryotype. Trends Genet., 22: 203–210.

[L]

L- 1 Lakra, W.S. and Rishi, K.K. (1991) Chromosomes of Indian fishes: an annotated list. Ind. J. Anim. Sci., 61: 342–349.

L- 2 Lakra, W.S. and Krishna, G. (1994) G-banded karyotypes of three species of Indian major carps. Chrom. Inform. Serv., 57: 23–25.

L- 3 Lakra, W.S. and Krishna, G. (1995) Karyology of grey mullet, *Mugil cephalus* (Teleostei, Mugiliformes) from India. Chrom. Inform. Serv., 58: 17–18.

L- 4 Lakra, W.S. (1996) Cytogenetic studies on endangered fish species: 1 Karyotypes of three species of mahseers, *Tor putitora*, *T. tor* and *T. khudree* (Cyprinidae: Pisces). Cytobios, 85: 205–218.

L- 5 Laliberte, M.F., Lafaurie, M., Lambert, J.C. and Ayraud, N. (1979) Etude preliminaire du caryotype de "*Mullus barbatus* Linne". Rapp. Comm. Int. Mer. Médit., 25/26: 125–130.

L- 6 Law, W.M., Ellender, R.D., Wharton, J.H. and Middlebrooks, B.L. (1978) Fish cell culture: properties of a cell line from the sheepshead, *Archosargus probatocephalus*. J. Fish. Res. Board Can., 35: 470–473.

L- 7 Le Comber, S.C. and Smith, C. (2004) Polyploidy in fishes: patterns and processes. Biol. J. Linn. Soc., 82: 431–442.

L- 8 Lee, M.H. and Loh, P.C. (1975) Some properties of an established fish cell line from the marine fish, *Caranx mate* (Omaka). Proc. Soc. Exper. Biol. Med., 150: 40–48.

L- 9 Lee, G.Y., So, J.N. and Kim, S.Z. (1982) Chromosomes and arm number of the acheilognathine fishes (Cyprinidae) in Korea. Basic Sci. Rev., Jeonbug Natn. Univ., 5: 61–69. (In Korean with English abstract)

L- 10 Lee, G.Y. (1982) Karyotypes of the acheilognathine fishes (Cyprinidae) in Korea (I). Ann. Rep. Biol. Res., Jeonbug Natn. Univ., 3: 19–24. (In Korean with English abstract)

L- 11 Lee, G.Y. (1983) Karyotypes of the acheilognathine fishes (Cyprinidae) in Korea (II). Ann. Rep. Biol. Res., Jeonbug Natn. Univ., 4: 1–9. (In Korean with English abstract)

L- 12 Lee, G.Y., Jang, S.I. and Yun, M.J. (1986) Karyotypes of nine species in the family Cyprinidae fishes from Korea. Korean J. Limnol., 19: 59–69. (In Korean with English abstract)

L- 13 Lee, G.Y., Jang, S.I., Yun, M.J., So, J.N., Jeon, S.R., Kang, Y.H., Itai, T. and Kimizuka, Y. (1987) Karyotypes of four species in genus *Moroco* (Pisces: Cyprinidae) from Korea and Japan. Korean J. Limnol., 20: 49–60. (In Korean with English abstract)

L- 14 Lee, G.Y., Jang, S.I., Yun, M.J. and So, J.N. (1987) Karyotypes of sixteen species of gobiid fishes from Korea. Basic Sci. Rev., Jeonbug Natn. Univ., 10: 1–14.

L- 15 Lee, H.Y., Yu, C.H., Jeon, S.K. and Lee, H.S. (1983) The karyotype analysis on 29 species of fresh water fish in Korea. Bull. Inst. Basic. Sci., Inha Univ., 4: 79–86. (In Korean with English abstract)

L- 16 Legendre, P. and Steven, D.M. (1969) Dénombrement des chromosomes chez quelques cyprins. Naturaliste can., 96: 913–918.

L- 17 LeGrande, W.H. (1975) Karyology of six species of Louisiana flatfishes (Pleuronectiformes: Osteichthyes). Copeia, 1975: 516–522.

L- 18 LeGrande, W.H. and Fitzsimons, J.M. (1976) Karyology of the mullets *Mugil curema* and *M. cephalus* (Perciformes: Mugilidae) from Louisiana. Copeia, 1976: 388–391.

L- 19 LeGrande, W.H. (1980) The chromosome complement of *Arius felis* (Siluriformes, Ariidae). Japan. J. Ichthyol., 27: 82–84.

L- 20 LeGrande, W.H. and Cavender, T.M. (1980) The chromosome complement of the stonecat madtom, *Noturus flavus* (Siluriformes: Ictaluridae), with evidence for the existence of a possible chromosomal race. Copeia, 1980: 341–344.

L- 21 LeGrande, W.H. (1981) Chromosomal evolution in North American catfishes (Siluriformes: Ictaluridae) with particular emphasis on the madtoms, *Noturus*. Copeia, 1981: 33–52.

L- 22 LeGrande, W.H., Dunham, R.A. and Smitherman, R.O. (1984) Karyology of three species of catfishes (Ictaluridae: *Ictalurus*) and four hybrid combinations. Copeia, 1984: 873–878.

L- 23 LeGrande, W.H. and Fitzsimons, J.M. (1988) Chromosome numbers of some Gulf coast sciaenid fishes. Copeia, 1988: 491–493.

L- 24 Lavoué, S., Miya, M., Poulsen, J.Y., Møller, P.R. and Nishida, M. (2008) Monophyly, phylogenetic position and inter-familial relationships of the Alepocephaliformes (Teleostei) based on whole mitogenome sequences. Mol. Phylogenet. Evol., 47: 1111–1121.

L- 25 Levan, A., Fredga, K. and Sandberg, A.A. (1964) Nomenclature for centromeric position on chromosomes. Hereditas, 52: 201–220.

L- 26 Levin, C.B. and Foster, N.R. (1972) Cytotaxonomic studies in Cyprinodontidae: multiple sex chromosomes in *Garmanella pulchra*. Notulae Naturae, 446: 1–5.

L- 27 Li, K., Li, Y.C. and Zhou, T. (1983) A comparative study of the karyotypes in two species of mud loaches. Zool. Res., 4(1): 75–81. (In Chinese with English abstract)

L- 28 Li, K., Li, Y.C., Zhou, M. and Zhou, T. (1983) Studies on the karyotypes of Chinese cyprinid fishes. II. Karyotypes of four species of Xenocyprininae. Acta Zool. Sinica, 29: 207–213. (In Chinese with English abstract)

L- 29 Li, K., Gui, J.F., Hong, Y.H., Li, Y.C. and Zhou, T. (1984) Studies on the karyotypes of Chinese cyprinid fishes. V. Karyotypes of species of gobionine fishes. J. Wuhan Univ. (Nat. Sci.), 1984: 113–122. (In Chinese with English abstract)

L- 30 Li, K., Li, Y.C. and Zhou, T. (1985) Cytological evidence for sex determination in *Leiobagrus marginatus*. Hereditas (Beijing), 7(3): 23–24. (In Chinese)

L- 31 Li, K., Li, Y.C. and Zhou, T. (1985) Studies on the karyotypes and C-banding patterns of three species in Channidae (Pisces). Acta Genet. Sinica, 12: 470–477. (In Chinese with English abstract)

L- 32 Li, K., Li, Y.C., Gui, J.F., Xie, X.Z. and Zhou, T. (1986) Studies on the karyotypes of Chinese cyprinid fishes. IX. Karyotypes of nine species of Abramidinae and one species of Xenocyprininae. Acta Hydrobiol. Sinica, 10: 189–193. (In Chinese)

L- 33 Li, M.F. and Clyburne, S. (1977) New cell line from the marine lumpfish, *Cyclopterus lumpus*. J. Fish. Res. Board Can., 34: 134–139.

L- 34 Lockwood, S.F., Seavey, B.T., Dillinger Jr., R.E. and Bickham, J.W. (1991) Variation in DNA content among age classes of broad whitefish (*Coregonus nasus*) from the Sagavanirktok River delta. Can. J. Zool., 69: 1335–1338.

L- 35 Li, S.S., Wang, R.F., Liu, G.Z., Wang, Y.X. and Li, C.Y. (1981) A karyotype study of *Coreglanis kishinouyei* and *Euchiloglanis davidi*. Trans. Chinese Ichthyol. Soc., 2: 153–156. (In Chinese with English abstract)

L- 36 Li, S.S., Wang, R.F., Liu, G.Z., Wang, Y.X. and Li, C.Y. (1981) Investigation on karyotype in *Liobagrus marginatus*. Hereditas (Beijing), 3(1): 23–24. (In Chinese)

L- 37 Li, S.S., Wang, R.F., Liu, G.Z., Wang, Y.X. and Li, C.Y. (1983) A karyotypical study of *Myxocyprinus asiaticus* (Bleeker). Zool. Res., 4: 182. (In Chinese)

L- 38 Li, S.S., Wang, R.F., Liu, G.Z., Wang, Y.X. and Li, C.Y. (1983) A karyotype study of eight species of teleost fishes. Hereditas (Beijing), 5(4): 25–28. (In Chinese)

L- 39 Li, Y.C., Li, K. and Zhou, T. (1982) A study of the karyotype of the rice-field eel. J. Wuhan Univ. (Nat. Sci.), 1982(1): 55–58, 116. (In Chinese with English abstract)

L- 40 Li, Y.C., Li, K. and Zhou, T. (1983) Studies on the karyotypes of Chinese cyprinid fishes. I. Karyotypes of ten species of Abramidinae. Acta Genet. Sinica, 10(3): 216–222. (In Chinese with English abstract)

L- 41 Li, Y.C., Li, K. and Zhou, T. (1983) Cellular DNA content of fourteen species of freshwater fishes. Acta Genet. Sinica, 10(5): 384–389. (In Chinese with English abstract)

L- 42 Li, Y.C., Li, K., Hong, Y.H., Gui, J.F. and Zhou, T. (1985) Studies on the karyotypes of Chinese cyprinid fishes. VII. Karyotypic analyses of seven species in the subfamily

Leuciscinae with a consideration for the phylogenetic relationships of some cyprinid fishes concerned. Acta Genet. Sinica, 12(5): 367-372. (In Chinese with English abstract)

L- 43 Li, Y.C., Li, K., Jiang, J.Q., Sun, Q.P. and Zhou, T. (1986) Studies on the karyotypes of Chinese cyprinid fishes. X. Karyotypes of five species of Barbinae and four species of Gobioninae. Zool. Res., 7(2): 183–189. (In Chinese with English abstract)

L- 44 Li, Y.C. and Li, K. (1987) Karyotypes of *Opsariichthys bidens* and *Misgurnus anguillicaudatus* (Pisces), with a consideration of the relationship between polymorphism of chromosome number and taxonomy of fish. J. Wuhan Univ. (Nat. Sci.), 1987(1): 107–112, 132. (In Chinese with English abstract)

L- 45 Li, Y.C., Li, K., Gui, J.F. and Zhou, T. (1987) Studies on the karyotypes of Chinese cyprinid fishes XI. Karyotypes of two species of Schizothoracinae and three species of Gobiobotinae. Acta Hydrobiol. Sinica, 11: 184–186. (In Chinese)

L- 46 Li, Y.C. and Gold, J.R. (1991) Cytogenetic studies in North American minnows (Cyprinidae). XXII. Chromosomal nucleolar organizer regions in the genus *Pimephales*. Can. J. Zool., 69: 2826–2830.

L- 47 Li, X.Q. and Peng, Y.D. (1994) Studies on karyotype of *Epinephelus fasciatomaculatus* and *Epinephelus fasciatus*. J. Zhanjiang Fish. Coll., 1994(2): 22–26. (In Chinese with English abstract)

L- 48 Libertini, A. and Meneghetti, F. (1998) Chromosome complement, Ag-NOR and C-banding in *Dentex dentex* L. (Sparidae, Perciformes). Chrom. Sci., 2: 117–118.

L- 49 Lieder, U. (1963) Über den gegenwärtigen Stand und die Methodik der Chromosomenuntersuchungen bei Fischen. Z. Fisch., 11: 673–684.

L- 50 Lieppman, M. and Hubbs, C. (1969) A karyological analysis of two cyprinid fishes, *Notemigonus crysoleucas* and *Notropis lutrensis*. Texas Rep. Biol. Med., 27: 427–435.

L- 51 Lima, N.R.W. and Galetti Jr., P.M. (1990) Chromosome characterization of the fish *Trichogenes longipinnis*. A possible basic karyotype of Trichomycteridae. Brazil. J. Genet., 13: 239–245.

L- 52 Lin, Y. and Ojima, Y. (1987) Studies on developing high resolution BG- and B-banding in fish chromosomes. La Kromosomo, II, 47/48: 1551–1556.

L- 53 Ling, J.X. (1982) Studies on the karyotypes of eight species of fish. J. Wuhan Univ. (Nat. Sci.), 1982(2): 109–112, 121–122. (In Chinese with English abstract)

L- 54 Liu, J. and Tian, M.C. (1996) A karyotype study on the mullet fish (*Mugil cephalus*). Chinese J. Zool., 31(1): 6–7. (In Chinese with English abstract)

L- 55 Liu, J.D., Yi, M.S., Zhao, G., Zhou, F., Wang, D.Q. and Yu, Q.X. (2002) Sex chromosomes in the spiny eel (*Mastacembelus aculeatus*) revealed by mitotic and meiotic analysis. Cytogenet. Genome Res., 98: 291–297.

L- 56 Liu, L.G., Zhao, J., Chen, X.L., Dong, C.W. and Cui, M. (2004) Karyotypic analysis from two clones of gynogenetic Pengze crucian carp (*Carassius auratus* of Pengze). Acta Genet. Sinica, 31: 780–786. (In Chinese with English abstract)

L- 57 Liu, L.S., Yang, J.H., Lin, J.H. and Hong, M.X. (1991) A study on the karyotype of *Sparus latus*. Chinese J. Zool., 26(1): 14–16. (In Chinese)

L- 58 Liu, L.Y. (1980) On karyotype of the grass-carp, *Ctenopharyngodon idellus*. Acta Zool. Sinica, 26: 126–131. (In Chinese with English abstract)

L- 59 Liu, L.Y. (1981) Karyotype analysis of *Aristichthys nobilis* by the leucocyte culture method. J. Beijing Normal Univ. (Nat. Sci.), 3: 79–83. (In Chinese with English abstract)

L- 60 Liu, L.Y. (1981) Karyotype analysis of *Hypophthalmichthys molitrix*. Acta Genet. Sinica, 8(3): 251–255. (In Chinese with English abstract)

L- 61 Liu, L.Y. (1983) Studies on the chromosome G-banding patterns in rice-field eels (*Monopterus albus* Zuiew). Acta Genet. Sinica, 10(3): 230–234. (In Chinese with English abstract)

L- 62 Liu, L.Y. (1984) Sister chromatid differentiation (SCD), sister chromatid exchanges (SCE) and dottid chromosomes (DC) in rice-field eel (*Monopterus albus* Zuiew). J. Beijing Normal Univ. (Nat. Sci.), 3: 85–90. (In Chinese with English abstract)

L- 63 Liu, S.Y. (1987) Studies on cytogenetics of *Ctenopharyngodon idellus*, *Megalobrama terminalis* and their triploid F_1 hybrid. Acta Hydrobiol. Sinica, 11: 52–58. (In Chinese with English abstract)

L- 64 Liu, S.Y. and Li, S.W. (1987) On the erythrocyte nucleus size and DNA contents of *Ctenopharyngodon idellus*, *Megalobrama terminalis* and F_1 hybrid of the triploid. Acta Genet. Sinica, 14(2): 142–148. (In Chinese with English abstract)

L- 65 Lodi, E. and Marchionni, V. (1980) Chromosome complement of the masked loach *Sabanejewia larvata* (De Fil.) (Pisces, Osteichthyes). Caryologia, 33: 435–440.

L- 66 Lodi, E. and Marchionni, V. (1980) The karyotypes of *Gambusia affinis holbrooki* Gir. from Viverone Lake (Italy) and from an artificial lake of Nicosia (Cyprus). Atti Soc. Ital. Sci. nat. Museo civ. Stor. nat. Milano, 121: 211–220.

L- 67 López, J.R., Alvarez, M.C., Thode, G. and Martinez, G. (1989) Karyotype divergence in *Symphodus melops* and *Symphodus roissali* (Labridae, Perciformes): C-banded and Ag-NOR karyotypes. Genome, 32: 35–39.

L- 68 Lou, Y., Zhang, K., Wu, Y. and Wang, Y. (1983) Studies on karyotype of black carp (*Mylopharyngodon piceus*). J. Fish. China, 7: 77–81. (In Chinese with English abstract)

L- 69 Lou, Y.D. (1997) Progress of fish karyotype studies in China. J. Fish. China, 21 (suppl.): 82–96. (In Chinese)

L- 70 Loudenslager, E.J. and Thorgaard, G.H. (1979) Karyotypic and evolutionary relationships of the Yellowstone (*Salmo clarki bouvieri*) and West-slope (*S. c. lewisi*) cuttthroat trout. J. Fish. Res. Board Can., 36: 630–635.

L- 71 Lourerio, M.A., Caetano, L.G. and Dias, A.L. (2000) Cytogenetic characterization of two species of the genus *Crenicichla* (Pisces, Cichlidae). Cytologia, 65: 57–63.

L- 72 Lozano, R., Ruiz Rejón, C. and Ruiz Rejón, M. (1991) An analysis of coho salmon chromatin by means of C-banding, AG- and fluorochrome staining, and *in situ* digestion with restriction endonucleases. Heredity, 66: 403–409.

L- 73 Lu, R.H., Li, Y.J. and Xu, K.S. (1984) A chromosome study of *Megalobrama amblycephala* Yih. Oceanol. Limnol. Sinica, 15: 487–492. (In Chinese with English abstract)

L- 74 Lueken, W. (1962) Chromosomenzahl bei *Orestias* (Pisces, Cyprinodontidae). Mitt. Hamburg. Zool. Mus. Inst., 60: 195–198.

L- 75 Lueken, W. and Foerster, W. (1969) Chromosomenuntersuchungen bei Fischen mit einer vereinfachten Zellkulturtechnik. Zool. Anz., 183: 168–176.

L- 76 Luo, J.L., Wang, Z.X., Lin, S.P. and Yang, J.H. (1986) Studies on the karyotype of *Clarias fuscus*. J. Fish. China, 10: 441–446. (In Chinese with English abstract)

L- 77 Lee, G.Y. (1984) Karyotypes of seven species of the Gobioninae fishes (Cyprinidae) in Korea. Korean J. Limnol., 17(3/4): 81–88. (In Korean with English abstract)

L- 78 Ludwig, A., Belfiore, N.M., Pitra, C., Svirsky, V. and Jenneckens, I. (2001) Genome duplication events and functional reduction of ploidy levels in sturgeon (*Acipenser*, *Huso* and *Scaphirhynchus*). Genetics, 158: 1203–1215.

L- 79 Lee, G.Y. (1986) Karyotypes of the family Gobiidae fishes in Korea (I). Korean J. Limnol., 19: 49–57. (In Korean with English abstract)

L- 80 Lee, G.Y (1986) Karyotypes of the family Gobiidae fishes from Korea. Zool. Sci., 3: 1115.

L- 81 Li, C.L., Cao, F.J., Liu, C.W. and Huang, X.H. (2005) Studies on the karyotypes of *Lutjanus johni*, *Lutjanus russelli* and *Lutjanus vitta*. Mar. Sci. Bull., 24(5): 23–26. (In Chinese with English abstract)

L- 82 Lamatsch, D.K., Steinlein, C., Schmid, M. and Schartl, M. (2000) Noninvasive determination of genome size and ploidy level in fishes by flow cytometry: detection of triploid *Poecilia formosa*. Cytometry, 39: 91–95.

L- 83 Libertini, A., Mandrioli, M., Colomba, M.S., Bertotto, D., Francescon, A. and Vitturi, R. (2002) A cytogenetic study of the common sole, *Solea solea*, from the northern Adriatic Sea. Chrom. Sci., 6: 63–66.

L- 84 Liu, S.M., Sezaki, K., Hashimoto, K., Kobayasi, H. and Nakamura, M. (1978) Simplified techniques for determination of polyploidy in ginbuna *Carassius auratus langsdorfi*. Bull. Japan. Soc. Sci. Fish., 44: 601–606.

L- 85 Leite, M.F. and Maistro, E.L. (2004) The karyotype of *Apareiodon affinis* (Pisces, Teleostei, Characiformes) from Sapucai River, Minas Gerais, Brazil. Cytologia, 69: 319–322.

L- 86 Lajus, D.L. (2007) Chromosomal analysis in population structuring and stock identification: Robertsonian polymorphism in the White Sea herring (*Clupea pallasi marisalbi*). In: Pisano, E., Ozouf-Costaz, C., Foresti, F. and Kapoor, B.G. eds. Fish cytogenetics. Science Publishers, Enfield, pp. 261–287.

*L- 87 Lajus, D.L. (1992) Khromosomnye nabory 13 vidov ryb Belogo morja. Problemy izuchenija, ratsional'nogo ispol'zovanija okhrany prirodnykh resursov Belogo morja. Reg. Konf. Thes. Dokl. Petrozavodsk, pp. 179–180. (After K-114)

L- 88 Lajus, D.L. (2001) Chromosomal polymorphism of the White Sea fishes. Abstract volume of the 10th European Congress of Ichthyology, Prague, p. 63.

L- 89 Larissa, P.B., Giuliano-Caetano, L. and Dias, A.L. (2008) Karyotype similarities among two populations of *Geophagus brasiliensis* (Perciformes, Cichlidae) from the Tibagi river basin/PR/Brazil. Caryologia, 61: 135–138.

L- 90 Lopes, P.A., Alberdi, A.J., Dergam, J.A. and Fenocchio, A.S. (1998) Cytotaxonomy of *Hoplias malabaricus* (Osteichthyes, Erythrinidae) in the Aguapey River (Province of Corrientes, Argentina). Copeia, 1998: 485–487.

L- 91 Leggatt, R.A. and Iwama, G.K. (2003) Occurrence of polyploidy in the fishes. Rev. Fish Biol. Fish., 13: 237–246.

[M]

M- 1 Mabuchi, K. and Nakabo, T. (1997) Revision of the genus *Pseudolabrus* (Labridae) from the East Asian waters. Ichthyol. Res., 44: 321–334.

M- 2 Mabuchi, K., Arai, R. and Nishida, M. (2002) Karyotypes and nuclear DNA contents of two *Pseudolabrus* species from the southern coast of Japan. Japan. J. Ichthyol., 49: 87–95.

M- 3 Machordom, A. and Doadrio, I. (2001) Evolutionary history and speciation modes in the cyprinid genus *Barbus*. Proc. R. Soc. Lond., B, 268: 1297–1306.

M- 4 Madeira, J.M., Collares-Pereira, M.J. and Elvira, B. (1992) Cytotaxonomy of Iberian loaches with some remarks on the karyological evolution of both families (Pisces, Cobitidae, Homalopteridae). Caryologia, 45: 273–281.

M- 5 Magtoon, W. and Uwa, H. (1985) Karyotype evolution and relationship of a small ricefish, *Oryzias minutillus*, from Thailand. Proc. Japan Acad., Ser. B, 61: 157–160.

M- 6 Magtoon, W. (1986) Distribution and phyletic relationships of *Oryzias* fishes in Thailand. In: Uyeno, T., Arai, R., Taniuchi, T. and Matsuura, K. eds. Indo-Pacific fish biology. Ichthyological Society of Japan, Tokyo, pp. 859–866.

M- 7 Magtoon, W. and Donsakul, T. (1987) Karyotypes of pangasiid catfishes, *Pangasius sutchi* and *P. larnaudii*, from Thailand. Japan. J. Ichthyol., 34: 396–398.

M- 8 Magtoon, W. and Arai, R. (1988) Karyotypes of bagrid catfishes, *Mystus wyckii* and *Bagroides macracanthus*, from Thailand. Bull. Natn. Sci. Mus. Tokyo, (A), 14: 113–117.

M- 9 Magtoon, W. and Arai, R. (1989) Karyotypes of five *Puntius* species and one *Cyclocheilichthys* species (Pisces, Cyprinidae) from Thailand. Bull. Natn. Sci. Mus. Tokyo, (A), 15: 167–175.

M- 10 Magtoon, W. and Arai, R. (1990) Karyotypes of three cyprinid fishes, *Osteochilus hasselti*, *O. vittatus*, and *Labiobarbus lineatus*, from Thailand. Japan. J. Ichthyol., 36: 483-487.

M- 11 Magtoon, W., Nadee, N., Higashitani, T., Takata, K. and Uwa, H. (1992) Karyotype evolution and geographical distribution of the Thai-medaka, *Oryzias minutillus*, in Thailand. J. Fish Biol., 41: 489–497.

M- 12 Magtoon, W. and Arai, R. (1993) Karyotypes and distribution of nucleolus organizer regions in cyprinid fishes from Thailand. Japan. J. Ichthyol., 40: 77–85.

M- 13 Maistro, E.L., Oliveira, C. and Foresti, F. (2002) Cytogenetic analysis of A- and B-chromosomes of *Rhamdia hilarii* (Teleostei, Pimelodidae): C-banding, silver nitrate and CMA_3 staining and restriction endonuclease banding. Cytologia, 67: 25–31.

M- 14 Makino, S. (1937) Notes on the chromosomes of some teleost fishes. Zool. Mag. Japan, 49: 75–76.

M- 15 Makino, S. (1941) A karyological study of gold-fish of Japan. Cytologia, 12: 96–111.

M- 16 Makoedov, A.N. (1987) The problem of stability of karyotype characters in the grayling fishes. Tsitologiya, 29: 490–496. (In Russian with English abstract)

M- 17 Mabuchi, K., Miya, M., Azuma, Y. and Nishida, M. (2007) Independent evolution of the specialized pharyngeal jaw apparatus in cichlid and labrid fishes. BMC Evol. Biol., 2007, 7: 10.

M- 18 Maldonado-Monroy, M.C., Uribe-Alcocer, M., Arreguin-Espinosa, J. and Castro-Perez, A. (1985) Karyological studies on *Dormitator maculatus* Bloch and *Gobiomorus dormitor* Lacépède (Gobiidae: Perciformes). Cytologia, 50: 663–669.

M- 19 Mandrioli, M., Cuoghi, B., Marini, M. and Manicardi, G.C. (2000) Cytogenetic analysis of the pufferfish *Tetraodon fluviatilis* (Osteichthyes). Chrom. Res., 8: 237–242.

M- 20 Manna, G.K. and Prasad, R. (1971) A new perspective in the mechanism on evolution of chromosome in fishes. Proc. First All Ind. Congr. Cytol. Genet., 1971: 237–240.

M- 21 Manna, G.K. and Prasad, R. (1973) Chromosomes in three species of fish (*Channa*). Nucleus, 16: 150–157.

M- 22 Manna, G.K. and Prasad, R. (1974) Somatic and germinal chromosomes of the tropical perch, *Anabas testudineus* (Bloch). Ind. J. Zool., 2: 11–14.

M- 23 Manna, G.K. and Prasad, R. (1974) Cytological evidence for two forms of *Mystus vittatus* (Bloch) as two species. Nucleus, 17: 4–8.

M- 24 Manna, G.K. and Prasad, R. (1974) Chromosome analysis in three species of fishes belonging to family Gobiidae. Cytologia, 39: 609–618.

M- 25 Manna, G.K. and Khuda-Bukhsh, A.R. (1974) Somatic chromosomes of two hybrid carps. Chrom. Inform. Serv., 16: 26–28.

M- 26 Manna, G.K. and Prasad, R. (1977) Chromosome analysis in five species of fresh water fishes. Nucleus, 20: 264–271.

M- 27 Manna, G.K. and Khuda-Bukhsh, A.R. (1977) Karyomorphology of cyprinid fishes and cytological evaluation of the family. Nucleus, 20: 119–127.

M- 28 Manna, G.K. and Khuda-Bukhsh, A.R. (1978) Karyomorphological studies in three species of teleostean fishes. Cytologia, 43: 69–73.

M- 29 Manna, G.K. (1984) Progress in fish cytogenetics. Nucleus, 27: 203–231.

M- 30 Manosroi, J., Meng-Umphan, K., Meevatee, U. and Manosroi, A. (2003) Chromosomal karyotyping from peripheral blood lymphocytes of the Mekong giant catfish (*Pangasianodon gigas*, Chevey). Asian Fish. Sci., 16: 241–246.

M- 31 Mao, L.J., Yang, L.B. and Qin, K.J. (1993) Karyotype studies on *Ernogrammus hexagrammus*. J. Ocean Univ. Qingdao, 23(4): 40–44. (In Chinese with English abstract)

M- 32 Mao, L.J., Yang, L.B. and Qin, K.J. (1993) Karyotypes of two gobiid species. J. Dalian Fish. College, 8(1): 1–7. (In Chinese)

M- 33 Mao, L.J. and Jin, X.M. (1994) Analysis of chromosome karyotype of *Enedrias nebulosus*. J. Dalian Fish. College, 9(4): 32–35. (In Chinese with English abstract)

M- 34 Mao, L.J. and Qiu, P. (1996) Karyotype of *Alectrias benjamini*. J. Dalian Fish. College, 11(3): 37–41. (In Chinese)

M- 35 Mao, L.J., Li, Y.J. and Qu, L. (1999) Analysis of karyotype of *Chaenogobius annularis*. J. Ocean Univ. Qingdao, 29(1): 42–46. (In Chinese with English abstract)

M- 36 Mao, L.J. and Li, Y.J. (2002) Karyotype analyses of five species of marine fishes. J. Dalian Fish. Univ., 17(2): 108–113. (In Chinese with English abstract)

M- 37 Margarido, V.P. and Galetti Jr., P.M. (1996) Chromosome studies in fish of the genus *Brycon* (Characiformes, Characidae, Bryconinae). Cytobios, 85: 219–228.

M- 38 Marian-Krasznai, T. and Krasznai, Z. (1978) Zytologische Untersuchungen bei der Familie Cyprinidae (Pisces). Biol. Zentralbl., 97: 205–214.

M- 39 Martinez, G., Thode, G., Alvarez, M.C. and López, J.R. (1989) C-banding and Ag-NOR reveal heterogeneity among karyotypes of serranids (Perciformes). Cytobios, 58: 53–60.

M- 40 Martinez, P., Viñas, A., Bouza, C., Arias, J., Amaro, R. and Sánchez, L. (1991) Cytogenetical characterization of hatchery stocks and natural populations of sea and brown trout from northwestern Spain. Heredity, 66: 9–17.

M- 41 Martins, I.C., Portella-Castro, A.L.B. and Julio Jr., H.F. (1995) Chromosome analysis of 5 species of the Cichlidae family (Pisces-Perciformes) from the Parana River. Cytologia, 60: 223–231.

M- 42 Martins, C., Giuliano-Caetano, L. and Dias, A.L. (1996) Occurrence of a B chromosome in *Cyphocharax modesta* (Pisces, Curimatidae). Cytobios, 85: 247–253.

M- 43 Martins, C. and Galetti Jr., P.M. (1997) Narrow chromosome diversity in fish of the genus *Schizodon* (Characiformes, Anostomidae). Cytobios, 92: 139–147.

M- 44 Martins, C., Venere, P.C., Mestriner, C.A., Cestari, M.M., Ferreira, R. and Galetti Jr., P.M. (2000) Chromosome relationships between Anostomidae and Chilodontidae fish (Characiformes). Cytologia, 65: 153–160.

M- 45 Martins, M.J., Collares-Pereira, M.J., Cowx, I.G. and Coelho, M.M. (1998) Diploids v. triploids of *Rutilus alburnoides*: spatial segregation and morphological differences. J. Fish Biol., 52: 817–828.

M- 46 Martins-Santos, I.C., Julio Jr., H.F. and Burin, I. (1996) Karyotypic studies of four species of the Sorubiminae subfamily (Pisces, Siluriformes). Caryologia, 49: 73–80.

M- 47 Matoso, D.A., Vicari, M.R., de Almeida, M.C., Shibatta, O.A., Moreira-Filho, O., Bertollo, L.A.C. and Artoni, R.F. (2002) Karyotypic studies in the Characidae fish, genus *Astyanax*. An endemic and highly isolated population of *Astyanax* sp. Cytologia, 67: 123–128.

M- 48 Matsuda, M. (1985) Karyotype of *Acrossocheilus hexagonolepis* (Cyprinidae). Ann. Rep. Biwako Bunkakan, 3: 1–3. (In Japanese)

M- 49 Matsuda, M. (1988) Karyotype of *Pseudorasbora pumila* subsp. (Cyprinidae). Ann. Rep. Biwako Bunkakan, 6: 31–33. (In Japanese)

M- 50 Matsuda, M. and Saitoh, K. (1989) Karyotype of a Japanese cyprinid fish, *Aphyocypris chinensis*. Ann. Rep. Biwako Bunkakan, 7: 21–23. (In Japanese)

M- 51 Matsumiya, T., Sakamoto, K. and Nishikawa, S. (1980) Chromosomes of four species of the family Hexagrammidae (Scorpaeniformes). Japan. J. Ichthyol., 27: 273–276.

M- 52 Matsuo, T. (1998) Karyotypes of the Oxudercinae (Teleostei, Gobiidae). Unpublished Master Thesis, Faculty of Fisheries, Nagasaki University. (In Japanese)

M- 53 Mayers, L.J. and Roberts, F.L. (1969) Chromosomal homogeneity of five populations of alewives, *Alosa pseudoharengus*. Copeia, 1969: 313–317.

M- 54 Mayr, B., Ráb, P. and Kalat, M. (1986) NORs and counterstain-enhanced fluorescence studies in Cyprinidae of different ploidy level. Genetica, 69: 111–118.

M- 55 Mayr, B., Ráb, P. and Kalat, M. (1986) Localisation of NORs and counterstain-enhanced fluorescence studies in *Salmo gairdneri* and *Salmo trutta* (Pisces, Salmonidae). Theor. Appl. Genet., 71: 703–707.

M- 56 Mayr, B., Kalat, M. and Ráb, P. (1990) Sequential counterstain-enhanced fluorescence chromosome banding in the fish *Anguilla anguilla*. Caryologia, 43: 277–281.

M- 57 Mazic, E.Y. and Toktosunov, A.T. (1984) The karyotypes of three species of loaches of the genus *Nemachilus* Hasselt (Cypriniformes, Cobitidae). Tsitologiya, 26: 960–962. (In Russian with English summary)

M- 58 Mazic, E.Y., Toktosunov, A.T. and Gnidenko, S.M. (1986) Comparative karyological analysis of the daces (*Leuciscus*, Cypriniformes, Cyprinidae) from the northern Tien-Shan. Zool. Zhurnal, 65: 1350–1355. (In Russian with English summary)

M- 59 Mazic, E.Y. and Toktosunov, A.T. (1986) Karyotype of the Amu Darya trout *Salmo trutta oxianus* (Salmoniformes, Salmonidae) from the Kyzylsu River. Zool. Zhurnal, 65: 1582–1586. (In Russian with English summary)

M- 60 Mazik, E.J., Toktosunov, A.T. and Ráb, P. (1989) Karyotype study of four species of the genus *Diptychus* (Pisces, Cyprinidae), with remarks on polyploidy of schizothoracine fishes. Folia Zool., 38: 325–332.

M- 61 McPhail, J.D. and Jones, R.L. (1966) A simple technique for obtaining chromosomes from teleost fishes. J. Fish. Res. Board Can., 23: 767–768.

M- 62 Merrilees, M.J. (1975) Karyotype of *Galaxias maculatus* from New Zealand. Copeia, 1975: 176–178.

M- 63 Mestriner, C.A., Galetti Jr., P.M., Valentini, S.R., Ruiz, I.R.G., Abel, L.D.S., Moreira-Filho, O. and Camacho, J.P.M. (2000) Structural and functional evidence that a B chromosome in the characid fish *Astyanax scabripinnis* is an isochromosome. Heredity, 85: 1–9.

M- 64 Mészáros, B. and Kató, I. (1976) On the evolution of fishes as seen from a karyological point of view. Acta Biol. Debrecina, 13: 255–274. (In Hungarian with English abstract)

M- 65 Michele, J.L. and Takahashi, C.S. (1977) Comparative cytology of *Tilapia rendalli* and *Geophagus brasiliensis* (Cichlidae, Pisces). Cytologia, 42: 535–537.

M- 66 Michele, J.L., Takahashi, C.S. and Ferrari, I. (1977) Karyotypic study of some species of the family Loricariidae (Pisces). Cytologia, 42: 539–546.

M- 67 Miller, I.N. (2000) Karyotype of the Far Eastern sculpin *Myoxocephalus stelleri* from Peter the Great Bay, Sea of Japan. Biol. Morya, 26: 121–124. (In Russian with English abstract)

M- 68 Miller, R.R. and Walters, V. (1972) A new genus of cyprinodontid fish from Nuevo Leon, Mexico. Contri. Sci. Nat. Hist. Mus. Los Angeles County, 233: 1–13.

M- 69 Miller, R.R. (1979) Ecology, habits and relationships of the Middle American Cuatro Ojos, *Anableps dowi* (Pisces: Anablepidae). Copeia, 1979: 82–91.

M- 70 Miller, R.R. and Uyeno, T. (1980) *Allodontichthys hubbsi*, a new species of goodeid fish from southwestern Mexico. Occ. Pap. Mus. Zool. Univ. Michigan, 692: 1–13.

M- 71 Miller, R.R. (1972) Classification of the native trouts of Arizona with the description of a new species, *Salmo apache*. Copeia, 1972: 401–422.

M- 72 Miller, R.R. and Fitzsimons, J.M. (1971) *Ameca splendens*, a new genus and species of goodeid fish from western México, with remarks on the classification of the Goodeidae. Copeia, 1971: 1–13.

M- 73 Mitrofanov, Y.A., Otradnova, V.V. and Pokozy, N.V. (1977) On the karyotype of the *Oncorhynchus keta* × *O. gorbusha* hybrid. Tsitologiya, 19: 798–808. (In Russian with English summary)

M- 74 Miyazawa, C.S. and Galetti Jr., P.M. (1994) First cytogenetical studies in *Characidium* species (Pisces: Characiformes, Characidiinae). Cytologia, 59: 73–79.

M- 75 Mizoguchi, S.M.H.N. and Martins-Santos, I.C. (1997) Macro- and microchromosomes B in females of *Astyanax scabripinnis* (Pisces, Characidae). Hereditas, 127: 249–253.

M- 76 Mizoguchi, S.M.H.N. and Martins-Santos, I.C. (1998) Activation patterns of the nucleolar organizer region in *Astyanax scabripinnis* (Pisces, Characidae). Cytologia, 63: 259–265.

M- 77 Molina, W.F., Schmid, M. and Galetti Jr., P.M. (1998) Heterochromatin and sex chromosomes in the neotropical fish genus *Leporinus* (Characiformes, Anostomidae). Cytobios, 94: 141–149.

M- 78 Molina, W.F., Maia-Lima, F.A. and Affonso, P.R.A.M. (2002) Divergence between karyotypical pattern and speciation events in Serranidae fish (Perciformes). Caryologia, 55: 299–305.

M- 79 Molina, W.F. and Galetti Jr., P.M. (2004) Karyotypic changes associated to the dispersive potential on Pomacentridae (Pisces, Perciformes). J. Exp. Mar. Biol. Ecol., 309: 109–119.

M- 80 Molina, W.F. (2005) Intraspecific karyotypical diversity in brackish water fishes of the Eleotridae family (Pisces, Perciformes). Cytologia, 70: 39–45.

M- 81 Mořan, P., Pendás, A.M., Garcia-Vázquez, E. and Linde, A.R. (1989) Chromosomal and morphological analysis of two populations of *Salmo trutta* sbp. *fario* employed in repopulation. J. Fish Biol., 35: 839–843.

M- 82 Moreira-Filho, O., Bertollo, L.A.C. and Galetti Jr., P.M. (1980) Evidences for a multiple sex chromosome system with female heterogamety in *Apareiodon affinis* (Pisces, Parodontidae). Caryologia, 33: 83–91.

M- 83 Moreira-Filho, O., Bertollo, L.A.C. and Galetti Jr., P.M. (1985) Karyotypic study of some species of family Parodontidae (Pisces-Cypriniformes). Caryologia, 38: 47–55.

M- 84 Moreira-Filho, O., Bertollo, L.A.C. and Galetti Jr., P.M. (1993) Distribution of sex chromosome mechanisms in neotropical fish and description of a ZZ/ZW system in *Parodon hilarii* (Parodontidae). Caryologia, 46: 115–125.

M- 85 Moreira-Filho, O., Fenocchio, A.S., Pastori, M.C. and Bertollo, L.A.C. (2001) Occurrence of a metacentric macrochromosome B in different species of the genus *Astyanax* (Pisces, Characidae, Tetragonopterinae). Cytologia, 66: 59–64.

M- 86 Morelli, S., Bertollo, L.A.C., Foresti, F., Moreira Filho, O. and Almeida Toledo, S. (1983) Cytogenetic considerations on the genus *Astyanax* (Pisces, Characidae). I. Karyotypic variability. Caryologia, 36: 235–244.

M- 87 Morelli, S., Bertollo, L.A.C. and Moreira Filho, O. (1983) Cytogenetic considerations on the genus *Astyanax* (Pisces, Characidae). II. Occurrence of natural triploidy. Caryologia, 36: 245–250.

M- 88 Morescalchi, A., Hureau, J.C., Olmo, E., Ozouf-Costaz, C., Pisano, E. and Stanyon, R. (1992) A multiple sex-chromosome system in Antarctic ice-fishes. Polar Biol., 11: 655–661.

M- 89 Morescalchi, A., Morescalchi, M.A., Odierna, G., Stingo, V. and Capriglione, T. (1996) Karyotype and genome size of zoarcids and notothenioids (Teleostei, Perciformes) from the Ross Sea: cytotaxonomic implications. Polar Biol., 16: 559–564.

M- 90 Morgan II, R.P., Goodfellow Jr., W.L. and Hocutt, C.H. (1984) Karyotype of *Nocomis micropogon*, *Rhinichthys cataractae* and their supposed hybrid, "*Rhinichthys bowersi*" (Pisces: Cyprinidae). Copeia, 1984: 990–992.

M- 91 Muramoto, J., Ohno, S. and Atkin, N.B. (1968) On the diploid state of the fish order Ostariophysi. Chromosoma, 24: 59–66.

M- 92 Muramoto, J. (1969) On the chromosomes of *Parasilurus asotus* (Linné). Chrom. Inform. Serv., 10: 18–19.

M- 93 Muramoto, J. and Igarashi, K. (1969) A preliminary note on the chromosomes and enzymatic patterns of three forms of sticklebacks. J. Fac. Sci. Hokkaido Univ., Ser. Zool., 17: 266–270.

M- 94 Muramoto, J., Igarashi, K., Itoh, M. and Makino, S. (1969) A study of the chromosomes and enzymatic patterns of sticklebacks of Japan. Proc. Japan Acad., 45: 803–807.

M- 95 Muramoto, J., Azumi, J. and Fukuoka, H. (1974) Karyotypes of 9 species of the Salmonidae. Chrom. Inform. Serv., 17: 20–23.

M- 96 Muramoto, J. (1975) A note on triploidy of the funa (Cyprinidae, Pisces). Proc. Japan Acad., 51: 583–587.

M- 97 Murofushi, M. and Yosida, T.H. (1979) Cytogenetical studies on fishes, II. Karyotypes of four carangid fishes. Japan. J. Genet., 54: 367–370.

M- 98 Murofushi, M. and Yosida, T.H. (1979) Cytogenetical studies on fishes, I. Karyotypes of four filefishes. Japan. J. Genet., 54: 191–195.

M- 99 Murofushi, M. and Aoki, H. (1979) Karyological study in *Scomber japonicus*. Rep. Mishima Res. Inst. Sci. Liv., Nihon Univ., 2: 37–40. (In Japanese with English abstract)

M–100 Murofushi, M., Oikawa, S., Nishikawa, S. and Yosida, T.H. (1980) Cytogenetical studies on fishes, III. Multiple sex chromosome mechanism in the filefish, *Stephanolepis cirrhifer*. Japan. J. Genet., 55: 127–132.

M–101 Murofushi, M., Oishi, M. and Nawa, N. (1980) Karyological studies in *Apogon semilineatus*. Rep. Mishima Res. Inst. Sci. Liv., Nihon Univ., 3: 47–50. (In Japanese with English abstract)

M–102 Murofushi, M., Oishi, M. and Nawa, N. (1980) Karyological studies in *Luciogobius guttatus*. Rep. Mishima Res. Inst. Sci. Liv., Nihon Univ., 3: 51–52. (In Japanese with English abstract)

M–103 Murofushi, M. and Nawa, N. (1981) A karyological study in yellowstriped butter fish *Labracoglossa argentiventris* Peters. Rep. Mishima Res. Inst. Sci. Liv., Nihon Univ., 4: 25–28. (In Japanese)

M–104 Murofushi, M. (1982) A karyological study of two beryciform fishes. Rep. Mishima Res. Inst. Sci. Liv., Nihon Univ., 5: 17–20. (In Japanese with English abstract)

M–105 Murofushi, M., Nishikawa, S. and Yosida, T.H. (1983) Cytogenetical studies on fishes. V. Multiple sex chromosome mechanism (XX-Y) found in two dragonet fishes. Proc. Japan Acad., Ser. B, 59: 58–61.

M–106 Murofushi, M., Nishikawa, S. and Yosida, T.H. (1983) Cytogenetical studies on fishes, IV. Karyotypes of six species in the sparoid fishes. Japan. J. Genet., 58: 361–367.

M–107 Murofushi, M., Nishikawa, S. and Yosida, T.H. (1984) Cytogenetical studies on fishes, IX. Karyological comparison of two species in snipefishes. Japan. J. Genet., 59: 155–158.

M–108 Murofushi, M. and Yosida, T.H. (1984) Cytogenetical studies on fishes. VIII. XX-Y sex chromosome mechanism newly found in the snake eel, *Muraenichthys gymnotus* (Anguilliformes, Pisces). Proc. Japan Acad., Ser. B, 60: 21–23.

M–109 Murofushi, M., Nishikawa, S. and Yosida, T.H. (1984) Cytogenetical studies on fishes, VII. Karyological comparison of two species in the sandperch fishes. La Kromosomo, II, 35/36: 1122–1125.

M–110 Murofushi, M., Nishikawa, S. and Yosida, T.H. (1984) Cytogenetical studies on fishes, VI. Karyological comparison of four species in the dragonet fish. La Kromosomo, II, 34: 1079–1084.

M–111 Murofushi, M. (1986) A study of karyotype classification and karyotype evolution in marine teleosts. Rep. Mishima Res. Inst. Sci. Liv., Nihon Univ., 9: 95–157. (In Japanese with English abstract)

M–112 Murofushi, M., Nakatsubo, T. and Smith, P.J. (1989) Karyological study on the New Zealand leatherjacket, *Parika scaber*, fish of the order Tetraodontiformes. Bull. Biogeogr. Soc. Japan, 44: 35–38.

M–113 Maistro, E.L., Jesus, C.M., Oliveira, C., Moreira-Filho, O. and Foresti, F. (2004) Cytogenetic analysis of A-, B-chromosomes and ZZ/ZW sex chromosomes of *Characidium gomesi* (Teleostei, Characiformes, Crenuchidae). Cytologia, 69: 181–186.

M–114 Maddock, M.B. and Schwartz, F.J. (1996) Elasmobranch cytogenetics: methods and sex chromosomes. Bull. Mar. Sci., 58: 147–155.

M–115 Morescalchi, M.A., Rocco, L. and Stingo, V. (2002) Cytogenetic and molecular studies in a lungfish, *Protopterus annectens* (Osteichthyes, Dipnoi). Gene, 295: 279–287.

M–116 Miller, P.J. and El-Tawil, M.Y. (1974) A multidisciplinary approach to a new species of *Gobius* (Teleostei: Gobiidae) from southern Cornwall. J. Zool. Lond., 174: 539–574.

M–117 Murakami, M. (2006) The origins of the gynogenetic triploid ginbuna (Japanese silver crucian carp, *Carassius auratus langsdorfi*). Fish Genet. Breed. Sci., 35: 87–92. (In Japanese with English abstract)

M–118 Murofushi, M., Yoda, K. and Deguchi, Y. (1987) Karyological study of four species in scorpionfishes. Rep. Mishima Res. Inst. Sci. Liv., Nihon Univ., 10: 37–42. (In Japanese with English abstract)

M–119 Morishima, K. and Arai, K. (2006) Natural clonal lines, polyploidy and atypical reproduction in the loach, *Misgurnus anguillicaudatus*. Fish Genet. Breed. Sci., 35: 93–100. (In Japanese with English abstract)

M–120 Mank, J.E. and Avise, J.C. (2006) Phylogenetic conservation of chromosome numbers in Actinopterygiian fishes. Genetica, 127: 321–327.

M–121 Molina, W.F. and Bacurau, T.O.F. (2006) Structural and numerical chromosome diversification in marine Perciformes (Priacanthidae and Gerreidae). Cytologia, 71: 237–242.

M–122 Mazzei, F., Ghigliotti, L., Bonillo, C., Coutanceau, J-P., Ozuouf-Costaz, C. and Pisano, E. (2004) Chromosomal patterns of major and 5S ribosomal DNA in six icefish species (Perciformes, Notothenioidei, Channichthyidae). Polar Biol., 28: 47–55.

M–123 Mazzei, F., Ghigliotti, L., Lecointre, G., Ozuouf-Costaz, C., Coutanceau, J-P., Detrich III, W. and Pisano, E. (2006) Karyotypes of basal lineages in notothenioid fishes: the genus *Bovichtus*. Polar Biol., 29: 1071–1076.

M–124 Mazzei, F., Ghigliotti, L., Coutanceau, J-P., Detrich III, H.W., Prirodina, V., Ozuouf-Costaz, C. and Pisano, E. (2008) Chromosomal characteristics of the temperate notothenioid fish *Eleginops maclovinus* (Cuvier). Polar Biol., 31: 629–634.

M–125 Majumdar, K.C. and McAndrew, B.J. (1986) Relative DNA content of somatic nuclei and chromosomal studies in three genera, *Tilapia*, *Sarotherodon*, and *Oreochromis* of the tribe Tilapiini (Pisces, Cichlidae). Genetica, 68: 175–188.

M–126 Morescalchi, A., Pisano, E., Stanyon, R. and Morescalchi, M.A. (1992) Cytotaxonomy of Antarctic teleosts of the *Pagothenia/Trematomus* complex (Nototheniidae, Perciformes). Polar Biol., 12: 553–558.

M–127 Morescalchi, M.A., Liguori, I., Rocco, L., Archimandritis, A. and Stingo, V. (2008) Karyotypic characterization and genomic organization of the 5S rDNA in *Polypterus senegalus* (Osteichthyes, Polypteridae). Genetica, 132: 179–186.

M–128 Martinez, E.R.M., Oliveira, C. and Foresti, F. (2004) Cytogenetic analyses of *Pseudopimelodus mangurus* (Teleostei: Siluriformes: Pseudopimelodidae). Cytologia, 69: 419–424.

M–129 Margarido, V.P. and Moreira-Filho, O. (2008) Karyotypic differentiation through chromosome fusion and number reduction in *Imparfinis hollandi* (Ostariophysi, Heptapteridae). Genet. Mol. Biol., 31(1) (suppl.): 235–238.

M–130 Martins, C. and Galetti Jr., P.M. (2001) Organization of 5S rDNA in species of the fish *Leporinus*: two different genomic locations are characterized by distinct nontranscribed spacers. Genome, 44: 903–910.

M–131 Moreira-Filho, O., Galetti Jr., P.M. and Bertollo, L.A.C. (2004) B chromosomes in the fish *Astyanax scabripinnis* (Characidae, Tetragonopterinae): an overview in natural populations. Cytogenet. Genome Res., 106: 230–234.

M-132 Medrano, L., Bernardi, G., Couturier, J., Dutrillaux, B. and Bernardi, G. (1988) Chromosome banding and genome compartmentalization in fishes. Chromosoma, 96: 178–183.

M-133 Mirsky, A.E. and Ris, H. (1951) The desoxyribonucleic acid content of animal cells and its evolutionary significance. J. General Physiol., 34: 451–462.

M-134 Miyaki, K., Tabeta, O. and Kayano, H. (1995) Karyotypes in six species of pufferfishes genus *Takifugu* (Tetraodontidae, Tetraodontiformes). Fish. Sci., 61: 594–598.

M-135 Maistro, E.L., Foresti, F., Oliveira, C. and Almeida-Toledo, L.F. (1992) Occurrence of macro B chromosomes in *Astyanax scabripinnis paranae* (Pisces, Characiformes, Characidae). Genetica, 87: 101–106.

M-136 Mazzuchelli, J. and Martins, C. (2009) Genomic organization of repetitive DNAs in the cichlid fish *Astronotus ocellatus*. Genetics, 136: 461–469.

M-137 Martinez, P., Castro, J., Pardo, B.G., Bouza, C., Hermida, M. and Vilas, R. (2009) High Ag-NOR-site variation associated to a secondary contact in brown trout from the Iberian Peninsula. Genetica, 136: 419–427.

M-138 Miya, M., Takeshima, H., Endo, H., Ishiguro, N.B., Inoue, J.G., Mukai, T., Satoh, T.P., Yamaguchi, M., Kawaguchi, A., Mabuchi, K., Shirai, S.M. and Nishida, M. (2003) Major patterns of higher teleostean phylogenies: a new perspective based on 100 complete mitochondrial DNA sequences. Mol. Phylogenet. Evol., 26: 121–138.

M-139 Miya, M., Satoh, T.P. and Nishida, M. (2005) The phylogenetic position of toadfishes (order Batrachoidiformes) in the higher ray-finned fish as inferred from partitioned Bayesian analysis of 102 whole mitochondrial genome sequences. Biol. J. Linn. Soc., 85: 289–306.

M-140 Miya, M., Holocroft, N.I., Satoh, T.P., Yamaguchi, M., Nishida, M. and Wiley, E.O. (2007) Mitochondrial genome and a nuclear gene indicate a novel phylogenetic position of deep-sea tube-eye fish (Stylephoridae). Ichthyol. Res., 54: 323–332.

M-141 Monteiro, R., Carvalho, C. and Collares-Pereira, M.J. (2009) Karyotype and genome size of *Iberochondrostoma almacai* (Teleostei, Cyprinidae) and comparison with the sister-species *I. lusitanicum*. Genet. Mol. Biol., 32: 268–275.

M-142 Martinez, E.R.M., Nirchio, M., Granado, A., Foresti, F. and Oliveira, C. (2008) Cytogenetic analysis of three catfish species of the family Pseudopimelodidae (Teleostei, Siluriformes). Genet. Mol. Biol., 31: 692–696.

M-143 Molina, W.F. and Galetti Jr., P.M. (2002) Robertsonian rearrangements in the reef fish *Chromis* (Perciformes, Pomacentridae) involving chromosomes bearing 5S rRNA genes. Genet. Mol. Biol., 25: 373–377.

M-144 Milhomem, S.S.R., Souza, A.C.P., Nascimento, A.L., Carvalho Jr., J.R., Feldberg, E., Pieczarka, J.C. and Nagamichi, C.Y. (2008) Cytogenetic studies in fishes of the genera *Hassar*, *Platydoras* and *Opsodoras* (Doradidae, Siluriformes) from Jari and Xingú rivers, Brazil. Genet. Mol. Biol., 31 (suppl.): 256–260.

M-145 Molina, W.F. and Galetti Jr., P.M. (2004) Multiple pericentric inversions and chromosomal divergence in the reef fishes *Stegastes* (Perciformes, Pomacentridae). Genet. Mol. Biol., 27: 543–548.

M-146 Mesquita, D.R., Porto, J.I.R. and Feldberg, E. (2008) Chromosomal variability in the wild ornamental species of *Symphysodon* (Perciformes: Cichlidae) from Amazon. Neotrop. Ichthyol., 6: 181–190.

M-147 Marques, D.K., Venere, P.C. and Galetti Jr., P.M. (2006) Chromosomal characterization of the bonytongue *Arapaima gigas* (Osteoglossiformes: Arapaimidae). Neotrop. Ichthyol., 4: 215–218.

M-148 Marcon, J.L., Feldberg, E., Porto, J.I.R. and Affonso, E.G. (1992) Estudos fisiologicos e citogeneticos em peixes do genero *Dianema* (Siluriformes, Callichthyidae) da Amazonia Central. IV. Simp. Citogenet. Evol. e Aplic. de Peixes Neotropicais, Rio de Janeiro, RJ, p. 34 (abstract). (In Portuguese)

*M-149 Martins-Santos, I.C., Silva, A.L.B. and Julio, H.F. (1990) Estudio cromossômico de *Aequidens plagiozonatus* (Perciformes, Cichlidae) do Rio Paraná. III. Simp. Citogenet. Evol. e Aplic. de Peixes Neotropicais, Botucatu, SP, p. 40 (abstract). (After K-114)

*M-150 Miller, R.R. and Echelle, A.A. (1975) *Cyprinodon tularosa*, a new cyprinodontid fish from the Tularosa basin, New Mexico. Southwest. Nat., 19: 365–377. (After K-114)

M-151 Margarido, V.P. and Galetti Jr., P.M. (1999) Heterochromatin patterns and karyotype relationships within and between the genera *Brycon* and *Salminus* (Pisces, Characidae). Genet. Mol. Biol., 22: 357–361.

M-152 Mestriner, C.A., Bertollo, L.A.C. and Galetti Jr., P.M. (1995) Chromosome banding and synaptonemal complexes in *Leporinus lacustris* (Pisces, Anostomidae): analysis of a sex system. Chrom. Res., 3: 440–443.

M-153 Molina, W.F., Margarido, V.P. and Galetti Jr., P.M. (2007) Natural triploidy in *Leporinus* cf. *elongatus* bearing sex chromosomes. Genet. Mol. Biol., 30: 567–569.

M-154 Martins, C. and Galetti Jr., P.M. (1998) Karyotype similarity between two sympatric *Schizodon* fish species (Anostomidae, Characiformes) from the Paraguay River basin. Genet. Mol. Biol., 21: 355–360.

M-155 Margarido, V.P. and Galetti Jr., P.M. (2000) Amplification of a GC-rich heterochromatin in the freshwater fish *Leporinus desmotes* (Characiformes, Anostomidae). Genet. Mol. Biol., 23: 569–573.

M-156 Molina, W.F. and Galetti Jr., P.M. (2007) Early replication banding in *Leporinus* species (Osteichthyes, Characiformes) bearing differentiated sex chromosomes (ZW). Genetica, 130: 153–160.

M-157 Moraes, V.P.O., Cereali, S.S., Froehlich, O. and Dias, A.L. (2007) Cytogenetic characterization of *Rhamdia quelen* (Siluriformes, Heptapteridae) from the Bodoquena Plateau, Mato Grosso do Sul, Brazil. Genet. Mol. Res., 6: 627–633.

M-158 Moraes, V.P.O., Souza Carneiro, J. and Dias, A.L. (2009) B chromosomes in four different populations of *Rhamdia quelen* (Siluriformes, Heptapteridae): a comparative study of frequency and distribution. Folia Biol. (Krakow), 57: 165–169.

M-159 Mizoguchi, S.M.H.K., Portela-Castro, A.L.B. and Martins-Santos, I.C. (2007) Cytogenetic characterization of *Crenicichla* (Pisces, Perciformes, Cichlidae) of the Iguaçu River. Genet. Mol. Res., 6: 650–656.

M-160 Mani, I., Kumar, R., Singh, M., Kushwaha, B., Nagpure, N.S., Srivastava, P.K., Murmu, K., Rao, D.S.K. and Lakra, W.S. (2009) Karyotypic diversity and evolution of seven mahseer species (Cyprinidae) from India. J. Fish Biol., 75: 1079–1091.

M-161 Mazzuchelli, J., Swarça, A.C. and Dias, A.L. (2007) Structural chromosome polymorphism in a *Pimelodus maculatus* La Cepède, 1803 population (Siluriformes, Pimelodidae) from the Paranapanema River basin, PR, Brazil. Brazil. J. Biol., 67 (special number): 631–637.

M-162 Mishra, A. (1998) *Gadusia variegata, Gonialosa manmina, Gagata cenia*. In: Ponniah, A.G. and John, G. eds. Fish chromosome atlas. National Bureau of Fish Genetic Resources, Lucknow, India, pp. 14–17, 202–203.

M-163 Meyer, M.K. and Förster, W. (1983) Eine neue *Ilyodon*-Art aus Guerrero, Mexico (Osteichthyes, Goodeidae). Zool. Abhand. Staat. Mus. Tier. Dresden, 38: 257–263.

M-164 Mariotto, M., Artoni, R.F. and Miyazawa, C.S. (2004) Occurrence of sexual chromosome, of the type ZZ/ZW, in *Ancistrus* cf. *dubius* (Loricariidae, Ancistrinae) of the Paraguay river basin, Mato Grosso, Brazil. Caryologia, 57: 327–331.

M-165 Mariotto, S. and Miyazawa, C.S. (2006) *Ancistrus* cf. *dubius* (Siluriformes, Ancistrinae), a complex of species. 1. Chromosomic characterization of four populations and occurrence of sexual chromosomes of type XX/XY, in the Pantanal basin of Mato Grosso, Brazil. Caryologia, 59: 299–304.

M–166 Magtoon, W. and Arai, R. (1994) Karyotypes of five *Rasbora* species and one *Danio* species (Cyprinidae) from Thailand. Proc. 4th Indo-Pacific Fish Conference, Bangkok, pp. 484–496.

*M–167 Mendonca, A.C.S., Molina, W.F., Gurgel, H.C.B. and Dias, A.L. (1990) Dados preliminares sobre os cromossomos de *Metynnis* cf. *roosevelti* (Eigenmann, 1915) (Characidae, Myleinae) da Lagoa Redonda, Rio Grande do Norte. III Simp. Citogenet. Evol. E Aplic. de Peixes Neotropicais, Botucatu, SP, p. 7 (abstract). (After K-114)

M–168 Maia, T.P. de A., Giuliano-Caetano, L., Rodriguez, M.S., Rubert, M., Takagui, F.H. and Dias, A.L. (2010) Chromosomal banding in three species of the genus *Rineloricaria* (Siluriformes, Loricariidae, Loricariinae). Ichthyol. Res., 57: 209–213.

M–169 Moysés, C.B., Daniel-Silva, M.F.Z., Lopes, C.E. and Almeida-Toledo, L.F. (2010) Cytotype-specific ISSR profiles and karyotypes in the Neotropical genus *Eigenmannia* (Teleostei: Gymnotiformes). Genetica, 138: 179–189.

[N]

N– 1 Nadamitsu, S. (1974) Karyological studies of two types of gobioid fish, "Yoshinobori". Bull. Hiroshima Women's Univ., 9: 3–9.

N– 2 Nascimento, A.L., Souza, A.C.P., Feldberg, E., Cavalho Jr., J.R., Barros, R.M.S., Pieczarka, J.C. and Nagamachi, C.Y. (2006) Cytogenetic analysis on *Pterophyllum scalare* (Perciformes, Cichlidae) from Jari river, Pará state. Caryologia, 59: 138–143.

N– 3 Nagpure, N.S., Kushwaha, B., Srivastava, S.K. and Ponniah, A.G. (2000) Karyomorphology of African catfish *Clarias gariepinus* (Clariidae). Chrom. Sci., 4: 43–44.

N– 4 Nagpure, N.S., Kushwaha, B., Srivastava, S.K. and Ponniah, A.G. (2001) Comparative cytogenetic studies in Indian major carps *Labeo rohita*, *Catla catla* and *Cirrhinus mrigala* (Cyprinidae, Pisces). Chrom. Sci., 5: 153–158.

N– 5 Nagy, A., Rajki, K., Horváth, L. and Csányi, V. (1978) Investigation on carp, *Cyprinus carpio* L. gynogenesis. J. Fish Biol., 13: 215–224.

N– 6 Nakai, Y., Kubota, S. and Kohno, S. (1991) Chromatin diminution and chromosome elimination in four Japanese hagfish species. Cytogenet. Cell Genet., 56: 196–198.

N– 7 Nakatsubo, T. and Murofushi, M. (1997) Karyological study of ocean sun fish *Mola mola*. In: Zhou, Y.Q. Zhou, H.Q., Yao, C.Q., Lu, Y., Hu, F.Y., Chi, H. and Din, F.H. eds. The fourth Asian fisheries forum. China Ocean Press, pp. 388–389.

N– 8 Nakayama, C.M., Feldberg, E. and Bertollo, L.A.C. (2008) Mapping of ribosomal genes and chromosomal markers in three species of the genus *Serrasalmus* (Characidae, Serrasalminae) from the Amazon basin. Genet. Mol. Biol., 31: 868–873.

N– 9 Nakayama, C., Jégu, M., Porto, J.I.R. and Feldberg, E. (2001) Karyological evidence for a cryptic species of piranha within *Serrasalmus rhombeus* (Characidae, Serrasalminae) in the Amazon. Copeia, 2001: 866–869.

N– 10 Nakayama, I. and Chourrout, D. (1993) Uniparental reproduction demonstrates the inheritance of Robertsonian variations in the rainbow trout, *Oncorhynchus mykiss*. Copeia, 1993: 553–557.

N– 11 Naran, D. (1997) Cytogenetic studies of *Pseudobarbus* and selected *Barbus* (Pisces: Cyprinidae) of southern Africa. M. Sc. Thesis, Rhodes Univ., 223 pp.

N– 12 Natarajan, R. and Subrahmanyam, K. (1968) A preliminary study on the chromosomes of *Tilapia mossambica* (Peters). Curr. Sci., 37: 262–263.

N– 13 Natarajan, R. and Subrahmanyam, K. (1974) A karyotype study of some teleosts from Portonovo waters. Proc. Ind. Acad. Sci., Sec. B, 79: 173–196.

N– 14 Navarrete, M.C. and Julio Jr., H.F. (1997) Cytogenetic analysis of four curimatids from the Paraguay basin, Brazil (Pisces: Characiformes: Curimatidae). Cytologia, 62: 241–247.

N- 15 Nayak, K. and Khuda-Bukhsh, A.R. (1987) Chromosomes of two species of fishes *Apocryptes* (Gobiidae: Pisces). Environ. Ecol., 5: 371–373.

N- 16 Nayak, K. and Khuda-Bukhsh, A.R. (1987) Karyotypes of two species of fishes. Environ. Ecol., 5: 318–320.

N- 17 Nayak, K. and Khuda-Bukhsh, A.R. (1988) Karyomorphology of a crocodile fish *Platycephalus tuberculatus* (Platycephalidae, Pisces). J. Inland Fish. Soc. India, 20(2): 35–37.

N- 18 Nayak, K. and Khuda-Bukhsh, A.R. (1991) Chromosome complements in *Mugil corsula* (Mugilidae, Pisces) with reference to somatic C-band localization. Nat. Acad. Sci. Letters, 14(4): 195–196.

N- 19 Nayyar, R.P. (1966) Karyotype studies in thirteen species of fishes. Genetica, 37: 78–92.

N- 20 Néo, D.M., Moreira Filho, O. and Camacho, J.P.M. (2000) Altitudinal variation for B chromosome frequency in the characid fish *Astyanax scabripinnis*. Heredity, 85: 136–141.

N- 21 Nielsen, J.L. and Sage, G.K. (2001) Microsatellite analyses of the trout of northwest Mexico. Genetica, 111: 269–278.

N- 22 Nikolskii, G.V. and Vasiliev, V.P. (1973) On several regularization in the distribution of chromosomes in fishes. Vopr. Ichthyol., 13: 3–22. (In Russian)

N- 23 Nirchio, M., Gomez, J.A. and Villalaz, J. (2001) Cariotipo del pez sapo *Batrachoides pacifici* (Batrachoididae: Teleostei) de la costa del Pacifico de Panamá. Saber, Univ. Oriente, Venezuela, 13: 81–83. (In Spanish with English abstract)

N- 24 Nirchio, M., Turner, B.J., Pérez, J.E., Gaviria, J.I. and Cequea, H. (2002) Karyotypes of three species of toadfish (Batrachoididae: Teleostei) from Venezuela. Sci. Mar., 66: 1–4.

N- 25 Nirchio, M., Fenocchio, A.S., Swarça, A.C., Pérez, J.E., Granado, A., Estrada, A. and Ron, E. (2003) Cytogenetic characterization of hybrids offspring between *Colossoma macropomum* (Cuvier, 1818) and *Piaractus brachypomus* (Cuvier, 1817) from Caicara del Orinoco, Venezuela. Caryologia, 56: 405–411.

N- 26 Nirchio, M., Fenocchio, A.S., Swarça, A.C. and Pérez, J.E. (2004) Karyology of the toadfish *Porichthys plectrodon* (Jordan and Gilbert, 1882) (Batrachoididae) from Margarita Island, Venezuela. Mar. Biol., 146: 161–165.

N- 27 Nirchio, M., Fenocchio, A.S., Swarça, A.C., Dias, A.L., Giuliano-Caetano, L., Ron, E., Gaviria, J.I. and Pérez, J.E. (2004) Cytogenetic characterization of *Thalassophryne maculosa* Günther, 1861 (Pisces: Batrachoididae) from Margarita Island, Venezuela. Caribbean J. Sci., 40: 218–222.

N- 28 Nishikawa, S., Amaoka, K. and Karasawa, T. (1971) On the chromosomes of two species of eels (*Anguilla*). Chrom. Inform. Serv., 12: 27–28.

N- 29 Nishikawa, S. and Karasawa, T. (1972) A comparative studies of the chromosomes in Japanese fishes–I. A study of the somatic chromosomes of three species of scups. J. Shimonoseki Univ. Fish., 20: 235–240. (In Japanese with English abstract)

N- 30 Nishikawa, S., Amaoka, K. and Karasawa, T. (1973) A preliminary study on the chromosomes of *Cichlasoma citrinella* (Cichlidae; Pisces). Chrom. Inform. Serv., 14: 32–33.

N- 31 Nishikawa, S., Amaoka, K. and Nakanishi, K. (1974) A comparative study of chromosomes of twelve species of gobioid fishes in Japan. Japan. J. Ichthyol., 21: 61–71.

N- 32 Nishikawa, S. and Sakamoto, K. (1977) Comparative studies on the chromosomes in Japanese fishes–III. Somatic chromosomes of three anguilliform fishes. J. Shimonoseki Univ. Fish., 25: 193–196. (In Japanese with English abstract)

N- 33 Nishikawa, S., Honda, M. and Wakatsuki, A. (1977) Comparative studies on the chromosomes in Japanese fishes–II. Chromosomes of eight species in scorpionfishes. J. Shimonoseki Univ. Fish., 25: 187–191. (In Japanese with English abstract)

N- 34 Nishikawa, S. and Sakamoto, K. (1978) Comparative studies on the chromosomes in Japanese fishes–V. Somatic chromosomes in Japanese blennioid fish, *Dictyosoma burgeri* Van der Hoven. J. Shimonoseki Univ. Fish., 27: 119–121. (In Japanese with English abstract)

N- 35 Nishikawa, S. and Sakamoto, K. (1978) Comparative studies on the chromosomes in Japanese fishes-IV. Somatic chromosomes of two lizardfishes. J. Shimonoseki Univ. Fish., 27: 113–117. (In Japanese with English abstract)

N- 36 Nishikawa, S. and Sakamoto, K. (1982) The karyotype of a hexagrammid, *Hexagrammos otakii* (Pisces: Scorpaeniformes). Chrom. Inform. Serv., 33: 17–18.

N- 37 Nogusa, S. (1960) A comparative study of the chromosomes in fishes with particular considerations on taxonomy and evolution. Mem. Hyogo Univ. Agricul., 3: 1–62.

N- 38 Nowruzfashkhami, M.R., Pourkazemi, M. and Baradarannoveiri, S. (2000) Chromosome study of Persian sturgeon *Acipenser persicus* B. Cytologia, 65: 197–202.

N- 39 Nygren, A., Nilsson, B. and Jahnke, M. (1968) Cytological studies in Atlantic salmon (*Salmo salar*). Ann. Acad. Reg. Sci. Upsalien., 12: 21–52.

N- 40 Nygren, A., Edlund, P., Hirsch, U. and Åhsgren, L. (1968) Cytological studies in perch (*Perca fluviatilis* L.), pike (*Esox lucius* L.), pike-perch (*Lucioperca lucioperca* L.), and ruff (*Acerina cernua* L.). Hereditas, 59: 518–524.

N- 41 Nygren, A., Nilsson, B. and Jahnke, M. (1971) Cytological studies in *Thymallus thymallus* and *Coregonus albula*. Hereditas, 67: 269–274.

N- 42 Nygren, A., Nilsson, B. and Jahnke, M. (1971) Cytological studies in the smelt (*Osmerus eperlanus* L.). Hereditas, 67: 283–286.

N- 43 Nygren, A., Nilsson, B. and Jahnke, M. (1971) Cytological studies in Hypotremata and Pleurotremata (Pisces). Hereditas, 67: 275–282.

N- 44 Nygren, A., Nilsson, B. and Jahnke, M. (1971) Cytological studies in *Salmo trutta* and *Salmo alpinus*. Hereditas, 67: 259–268.

N- 45 Nygren, A., Leijon, U., Nilsson, B. and Jahnke, M. (1971) Cytological studies in *Coregonus* from Sweden. Ann. Acad. Reg. Sci. Upsalien., 15: 5–20.

N- 46 Nygren, A. and Jahnke, M. (1972) Microchromosomes in primitive fishes. Swedish J. Agric. Res., 2: 229–238.

N- 47 Nygren, A. and Jahnke, M. (1972) Cytological studies in *Myxine glutinosa* (Cyclostomata) from the Gullmaren Fjord in Sweden. Swedish J. Agric. Res., 2: 83–88.

N- 48 Nygren, A., Nilsson, B. and Jahnke, M. (1972) Cytological studies in Atlantic salmon from Canada, in hybrids between Atlantic salmon from Canada and Sweden and in hybrids between Atlantic salmon and sea trout. Hereditas, 70: 295–306.

N- 49 Nygren, A., Bergkvist, G., Windahl, T. and Jahnke, G. (1974) Cytological studies in Gadidae (Pisces). Hereditas, 76: 173–178.

N- 50 Nygren, A., Andersson, J., Jonsson, L. and Jahnke, G. (1975) Cytological studies in Cyprinidae (Pisces). Hereditas, 81: 165–172.

N- 51 Nygren, A., Nyman, L., Svensson, K. and Jahnke, G. (1975) Cytological and biochemical studies in back-crosses between the hybrid Atlantic salmon × sea trout and its parental species. Hereditas, 81: 55–62.

N- 52 Nirchio, M., Cervigón, F., Porto, J.I.R., Pérez, J.E., Gómez, J.A. and Villalaz, J. (2003) Karyotype supporting *Mugil curema* Valenciennes, 1836 and *Mugil gaimardianus* Desmarest, 1831 (Mugilidae: Teleostei) as two valid nominal species. Sci. Mar., 67: 113–115.

N- 53 Nirchio, M., Ron, E. and Rossi, A.R. (2005) Karyological characterization of *Mugil trichodon* Poey, 1876 (Pisces: Mugilidae). Sci. Mar., 69: 525–530.

N- 54 Nagata, Y., Nagata, A., Fujioka, Y. and Sakai, H. (1994) A karyotype study on two species of *Zacco* (type A, type B) (Pisces, Cyprinidae) from Yamaguchi Prefecture. J. Shimonoseki Univ. Fish., 43: 17–20. (In Japanese with English abstract)

N- 55 Nagpure, N.S., Kumar, R., Srivastava, S.K., Kushwaha, B., Singh, P.J., Gopalakrishnan, A. and Baseer, V.S. (2005) A report on high number of chromosomes in three endemic fish species from the Western Ghats of India. Nucleus, 48: 165–170.

N- 56 Nagpure, N.S., Kushwaha, B., Srivastava, S.K., Kumar, R., Gopalakrishnan, A., Baseer, V.S. and Verma, M.S. (2003) Characterization of three endemic fish species from Western Ghats using cytogenetic markers. Nucleus, 46: 110–114.

N- 57 Nagpure, N.S., Kumar, R., Srivastava, S.K., Gopalakrishnan, A., Verma, M.S. and Baseer, V.S. (2004) Cytogenetic studies of fish species *Horabagrus nigricollaris*, *Puntius denisonii* and *Puntius sarana subnasutus* endemic to the Western Ghats. Nucleus, 47: 143–148.

N- 58 Nagpure, N.S., Kumar, R., Verma, M.S., Srivastava, S.K., Anna Mercy, T.V., Gopalakrishnan, A. and Basheer, V.S. (2004) Characterization of fish, *Pristolepis marginata* Jerdon using cytogenetic markers. J. Cytol. Genet., 5 (NS): 167–172.

N- 59 Nirchio, M., Gaviria, J.I., Oliveira, C., Ferreira, I.A. and Martins, C. (2007) Cytogenetic analysis of three species of the genus *Haemulon* (Teleostei: Haemulinae) from Margarita Island, Venezuela. Genetica, 131: 135–140.

N- 60 Nirchio, M., Granado, A., Ron, E. and Pérez, J.E. (2002) Karyotype and nucleolar organizer regions in *Serrasalmus rhombeus* (Serrasalminae) from Caicara del Orinoco, Venezuela. Interciencia, 27: 676–678.

N- 61 Nirchio, M., Cequea, H. and Turner, B.J. (2003) Karyotypic characterization and nucleolus organizer regions in *Cyprinodon dearborni* (Meek, 1909) from Venezuela. Interciencia, 28: 352–354.

N- 62 Nakai, Y., Kubota, S., Goto, Y., Ishibashi, T., Davison, W. and Kohno, S. (1995) Chromosome elimination in three Baltic, south Pacific and north-east Pacific hagfish species. Chrom. Res., 3: 321–330.

N- 63 Naran, D., Skelton, P.H. and Villet, M.H. (2006) Karyology of the redfin minnows, genus *Pseudobarbus* Smith, 1841 (Teleostei: Cyprinidae): one of the evolutionarily tetraploid lineages of South African barbines. Afr. Zool., 41: 178–182.

N- 64 Nirchio, M., Oliveira, C., Ferreira, I.A., Martins, C., Rossi, A.R. and Sola, L. (2009) Classical and molecular cytogenetic characterization of *Agonostomus monticola*, a primitive species of Mugilidae (Mugiliformes). Genetica, 135: 1–5.

N- 65 Nagpure, N.S., Kumar, R., Srivastava, S.K., Kushwaha, B., Gopalakrishnan, A. and Basheer, V.S. (2006) Cytogenetic characterization of two marine ornamental fishes, *Chaetodon collare* and *Stegastes insularis*. J. Mar. Biol. Ass. India, 48: 267–269.

N- 66 Nirchio, M., Oliveira, C., Ferreira, I.A., Pérez, J.E., Gaviria, J.I., Harrison, I., Rossi, A.R. and Sola, L. (2007) Comparative cytogenetic and allozyme analysis of *Mugil rubrioculus* and *M. curema* (Teleostei: Mugilidae) from Venezuela. Interciencia, 32: 757–762.

N- 67 Naran, D., Skelton, P.H. and Villet, M.H. (2007) Karyology of three evolutionarily hexaploid southern African species of yellowfish, *Labeobarbus* Rüppel, 1836 (Cyprinidae). African Zool., 42: 254–260.

N- 68 Nelson, J.S. (2006) Fishes of the World. Fourth edition. John Wiley & Sons, New Jersey, USA.

N- 69 Nirchio, M., Oliveira, C., Ferreira, I.A., Granado, A. and Ron, E. (2007) Extensive polymorphism and chromosomal characteristics of ribosomal DNA in the characid fish *Triportheus venezuelensis* (Characiformes, Characidae). Genet. Mol. Biol., 30: 25–30.

N- 70 Nirchio, M., Rondón, R., Oliveira, C., Ferreira, I.A., Martins, C., Pérez, J., Sola, L. and Rossi, A.R. (2008) Cytogenetic studies in three species of *Lutjanus* (Perciformes: Lutjanidae: Lutjaninae) from the Isla Margarita, Venezuela. Neotrop. Ichthyol., 6: 101–108.

N- 71 Nakayama, C.M., Porto, J.I.R. and Feldberg, E. (2002) A comparative cytogenetic study of five piranha species (*Serrasalmus*, Serrasalminae) from the Amazonian basin. Genetica, 114: 231–236.

N- 72 Nagpure, N.S., Kumar, R., Kushwaha, B., Srivastava, S.K., Sigh, P.J. and Lakra, W.S. (2006) Karyological studies of fish species *Garra surendranathanii* and *Barilius gatensis* from the Western Ghats of India. Nucleus, 49: 171–175.

N- 73 Nagpure, N.S., Kumar, R., Srivastava, S.K., Gopalakrishnan, A., Basheer, V.S. and Kapoor, D. (2006) Cytogenetic studies of two cyprinids *Garra mullya* and *Barilius bakeri* from the Western Ghats of India. J. Cytol. Genet., 7 (NS): 69–77.

N- 74 Nirchio, M., Cipriano, R., Cestari, M. and Fenocchio, A. (2005) Cytogenetical and morphological features reveal significant differences among Venezuelan and Brazilian samples of *Mugil curema* (Teleostei: Mugilidae). Neotrop. Ichthyol., 3: 107–110.

N- 75 Nakatani, Y., Takeda, H., Kohara, Y. and Morishita, S. (2007) Reconstruction of the vertebrate ancestral genome reveals dynamic genome reorganization in early vertebrates. Genome Res., 17: 1254–1265.

N- 76 Nirchio, M., Oliveira, C., Ferreira, D.C., Rondón, R., Pérez, J.E., Hett, A.K., Rossi, A.R. and Sola, L. (2009) Cytogenetic characterization of *Rhomboplites aurorubens* and *Ocyurus chrysurus*, two monotypic genera of Lutjaninae from Cubagua Island, Venezuela, with a review of the cytogenetics of Lutjanidae (Teleostei: Perciformes). Neotrop. Ichthyol., 7: 587–594.

N- 77 Naruse, K., Tanaka, M., Mita, K., Shima, A., Postlethwait, J. and Mitani, H. (2004) A medaka gene map: the trace of ancestral vertebrate proto-chromosomes revealed by comparative gene mapping. Genome Res., 14: 820–828.

[O]

O- 1 Oberdorff, T., Ozouf-Costaz, C. and Agnese, J.F. (1990) Chromosome banding in African catfishes: nucleolar organizer regions in five species of the genus *Synodontis* and one of the genus *Hemisynodontis* (Pisces, Mochokidae). Caryologia, 43: 9–16.

O- 2 Oellermann, L.K. and Skelton, P.H. (1990) Hexaploidy in yellowfish species (*Barbus*, Pisces, Cyprinidae) from southern Africa. J. Fish Biol., 37: 105–115.

O- 3 Ohno, S., Stenius, C., Faisst, E. and Zenzes, M.T. (1965) Post-zygotic chromosomal rearrangements in rainbow trout (*Salmo irideus* Gibbons). Cytogenetics, 4: 117–129.

O- 4 Ohno, S. and Atkin, N.B. (1966) Comparative DNA values and chromosome complements of eight species of fishes. Chromosoma, 18: 455–466.

O- 5 Ohno, S., Muramoto, J., Christian, L. and Atkin, N.B. (1967) Diploid-tetraploid relationship among Old-World members of the fish family Cyprinidae. Chromosoma, 23: 1–9.

O- 6 Ohno, S., Wolf, U. and Atkin, N.B. (1968) Evolution from fish to mammals by gene duplication. Hereditas, 59: 169–187.

O- 7 Ohno, S., Muramoto, J., Stenius, C., Christian, L., Kittrell, W.A. and Atkin, N.B. (1969) Microchromosomes in holocephalian, chondrostean and holostean fishes. Chromosoma, 26: 35–40.

O- 8 Ohno, S., Muramoto, J., Klein, J. and Atkin, N.B. (1969) Diploid-tetraploid relationship in clupeoid and salmonoid fish. In: Darlington, C.D. and Lewis, K.R. eds. Chromosomes today. Vol. 2. Oliver & Boyd, Edinburgh, pp. 139–147.

O- 9 Ohno, S. (1970) The enormous diversity in genome sizes of fish as a reflexion of nature's extensive experiments with gene duplication. Trans. Amer. Fish. Soc., 99: 120–130.

O- 10 Ohno, S., Christian, L., Romero, M., Dofuku, R. and Ivey, C. (1973) On the question of American eels, *Anguilla rostrata* versus European eels, *Anguilla anguilla*. Experientia, 29: 891.

O- 11 Ohno, S. (1974) Animal cytogenetics. Vol. 4: Chordata 1. Protochordata, Cyclostomata and Pisces. Gebrueder Borntraeger, Berlin.

O- 12 Ojima, Y., Hitotsumachi, S. and Makino, S. (1966) Cytogenetic studies in lower vertebrates. I. A preliminary report on the chromosomes of the funa (*Carassius auratus*) and gold-fish (a revised study). Proc. Japan Acad., 42: 62–66.

O- 13 Ojima, Y. and Hitotsumachi, S. (1967) Cytogenetic studies in lower vertebrates. IV. A note on the chromosomes of the carp (*Cyprinus carpio*) in comparison with those of the funa and the goldfish (*Carassius auratus*). Japan. J. Genet., 42: 163–167.

O- 14 Ojima, Y. and Hitotsumachi, S. (1969) Cytogenetical studies in loaches (Pisces, Cobitidae). Zool. Mag. Japan, 78: 139–141. (In Japanese with English abstract)

O- 15 Ojima, Y. and Hitotsumachi, S. (1969) The karyotype of the medaka, *Oryzias latipes*. Chrom. Inform. Serv., 10: 15–16.

O- 16 Ojima, Y., Hayashi, M. and Ueno, K. (1970) Cytogenetic studies in lower vertebrates, IX. A study of the chromosomes of five species of cyprinid fishes of Japan. Chrom. Inform. Serv., 11: 27–29.

O- 17 Ojima, Y., Hitotsumachi, S. and Hayashi, M. (1970) A blood culture method for fish chromosomes. Japan. J. Genet., 45: 161–162.

O- 18 Ojima, Y., Hayashi, M. and Ueno, K. (1972) Cytogenetic studies in lower vertebrates. X. Karyotype and DNA studies in 15 species of Japanese Cyprinidae. Japan. J. Genet., 47: 431–440.

O- 19 Ojima, Y., Takayama, S. and Yamamoto, K. (1972) Chromosome preparation from cultured scale epithelium of teleost fish. Japan. J. Genet., 47: 445–446.

O- 20 Ojima, Y., Ueno, K. and Hayashi, M. (1973) Karyotypes of the acheilognathine fishes (Cyprinidae) of Japan with a discussion of phylogenetic problems. Zool. Mag. Japan, 82: 171–177. (In Japanese with English abstract)

O- 21 Ojima, Y., Hayashi, M. and Ueno, K. (1975) Triploidy appeared in the back-cross offspring from funa-carp crossing. Proc. Japan Acad., 51: 702–706.

O- 22 Ojima, Y., Ueno, K. and Hayashi, M. (1976) A review of the chromosome numbers in fishes. La Kromosomo, II, 1: 19–47.

O- 23 Ojima, Y. and Asano, N. (1977) A cytological evidence for gynogenetic development of the ginbuna (*Carassius auratus langsdorfii*). Proc. Japan Acad., Ser. B, 53: 138–142.

O- 24 Ojima, Y. and Ueda, T. (1978) New C-banded marker chromosomes found in carp-funa hybrids. Proc. Japan Acad., Ser. B, 54: 15–20.

O- 25 Ojima, Y. and Makino, S. (1978) Triploidy induced by cold shock in fertilized eggs of the carp. Proc. Japan Acad., Ser. B, 54: 359–362.

O- 26 Ojima, Y., Ueda, T. and Narikawa, T. (1979) A cytogenetic assessment on the origin of the gold-fish. Proc. Japan Acad., Ser. B, 55: 58–63.

O- 27 Ojima, Y. and Kashiwagi, E. (1979) A karyotype study of eleven species of labrid fishes from Japan. Proc. Japan Acad., Ser. B, 55: 280–285.

O- 28 Ojima, Y. and Takai, A. (1979) The occurrence of spontaneous polyploid in the Japanese common loach, *Misgurnus anguillicaudatus*. Proc. Japan Acad., Ser. B, 55: 487–491.

O- 29 Ojima, Y. and Takai, A. (1979) Further cytogenetical studies on the origin of the goldfish. Proc. Japan Acad., Ser. B, 55: 346–350.

O- 30 Ojima, Y. and Yamano, T. (1980) The assignment of the nucleolar organizer in the chromosomes of the funa (*Carassius*, Cyprinidae, Pisces). Proc. Japan Acad., Ser. B, 56: 551–556.

O- 31 Ojima, Y. and Kashiwagi, E. (1980) Further studies of the chromosomes of the Labridae (Pisces). A preliminary note. Proc. Japan Acad., Ser. B, 56: 328–331.

O- 32 Ojima, Y. and Yamamoto, T. (1980) A chromosome study of the holostean long nose gar, *Lepisosteus osseus*. Chrom. Inform. Serv., 28: 7–8.

O- 33 Ojima, Y., Maeki, K., Takayama, S. and Nogusa, S. (1963) A cytotaxonomic study on the Salmonidae. Nucleus, 6: 91–98.

O- 34 Ojima, Y. and Kurishita, A. (1980) A new method to increase the number of mitotic cells in the kidney tissue for fish chromosome studies. Proc. Japan Acad., Ser. B, 56: 610–615.

O- 35 Ojima, Y. and Takai, A. (1981) A karyotype study of colored-carp (*Cyprinus carpio*). A preliminary report. Proc. Japan Acad., Ser. B, 57: 7–12.

O- 36 Ojima, Y. and Kashiwagi, E. (1981) Chromosomal evolution associated with Robertsonian fusion in the genus *Dascyllus* (Chrominae, Pisces). Proc. Japan Acad., Ser. B, 57: 368–370.

O- 37 Ojima, Y. (1982) Evolution of the sex chromosomes in Pisces. A review. La Kromosomo, II, 27/28: 849–859. (In Japanese)

O- 38 Ojima, Y. (1982) Methods in fish cytogenetics. Nucleus, 25: 1–7.

O- 39 Ojima, Y. and Ueda, H. (1982) A karyotypical study of the conger eel (*Conger myriaster*) in *in vitro* cells, with special regard to the identification of the sex chromosome. Proc. Japan Acad., Ser. B, 58: 56–59.

O- 40 Ojima, Y., Ueda, H. and Takai, A. (1984) Sex chromosome differentiation in *Parapercis sexfasciata* (Perciformes, Pisces). Proc. Japan Acad., Ser. B, 60: 137–140.

O- 41 Ojima, Y. and Kojima, T. (1985) Chromosomal polymorphisms in Apogonidae fishes. Proc. Japan Acad., Ser. B, 61: 79–82.

O- 42 Ojima, Y., Takarada, Y. and Takai, A. (1986) Microcell-mediated transfer of fish chromosomes into mouse cells. Proc. Japan Acad., Ser. B, 62: 91–94.

O- 43 Ojima, Y. and Kikuno, T. (1986) A heteromorphic chromosome of *Beryx splendens*, Berycidae (Pisces). Proc. Japan Acad., Ser. B, 62: 317–320.

O- 44 Ojima, Y., Nakanishi, Y. and Takai, A. (1986) Chromosomal studies of cultured cells from the hybrids between *Huso huso* and *Acipenser ruthenus*. Proc. Japan Acad., Ser. B, 62: 87–90.

O- 45 Ojima, Y. (1987) The origin of natural monument, scarlet crucian carp (hibuna) in the Harutori Lake, Kushiro, Hokkaido, Japan. Proc. Japan Acad., Ser. B, 63: 365–368.

O- 46 Ojima, Y. and Kikuno, T. (1987) Karyotypes of a gobiesociform and two perciform fishes (Teleostei). Proc. Japan Acad., Ser. B, 63: 201–204.

O- 47 Ojima, Y., Takai, A. and Yamamoto, K. (1988) Production and chromosomal features of triploid hybrids between *Rhodeus ocellatus ocellatus* and *Acheilognathus limbatus* (Cyprinidae, Pisces). Proc. Japan Acad., Ser. B, 64: 225–228.

O- 48 Ojima, Y. and Yamamoto, K. (1990) Cellular DNA contents of fishes determined by flow cytometry. La Kromosomo, II, 57: 1871–1888.

O- 49 Ojima, Y. (1983) Fish cytogenetics. In: Sharma, A.K. and Sharma, A. eds. Chromosomes in evolution of eukaryotic groups. Vol. I. CRC press, Florida, pp. 111–145.

O- 50 Oliveira, C., Almeida-Toledo, L.F., Foresti, F., Britsky, H.A. and Toledo-Filho, S.A. (1988) Chromosome formulae of neotropical freshwater fishes. Brazil. J. Genet., 11: 577–624.

O- 51 Oliveira, C., Almeida-Toledo, L.F., Foresti, F. and Toledo-Filho, S.A. (1988) Supernumerary chromosomes, Robertsonian rearrangement and multiple NORs in *Corydoras aeneus* (Pisces, Siluriformes, Callichthyidae). Caryologia, 41: 227–236.

O- 52 Oliveira, C., Almeida-Toledo, L.F. and Toledo-Filho, S.A. (1990) Comparative cytogenetic analysis of three cytotypes of *Corydoras nattereri* (Pisces, Siluriformes, Callichthyidae). Cytologia, 55: 21–26.

O- 53 Oliveira, C., Andreata, A.A., Almeida-Toledo, L.F. and Toledo-Filho, S.A. (1991) Karyotype and nucleolus organizer regions of *Pyrrhulina* cf. *australis* (Pisces, Characiformes, Lebiasinidae). Brazil. J. Genet., 14: 685–690.

O- 54 Oliveira, C., Almeida-Toledo, L.F., Mori, L. and Toledo-Filho, S.A. (1992) Extensive chromosomal rearrangements and nuclear DNA content changes in the evolution of the armoured catfishes genus *Corydoras* (Pisces, Siluriformes, Callichthyidae). J. Fish Biol., 40: 419–431.

O- 55 Oliveira, C., Almeida-Toledo, L.F., Mori, L. and Toledo-Filho, S.A. (1993) Cytogenetic and DNA content studies of armoured catfishes of the genus *Corydoras* (Pisces, Siluriformes, Callichthyidae) from the southeast coast of Brazil. Brazil. J. Genet., 16: 617–629.

O- 56 Oliveira, C., Almeida-Toledo, L.F., Mori, L. and Toledo-Filho, S.A. (1993) Cytogenetic and DNA content in six genera of the family Callichthyidae (Pisces, Siluriformes). Caryologia, 46: 171–188.

O- 57 Oliveira, C., Torres, R.A., Favorito, S. and Foresti, F. (1997) Cytogenetic studies of *Mastacembelus armatus* (Pisces, Mastacembelidae). Cytobios, 92: 83–89.

O- 58 Oliveira, C., Saboya, S.M.R., Foresti, F., Senhorini, J.A. and Bernardino, G. (1997) Increased B chromosome frequency and absence of drive in the *Prochilodus lineatus*. Heredity, 79: 473–476.

O- 59 Oliveira, C. and Wright, J.M. (1998) Molecular cytogenetic analysis of heterochromatin in the chromosomes of tilapia, *Oreochromis niloticus* (Teleostei: Cichlidae). Chrom. Res., 6: 205–211.

O- 60 Oliveira, C. and Gosztonyi, A.E. (2000) A cytogenetic study of *Diplomystes mesembrinus* (Teleostei, Siluriformes, Diplomystidae) with a discussion of chromosome evolution in siluriforms. Caryologia, 53: 31–37.

O- 61 Olmo, E., Stingo, V., Odierna, G. and Capriglione, T. (1980) Cryptic polyploidy in sharks and rays as revealed by DNA renaturation kinetics. Acad. Nat. Lincei, ser. 8, 68: 555–560.

O- 62 Olmo, E., Stingo, V., Odierna, G. and Capriglione, T. (1982) Sequence organization in the DNA of three selachians. Experientia, 38: 339–340.

O- 63 Osinov, A.G., Vasilieva, E.D. and Vasiliev, V.P. (1983) Hybrid origin of unisexual triploid form of the genus *Cobitis* (Cobitidae, Pisces). Dokl. Akad. Sci. USSR, 272: 716–718. (In Russian)

O- 64 Ostberg, C.O. and Thorgaard, G.H. (1999) Geographic distribution of chromosome and microsatellite DNA polymorphisms in *Oncorhynchus mykiss* native to western Washington. Copeia, 1999: 287–298.

O- 65 Ota, K., Kobayashi, T., Ueno, K. and Gojobori, T. (2000) Evolution of heteromorphic sex chromosomes in the order Aulopiformes. Gene, 259: 25–30.

O- 66 Oyhenart-Perera, M.F., Luengo, J.A. and Brum-Zorrilla, N. (1975) Estudio, citogenetico de *Cichlasoma facetum* (Jenyns) y *Crenicichla sexatilis* (Linn.) (Teleostei, Cichlidae). Rev. Biol. Uruguay, 3: 29–36.

O- 67 Ozouf-Costaz, C. and Bazignan, M.D. (1987) Chromosome relationships among 15 species of Nototheniidae. Proc. 5th Congr. Europ. Ichthyologists, pp. 413–419.

O- 68 Ozouf-Costaz, C. (1987) Karyotypes of *Chaenodraco wilsoni* and *Chionodraco myersi* (Channichthyidae) from Prydz Bay, Antarctica. Copeia, 1987: 503–505.

O- 69 Ozouf-Costaz, C., Teugels, G.G. and Legendre, M. (1990) Karyological analysis of three strains of the African catfish, *Clarias gariepinus* (Clariidae), used in aquaculture. Aquaculture, 87: 271–277.

O- 70 Ozouf-Costaz, C., Hureau, J.C. and Beaunier, M. (1991) Chromosome studies on fish of the suborder Notothenioidei collected in the Weddell Sea during Epos 3 Cruise. Cybium, 15: 271–289.

O- 71 Ozouf-Costaz, C., Pisano, E., Thaeron, C. and Hureau, J.-C. (1997) Antarctic fish chromosome banding: significance for evolutionary studies. Cybium, 21: 399–409.

O- 72 Oliveira, C., Alves, A.L. and Foresti, F. (2006) The karyotype of *Scoloplax distolothrix* (Teleostei: Siluriformes: Scoloplacidae). Cytologia, 71: 257–259.

O- 73 Ojima, Y. (1983) Fish Cytogenetics. Suikosha, Tokyo. (In Japanese)

O- 74 Olmo, E., Stingo, V., Cobror, O., Capriglione, T. and Odierna, G. (1982) Repetitive DNA and polyploidy in selachians. Comp. Biochem. Physiol., 73B: 739–745.

O- 75 Osinov, A.G., Vasilieva, Y.D. and Vasiliev, V.P. (1990) The problem of reticulate speciation in vertebrates: diploid-triploid-tetraploid complexes in the genus *Cobitis* (Cobitidae). II. Characteristics of the triploid form. J. Ichthyol., 30(4): 59–67.

O- 76 Oliveira, C. and Almeida Toledo, L.F. (2006) Evidence of an XX/XY sex chromosome system in the fish *Dormitator maculatus* (Teleostei, Eleotrididae). Genet. Mol. Biol., 29: 653–655.

O- 77 Ocalewicz, K., Hliwa, P., Krol, J., Rábová, M., Stabinski, R. and Ráb, P. (2007) Karyotype and chromosomal characteristics of Ag-NOR sites and 5S rDNA in European smelt, *Osmerus eperlanus*. Genetica, 131: 29–35.

O- 78 Oliveira, C., Nirchio, M., Granado, Á. and Levy, S. (2003) Karyotypic characterization of *Prochilodus mariae*, *Semaprochilodus kneri* and *S. laticeps* (Teleostei: Prochilodontidae) from Caicara del Orinoco, Venezuela. Neotrop. Ichthyol., 1: 47–52.

O- 79 Oliveira, R.R., Souza, I.L. and Venere, P.C. (2006) Karyotype description of three species of Loricariidae (Siluriformes) and occurrence of the ZZ/ZW sexual system in *Hemiancistrus spilomma* Cardoso & Lucinda, 2003. Neotrop. Ichthyol., 4: 93–97.

O- 80 Oliveira, R.R., Feldberg, E., Anjos, M.B. and Zuanon, J. (2007) Karyotype characterization and ZZ/ZW sex chromosome heteromorphism in two species of the catfish genus *Ancistrus* Kner, 1854 (Siluriformes: Loricariidae) from the Amazon basin. Neotrop. Ichthyol., 5: 301–306.

O- 81 Oliveira, C., Foresti, F. and Pedroso, G. (1992) Análise da distribuiçao da heterocromatina constitutiva em *Salminus maxillosus* (Dourado) e *S. hilarii* (Tabarana). IV. Simp. Citogenet. Evol. e Aplic. de Peixes Neotropicais, Rio de Janeiro, RJ, p. 27 (abstract). (In Portuguese)

*O- 82 Oliveira, M.I.B., Sanguino, E.C.B. and Falcao, J.N. (1992) Estudos citogenéticos em *Pyrrhulina* sp. (Teleostei, Characiformes, Lebiasinidae). IV. Simp. Citogenet. Evol. E Aplic. De Peixes Neotropicais, Rio de Janeiro, RJ, p. 13 (abstract). (After K-114)

*O- 83 Ozouf-Costaz, C., Parrent, M., Teugels, G.G. and Van der Berg, E. (1994) Further karyological data on African clariid catfishes (Ostariophysi, Siluroidei). Abstr. VIII Congr. Soc. Europ. Ichthyol., Oviedo, p. 56. (After K-114)

O- 84 Oliveira, R.R., Feldberg, E., Dos Anjos, M.B. and Zuanon, J. (2008) Occurrence of multiple sexual chromosomes (XX/XY_1Y_2 and $Z_1Z_1Z_2Z_2/Z_1Z_2W_1W_2$) in catfishes of the genus *Ancistrus* (Siluriformes: Loricariidae) from the Amazon basin. Genetica, 134: 243–249.

O- 85 Oliveira, R.R., Feldberg, E., Dos Anjos, M.B. and Zuanon, J. (2009) Mechanisms of chromosomal evolution and its possible relation to natural history characteristics in *Ancistrus* catfishes (Siluriformes: Loricariidae). J. Fish Biol., 75: 2209–2225.

O- 86 Oliveira, C., Almeida-Toledo, L.F. and Foresti, F. (2007) Karyotypic evolution in neotropical fishes. In: Pisano, E., Ozouf-Costaz, C., Foresti, F. and Kapoor, B.G. eds. Fish cytogenetics. Science Publishers, Enfield, pp. 111–164.

[P]

P- 1 Pacheco, R.B., Giuliano-Caetano, L. and Dias, A.L. (2001) Cytotypes and multiple NORs in an *Astyanax altiparanae* population (Pisces, Tetragonopterinae). Chrom. Sci., 5: 109–114.

P- 2 Paintner-Marques, T.R., Giuliano-Caetano, L. and Dias, A.L. (2002) Karyotypic diversity in a *Bryconamericus* aff. *exodon* population (Characidae, Tetragonopterinae). Cytologia, 67: 397–402.

P- 3 Palazón, J.L., Nirchio, M. and Sarasquete, C. (2003) Conventional karyotype and nucleolar organiser regions of the toadfish *Halobatrachus didactylus* (Schneider, 1801) (Pisces: Batrachoididae). Sci. Mar., 67: 445–449.

P- 4 Pan, W.M. (1996) The karyotype and the rate of chromosome spontaneous generation aberration of *Sebasticus marmoratus* Cuvier and Valenciennes. J. Fish. China, 20: 175–177. (In Chinese)

P- 5 Petrov, D.A. (2001) Evolution of genome size: new approaches to an old problem. Trends Genet., 17(1): 23–28.

P- 6 Park, E.H. (1974) A list of the chromosome numbers of fishes. College Rev. College Liberal Arts Sci., Seoul Natn. Univ., 20: 346–372.

P- 7 Park, E.H. and Kang, Y.S. (1976) Karyotype conservation and difference in DNA amount in anguilloid fishes. Science, 193: 64–66.

P- 8 Park, E.H. and Park, S.D. (1979) A culture technique using marine fish kidney to obtain chromosomes. J. Fish. Res. Board Can., 36: 458–461.

P- 9 Park, E.H. and Kang, Y.S. (1979) Karyological confirmation of conspicuous ZW sex chromosomes in two species of Pacific anguilloid fishes (Anguilliformes: Teleostomi). Cytogenet. Cell Genet., 23: 33–38.

P- 10 Park, E.H. and Grimm, H. (1981) Distribution of C-band heterochromatin in the ZW sex chromosomes of European and American eels (Anguillidae, Teleostomi). Cytogenet. Cell Genet., 31: 167–174.

P- 11 Park, E.H. and Kang, Y.S. (1981) Karyotype and genome size of two variants of mandarine fish, *Siniperca scherzeri* (Teleostei; Serranidae). Korean J. Genet., 3: 63–68.

P- 12 Park, I.S., Kim, C.H. and Kim, D.S. (1999) Karyotypes and cellular DNA contents of two species in the genus, *Lateolabrax* from Korea. Fish. Sci., 65: 488–489.

P- 13 Passakas, T. (1976) Further investigations on the chromosomes of *Anguilla anguilla*. Folia Biol. (Krakow), 24: 239–244.

P- 14 Passakas, T. (1981) Comparative studies on the chromosomes of the European eel (*Anguilla anguilla* L.) and the American eel (*Anguilla rostrata* Le Sueur). Folia Biol. (Krakow), 29: 41–57.

P- 15 Pastori, M.C., Cano, J., Bertollo, C. and Fenocchio, A.S. (1998) Cytogenetic study of two species of needlefish (Belonidae) from Argentina. Ital. J. Zool., 65 suppl.: 57–60.

P- 16 Patro, R. and Prasad, R. (1979) Chromosomes of six marine percoids from the Indian Sea. Ind. Biologist, 11: 9–12.

P- 17 Patro, R. and Prasad, R. (1981) Chromosomal studies in five Indian flatfishes. Copeia, 1981: 498–503.

P- 18 Paugy, D., Guégan, J.F. and Agnèse, J.F. (1990) Three simultaneous and independent approaches to the characterization of a new species of *Labeo* (Teleostei, Cyprinidae) from West Africa. Can. J. Zool., 68: 1124–1131.

P- 19 Pauls, E. and Bertollo, L.A.C. (1983) Evidence for a system of supernumerary chromosomes in *Prochilodus scrofa* Steindachner, 1881 (Pisces, Prochilodontidae). Caryologia, 36: 307–314.

P- 20 Pauls, E. and Bertollo, L.A.C. (1990) Distribution of a supernumerary chromosome system and aspects of karyotypic evolution in the genus *Prochilodus* (Pisces, Prochilodontidae). Genetica, 81: 117–123.

P- 21 Pazza, R. and Júlio Jr., H.F. (2003) Occurrence of three sympatric cytotypes of *Hoplias malabaricus* (Pisces, Erythrinidae) in the Upper Paraná River foodplain (Brazil). Cytologia, 68: 159–163.

P- 22 Peňaz, M., Ráb, P. and Prokeš, M. (1979) Cytological analysis, gynogenesis and early development of *Carassius auratus gibelio*. Acta Sci. Nat. Brno, 13(7): 1–33.

P- 23 Perry, J., Slater, H.R. and Choo, K.H.A. (2004) Centric fission—simple and complex mechanisms. Chrom. Res., 12: 627–640.

P- 24 Pezold, F. (1984) Evidence for multiple sex chromosomes in the freshwater goby, *Gobionellus shufeldti* (Pisces: Gobiidae). Copeia, 1984: 235–238.

P- 25 Pfister, S.C., Moreira Filho, O. and Bertollo, L.A.C. (1997) Cytogenetic studies in *Orthospinus francisensis* (Pisces, Characidae, Stethaprioninae) and some considerations about karyotype evolution of the group. Cytobios, 91: 97–101.

P- 26 Phan, V.N., Gomes, V. and Suzuki, H. (1986) Estudos citogenéticos de peixes Antárticos. I. Cariótipos de *Notothenia gibberifrons* (Lönnberg, 1905), *Trematomus bernacchii* (Boulenger, 1902) e *T. hansoni* (Boulenger, 1902) (Perciformes, Nototheniidae). An. Acad. Brasil. Cienc., 58 (suppl.): 23–27. (In Portuguese with English summary)

P- 27 Phan, V.N., Gomes, V., Suzuki, H. and Rocha Passos, M.J.A.C. (1987) Karyotypes of two Antarctic fishes, *Notothenia gibberifrons* and *Notothenia coriiceps neglecta*. Japan. J. Ichthyol., 33: 384–387.

P- 28 Phillips, R.B., Zajicek, K.D. and Utter, F.M. (1985) Q band chromosomal polymorphisms in chinook salmon (*Oncorhynchus tshawytscha*). Copeia, 1985: 273–278.

P- 29 Phillips, R.B. and Kapuscinksi, A.R.D. (1987) A Robertsonian polymorphism in pink salmon (*Oncorhynchus gorbuscha*) involving the nucleolar organizer region. Cytogenet. Cell Genet., 44: 148–152.

P- 30 Phillips, R.B., Pleyte, K.A. and Hartley, S.E. (1988) Stock-specific differences in the number and chromosome positions of the nucleolar organizer regions in arctic char (*Salvelinus alpinus*). Cytogenet. Cell Genet., 48: 9-12.

P- 31 Phillips, R.B., Pleyte, K.A. and Ihssen, P.E. (1989) Patterns of chromosomal nucleolar organizer region (NOR) variation in fishes of the genus *Salvelinus*. Copeia, 1989: 47–53.

P- 32 Phillips, R.B., Zajicek, K.D. and Ihssen, P.E. (1989) Population differences in chromosome-banding polymorphisms in lake trout. Trans. Amer. Fish. Soc., 118: 64–73.

P- 33 Phillips, R.B., Reed, K.M. and Ráb, P. (1996) Revised karyotypes and chromosome banding of coregonid fishes from the Laurentian Great Lakes. Can. J. Zool., 74: 323–329.

P- 34 Phillips, R.B. and Ráb, P. (2001) Chromosome evolution in the Salmonidae (Pisces): an update. Biol. Rev., 76: 1–25.

P- 35 Pijnacker, L.P. and Ferwerda, M.A. (1995) Zebrafish chromosome banding. Genome, 38: 1052–1055.

P- 36 Pisano, E., Ghigliotti, L., Mazzei, F. and Ozouf-Costaz, C. (2003) Cytogenetic features of *Notothenia angustata* Hutton, 1875, an Antarctic fish living in non-Antarctic waters. In: Huiskes, A.H.L., Gieskes, W.W.C., Rozema, J., Schorno, R.M.L., van der Vies, S.M. and Wolff, W.J. eds. Antarctic biology in a global context. Backhuys Publisher, Leiden, pp. 117–120.

P- 37 Popov, P. and Zekov, A. (1985) A study on the karyotype of *Tinca tinca*. Anim. Sci., 22: 43–49. (In Russian with English summary)

P- 38 Portela, A.L.B.S., Galetti Jr., P.M. and Bertollo, L.A.C. (1988) Considerations on the chromosome evolution of Tetragonopterinae (Pisces, Characidae). Brazil. J. Genet., 11: 307–316.

P- 39 Portela-Castro, A.L.B. and Júlio Jr., H.F. (2002) Karyotype relationships among species of subfamily Tetragonopterinae (Pisces, Characidae): cytotaxonomy and evolution aspects. Cytologia, 67: 329–336.

P- 40 Porto, J.I.R., Feldberg, E., Falcão, J.N. and Nakayama, C.M. (1993) Cytogenetic studies in Hemiodontidae (Ostariophysi, Characiformes) fishes from the Central Amazon. Cytologia, 58: 397–402.

P- 41 Porto-Foresti, F., Oliveira, C. and Foresti, F. (2000) Chromosomes and nuclear DNA content of *Diplectrum radiale* (Perciformes, Serranidae). Chrom. Sci., 4: 95–97.

P- 42 Porto-Foresti, F., Andreata, A.A., Oliveira, C. and Foresti, F. (2000) The karyotype of *Pseudoplatystoma fasciatum* (Teleostei, Siluriformes) from the Rio Paraguay basin. Chrom. Sci., 4: 99–102.

P- 43 Porto-Foresti, F., Oliveira, C. and Foresti, F. (2005) First chromosome characterization in the neotropical eel, *Gymnothorax ocellatus* (Pisces, Muraenidae). Cytologia, 70: 283–286.

P- 44 Post, A. (1965) Vergleichende Untersuchungen der Chromosomenzahlen bei Susswasser-Teleosteern. Z. Zool. Syst. Evol.-Forsch., 3: 47–93.

P- 45 Post, A. (1972) Ergebnisse der Forschungsreisen des FFS "Walther Herwig" nach Südamerika. XXIV. Die Chromosomenzahlen einiger atlantischer Myctophidenarten (Osteichthyes, Myctophoidei, Myctophidae). Arch. FischWiss., 23: 89–93.

P- 46 Post, A. (1974) Ergebnisse der Forschungsreisen des FFS "Walther Herwig" nach Südamerika. XXXIV. Die Chromosomen von drei Arten aus der Familie Gonostomatidae (Osteichthyes, Stomiatoidei). Arch. FischWiss., 25: 51–55.

P- 47 Powers, P.K. and Gold, J.R. (1992) Cytogenetic studies in North American minnows (Cyprinidae). XX. Chromosomal NOR variation in the genus *Luxilus*. Copeia, 1992: 332–343.

P- 48 Prasad, R. and Manna, G.K. (1974) Somatic and germinal chromosomes of a live-fish, *Heteropneustes fossilis* (Bloch). Caryologia, 27: 217–223.

P- 49 Prasad, R. and Manna, G.K. (1976) Chromosomes of the fishes, *Tilapia mossambica* and *Notopterus notopterus*. Chrom. Inform. Serv., 21: 11–13.

P- 50 Prirodina, V.P. (1978) Karyotype of the New Zealand spiny dogfish of the genus *Squalus* L. (fam. Squalidae). In: Skarlato, O.A., Andoiyashev, A.P., Barsukov, V.V., Dorofeeva, E.A., Korovina, V.M. and Neelov, A.V. eds. Morphology and systematics of fish. Akademiya Nauk SSSR, Leningrad, pp. 53–56. (In Russian)

P- 51 Prirodina, V.P. (1984) Karyotypes of three species of the nototheniid fishes. Biol. Morya, Vladivostok, 1984(3): 74–76. (In Russian)

P- 52 Prirodina, V.P. and Neyelov, A.V. (1984) Karyotypes in two species of the genus *Notothenia* s. str. (fam. Nototheniidae) from the West Antarctica. Trudy Zool. Inst., 127: 32–37. (In Russian)

P- 53 Phillips, R.B. (2007) Application of fluorescence *in situ* hybridization (FISH) to genome mapping in fishes. In: Pisano, E., Ozouf-Costaz, C., Foresti, F. and Kapoor, B.G. eds. Fish cytogenetics. Science Publishers, Enfield, pp. 455–471.

P- 54 Phillips, R.B. and Ihssen, P.E. (1985) Chromosome banding in salmonid fish: nucleolar organizer regions in *Salmo* and *Salvelinus*. Can. J. Genet. Cytol., 27: 433–440.

P- 55 Prirodina, V.P. (1986) Karyotype of *Cottoperca gobio* (Bovichthyidae, Notothenioidei) as compared to karyotypes of other Notothenioidei. Morphology and distribution of fishes of the Southern Ocean. Proc. Zool. Inst. Leningrad, 153: 67–71. (In Russian)

P- 56 Prirodina, V.P. and Ozouf-Costaz, C. (1995) Description of karyotypes of species of *Harpagifer* (Harpagiferidae, Notothenioidei) from the South Orkney Islands and Macquarie Island. J. Ichthyol., 35(9): 341–344.

P- 57 Prirodina, V.P. (1995) Chromosomal complement of *Artedidraco mirus* and *Pogonophryne dolichobranchiata* (Artedidraconidae, Notothenioidei) from the Scotia Sea. J. Ichthyol., 35(9): 395–399.

P- 58 Pisano, E., Ozouf-Costaz, C., Hureau, J-C. and Williams, R. (1995) Chromosome differentiation in the subantarctic Bovichtidae species *Cottoperca gobio* (Günther, 1861) and *Pseudaphritis urvillii* (Valenciennes, 1832) (Pisces, Perciformes). Antarctic Sci., 7: 381–386.

P- 59 Potter, I.C., Robinson, E.S. and Walton, S.M. (1968) The mitotic chromosomes of the lamprey *Mordacia mordax* (Agnatha: Petromyzonidae). Experientia, 24: 966–967.

P- 60 Potter, I.C. and Rothwell, B. (1970) The mitotic chromosomes of the lamprey, *Petromyzon marinus* L. Experientia, 26: 429–430.

P- 61 Pie, M.R., Torres, R.A. and Brito, D.M.A. (2007) Evolution of genome size in fishes: a phylogenetic test of the Hinegardner and Rosen hypothesis. Genetica, 131: 51–58.

P- 62 Phillips, R.B., DeKoning, J., Morasch, M.R., Park, L.K. and Devlin, R.H. (2007) Identification of the sex chromosome pair in chum salmon (*Oncorhynchus keta*) and pink salmon (*Oncorhynchus gorbuscha*). Cytogenet. Genome Res., 116: 298–304.

P- 63 Pardo, B.G., Bouza, C., Castro, J., Martinez, P. and Sánchez, L. (2001) Localization of ribosomal genes in Pleuronectiformes using Ag-, CMA_3-banding and *in situ* hybridization. Heredity, 86: 531–536.

P- 64 Pleyte, K.A., Phillips, R.B. and Hartley, S.E. (1989) Q-band chromosomal polymorphisms in arctic char (*Salvelinus alpinus*). Genome, 32: 129–133.

P- 65 Portela-Castro, A.L.B., Julio Jr., H.F., Santos, I.C.M. and Pavanelli, C.S. (2008) Occurrence of two cytotypes in *Bryconamericus* aff. *iheringii* (Characidae): karyotype analysis by C- and G-banding and replication bands. Genetica, 133: 113–118.

P- 66 Pazza, R. and Kavalco, K.F. (2007) Chromosomal evolution in the neotropical characin *Astyanax* (Teleostei, Characidae). Nucleus, 50: 523–547.

P- 67 Park, I.S., Kim, H.B. and Lee, Y.D. (1995) Karyotypic analysis of four labrid fishes from Korea. Korean J. Ichthyol., 7: 79–83.

P- 68 Park, I.S. and Lee, C.L. (1996) Cytogenetic analysis of bagrid catfish, *Pseudobagrus fulvidraco* (Teleostomi: Siluriformes). Korean J. Ichthyol., 8(2): 10–15.

P- 69 Park, I.S., Choi, Y., Kim, Y.H., Nam, Y.K. and Kim, D.S. (2000) Flow cytometric and cytogenetic studies in *Rhynchocypris oxycephalus* and *R. steindachneri*. J. Aquaculture (Korea), 13: 193–196. (In Korean with English abstract)

P- 70 Park, I.S. and Lee, J.S. (2005) Cytogenetical study of fishes from coastal area in Jo Island, Busan, Korea. I. *Parapercis sexfasciata* (Temminck et Schlegel), *Sebastiscus marmoratus* (Cuvier) and *Pleuronectes yokohamae* (Günther). Korean J. Ichthyol., 27: 41–45.

P- 71 Park, M.K., Kim, I.S. and Park, J.Y. (2006) Morphological characters and geographical distributions of Korean ricefish (Pisces, Adrianichthyidae), *Oryzias*. Korean J. Ichthyol., 18: 163–169.

P- 72 Pedersen, R.A. (1971) DNA content, ribosomal gene multiplicity, and cell size in fish. J. Exp. Zool., 177: 65–78.

P- 73 Portela-Castro, A.L.B., Júlio Jr., H.F. and Nishiyama, P.B. (2001) New occurrence of microchromosomes B in *Moenkhausia sanctaefilomenae* (Pisces, Characidae) from the Paraná River of Brazil: analysis of the synaptonemal complex. Genetica, 110: 277–283.

P- 74 Post, A. (1973) Chromosomes of two fish-species of the genus *Diretmus* (Osteichthyes, Beryciformes: Diretmidae). In: Schröder, J.H. ed. Genetics and mutagenesis of fish. Springer, Berlin, pp. 103–111.

P- 75 Putnam, N.H. and 36 others (2008) The amphioxus genome and the evolution of the chordate karyotype. Nature, 453: 1064–1072.

P- 76 Pereira, C., Neto, A. and Collares-Pereira, M.J. (2009) Cytogenetic survey of species of two distinct genera of Iberian nases (Cyprinidae, Leuciscinae) that hybridize extensively in nature. I. Evidence of a similar and conserved chromosome pattern with some few species-specific markers at macro-structural level. Genetica, 137: 285–291.

P- 77 Peres, M., Adriano, W., Bertollo, L.A.C. and Moreira-Filho, O. (2007) Karyotypic characterization of two species of the genus *Serrapinnus* (Characiformes, Characidae), with the description of a structural polymorphism in *S. heterodon*. Caryologia, 60: 319–324.

P- 78 Phillips, R.B., Morasch, M.R., Wheeler, P.A. and Thorgaard, G.H. (2005) Rainbow trout (*Oncorhynchus mykiss*) of Idaho and Alaskan origin (2n = 58) share a chromosome fusion relative to trout of California origin (2n = 60). Copeia, 2005: 661–664.

P- 79 Pazza, R., Kavalco, K.F. and Bertollo, L.A.C. (2006) Chromosome polymorphism in *Astyanax fasciatus* (Teleostei, Characidae). 1. Karyotype analysis, Ag-NORs and mapping of the 18S and 5S ribosomal genes in sympatric karyotypes and their possible hybrid forms. Cytogenet. Genome Res., 112: 313–319.

P- 80 Pazza, R., Kavalco, S.A.F., Penteado, P.R., Kavalco, K.F. and Almeida-Toledo, L.F. (2008) The species complex *Astyanax fasciatus* Cuvier (Teleostei, Characiformes)—A multidisciplinary approach. J. Fish Biol., 72: 2002–2010.

P- 81 Pastori, M.C., Fenocchio, A.S. and Lopez, P.A. (1997) First description of microchromosome in the Anostomidae fish *Schizodon nasutus* from Argentina. Brazil. J. Genet., 20: 425–427.

P- 82 Porto-Foresti, F., Oliveira, C., Maistro, E.L. and Foresti, F. (1997) Estimated frequency of B-chromosomes and population density of *Astyanax scabripinnis paranae* in a small stream. Brazil. J. Genet., 20: 377–380.

P- 83 Pauls, E., Affonso, P.R.A.M., Netto, M.R.C.B. and Pacheco, M.L. (1996) Supernumerary chromosomes on marine fish *Upeneus parvus* (Poey 1853, Mullidae) from Atlantic Ocean. Arch. Zootec., 45: 295–299.

*P- 84 Prado, C.H.B.A. and Galetti jr., P.M. (1986) Consideraçoes iniciais sobre a estrutura cromossomica no gênero *Pigocentrus* (Serrasalminae, Characiformes). I. Simp.

Citogenet. Evol. e Aplic. de Peixes Neotropicais, Sao Carlos, SP, p. 56 (abstract). (After K-114)

P- 85 Porto, J.I.R. and Feldberg, E. (1992) Comparative cytogenetic study of the armored catfishes of the genus *Hoplosternum* (Siluriformes, Callichthyidae). Brazil. J. Genet., 15: 359–367.

*P- 86 Pedraca, E.B., Valentim, F.C. and Feldberg, E. (1994) Caracterizaçao cromossômica no gênero *Plagioscion* (Perciformes, Sciaenidae) da Bacia Amazônica Central. V. Simp. Citogenet. Evol. e Aplic. de Peixes Neotropicais, Botucatu, SP, p. 48 (abstract). (After K-114)

*P- 87 Peixoto, R.M. and Erdtmann, B. (1988) Estudio citigenéticos no gênero *Gymnogeophagus* (Pisces, Perciformes, Cichlidae). II. Simp. Citogenet. Evol. e Aplic. de Peixes Neotropicais, Sao Carlos, SP, p. 27 (abstract). (After K-114)

P- 88 Porto, J.I.R., Feldberg, E., Nakayama, C.M. and Falcao, J.N. (1992) A checklist of chromosome numbers and karyotypes of Amazonian freshwater fishes. Rev. Hydrobiol. Trop., 25: 287–299.

P- 89 Porto, J.I.R. and Feldberg, E. (1993) Is *Callichthys* Linné (Ostariophysi, Siluriformes, Callichthyidae) a monotypic genus? Acta Amazonica, 23: 311–314.

P- 90 Paintner-Marques, T.R., Giuliano-Caetano, L. and Dias, A.L. (2003) Cytogenetic characterization of a population of *Bryconamericus* aff. *iheringii* (Characidae, Tetragonopterinae). Genet. Mol. Biol., 26: 145–149.

P- 91 Peres, W.A.M., Buckup, P.A., Kantek, D.L.Z., Bertollo, L.A.C. and Moreira-Filho, O. (2009) Chromosomal evidence of downstream dispersal of *Astyanax fasciatus* (Characiformes, Characidae) associated with river shed interconnection. Genetica, 137: 305–311.

P- 92 Prasad, R. (1998) *Clarias batrachus*. In: Ponniah, A.G. and John, G. eds. Fish chromosome atlas. National Bureau of Fish Genetic Resources, Lucknow, India, pp. 214–215.

P- 93 Phillips, R.B. (1983) Chromosomal location of nucleolar organizer regions (NORs) in salmonids. Genetica, 104, no. 1, pt. 2, supplement: s56–s57.

P- 94 Pisano, E., Mazzei, F., Derome, N., Ozouf-Costaz, C., Hureau, J.-C. and di Prisco, G. (2001) Cytogenetics of the bathydraconid fish *Gymnodraco acuticeps* (Perciformes, Notothenioidei) from Terra Nova Bay, Ross Sea. Polar Biol., 24: 846–852.

P- 95 Poletto, A.B., Ferreira, I. and Martins, C. (2010) The B chromosomes of the African cichlid fish *Haplochromis obliquidens* harbour 18S rRNA gene copies. BMC Genetics, 2010, 11: 1.

P- 96 Ponniah, A.G. and John, G. (eds). (1998) Fish chromosome atlas. National Bureau of Fish Genetic Resources (NBFGR), special publication, no. 1. Lucknow, India, pp. 323 pp.

P- 97 Peichel, C.L., Ross, J.A., Matson, C.K., Dickson, M., Grimwood, J., Schmutz, J., Myers, R.M., Mori, S., Schluter, D. and Kingsley, D.M. (2004) The master sex-determination locus in threespine sticklebacks is on a nascent Y chromosome. Curr. Biol., 14: 1416–1424.

[R]

R- 1 Ráb, P. (1981) Karyotype of European catfish *Silurus glanis* (Siluridae, Pisces), with remarks on cytogenetics of siluroid fishes. Folia Zool., 30: 271–286.

R- 2 Ráb, P. (1981) Karyotype of European mudminnow, *Umbra krameri*. Copeia, 1981: 911–913.

R- 3 Ráb, P. (1981) Evolutionary polyploidy of fishes. In: Peňáz, M. and Prokeš, H. eds. Tropical problems of ichthyology. Proc. Symp. Trop. Prob. Ichthyol., Brno, pp. 107–111.

R- 4 Ráb, P. (1981) Karyotypes of two African barbels *Barbus bariloides* and *Barbus holotaenia*. Folia Zool., 30: 181–190.

R- 5 Ráb, P. (1982) Chromosome study of the Danube salmon *Hucho hucho* (Pisces, Salmonidae). Folia Zool., 31: 181–190.

R- 6 Ráb, P., Liehman, P. and Prokes, M. (1983) Karyotype of *Cichlasoma tetracanthum* (Pisces, Cichlidae) from Cuba. Folia Zool., 32: 185–188.

R- 7 Ráb, P. (1984) Chromosome study of four poeciliid fishes from Cuba. Folia Zool., 33: 229–234.

R- 8 Ráb, P. (1985) Karyotype of the Danube goby, *Proterorhinus marmoratus* (Pisces, Gobiidae). Folia Zool., 34: 329–334.

R- 9 Ráb, P. (1986) Karyotype of the European burbot, *Lota lota* (L.) (Gadidae). J. Ichthyol., 1986(2): 127–131.

R- 10 Ráb, P. (1986) A note on the karyotype of the sterlet, *Acipenser ruthenus* (Pisces, Acipenseridae). Folia Zool., 35: 73–78.

R- 11 Ráb, P., Roth, P. and Mayr, B. (1987) Karyotype study of eight species of European percid fishes (Pisces, Percidae). Caryologia, 40: 307–318.

R- 12 Ráb, P. and Mayr, B. (1987) Chromosome banding studies in European esocoid fishes: localization of nucleolar organizer regions in *Umbra krameri* and *Esox lucius*. Copeia, 1987: 1062–1067.

R- 13 Ráb, P. and Roth, P. (1989) Chromosome studies in European leuciscine fishes (Pisces, Cyprinidae). Aneuploidy due to a B-chromosome in *Rutilus rutilus*. Folia Zool., 38: 333–337.

R- 14 Ráb, P., Pokorny, J. and Roth, P. (1989) Chromosome studies of the common carp, *Cyprinus carpio*. I. Karyotype of Amurian carp, *C. carpio haematopterus*. Caryologia, 42: 27–36.

R- 15 Ráb, P. and Roth, P. (1989) Chromosome studies of European leuciscine fishes (Pisces: Cyprinidae). Karyotype of *Rutilus pigus virgo* and *R. rutilus*. Folia Zool., 38: 239–245.

R- 16 Ráb, P., Roth, P. and Arefjev, V.A. (1990) Chromosome studies of European leuciscine fishes (Pisces, Cyprinidae). Karyotype of *Aspius aspius*. Caryologia, 43: 249–255.

R- 17 Ráb, P. (1991) Fish cytogenetics and its application in fish reproductive biology. Bull. Inst. Zool. Academia Sinica, Monograph, 16: 357–373.

R- 18 Ráb, P., Mayr, B. and Roth, P. (1991) Chromosome banding study of European catfish, *Silurus glanis* (Pisces, Siluridae). Genetica, 83: 153–157.

R- 19 Ráb, P., Roth, P. and Vasiljeva, E.D. (1991) Chromosome banding study of the golden loach, *Sabanejewia auratus balcanica* from Slovakia (Cobitidae). Japan. J. Ichthyol., 38: 141–146.

R- 20 Ráb, P. (1991) The karyotype of the cyprinid fish *Pseudaspius leptocephalus*. Japan. J. Ichthyol., 38: 329–331.

R- 21 Ráb, P., Machordom, A., Perdices, A. and Guegan, J.-F. (1995) Karyotypes of three « small » *Barbus* species (Cyprinidae) from Republic of Guinea (Western Africa) with a review on karyology of African small *Barbus*. Caryologia, 48: 299–307.

R- 22 Ráb, P. and Collares-Pereira, J. (1995) Chromosomes of European cyprinid fishes (Cyprinidae, Cypriniformes): a review. Folia Zool., 44: 193–214.

R- 23 Ráb, P., Karakousis, Y., Rábová, M. and Economidis, P.S. (1996) Banded karyotype of the cyprinid fish *Leuciscus borysthenicus*. Ichthyol. Res., 43: 463–468.

R- 24 Ráb, P., Karakousis, Y. and Rábová, M. (1996) Karyotype, NOR phenotype and C-banding study of *Barbus cyclolepis* from Greece. Folia Zool., 45 (suppl. 1): 77–83.

R- 25 Ráb, P., Rábová, M., Economidis, P.S. and Triantaphyllidis, C. (2000) Banded karyotype of the Greek endemic cyprinid fish, *Pachychilon macedonicum*. Ichthyol. Res., 47: 107–110.

R- 26 Ráb, P., Machordom, A., Rábová, M. and Doadrio, I. (2000) Karyotype of African bariliine fish *Raiamas steindachneri* (Osteichthyes, Cyprinidae). Folia Zool., 49: 75–80.

R- 27 Rachlin, J.W., Beck, A.P. and O'Connor, J.M. (1978) Karyotypic analysis of the Hudson river striped bass, *Morone saxatilis*. Copeia, 1978: 343–345.

R- 28 Radda, A.C. and Scheel, J.J. (1974) *Aphyosemion puerzli* nov. spec. und *Aphyosemion robertsoni* nov. spec., zwei neue Rivulinen aus Kamerun. Aquarium J., 4(3): 33–41.

R- 29 Radda, A. (1975) Studien zur Cytogenetik, Taxonomie und Verbreitung der « Gardneri-Gruppe » der Gattung *Aphyosemion* (Osteichthyes, Cyprinodontidae) aus dem südlichen Westkamerun. Nat. Hist. Mus. Stadt Bern, Jahrbuch, 5: 189–199.

R- 30 Radda, A.C. and Scheel, J.J. (1975) Neue Formen der Gattung *Aphyosemion* Myers aus dem südwestlichen Kamerun. Aquarium, 69: 97–103.

R- 31 Rafique, M., Ahmad, M. and Saeed-Ul-Haque (2000) Comparative karyomorphology of three fish species belonging to the genus *Barilius* (Cyprinidae) from Pakistan. Pakistan J. Zool., 32: 337–342.

R- 32 Raicu, P. and Taisescu, E. (1972) *Misgurnus fossilis*, a tetraploid fish species. J. Heredity, 63: 92–94.

R- 33 Raicu, P., Taisescu, E. and Cristian, A. (1972) Diploid chromosome complement of the carp. Cytologia, 37: 355–358.

R- 34 Raicu, P., Taisescu, E. and Bănărescu, P. (1973) A comparative study of the karyotype in the genus *Gobio* (Pisces, Cyprinidae). Cytologia, 38: 731–736.

R- 35 Raicu, P. and Taisescu, E. (1977) Cytogenetic study in *Salmo irideus* and *S. trutta fario*. Cytologia, 42: 311–314.

R- 36 Raicu, P., Taisescu, E. and Bănărescu, P. (1981) *Carassius carassius* and *C. auratus*, a pair of diploid and tetraploid representative species (Pisces, Cyprinidae). Cytologia, 46: 233–240.

R- 37 Ramirez, S.A. (1980) A modified technique for fish karyotype analysis using scale epithelium. Copeia, 1980: 543–545.

R- 38 Ratty, F.J., Song, Y.C. and Laurs, R.M. (1986) Chromosomal analysis of albacore, *Thunnus alalunga*, and yellowfin, *Thunnus albacares*, and skipjack, *Katsuwonus pelamis*, tuna. Fish. Bull., 84: 469–476.

R- 39 Ravedutti, C.G. and Júlio Jr., H.F. (2001) Cytogenetic analysis of three species of the neotropical family Auchenipteridae (Pisces, Siluriformes) from the Paraná River basin, Brazil. Cytologia, 66: 65–70.

R- 40 Refstie, T., Vassvik, V. and Gjedrem, T. (1977) Induction of polyploidy in salmonids by cytochalasin B. Aquaculture, 10: 65–74.

R- 41 Regan, J.D., Sigel, M.M., Lee, W.H., Llamas, K.A. and Beasley, A.R. (1968) Chromosomal alterations in marine fish cells *in vitro*. Can. J. Genet. Cytol., 10: 448–453.

R- 42 Ren, X.H., Cui, J.X. and Yu, Q.X. (1991) The heterogeneities of heterochromatins and NORs in the genomes of Chinese cyprinid fishes. Cytologia, 56: 673–677.

R- 43 Ren, X.H. and Yu, X.J. (1993) Characterization of nucleolar organizer regions of twelve species of Chinese cyprinid fishes. Caryologia, 46: 201–207.

R- 44 Rio, G.J., Magnavita, F.J., Rubin, J.A. and Beckert, W.H. (1973) Characteristics of an established goldfish *Carassius auratus* (L.) cell line. J. Fish Biol., 5: 315–321.

R- 45 Rishi, K.K. (1973) A preliminary report on the karyotypes of eighteen marine fishes. Res. Bull. Panjab Univ., (N.S.), 24(3/4): 161–162.

R- 46 Rishi, K.K. (1973) Somatic karyotypes of three teleosts. Genen Phaenen, 16(3): 101–107.

R- 47 Rishi, K.K. (1975) Somatic and meiotic chromosomes of *Trichogaster fasciatus* (Bl. & Sch.) (Teleostei, Perciformes: Osphronemidae). Genen Phaenen, 18: 49–53.

R- 48 Rishi, K.K. and Gaur, P. (1976) Cytological female heterogamety in jet-black molly, *Mollienesia sphenops*. Curr. Sci., 45: 669–670.

R- 49 Rishi, K.K. (1976) Mitotic and meiotic chromosomes of a teleost, *Callichrous bimaculatus* (Bloch) with indications of male heterogamety. Cien. Cultura, 28: 1171–1173.

R- 50 Rishi, K.K. (1976) Karyotypic studies on four species of fishes. Nucleus, 19: 95–98.

R- 51 Rishi, K.K., Sharma, M.P. and Mankotia, R. (1977) Somatic chromosomes of three Indian teleosts. Matsya, 3: 6–9.

R- 52 Rishi, K.K. (1978) Chromosomal homogeneity in two Indian catfishes. Geobios, 5: 121–123.

R- 53 Rishi, K.K. (1978) Giemsa-banding in fish chromosomes. Curr. Sci., 47: 393–394.

R- 54 Rishi, K.K. (1979) Somatic G-banded chromosomes of *Colisa fasciatus* (Perciformes: Belontidae) and confirmation of female heterogamety. Copeia, 1979: 146–149.

R- 55 Rishi, K.K. (1981) Chromosomal studies on four cyprinid fishes. Intl. J. Acad. Ichthyol., 2: 1–4.

R- 56 Rishi, K.K., Haobam, M.S. and Singh, J. (1981) A new karyotype in a teleost fish. Curr. Sci., 50: 244–245.

R- 57 Rishi, K.K. and Singh, J. (1982) Karyological studies on five estuarine fishes. Nucleus, 25: 178–180.

R- 58 Rishi, K.K. and Singh, J. (1983) Karyological studies on two Indian estuarine catfishes, *Plotosus caninus* Ham. and *Pseudeutropius atherinoides* (Bloch). Caryologia, 36: 139–144.

R- 59 Rishi, K.K., Singh, J. and Haobam, M.S. (1983) Karyological study on a marine catfish, *Arius dussumieri* (Val.) (Ariidae: Siluriformes). Chrom. Inform. Serv., 34: 7–9.

R- 60 Rishi, K.K. and Singh, J. (1983) Chromosomes of *Notopterus notopterus* (Pallas) (Notopteridae: Clupeiformes). Chrom. Inform. Serv., 34: 9–10.

R- 61 Rishi, K.K. and Singh, J. (1983) Chromosomal analysis of the Indian silurid, *Wallago attu* (Schneider) (family: Siluridae). Chrom. Inform. Serv., 34: 10–11.

R- 62 Rishi, K.K., Singh, J. and Kaul, M.M. (1983) Chromosomal analysis of *Schizothoraichthys progastus* (McClell.) (Cyprinidae: Cypriniformes). Chrom. Inform. Serv., 34: 12–13.

R- 63 Rishi, K.K. and Haobam, M.S. (1984) Karyological analysis of two marine fishes. Perspect. Cytol. Genet., 4: 425–428.

R- 64 Rishi, K.K. and Haobam, M.S. (1984) Karyotypic study on two freshwater mud-eels. Perspect. Cytol. Genet., 4: 429–432.

R- 65 Rishi, K.K. and Haobam, M.S. (1984) Karyotypes of three forms of fishes having high chromosome number. Intl. J. Acad. Ichthyol., 5: 139–144.

R- 66 Rishi, K.K. and Haobam, M.S. (1984) Somatic chromosomes in a marine fish, *Megalops cyprinoides* (Broussonet) (Megalopidae: Elopiformes). Chrom. Inform. Serv., 36: 22–24.

R- 67 Rishi, K.K. and Mandhan, R.P. (1990) Analysis of C-banded heterochromatin in the chromosomes of *Labeo rohita* (Ham.) (Cyprinidae). Chrom. Inform. Serv., 48: 3–4.

R- 68 Rishi, K.K., Mandhan, R.P. and Lakra, W.S. (1991) Ag-NOR location in the karyotype of *Anabas testudineus* (Bloch) (Anabantidae, Perciformes, Pisces). Chrom. Inform. Serv., 52: 34–35.

R- 69 Rishi, K.K. and Manjusha (1991) Ag-NOR study in the chromosomes of *Heteropneustes fossilis* (Bloch) (Heteropneustidae, Siluriformes, Pisces). Chrom. Inform. Serv., 50: 17–18.

R- 70 Rishi, K.K., Rishi, S., Mandhan, R.P. and Thind, A.D.K. (1993) Ag-NOR localization and C-heterochromatin in *Channa punctatus* (Bloch) (Ophiocephalidae: Teleostei). Chrom. Inform. Serv., 55: 7–9.

R- 71 Rishi, K.K. and Shashikala (1994) High chromosome numbers in two Indian hill-stream cyprinids—*Tor tor* and *Tor putitora*. Chrom. Inform. Serv., 57: 25–26.

R- 72 Rishi, K.K. and Thind, A.D.K. (1994) C-heterochromatin and NOR-localization in the chromosomes of *Notopterus notopterus* (Notopteridae: Clupeiformes: Osteichthyes). Chrom. Inform. Serv., 57: 20–21.

R- 73 Rishi, K.K., Shashikala and Rishi, S. (1997) Karyotype analysis by restriction enzyme banding in two Indian fishes, *Channa punctatus* (Ophiocephalidae) and *Labeo dero* (Cyprinidae). Chrom. Sci., 1: 105–107.

R- 74 Rishi, K.K., Shashikala and Rishi, S. (1998) Karyotype study on six Indian hill-stream fishes. Chrom. Sci., 2: 9–13.

R- 75 Rishi, K.K. and Rishi, S. (1981) Giemsa banding in fish chromosomes. Perspect. Cytol. Genet., 3: 103–106.

R- 76 Rishi, S., Gill, P.P.K. and Rishi, K.K. (1994) C-banding and Ag-NOR staining in the spiny eel, *Macrognathus aculeatus* (Bloch) (Mastacembelidae, Mastacembeliformes). Chrom. Inform. Serv., 57: 22–23.

R- 77 Rishi, S., Gill, P.K. and Rishi, K.K. (1997) Karyotypic studies in eight species of fishes belonging to family Belontiidae (Pisces, Perciformes). Chrom. Sci., 1: 55–59.

R- 78 Rishi, S., Gill, P.K. and Rishi, K.K. (2001) Heterochromatin localization in the chromosomes of five belontid fishes (Belontiidae, Perciformes). Chrom. Sci., 5: 35–38.

R- 79 Rivlin, K.A., Dale, G. and Rachlin, J.W. (1986) Karyotypic analysis of three species of cardinal fish (Apogonidae) and its implications for the taxonomic status of the genera *Apogon* and *Phaeoptyx*. Ann. New York Acad. Sci., 463: 211–213.

R- 80 Roberts, F.L. (1964) A chromosome study of twenty species of Centrarchidae. J. Morphol., 115: 401–409.

R- 81 Roberts, F.L. (1965) A chromosome study of *Gambusia affinis holbrookii*. Copeia, 1965: 238–239.

R- 82 Roberts, F.L. (1966) Cell culture of fibroblasts from *Clupea harengus* gonads. Nature, 212: 1592–1593.

R- 83 Roberts, F.L. (1970) Atlantic salmon (*Salmo salar*) chromosomes and speciation. Trans. Amer. Fish. Soc., 99: 105–111.

R- 84 Robinson, E.S. and Potter, I.C. (1969) Meiotic chromosomes of *Mordacia praecox* and a discussion of chromosome numbers in lampreys. Copeia, 1969: 824–828.

R- 85 Robinson, E.S., Potter, I.C. and Webb, C.J. (1974) Homogeneity of Holarctic lamprey karyotypes. Caryologia, 27: 443–454.

R- 86 Robinson, E.S. and Potter, I.C. (1981) The chromosomes of the southern hemispheric lamprey, *Geotria australis* Gray. Experientia, 37: 239–240.

R- 87 Rocco, L., Morescalchi, M.A., Costagliola, D. and Stingo, V. (2002) Karyotype and genome characterization in four cartilaginous fishes. Gene, 295: 289–298.

R- 88 Rocco, L., Costagliola, D., Fiorillo, M., Tinti, F. and Stingo, V. (2005) Molecular and chromosomal analysis of ribosomal cistrons in two cartilaginous fish, *Taeniura lymma* and *Raja montagui* (Chondrichthyes, Batoidea). Genetica, 123: 245–253.

R- 89 Rodorigues-Daga, R., Amores, A. and Thode, G. (1993) Karyotype and nucleolus organizer regions in *Epinephelus caninus* (Pisces, Serranidae). Caryologia, 46: 71–76.

R- 90 Rong, S.B., Zhang, T.G., Zhang, W.H. and Chen, J.D. (1991) The karyotype of *Gymnothorax reevesi*. Tropic Oceanology, 10(2): 98–100. (In Chinese with English abstract)

R- 91 Ross, M.R. (1973) A chromosome study of five species of etheostomine fishes (Percidae). Copeia, 1973: 163–165.

R- 92 Rossi, A.R., Crosetti, D., Gornung, E. and Sola, L. (1996) Cytogenetic analysis of global populations of *Mugil cephalus* (striped mullet) by different staining techniques and fluorescent *in situ* hybridization. Heredity, 76: 77–82.

R- 93 Rossi, A.R., Gornung, E., Crosetti, D., De Innocentiis, S. and Sola, L. (2000) Cytogenetic analysis of *Oedalechilus labeo* (Pisces: Mugilidae), with a report of NOR variability. Mar. Biol., 136: 159–162.

R- 94 Rossi, A.R., Gornung, E., Sola, L. and Nirchio, M. (2005) Comparative molecular cytogenetic analysis of two congeneric species, *Mugil curema* and *M. liza* (Pisces, Mugiliformes), characterized by significant karyotype diversity. Genetica, 125: 27–32.

R- 95 Rudek, Z. (1974) Karyological investigations of two forms of *Vimba vimba* (Linnaeus 1758) occurring in Poland. Folia Biol. (Krakow), 22: 211–215.

R- 96 Ruiz-Carus, R. and Uribe-Alcocer, M. (2003) Karyotype analysis of *Eucinostomus argenteus*, *E. gula*, *E. harengulus*, and *Eugerres plumieri* (Teleostei, Gerreidae) from Florida and Puerto Rico. Environ. Biol. Fishes, 67: 269–276.

R- 97 Rukhkyan, R.G. and Arakelyan, G.L. (1980) Karyological bases of hybrid origin of Sevan whitefish *Coregonus lavaretus*. In: Kirpichnikov, V.S. ed. Karyological variability, mutagenesis and gynogenesis in fishes. Inst. Cytol. Acad. Sci. USSR, Leningrad, pp. 34–42. (In Russian with English summary)

R- 98 Rukhkyan, R.G. (1982) A comparative analysis of the karyotypes of the Sevan trout *Salmo ischchan* Kessler. Tsitologiya, 24: 66–77. (In Russian with English summary)

R- 99 Rukhkyan, R.G. (1984) The origin and speciation of the Alabalakh trout (genus *Salmo*, Salmonidae) in the light of its karyological peculiarities. J. Ichthyol., 24(2): 146–151.

R- 100 Russo, C., Rocco, L., Stingo, V., Aprea, G. and Odierna, G. (1999) A cytogenetic analysis of *Gambusia holbrooki* (Cyprinodontiformes, Poecilidae) from the River Sarno. Ital. J. Zool., 66: 291–296.

R- 101 Rock, J., Eldridge, M., Champion, A., Johnston, P. and Joss, J. (1996) Karyotype and nuclear DNA content of the Australian lungfish, *Neoceratodus forsteri* (Ceratodidae: Dipnoi). Cytogenet. Cell Genet., 73: 187–189.

R- 102 Ruma, F., Ahmed, A.T.A. and Alam, S.S. (2006) Karyotype analysis of *Channa punctata* Bloch and *Channa orientalis* Schneider with Giemsa, CMA and DAPI. Cytologia, 71: 425–430.

R- 103 Ryazanova, I.N. (2005) Karyotype of the snow sculpin *Myoxocephalus brandti* Steindachner (Cottidae) from Peter the Great Bay, Sea of Japan. Russian J. Mar. Biol., 31: 238–242.

R- 104 Rabova, M., Ráb, P. and Ozouf-Costaz, C. (2001) Extensive polymorphism and chromosomal characteristics of ribosomal DNA in a loach fish, *Cobitis vardarensis* (Ostariophysi, Cobitidae) detected by different banding techniques and fluorescence *in situ* hybridization (FISH). Genetica, 111: 413–422.

R- 105 Ren, X.H., Cui, J. and Yu, Q. (1992) Chromosomal heterochromatin differentiation of cyprinid fishes. Genetica, 87: 47–51.

R- 106 Ron, E. and Nirchio, M. (2005) Cytogenetic characterization of *Haemulon flavolineatum* Desmarest, 1823 (Pisces: Haemulidae) from Margarita Island, Venezuela. Bol. Inst. Oceanogr., Univ. Oriente, 44: 35–40. (In Spanish with English abstract)

R- 107 Roncati, H.A., Pastori, M.C. and Fenocchio, A.S. (2007) Cytogenetic studies and evolutive considerations on fishes of the family Cichlidae (Perciformes) from Parana River (Argentina). Cytologia, 72: 379–384.

R- 108 Rodrigues, M.M., Baroni, S. and Almeida-Toledo, L.F. (2007) Karyological characterization of four marine fish species, genera *Trachinotus* and *Selene* (Perciformes: Carangidae), from the southeast Brazilian coast. Cytologia, 72: 95–99.

R- 109 Rivlin, K.A., Rachlin, J.W. and Dale, G. (1987) Intraspecific chromosomal variation in *Apogon binotatus* (Perciformes: Apogonidae) from the Florida Keys and St. Croix. Ann. New York Acad. Sci., 494: 263–265.

R- 110 Rivlin, K.A., Rachlin, J.W. and Warkentine, B.E. (1988) G-banding of the chromosomes of *Apogon maculatus* and *A. pseudomaculatus* (Perciformes: Apogonidae). Ann. New York Acad. Sci., 529: 160–163.

R- 111 Rocha, É.C. and Molina, W.F. (2008) Cytogenetic analysis in western Atlantic snappers (Perciformes, Lutjanidae). Genet. Mol. Biol., 31: 461–467.

R- 112 Rocco, L., Liguori, I., Costagliola, D., Morescalchi, M.A., Tinti, F. and Stingo, V. (2007) Molecular and karyological aspects of Batoidea (Chondrichthyes, Elasmobranchi) phylogeny. Gene, 389: 80–86.

R- 113 Ryazanova, I.N. (2007) Karyotype of the sculpin *Myoxocephalus jaok* (Pisces: Cottidae) from Peter the Great Bay, Sea of Japan. Russian J. Mar. Biol., 33: 267–270.

R- 114 Ryazanova, I.N., Borisenko, S.A. and Frolov, S.V. (2008) Comparative analysis of the karyotype of the sculpin *Myoxocephalus ochotensis* (Pisces: Cottidae) from the Odyan Bay of the Sea of Okhotsk. Russian J. Mar. Biol., 34: 195–198.

R- 115 Ráb, P. and Crossman, E.J. (1994) Chromosomal NOR phenotypes in North American pikes and pickerels, genus *Esox*, with notes on the Umbridae (Euteleostei: Esocae). Can. J. Zool., 72: 1951–1956.

R- 116 Robinson, E.S., Potter, I.C. and Atkin, N.B. (1975) The nuclear DNA content of lampreys. Experientia, 31: 912–913.

R- 117 Ravi, V. and Venkatesh, B. (2008) Rapidly evolving fish genomes and teleost diversity. Curr. Opinion Genet. Develop., 18: 544–550.

R- 118 Rocco, L. (2007) Molecular markers in cartilaginous fish cytogenetics. In: Pisano, E., Ozouf-Costaz, C., Foresti, F. and Kapoor, B.G. eds. Fish cytogenetics. Science Publishers, Enfield, pp. 473–490.

R- 119 Rubert, M., Zawadzki, C.H. and Giuliano-Caetano, L.(2008) Cytogenetic characterization of *Hypostomus nigromaculatus* (Siluriformes: Loricariidae). Neotrop. Ichthyol., 6: 93–100.

R- 120 Ráb, P., Rábová, M., Reed, K.M. and Phillips, R.B. (1999) Chromosomal characteristics of ribosomal DNA in the primitive semionotiform fish, longnose gar *Lepisosteus osseus*. Chrom. Res., 7: 475–480.

R- 121 Ráb, P., Karakousis, Y. and Peios, C. (1994) Karyotype of *Silurus aristotelis* with reference to the cytotaxonomy of the genus *Silurus* (Pisces, Siluridae). Folia Zool., 43: 75–81.

*R–122 Rishi, S., Rishi, K.K. and Thind, A.D.K. (1994) NOR studies in six species of the genus *Mystus* (Bagridae, Osteichthyes). Abstr. VIII Congr. Soc. Europ. Ichthyol., Oviedo, p. 59. (After K-114)

*R–123 Reggi, R., Perico, E., Suminsky, M. and Camillo, J.C.A. (1986) Estudos citogeneticos em Papa-Terra, *Menticirrhus litoralis* (Perciformes, Sciaenidae). I. Simp. Citogenet. Evol. e Aplic. de Peixes Neotropicais, Sao Carlos, SP, p. 57 (abstract). (After K-114)

R- 124 Ribeiro, L.B., Matoso, D.A., Almeida, M.C., Vicari, M.R., Moraes-Neto, A., Svidnicki, M.C.C.M. and Artoni, R.F. (2008) Karyotypic variability in *Iheringichthys labrosus* (Teleostei, Pimelodidae) from the Tibagi River basin (Paraná State, Brazil). Genet. Mol. Res., 7: 718–724.

*R–125 Raghunath, P. and Prasad, R. (1980) Chromosomes of six marine percoids from the Indian Sea. Ind. Biologist, 11: 9–12. (After A-124)

[S]

S- 1 Safar, P., Mahmood, K., Bahram, K. and Masoud, S. (2000) Karyological study of two populations of *Capoeta capoeta* from North Iran. Caryologia, 65: 231–234.

S- 2 Sahoo, P.K., Barat, A. and Ponniah, A.G. (1997) Replication banding in *Channa punctatus* (Channidae, Pisces). Chrom. Sci., 1: 123–125.

S- 3 Saitoh, K., Takai, A. and Ojima, Y. (1984) Chromosomal study on the three local races of the striated spined loach (*Cobitis taenia striata*). Proc. Japan Acad., Ser. B, 60: 187–190.

S- 4 Saitoh, K. (1989) Multiple sex-chromosome system in a loach fish. Cytogenet. Cell Genet., 52: 62–64.

S- 5 Sakamoto, K. and Nishikawa, S. (1980) Chromosomes of three flatfishes (Pleuronectiformes). Japan. J. Ichthyol., 27: 268–272.

S- 6 Salas, E. and Boza, J. (1991) Citotaxonomia comparativa de tres especies de *Cichlasoma* (Pisces: Cichlidae) nativas de Costa Rica. Rev. Biol. Trop., 39: 219–224.

S- 7 Salekhova, L.P. and Vasiliev, V.P. (1980) Ecological and karyological evidence of species independence of *Spicara flexuosa* Raf., and *Spicara maena* (L.) (Fam. Centracanthidae).

In: Kirpichnikov, V.S. ed. Karyological variability, mutagenesis and gynogenesis in fishes. Inst. Cytol. Acad. Sci. USSR, Leningrad, pp. 50–54. (In Russian with English summary)

S– 8 Salvador, L.B. and Moreira-Filho, O. (1992) B chromosomes in *Astyanax scabripinnis* (Pisces, Characidae). Heredity, 69: 50–56.

S– 9 Salvadori, S., Coluccia, E., Cannas, R., Cau, A. and Deiana, A.M. (2003) Replication banding in two Mediterranean moray eels: chromosomal characterization and comparison. Genetica, 119: 253–258.

S– 10 Sánchez, S. and Fenocchio, A.S. (1996) Karyotypic analysis in three populations of the South-American eel like fish *Synbranchus marmoratus*. Caryologia, 49: 65–71.

S– 11 Sánchez, S. and Fenocchio, A.S. (1996) Karyotypic studies and cytotaxonomic considerations on *Callichthys callichthys* (Pisces, Siluroidei) from Argentina. Cytologia, 61: 247–252.

S– 12 Sánchez, S. and Jorge, L.C. (1999) A new report of the ZZ/ZW sex chromosome system in the genus *Triportheus* (Pisces, Triporthinae). Cytologia, 64: 395–400.

S– 13 Sasaki, M. and Hitotsumachi, S. (1967) Notes on the chromosomes of a fresh water lamprey, *Entosphenus reissneri* (Cyclostomata). Chrom. Inform. Serv., 8: 22–24.

S– 14 Sasaki, M., Hitotsumachi, S., Makino, S. and Terao, T. (1968) A comparative study of the chromosomes in the chum salmon, the kokanee salmon and their hybrids. Caryologia, 21: 389–394.

S– 15 Sasaki, T. and Sakamoto, K. (1977) Karyotype of the rockfish, *Sebastes taczanowskii* (Steindachner). Chrom. Inform. Serv., 22: 7–8.

S– 16 Sawada, Y. and Sakamoto, K. (1980) Chromosomes of a rare callionymid, *Draculo mirabilis*. Japan. J. Ichthyol., 26: 367–368.

S– 17 Sawada, Y. (1976) Chromosomes of *Leptobotia curta* (Cobitidae, Cypriniformes). Japan. J. Ichthyol., 23: 175–177.

S– 18 Scheel, J.J. (1966) Taxonomic studies of African and Asian tooth-carps (Rivulinae) based on chromosome numbers, haemoglobin patterns, some morphological traits and crossing experiments. Vidensk. Medd. Dansk naturh. Foren., 129: 123–148.

S– 19 Scheel, J.J. (1968) Zoogeographie und verwandtschaftliche Beziehungen von Rivulinae des Atlantischen Afrika, 1–3. Aquarien Terrarien, 15: 1–4, 6–12.

S– 20 Scheel, J.J. (1968) Zoogeographie und verwandtschaftliche Beziehungen von Rivulinae des Atlantischen Afrika IV. Aquarien Terrarien, 15: 118–121, 156–158, 200.

S– 21 Scheel, J.J. (1968) Zoogeographie und verwandtschaftliche Beziehungen von Rivulinae des Atlantischen Afrika 6. Aquarien Terrarien, 15: 201.

S– 22 Scheel, J.J. (1968) Zoogeographie und verwandtschaftliche Beziehungen von Rivulinae des Atlantischen Afrika 7. Aquarien Terrarien, 15: 226–227.

S– 23 Scheel, J.J. (1968) Zoogeographie und verwandtschaftliche Beziehungen von Rivulinae des Atlantischen Afrika 8. Aquarien Terrarien, 15: 304–305.

S– 24 Scheel, J.J. (1972) Rivuline karyotypes and their evolution (Rivulinae, Cyprinodontidae, Pisces). Z. Zool. Syst. Evol. Forsch., 10: 180–209.

S– 25 Scheel, J.J. (1972) Cichlid chromosomes: the nominal genus *Pelmatochromis*. Advanced Aquarists Mag., 1972, May: 6–7.

S– 26 Scheel, J.J. (1972) Rivulin chromosomes: The genus *Pterolebias* Garman. Advanced Aquarists Mag., 1972, May: 4–5.

S– 27 Scheel, J.J., Simonsen, V. and Gyldenholm, A.O. (1972) The karyotypes and some electrophoretic patterns of fourteen species of the genus *Corydoras*. Z. Zool. Syst. Evol. Forsch., 10: 144–152.

S– 28 Scheel, J.J. (1972) Cytotaxonomic studies: the *Aphyosemion elegans* group. Z. Zool. Syst. Evol. Forsch., 10: 122–127.

S- 29 Scheel, J.J. (1972) Die Chromosomen der drei Neon-Tetras. Aquarien Terrarien, 19(9): 307–309.
*S- 30 Scheel, J.J. (1973) Fish chromosomes and their evolution. Internal Report of Danmarks Akvarium, Charlottenlund, Denmark. (After O-50)
S- 31 Scheel, J.J. and Radda, A.C. (1974) Beschreibung zweier neuer Cyprinodontidae aus dem tropischen Afrika. Aquaria, 21: 157–162.
S- 32 Scheel, J.J. (1974) Rivuline studies. Taxonomic studies of rivuline cyprinodonts from tropical Atlantic Africa (Rivulinae, Cyprinodontidae, Atheriniformes, Pisces). Ann. Mus. Royal Afr. Centrale, 211: 1–150.
S- 33 Scheel, J.J. (1975) A review of *Aphyosemion arnoldi* and its allied. BKA-Killi News, May, 13 pp.
S- 34 Scheel, J.J. (1975) A review of the *Aphyosemion scheeli* group. BKA-Killi News, Oct. 4 pp.
S- 35 Scheel, J.J. (1975) The karyotypic variation in *Aphyosemion*. BKA-Killi News, June, 4 pp.
S- 36 Scheel, J.J. (1975) A review of *Aphyosemion gardneri* and its allied. BKA-Killi News, Aug. 13 pp.
S- 37 Scheel, J.J. (1973) A review of *Aphyosemion calliurum* (Blgr.). AKA-KN, 6(8): 5–14.
S- 38 Scheel, J.J. and Radda, A.C. (1974) Description of two new Cyprinodontidae from tropical Africa. BKA-Killi News, Dec. 6 pp.
S- 39 Schmidtke, J., Schmitt, E., Matzke, E. and Engel, W. (1979) Non-repetitive DNA sequence divergence in phylogenetically diploid and tetraploid teleostean species of the family Cyprinidae and the order Isospondyli. Chromosoma, 75: 185–198.
S- 40 Schwartz, F.J. and Maddock, M.B. (1986) Comparisons of karyotypes and cellular DNA contents within and between major lines of elasmobranchs. In: Uyeno, T., Arai, R., Taniuchi, T. and Matsuura, K. eds. Indo-Pacific fish biology. Ichthyological Society of Japan, Tokyo, pp. 148–157.
S- 41 Schwartz, F.J. and Maddock, M.B. (2002) Cytogenetics of the elasmobranchs: genome evolution and phylogenetic implications. Mar. Freshwater Res., 53: 491–502.
S- 42 Setzer, P.Y. (1970) An analysis of a natural hybrid swarm by means of chromosome morphology. Trans. Amer. Fish. Soc., 99: 139–146.
S- 43 Severin, S.O. (1979) The karyotype of the European grayling, *Thymallus thymallus*. J. Ichthyol., 19 (2): 44–49.
S- 44 Severin, S.O. and Zinov'yev, Y.A. (1982) Karyotypes of isolated populations of *Thymallus arcticus* from the Ob River basin. J. Ichthyol., 22(1): 13–22.
S- 45 Severin, S.O. (1985) Karyological analysis of three populations of grayling, *Thymallus thymallus*, in the Kama basin. J. Ichthyol., 25(6): 135–146.
S- 46 Shan, S.X. and Jiang, Y.G. (1988) Karyotype studies of crucian carp, *Carassius auratus gibelio*. Acta Hydrobiol. Sinica, 12: 381–384. (In Chinese)
S- 47 Salvadori, S., Cau, A., Coluccia, E., Milia, A. and Deiana, A.M. (1994) Karyotype, C- and G-banding, and nucleolar organizer regions of *Conger conger* (Osteichthyes, Anguilliformes). Boll. Zool., 61: 59–63.
S- 48 Sharma, K.K. and Sharma, O.P. (1995) Somatic and meiotic chromosomes of *Amblyceps mangois* (Siluriformes: Amblyceptidae). Chrom. Inform. Serv., 58: 20–21.
S- 49 Sharma, O.P. and Agarwal, A. (1981) The somatic and meiotic chromosomes of *Puntius conchonius* (Cyprinidae) from the Jammu and Kashmir state, India. Genetica, 56: 235–237.
S- 50 Sharma, O.P. and Tripathi, N.K. (1981) Somatic and meiotic chromosomes of a fresh water teleost *Barilius bendelisis* (Pisces: Cyprinidae). Chrom. Inform. Serv., 31: 20–22.
S- 51 Sharma, O.P. and Tripathi, N.K. (1981) Studies on the somatic and meiotic chromosomes of a fresh water teleost *Xenentodon cancila* (Belonidae: Pisces). Chrom. Inform. Serv., 30: 11–13.

S- 52 Sharma, O.P. and Tripathi, N.K. (1981) Chromosome number and karyotype of the catfish *Gagata viridescens* (Siluridae) from Jammu (J & K), India. Chrom. Inform. Serv., 30: 13–15.

S- 53 Sharma, O.P. and Tripathi, N.K. (1982) Somatic and germ cell chromosomes of *Gambusia affinis holbrooki*. Chrom. Inform. Serv., 32: 24–25.

S- 54 Sharma, O.P. and Tripathi, N.K. (1984) Studies on the chromosomes of *Nandus nandus* and *Badis badis* from the J and K State, India. Cytologia, 49: 73–79.

S- 55 Sharma, O.P. and Tripathi, N.K. (1986) Karyotypic diversity in genus *Mystus* (Bagridae: Pisces). Cytologia, 51: 1–9.

S- 56 Sharma, O.P. and Tripathi, N.K. (1986) On the chromosomes of two cyprinid fishes. Chrom. Inform. Serv., 41: 12–15.

S- 57 Sharma, O.P. and Tripathi, N.K. (1988) Female heterogamety in two teleostean fishes. Cytologia, 53: 81–86.

S- 58 Sharma, O.P., Sharma, K.K., Tripathi, N.K. and Gupta, S.C. (1995) Localization of C-heterochromatin in two cyprinid fishes *Puntius ticto* and *Schizothorax niger*. Chrom. Inform. Serv., 58: 19–20.

S- 59 Sharpnack, L.A. and Fitzsimons, J.M. (1980) Karyology of four species of cyprinodont fishes from the Louisiana Gulf Coast. Proc. Louisiana Acad. Sci., 43: 39–45.

S- 60 Shelenkova, N.Y. (1986) Karyotypes of two isolated populations of the sockeye salmon *Oncorhynchus nerka* Walb. Tsitologiya, 28: 735–739. (In Russian with English summary)

S- 61 Shelenkova, N.Y. (1987) A study of karyotypes of two Pacific salmon species, *Oncorhynchus nerka* Walbaum and *O. kisutch* W., by the C-banding method. Tsitologiya, 29: 100–104. (In Russian with English summary)

S- 62 Shen, J.B., Fan, Z.T. and Wang, G.R. (1983) Karyotype studies of male triploid crucian carp (Fangzheng crucian carp) in Heilongjiang. Acta Genet. Sinica, 10(2): 133–136. (In Chinese with English abstract)

S- 63 Shen, J.B., Fan, Z.T. and Wang, G.R. (1983) Karyotype of *Pseudobagrus fulvidraco*. Hereditas (Beijing), 5(2): 23–24. (In Chinese)

S- 64 Shen, J.B., Wang, G.R. and Fan, Z.T. (1983) The ploidy of crucian carp and its geographical distribution in Heilongjiang main river. J. Fish. China, 7(2): 87–94. (In Chinese with English abstract)

S- 65 Shen, J.B., Wang, G.R., Fan, Z.T. and Li, J. (1984) Chromosome numbers of 16 fishes from Heilongjiang Province, China. Zool. Res., 5(1) suppl.: 90. (In Chinese)

S- 66 Shen, J.B., Fan, Z.T., Li, S.W., Cheng, R.X. and Xue, S.B. (1984) Comparative studies of the somatic cell and spermatozoon DNA contents and ploidy of Fangzheng crucian carp and Zha-long Lake goldfish. Acta Zool. Sinica, 30: 7–13. (In Chinese with English abstract)

S- 67 Shi, A.J., Feng, K.M., Hu, G.Z. and Xie, S.R. (1986) Comparison between artificially induced *Cyprinus carpio* gynogenetic offspring and their parents with reference to karyotype and enzymic histochemistry. Acta Hydrobiol. Sinica, 10: 127–134. (In Chinese)

S- 68 Shichiri, M., Kikuma, Y., Kuo, C.-H., Liu, L.-L., Kubota, S. and Kohno, S. (1997) Chromosome elimination and germ line-restricted microchromosomes in *Paramyxine sheni* from Taiwan. Chrom. Sci., 1: 49–53.

S- 69 Shimabukuro-Dias, C.K., Oliveira, C. and Foresti, F. (2004) Karyotype valiability in eleven species of the catfish genus *Corydoras* (Siluriformes: Callichthyidae). Ichthyol. Explor. Freshwaters, 15: 135–146.

S- 70 Shinderite, V.S. and Vasiliev, V.P. (1976) The karyotype of the Vimba *Vimba vimba*. Vopr. Ichthyol., 16: 366–369. (In Russian)

S- 71 Simon, R.C. (1963) Chromosome morphology and species evolution in the five North American species of Pacific salmon (*Oncorhynchus*). J. Morphol., 112: 77–95.

S– 72 Skvortsova, T.A. (1975) Chromosome complexes of the White Sea herring (*Clupea harengus pallasi* n. *maris-albi* Berg) and the Baltic herring (*Clupea harengus harengus* n. *membras* L.). Issled. Fauny Morei, 16: 104–108. (In Russian)

S– 73 Smith, M.L. and Miller, R.R. (1987) *Allotoca goslinei*, a new species of goodeid fish from Jalisco, Mexico. Copeia, 1987: 610–616.

S– 74 Sobita, N., Bhagirath, T.H. and Dhananjoy, C.H. (2004) Chromosome complements of a cyprinid fish *Puntius manipurensis*. Nucleus, 47: 59–63.

S– 75 Sofradžija, A., Berberović, Lj. and Pavlović, B. (1972) Chromosome number in the cells of the testes of *Paraphoxinus pstrossi* Steind. Genetika, Acta Biol. Iugoslavica, 4: 17–23. (In Serbo-Croatian with English summary)

S– 76 Sofradžija, A. and Berberović, Lj. (1973) The chromosome number of *Barbus meridionalis petenyi* Heckel (Cyprinidae, Pisces). Bull. Sci. Yougoslavie, Sec. A, 18: 77–78.

S– 77 Sofradžija, A., Hadžiselimović, R. and Marić, C. (1975) Basic data about the chromosome complement of *Rhodeus sericeus amarus* (Bloch 1782), Cyprinidae, Pisces. Bull. Sci. Yougoslavie, Sec. A, 20: 7–8.

S– 78 Sofradžija, A. and Berberović, Lj. (1975) Basic data about chromosome complement of the species *Gobio gobio* (Linaeus, 1758), Cyprinidae, Pisces. Ichthyologia, 7: 53–59. (In Serbo-Croatian with English summary)

S– 79 Sofradžija, A. (1977) Caryology and cytotaxonomy of the *Leuciscus* species from the waters in Bosnia and Herzegovina. God. Biol. Inst. Univ. Sarajevu, 30: 113–211. (In Serbo-Croatian with English summary)

S– 80 Sofradžija, A., Berberović, Lj. and Hadžiselimović, R. (1978) Chromosome sets of *Carassius carassius* and *Carassius auratus gibelio*. Ichthyologia, 10: 135–142. (In Serbo-Croatian with English summary)

S– 81 Sofradžija, A. and Berberović, Lj. (1978) Diploid-triploid sexual dimorphism in *Cobitis taenia taenia* L. (Cobitidae, Pisces). Genetika, Acta Biol. Iugoslavica, 10: 389–397. (In Serbo-Croatian with English summary)

S– 82 Sofradžija, A., Berberović, Lj. and Hadžiselimović, R. (1979) Chromosome complements of *Alburnus alburnus* (L. 1758) and *Alburnoides bipunctatus* (Bloch, 1782), Cyprinidae, Pisces. Ichthyologia, 11: 33–41. (In Serbo-Croatian with English summary)

S– 83 Sofradžija, A. and Vuković, T. (1979) Chromosome complement of *Nemachilus barbatulus* (Linnaeus, 1758), Cobitidae, Pisces. Ichthyologia, 11: 43–49. (In Serbo-Croatian with English summary)

S– 84 Sofradžija, A., Berberović, Lj. and Sahinović, N. (1979) The chromosomes of [*Scardinius erythrophthalmus* (L. 1758)], Cyprinidae, Pisces. Radovi Anubih, 63: 131–142. (In Serbo-Croatian with English summary)

S– 85 Sofradžija, A. and Hadžiselimović, R. (1981) Chromosome complement of the species *Chondrostoma nasus* (Cyprinidae, Pisces). God. Biol. Inst. Univ. Sarajevu, 34: 107–115. (In Serbo-Croatian with English summary)

S– 86 Sofradžija, A. (1985) Cytogenetic investigations on Adriatic fishes *Uranoscopus scaber* L., *Solea lutea* (Risso) and *Serranus scriba* (L.). Bilješke notes, Inst. Oceanograf. Ribarstyo, Yugoslavija, 64: 1–8. (In Serbo-Croatian with English summary)

S– 87 Sola, L., Gentili, G. and Cataudella, S. (1980) Eel chromosomes: cytotaxonomical interrelationships and sex chromosomes. Copeia, 1980: 911–913.

S– 88 Sola, L., Cataudella, S. and Capanna, E. (1981) New developments in vertebrate cytotaxonomy. III. Karyology of bony fishes: a review. Genetica, 54: 285–328.

S– 89 Sola, L., Natili, G.L. and Cataudella, S. (1989) Chromosome variability in introduced whitefishes (genus *Coregonus*) from two lakes in central Italy. Copeia, 1989: 189–192.

S– 90 Sola, L., Bressanello, S., Rossi, A.R., Iaselli, V., Crosetti, D. and Cataudella, S. (1993) A karyotype analysis of the genus *Dicentrarchus* by different staining techniques. J. Fish Biol., 43: 329–337.

S- 91 Sola, L., Cipelli, O., Gornung, E., Rossi, A.R., Andaloro, F. and Crosetti, D. (1997) Cytogenetic characterization of the greater amberjack, *Seriola dumerili* (Pisces: Carangidae), by different staining techniques and fluorescence *in situ* hybridization. Mar. Biol., 128: 573–577.

S- 92 Sola, L, Marzovillo, M., Rossi, A.R., Gornung, E., Bressanello, S. and Turner, B.J. (1997) Cytogenetic analysis of a self-fertilizing fish, *Rivulus marmoratus*: remarkable chromosomal constancy over a vast geographic range. Genome, 40: 945–949.

S- 93 Sola, L., Gornung, E., Naoi, H., Gunji, R., Sato, C., Kawamura, K., Arai, R. and Ueda, T. (2003) FISH-mapping of 18S ribosomal RNA genes and telomeric sequences in the Japanese bitterlings *Rhodeus ocellatus kurumeus* and *Tanakia limbata* (Pisces, Cyprinidae) reveals significant cytogenetic differences in morphologically similar karyotypes. Genetica, 119: 99–106.

S- 94 Son, Y.M. and Lee, J.H. (1989) Karyotypes of genus *Liobagrus* (Pisces: Amblycipitidae) in Korea. Korean J. Ichthyol., 1: 64–72.

S- 95 Song, Y.C., Ratty, F.J. and Laurs, R.M. (1986) Transmission electron microscope observations of C-banded skipjack tuna chromosomes. J. Fish Biol., 28: 419–424.

S- 96 Song, Y.C. (1987) Studies on karyotypes and C-banding patterns of three fish species in Scombridae. Oceanol. Limnol. Sinica, 18: 352–356. (In Chinese with English abstract)

S- 97 Song, Z. and Zan, R.G. (1985) Morphological changes of chromosomes during the embryo development of fishes. Acta Biol. Exp. Sinica, 18: 77–87. (In Chinese with English abstract)

S- 98 Souza, I.L., Moreira-Filho, O. and Bertollo, L.A.C. (1995) Karyotypic characterization of *Aphyocharax difficilis* (Pisces, Characidae): C-banding, Ag-NORs and occurrence of diplochromosomes. Cytobios, 83: 33–39.

S- 99 Souza, I.L. and Moreira-Filho, O. (1995) Cytogenetic diversity in the *Astyanax scabripinnis* species complex (Pisces, Characidae). I. Allopatric distribution in a small stream. Cytologia, 60: 1–11.

S- 100 Souza, I.L., Moreira-Filho, O. and Bertollo, L.A.C. (1995) Cytogenetic diversity in the *Astyanax scabripinnis* (Pisces, Characidae) complex. II. Different cytotypes living in sympatry. Cytologia, 60: 273–281.

S- 101 Souza, I.L., Galián, J., Rúa, P.D.L., Bertollo, L.A.C. and Moreira-Filho, O. (2001) Nonrandom distribution of the GC-rich heterochromatin and nucleolar rDNA sites on *Astyanax scabripinnis* chromosomes. Cytologia, 66: 85–91.

S- 102 Souza, L., Giuliano-Caetano, L. and Dias, A.L. (2003) Karyotypic study of three species of *Pimelodus* (Pisces, Pimelodidae) from the Paraguai river basin. Cytologia, 68: 345–350.

S- 103 Sriramulu, V. (1959) A note on the chromosomes of *Oryzias melastigma* (McClelland). Curr. Sci., 1959: 117–118.

S- 104 Srivastava, M.D.L. and Das, B. (1968) Somatic chromosomes of *Clarias batrachus* (L.) (Clariidae: Teleostomi). Caryologia, 21: 349–352.

S- 105 Srivastava, M.D.L. and Das, B. (1969) Somatic chromosomes of teleostean fish. J. Heredity, 60(2): 57–58.

S- 106 Stange, E.A.R. and Almeida-Toledo, L.F. (1993) Supernumerary B chromosomes restricted to males in *Astyanax scabripinnis* (Pisces, Characidae). Brazil. J. Genet., 16: 601–615.

S- 107 Starmach, J. (1967) Die Chromosomen von *Cottus poecilopus* Heckel und *Cottus gobio* L. Acta Hydrobiol., 9: 301–307.

S- 108 Stevenson, M.M. (1981) Karyomorphology of several species of *Cyprinodon*. Copeia, 1981: 494–498.

S- 109 Sato, L.R., Oliveira, C. and Foresti, F. (2004) Karyotype description of five species of *Trichomycterus* (Teleostei: Siluriformes: Trichomycteridae). Genet. Mol. Biol., 27: 45–50.

S- 110 Sá-Gabriel, L.G. and Molina, W.F. (2005) Karyotype diversification in fishes of the Balistidae, Diodontidae and Tetraodontidae (Tetraodontiformes). Caryologia, 58: 229–237.

S- 111 Stingo, V. (1979) New developments in vertebrate cytotaxonomy. II. The chromosomes of the cartilaginous fishes. Genetica, 50: 227–239.

S- 112 Stingo, V., du Buit, M. and Odierna, G. (1980) Genome size of some selachian fishes. Boll. Zool., 47: 129–137.

S- 113 Stingo, V. and Capriglione, T. (1986) DNA and chromosomal evolution in cartilaginous fish. In: Uyeno, T., Arai, R., Taniuchi, T. and Matsuura, K. eds. Indo-Pacific fish biology. Ichthyological Society of Japan, Tokyo, pp. 140–147.

S- 114 Stingo, V., Rocco, L., Odierna, G. and Bellitti, M. (1995) NOR and heterochromatin analysis in two cartilaginous fishes by C-, Ag- and RE (restriction endonuclease)-banding. Cytogenet. Cell Genet., 71: 228–234.

S- 115 Stingo, V. and Rocco, L. (2001) Selachian cytogenetics: a review. Genetica, 111: 329–347.

S- 116 Subrahmanyam, K. and Natarajan, R. (1970) A study on the somatic chromosomes of *Therapon* Cuvier (Teleostei: Perciformes). Proc. Ind. Acad. Sci., 72B: 288–294.

S- 117 Subrahmanyam, K. (1970) Effect of rotenone on fish chromosomes. J. Annamalai Univ. Sci., 28: 243–245.

S- 118 Subrahmanyam, K. and Ramamoorthi, K. (1971) A karyotype study in the estuarine worm eel, *Moringua linearis* (Gray). Sci. Culture, 37: 201–202.

S- 119 Suciu, R. and Ráb, P. (1992) A note on the karyotype of *Percarina demidoffi* (Pisces, Percidae). Folia Zool., 41: 93–95.

S- 120 Sumner, A.T. (1972) A simple technique for demonstrating centromeric heterochromatin. Exp. Cell Res., 75: 304–306.

S- 121 Sun, G.I. and Liu, L. (1990) The karyotype in embryos of the icefish, *Neosalanx taihuensis* Chen. J. Fish. China, 14: 357–358. (In Chinese)

S- 122 Suzuki, T. (1996) *Hypseleotris cyprinoides*. In: Fisheries Agency of Japan ed. Basic data on rare Japanese wild aquatic species. (III). Nihonsuisanshigenhogokyokai, Tokyo, pp. 215–221. (In Japanese)

S- 123 Suzuki, A. and Taki, Y. (1981) Karyotype of tetraploid origin in a tropical Asian cyprinid, *Acrossocheilus sumatranus*. Japan. J. Ichthyol., 28: 173–176.

S- 124 Suzuki, A. and Taki, Y. (1982) Karyotype of a noemacheiline loach, *Lefua echigonia*. Japan. J. Ichthyol., 29: 303–304.

S- 125 Suzuki, A., Taki, Y. and Urushido, T. (1982) Karyotypes of two species of Arowana, *Osteogrossum bicirrhosum* and *O. ferreirai*. Japan. J. Ichthyol., 29: 220–222.

S- 126 Suzuki, A. and Taki, Y. (1986) Chromosomes and DNA values of two cyprinid fishes of the subfamily Barbinae. Japan. J. Ichthyol., 32: 459–462.

S- 127 Suzuki, A. and Taki, Y. (1988) Karyotype and DNA content in the cyprinid *Catlocarpio siamensis*. Japan. J. Ichthyol., 35: 389–391.

S- 128 Suzuki, A. and Yamanaka, K. (1988) Chromosomes of an African lungfish, *Protopterus annectens*. Proc. Japan Acad., Ser. B, 64: 119–121.

S- 129 Suzuki, A., Taki, Y., Takeda, M. and Akatsu, S. (1988) Multiple sex chromosomes in a monodactylid fish. Japan. J. Ichthyol., 35: 98–101.

S- 130 Suzuki, A., Takeda, M., Tanaka, H. and Yoo, M.S. (1988) Chromosomes of *Scatophagus argus* and *Selenotoca multifasciata* (Scatophagidae). Japan. J. Ichthyol., 35: 102–104.

S- 131 Suzuki, A., Yamanaka, K., Urushido, T. and Kondo, N. (1988) A banding chromosome study of African fish family Polypteridae (Pisces: Polypteriformes). Proc. Japan Acad., Ser. B, 64: 299–302.

S- 132 Suzuki, A., Yamanaka, K., Urushido, T. and Kondo, N. (1989) N-banding chromosomes of two polypterids, *Polypterus senegalus* and *Calamoichthys calabaricus* (Polypteriformes, Pisces). Chrom. Inform. Serv., 46: 23–25.

S- 133 Suzuki, A. (1989) Chromosomes of European rudd, *Scardinius erythrophthalmus* (Cypriniformes, Pisces). Chrom. Inform. Serv., 46: 21–23.

S- 134 Suzuki, A. and Takeda, M. (1991) A preliminary report on the stickleback species, *Pungitius pungitius* (Linnaeus) and *Pungitius* sp. sensu Nakamura (Musashitomiyo) of the family Gasterosteidae. Chrom. Inform. Serv., 50: 37–41.

S- 135 Suzuki, A. and Hirata, J. (1991) Chromosomes and DNA contents of *Amia calva*. Chrom. Inform. Serv., 50: 34–37.

S- 136 Suzuki, A. (1991) Chromosomes of a Chinese cyprinid fish, *Puntius semifasciolatus*. Chrom. Inform. Serv., 51: 8–10.

S- 137 Suzuki, A. (1992) Chromosomes and DNA studies of eight species in the family Cobitidae (Pisces, Cypriniformes). La Kromosomo, II, 67/68: 2275–2282.

S- 138 Suzuki, A. (1992) Karyotype and DNA contents of the Chinese catostomid fish, *Myxocyprinus asiaticus*. Chrom. Inform. Serv., 53: 25–27.

S- 139 Suzuki, A. and Taki, Y. (1992) Karyological and cytochemical studies of *Pseudogastromyzon myersi* and *Homaloptera hoffmanni* in the order Cypriniformes. Chrom. Inform. Serv., 52: 32–34.

S- 140 Suzuki, A. and Tonosaki, H. (1993) Tandem translocation chromosomes in *Allocyttus verrucossus* (Pisces; Zeiformes). Bull. Biogeograph. Soc. Japan, 48: 69–72.

S- 141 Suzuki, A., Taki, Y., Mochizuki, M. and Hirata, J. (1995) Chromosomal speciation in Eurasian and Japanese Cyprinidae (Pisces, Cypriniformes). Cytobios, 83: 171–186.

S- 142 Suzuki, A. and Tonosaki, H. (1995) Karyotypes by Ag-NOR staining in three species of pleuronectid fishes. Chrom. Inform. Serv., 59: 11–12.

S- 143 Suzuki, A. (1996) Chromosomal evolution of the suborder Cyprinoidei, with special reference to topical problems of phylogenetic relationships of their families. Cytobios, 87: 169–179.

S- 144 Suzuki, A. and Taki, Y. (1996) Tetraploidization in the cobitid subfamily Botinae (Pisces, Cypriniformes). Cytobios, 85: 229–245.

S- 145 Suzuki, A. (2003) Cytogenetical studies on three species of Noemacheilinae and Gastromyzoninae fishes (Cobitidae; Cypriniformes). Folia Biol. (Krakow), 51 (suppl.): 125–128.

S- 146 Swarça, A.C., Giuliano-Caetano, L. and Dias, A.L. (1999) Cytogenetic characterization through chromosomic banding of *Pinirampus pirinampu* (Pisces, Pimelodidae) from the Tibagi river basin PR/Brazil. Caryologia, 52: 31–35.

S- 147 Swarça, A.C., Giuliano-Caetano, L., Vanzela, A.L.L. and Dias, A.L. (2001) Heteromorphism of rDNA size in *Pinirampus pirinampu* (Pisces: Pimelodidae) detected by *in situ* hybridization. Cytologia, 66: 275–278.

S- 148 Swarça, A.C., Cestari, M.M., Giuliano-Caetano, L. and Dias, A.L. (2001) Cytogenetic characterization of the large South American siluriform fish species *Zungaro zungaro* (Pisces, Pimelodidae). Chrom. Sci., 5: 51–55.

S- 149 Swarça, A.C., Vidotto, A.P. and Dias, A.L. (2003) Cytogenetic characterization of *Pimelodella* aff. *avanhandavae* (Siluriformes, Pimelodidae) from Tibagi River (Paraná State, Brazil). Caryologia, 56: 421–425.

S- 150 Swarça, A.C., Fenocchio, A.S., Cestari, M.M. and Dias, A.L. (2005) Karyotype divergence among populations of giant catfish *Pseudoplatystoma corruscans* (Teleostei: Pimelodidae) indicates higher species diversity. Ichthyol. Explor. Freshwaters, 16: 325–330.

S- 151 Swarup, H. (1959) Production of triploidy in *Gasterosteus aculeatus* (L.). J. Genet., 56: 129–142.

S- 152 Scheel, J.J. (1972) The chromosomes of the third neon tetra. TFH, 1972, July: 60–65.

S- 153 Scheel, J.J. and Christensen, B. (1970) The chromosomes of the two common neon tetras. TFH, 19(2): 24–31.

S- 154 Stivari, M.K. and Martins-Santos, I.C. (2004) Karyotype diversity in two populations of *Rhamdia quelen* (Pisces, Heptapteridae). Cytologia, 69: 25–34.

S- 155 Sánchez, S., Laudicina, A. and Jorge, L.C. (2004) A new report of multiple sex chromosome system in the order Gymnotiformes (Pisces). Cytologia, 69: 155–160.

S- 156 Sekiya, S. and Honma, Y. (1975) Morphological and cytological studies on the ironfish occurring in Niigata and Sado Island. Proc. Japan. Soc. Syst. Zool., 11: 53–60. (In Japanese with English abstract)

S- 157 Slechtova, V., Bohlen, J., Freyhof, J. and Ráb, P. (2006) Molecular phylogeny of the Southeast Asian freshwater fish family Botiidae (Teleostei: Cobitoidea) and the origin of polyploidy in their evolution. Mol. Phylogenet. Evol., 39: 529–541.

S- 158 Scheel, J.J. (1990) Atlas of killifishes of the Old World. T.F.H. Publications, Neptune, NJ, USA.

S- 159 Scheel, J.J. (1975) Rivulins of the Old World. T.F.H. Publications, Neptune, NJ, USA.

S- 160 Smith, M.L. and Miller, R.R. (1980) *Allotoca maculata*, a new species of goodeid fish from western México, with comments on *Allotoca dugesi*. Copeia, 1980: 408–417.

S- 161 Schultz, R.J. (1967) Gynogenesis and triploidy in the viviparous fish *Poeciliopsis*. Science, 157: 1564–1567.

S- 162 Stevenson, M.M. (1975) A comparative chromosome study of the pupfish genus *Cyprinodon* (Teleostei: Cyprinodontidae). Unpubl. Ph D Diss., University of Oklahoma, Norman, USA.

S- 163 Santini, F., Harmon, L.J., Carnevale, G. and Alfaro, M.E. (2009) Did genome duplication drive the origin of teleosts? A comparative study of diversification in ray-finned fishes. BMC Evol. Biol., 2009, 9: 194. doi: 10.1186/1471-2148-9-194.

S- 164 Schmid, M., Ziegler, C.G., Steinlein, C., Nanda, I. and Schartl, M. (2006) Cytogenetics of the bleak (*Alburnus alburnus*), with special emphasis on the B chromosomes. Chrom. Res., 14: 231–242.

S- 165 Swarça, A.C., Fenocchio, A.S., Cestari, M.M., Bertollo, L.A.C. and Dias, A.L. (2006) Heteromorphic sex chromosome system with an exceptionally large Y chromosome in a catfish *Steindachneridion* sp. (Pimelodidae). Cytogenet. Genome Res., 112: 325–328.

S- 166 Swarça, A.C., Giuliano-Caetano, L. and Dias, A.L. (2000) Cytogenetics of species of the families Pimelodidae and Rhamdiidae (Siluriformes). Genet. Mol. Biol., 23: 589–593.

S- 167 Sideleva, V.G., Prirodina, V.P. and Khanayev, I.V. (1995) Karyotypes of benthic Lake Baikal cottoids, with remarks about karyological evolution in relation to bathymeric distribution. J. Ichthyol., 35(9): 295–304.

S- 168 Sideleva, V.G. (2003) The endemic fish of Lake Baikal. Buckhuys Publishers, Leiden.

S- 169 Swarça, A.C., Fenocchio, A.S., Cestari, M.M. and Dias, A.L. (2003) Analysis of heterochromatin by combination of C-banding and CMA_3 and DAPI staining in two fish species (Pimelodidae, Siluriformes). Genetica, 119: 87–92.

S- 170 Sola, L., Natili, G.L. and Cataudella, S. (1988) Cytogenetical characterization of *Odontesthes bonariensis* (Pisces, Atherinidae), an Argentine species introduced in Italy. Genetica, 77: 217–224.

S- 171 Sahoo, P.K., Nanda, P. and Barat, A. (2007) Karyotypic diversity among three species of *Garra* (family Cyprinidae) from River Dikrong, Arunachal Pradesh, India. Cytologia, 72: 259–263.

S- 172 Sonnenberg, R. (2007) Two new species of *Chromaphyosemion* (Cyprinodontiformes: Nothobranchiidae) from the coastal plain of Equatorial Guinea. Ichthyol. Explor. Freshwaters, 18: 359–373.

S- 173 Sahoo, P.K., Nanda, P. and Barat, A. (2007) Karyotypic analysis of *Neolissocheilus hexagonolepis* (McClelland), *Puntius ticto* (Ham.) and *P. chola* (Ham.) (family: Cyprinidae, Pisces). Cytologia, 72: 409–413.

S- 174 Souza, I.L., Venere, P.C. and Moreira-Filho, O. (2007) Constitutive heterochromatin and Ag-NOR polymorphisms in the small characid fish *Astyanax scabripinnis* (Jenyns, 1842). Cytologia, 72: 63–69.

S- 175 Sobita, N. and Bhagirath, T. (2007) *Colisa fasciata* (Schneider) (Perciformes, Anabantoidei, Belontidae) karyotype is divergent and ZW sex chromosomes are allocyclic and asynaptic. Nucleus, 50: 45–49.

S- 176 Scheel, J.J. (1969) Notes on the taxonomy of *Austrofundulus dolichopterus* and other annual rivuline species of the New World. JAKA, 6(1): 8–16.

S- 177 Subrahmanyam, K. (1969) A karyotypic study of the estuarine fish *Boleophthalmus boddaerti* (Pallas) with calcium treatment. Curr. Sci., 18: 437–439.

S- 178 Sena, D.C.S. and Molina, W.F. (2007) Robertsonian rearrangements and pericentric inversions in Scaridae fish (Perciformes). Genet. Mol. Res., 6: 575–580.

S- 179 Sola, L., Nirchio, M. and Rossi, A.R. (2008) The past and the future of cytogenetics of Mugilidae: an updated overview. Bol. Inst. Oceanogr. Venezuela, 47: 27–33.

S- 180 Swarça, A.C., Fenocchio, A.S. and Dias, A.L. (2007) An update cytogenetic review for species of the families Pseudopimelodidae, Pimelodidae and Heptapteridae (Pisces, Siluriformes). Suggestion of a cytotaxonomical classification. Caryologia, 60: 338–348.

S- 181 Sá-Gabriel, L.G. and Molina, W.F. (2004) Sex chromosomes in *Stephanolepis hispidus* (Monacanthidae, Tetraodontiformes). Cytologia, 69: 447–452.

S- 182 Salvadori, S., Coluccia, E., Cannas, R., Cau, A. and Deiana, A.M. (2009) A ZZ-ZW sex chromosome system in the finless eel *Dalophis imberbis* (Anguilliformes, Ophichthidae). Genetica, 135: 283–288.

S- 183 Schultz, R.J. (1971) Special adaptive problems associated with unisexual fishes. Amer. Zool., 11: 351-360.

S- 184 Scheel, J.J. (1982, personal communication)

S- 185 Stingo, V., Capriglione, T., Rocco, L., Improta, R. and Morescalchi, A. (1989) Genome size and A-T rich DNA in selachians. Genetica, 79: 197–205.

S- 186 Scavone, M.D.P. and Júlio Jr., H.F. (1995) Cytogenetic analysis and heterochromatin distribution in ZZ/ZW sex chromosomes of the mailed catfish *Loricariichthys platymetopon* (Loricariidae: Siluriformes). Braz. J. Genet., 18: 31–35.

S- 187 Sola, L., Rossi, A.R., Iaselli, V., Rasch, E.M. and Monaco, P.J. (1992) Cytogenetics of bisexual/unisexual species of *Poecilia*. II. Analysis of heterochromatin and nucleolar organizer regions in *Poecilia mexicana mexicana* by C-banding and DAPI, quinacrine, chromomycin A_3, and silver staining. Cytogenet. Cell Genet., 60: 229–235.

S- 188 Sola, L., Bressanello, S., Rasch, E.M. and Monaco, P.J. (1993) Cytogenetics of bisexual/unisexual species of *Poecilia*. IV. Sex chromosomes, sex chromatin composition and Ag-NOR polymorphisms in *Poecilia latipinna*: a population from Mexico. Heredity, 70: 67–71.

S- 189 Singh, M., Kumar, R., Nagpure, N.S., Kushwaha, B., Gond, I. and Lakra, W.S. (2009) Chromosomal localization of 18S and 5S rDNA using FISH in the genus *Tor* (Pisces, Cyprinidae). Genetica, 137: 245–252.

S- 190 Shimabukuro-Dias, C.K., Oliveira, C. and Foresti, F. (2004) Cytogenetic analysis of five species of the subfamily Corydoradinae (Teleostei: Siluriformes: Callichthyidae). Genet. Mol. Biol., 27: 549–554.

S- 191 Smith, E.M. and Gregory, T.R. (2009) Patterns of genome size diversity in the ray-finned fishes. Hydrobiologia, 625: 1–25.

*S–192 Stange, E.A.R., Silva, V.E. and Dutra, S.V. (1986) Citogenética no género *Astyanax* (Pisces, Characidae) das bacias dos rios Benevente e Jucu. I Simp. Citogenet. Evol. e Aplic. de Peixes Neotropicais, Sao Carlos, SP, p. 58 (abstract). (After K-114)

*S–193 Skelton, P.H. and Naran, D. (1995) Polyploidy in southern African barbine cyprinids. Abstr. VIII Congr. Soc. Europ. Ichthyol., Oviedo, p. 62. (After K-114)

*S–194 Simovic, S.Z., Markovic, G.S. and Simonovic, P.D. (1994) New data of the chromosome complements of the five species of family Cyprinidae. Abstr. VIII Congr. Soc. Europ. Ichthyol., Oviedo, p. 62. (After K-114)

S- 195 Sofradžija, A. (1984) First data on the chromosomes of the three Adriatic fish species (*Scorpaena porcus*, *S. ustulata* and *Corvina nigra*). Ichthyologia, 16: 57–61. (In Serbo-Croatian with English summary)

*S-196 Stange, E.A.R. and Passamani, M. (1986) Caracterizaçao cromossômica de *Awaous tajasica* (Lichtenstein, 1822) (Pisces, Gobiidae). I. Simp. Citogenet. Evol. e Aplic. de Peixes Neotropicais, Sao Carlos, SP, p. 59 (abstract). (After K-114)

S- 197 Souza, L., Swarca, A.C. and Dias, A.L. (2004) Analysis of the nucleolus organizer regions in 5 species of the genus *Pimelodus* (Siluriformes, Pimelodidae), using $AgNO_3$, CMA_3 and FISH with the 18S rDNA probe. Caryologia, 57: 145–151.

S- 198 Stolf, R., Swarca, A.C., Giuliano-Caetano, L. and Dias, A.L. (2004) Analyses of karyotype and nucleolus organizer regions of *Imparfinis* aff. *schubarti* (Siluriformes, Pimelodidae) of the Tibagi river basin, Paraná, Brazil. Caryologia, 57: 348–352.

S- 199 Swarça, A.C., Fenocchio, A.S., Cestari, M.M. and Dias, A.L. (2005) First chromosome data on *Steindachneridion scripta* (Pisces, Siluriformes, Pimelodidae) from Brazilian rivers: Giemsa, CBG, G-, and RE banding. Genet. Mol. Res., 4: 734–741.

S- 200 Swarça, A.C., Fenocchio, A.S., Cestari, M.M. and Dias, A.L. (2008) Analyses of the structure of NORs in two species of South American Sorubiminae fishes (Siluriformes) by means of several cytogenetic techniques. Folia Biol. (Krakow), 56: 31–35.

S- 201 Santi-Rampazzo, A.P., Nishiyama, P.B., Ferreira, P.E.B. and Martins-Santos, I.C. (2007) Cytogenetic analysis and description of the sexual chromosome determination system ZZ/ZW of species of the fish genus *Serrapinnus* (Characidae, Cheirodontinae). Genet. Mol. Res., 6: 504–509.

*S-202 Sharma, O.P. and Tripathi, N.K. (1984) Somatic chromosome analysis of *Wallago attu*, a siluroid fish from Jammu (J & K), India. Perspect. Cytol. Genet., 4: 437–439. (After P-96)

S- 203 Singh, M., Kumar, R., Nagpure, N.S., Kushwaha, B., Mani, I., Chauhan, U.K. and Lakra, W.S. (2009) Population distribution of 45S and 5S rDNA in golden mahseer, *Tor putitora*: population-specific FISH marker. J. Genet., 88: 315–320.

S- 204 Shields, B.A., Guise, K.S. and Underhill, J.C. (1990) Chromosomal and mitochondrial DNA characterization of a population of dwarf cisco (*Coregonus artedii*) in Minnesota. Can. J. Fish. Aquat. Sci., 47: 1562–1569.

S- 205 Sato, Y. and Nishida, M. (2010) Teleost fish with specific genome duplication as unique models of vertebrate evolution. Environ. Biol. Fishes, 88: 169–188.

S- 206 Silva, D.S., Milhomem, S.S.R., Pieczarka, J.C. and Nagamachi, C.Y. (2009) Cytogenetic studies in *Eigenmannia virescens* (Sternopygidae, Gymnotiformes) and new inferences on the origin of sex chromosomes in the *Eigenmannia* genus. BMC Genetics, 2009, 10: 74.

S- 207 Schultz, R.J. (1961) Reproductive mechanism of unisexual and bisexual strains of the viviparous fish *Poeciliopsis*. Evolution, 15: 302–325.

[T]

T- 1 Takahashi, J. and Oka, M. (1976) Karyotypes and electrophoretic patterns of hemoglobins in loaches of the genus *Cobitis*. Japan. J. Ichthyol., 23: 114–117. (In Japanese with English abstract)

T- 2 Takai, A. and Ojima, Y. (1982) The assignment of the nucleolus organizer regions in the chromosomes of the carp, the funa and their hybrids (Cyprinidae, Pisces). Proc. Japan Acad., Ser. B, 58: 303–306.

T- 3 Takai, A. and Ojima, Y. (1983) Tetraploidy appeared in the offspring of triploid Ginbuna, *Carassius auratus langsdorfii* (Cyprinidae, Pisces). Proc. Japan Acad., Ser. B, 59: 347–350.

T- 4 Takai, A. and Ojima, Y. (1984) Some features on the nucleolus organizer regions in the chromosomes of the cyprinid fishes. A preliminary report. Proc. Japan Acad., Ser. B, 60: 410–413.

T- 5 Takai, A. and Ojima, Y. (1985) Karyotypic studies of five species of Anguilliformes (Pisces). Proc. Japan Acad., Ser. B, 61: 253–256.

T- 6 Takai, A. and Ojima, Y. (1986) Some features on nucleolus organizer regions in fish chromosomes. In: Uyeno, T., Arai, R., Taniuchi, T. and Matsuura, K. eds. Indo-Pacific fish biology. Ichthyological Society of Japan, Tokyo, pp. 899–909.

T- 7 Takai, A. and Ojima, Y. (1987) Comparative studies of karyotypes and distributions of nucleolus organizer regions in pomacentrid fishes. I. Proc. Japan Acad., Ser. B, 63: 17–20.

T- 8 Takai, A., Kikuno, T. and Ojima, Y. (1987) Karyotype, C-bands and NORs in *Anago anago* (Congridae, Pisces). Chrom. Inform. Serv., 42: 30–32.

T- 9 Takai, A. and Ojima, Y. (1987) Comparative chromosomal studies in three balistid fishes. La Kromosomo, II, 47/48: 1545–1550.

T- 10 Takai, A. and Ojima, Y. (1988) Karyotype and banding analyses in a balisted fish, *Balistoides viridescens* "gomamongara". Chrom. Inform. Serv., 45: 25–27.

T- 11 Takai, A. and Ojima, Y. (1988) Chromosomal distribution of nucleolus organizer regions in six species of Rhodeinae (Cyprinidae, Pisces). Proc. Japan Acad., Ser. B, 64: 229–232.

T- 12 Takai, A. and Ojima, Y. (1988) Chromosomal distribution of C-banded heterochromatin in cyprinid fishes. Proc. Japan Acad., Ser. B, 64: 49–52.

T- 13 Takai, A. and Ojima, Y. (1989) Conventional and C-banding karyotypes and electrophoretic pattern of muscle protein in fish of genus *Carassius* from Ishigaki Island (Cyprinidae, Pisces). La Kromosomo, II, 54: 1782–1786.

T- 14 Takai, A. and Ojima, Y. (1990) A comparative study of chromosomal C-band distribution in the carp, the funa and their hybrids. Chrom. Inform. Serv., 49: 8–10.

T- 15 Takai, A. and Ojima, Y. (1990) Characteristics of chromosomes and electrophoretic patterns of muscle protein and lactate dehydrogenase in fish of genus *Carassius* (Cypriniformes). Cytologia, 55: 255–263.

T- 16 Takai, A. and Ojima, Y. (1991) Comparative studies of karyotypes and distribution of nucleolus organizer regions in pomacentrid fish. 2. Cytobios, 65: 199–205.

T- 17 Takai, A. and Ojima, Y. (1992) Chromosomal distribution of nucleolus organizer regions in Japanese cyprinid fish. Cytobios, 71: 7–17.

T- 18 Takai, A. and Ojima, Y. (1993) G-banding chromosomes of Chinese crucian carp from cultured fin cells. Cytologia, 58: 211–215.

T- 19 Takai, A. and Ojima, Y. (1995) Chromosome evolution associated with Robertsonian rearrangements in pomacentrid fish (Perciformes). Cytobios, 84: 103–110.

T- 20 Takai, A. and Ojima, Y. (1995) A chromosomal study of a serranid fish, *Chromileptes altivelis* (Perciformes), using fin cultures. Chrom. Inform. Serv., 59: 9–10.

T- 21 Takai, A. and Ueno, K. (1997) Karyotypes and chromosomal distribution of constitutive heterochromatin and nucleolus organizer regions in three kyphosid fish (Perciformes). Cytobios, 89: 141–148.

T- 22 Takai, A. and Ojima, Y. (1998) C-banded karyotype and nucleolus organizer regions of a notopterid fish, *Notopterus chitala* (Notopteridae, Osteoglossiformes). Chrom. Sci., 2: 35–38.

T- 23 Takai, A. and Ojima, Y. (1999) Constitutive heterochromatin distribution in the chromosomes of pomacentrid fishes (Perciformes). Cytologia, 64: 87–91.

T- 24 Takai, A., Yoshikawa, T. and Ojima, Y. (2002) C-banded karyotype and nucleolus organizer regions in a discus fish, *Symphysodon aequifasciata axelrodi* (Cichlidae, Perciformes). Chrom. Sci., 6: 53–55.

T- 25 Takarada, Y., Ojima, Y. and Takai, A. (1989) Establishment of a fish cell line transformed by MNNG, RFM. Proc. Japan Acad., Ser. B, 65: 108–111.

T- 26 Takarada, Y., Ojima, Y. and Takai, A. (1989) Gene mapping of the gold-fish using mouse-fish cell hybrids. Chrom. Inform. Serv., 47: 30–32.

T- 27 Tatewaki, R., Kagohashi, Y. and Kitada, J.-I. (1999) High incidence of NOR associations in metaphases of *Lateolabrax latus* (Perciformes, Pisces). Chrom. Sci., 3: 75–77.

T- 28 Taylor, K.M. (1967) The chromosomes of some lower chordates. Chromosoma, 21: 181–188.

T- 29 Teugels, G.G., Ozouf-Costaz, C., Legendre, M. and Parrent, M. (1992) A karyological analysis of the artificial hybridization between *Clarias gariepinus* (Burchell, 1822) and *Heterobranchus longifilis* Valenciennes, 1840 (Pisces; Clariidae). J. Fish Biol., 40: 81–86.

T- 30 Thode, G., Cano, J. and Alvarez, M.C. (1983) Association of nucleolus organizer chromosomes shown by silver staining in *Gobius fallax*. J. Heredity, 74: 480–482.

T- 31 Thode, G. and Alvarez, M.C. (1983) The chromosome complements of two species of *Gobius* (Teleostei, Perciformes). Experientia, 39: 1312–1314.

T- 32 Thode, G., Cano, J. and Alvarez, M.C. (1983) A karyological study on four species of Mediterranean gobiid fishes. Cytologia, 48: 131–138.

T- 33 Thode, G., Giles, V. and Alvarez, M.C. (1985) Multiple chromosome polymorphism in *Gobius paganellus* (Teleostei, Perciformes). Heredity, 54: 3–7.

T- 34 Thode, G., Alvarez, M.C., Garcia, E. and Giles, V. (1985) Variations in C-banding patterns and DNA values in two scorpion-fishes (*Scorpaena porcus* and *S. notata*, Teleostei). Genetica, 68: 69–74.

T- 35 Thode, G., Alvarez, M.C., Giles, V. and Garcia, E. (1985) Chromosome complement, C-banding and Ag-NOR location in *Ophysurus serpens* (Ophichthidae, Auguilliformes). Cytobios, 43: 73–77.

T- 36 Thode, G. (1987) Karyotype analysis of the clingfish, *Lepadogaster candollei* Risso (Gobiesociformes). Cytobios, 51: 163–169.

T- 37 Thode, G., Martinez, G., Ruiz, J.L. and Lopéz, J.R. (1988) A complex chromosomal polymorphism in *Gobius fallax* (Gobiidae, Perciformes). Genetica, 76: 65–71.

T- 38 Thode, G., Amores, A. and Martinez, G. (1994) The karyotype of *Balistes carolinensis* Gmelin (Pisces, Tetraodontiformes), a specialized species. Caryologia, 47: 257–263.

T- 39 Thompson, K.W. (1976) Some aspects of chromosomal evolution of the Cichlidae (Teleostei: Perciformes) with emphasis on Neotropical forms. Unpublished Ph. D. Dissertation, The University of Texas at Austin.

T- 40 Thompson, K.W., Hubbs, C. and Edwards, R.J. (1978) Comparative chromosome morphology of the blackbasses. Copeia, 1978: 172–175.

T- 41 Thompson, K.W. (1979) Cytotaxonomy of 41 species of neotropical Cichlidae. Copeia, 1979: 679–691.

T- 42 Thompson, K.W. (1981) Karyotypes of six species of African Cichlidae (Pisces: Perciformes). Experientia, 37: 351–352.

T- 43 Thompson, K.W. (1991) Cytotaxonomy of the Bantam sunfish *Lepomis symmetricus* (Centrarchidae). I. Robertsonian variation in a population from the Yazoo river basin in Mississippi (U.S.A.). Caryologia, 44: 109–114.

T- 44 Thorgaard, G.H. (1976) Robertsonian polymorphism and constitutive heterochromatin distribution in chromosomes of the rainbow trout (*Salmo gairdneri*). Cytogenet. Cell Genet., 17: 174–184.

T- 45 Thorgaard, G.H. (1977) Heteromorphic sex chromosomes in male rainbow trout. Science, 196: 900–902.

T- 46 Thorgaard, G.H. (1978) Sex chromosomes in the sockeye salmon: a Y-autosome fusion. Can. J. Genet. Cytol., 20: 349–354.

T- 47 Thorgaard, G.H. (1983) Chromosomal differences among rainbow trout populations. Copeia, 1983: 650–662.

T- 48 Torres, R.A., Foresti, F. and Oliveira, C. (2002) Occurrence of karyotypical mosaicism in *Trichomycterus paolence* (Teleostei, Trichomycteridae). Caryologia, 55: 283–287.

T- 49 Tripathi, N.K. and Das, C.C. (1980) Chromosomes in three species of Asian catfish. Copeia, 1980: 916–918.

T- 50 Tripathi, N.K. and Das, C.C. (1981) The karyotype of *Trichogaster fasciatus* (Osphranemidae, Pisces). Curr. Sci., 20: 916–917.

T- 51 Tripathi, N.K. and Sharma, O.P. (1986) Karyotypic studies on an air breathing fish, *Heteropneustes fossilis* (Heteropneustidae: Siluridae). Chrom. Inform. Serv., 40: 14–16.

T- 52 Tripathi, N.K. and Sharma, O.P. (1987) Cytological studies on six cyprinid fishes. Genetica, 73: 243–246.

T- 53 Tripathi, N.K. and Das, C.C. (1988) Karyotypes of five Indian perciform fishes. Copeia, 1988: 231–233.

T- 54 Tsingenopoulos, C. and Karakousis, Y. (1996) Phylogenetic relationships of *Leuciscus keadicus*, an endemic cyprinid species from Greece, with other Greek species in the genus *Leuciscus*. Folia Zool., 45: 87–93.

T- 55 Tsigenopoulos, C.S., Ráb, P., Naran, D. and Berrebi, P. (2002) Multiple origins of polyploidy in the phylogeny of southern African barbs (Cyprinidae) as inferred from mtDNA markers. Heredity, 88: 466–473.

T- 56 Turner, B.J. and Liu, R.K. (1977) Extensive interspecific genetic compatibility in the New World killifish genus *Cyprinodon*. Copeia, 1977: 259–269.

T- 57 Turner, B.J., Grudzien, T.A., Adkisson, K.P. and Worrell, R.A. (1985) Extensive chromosomal divergence within a single river basin in the goodeid fish, *Ilyodon furcidens*. Evolution, 39: 122–134.

T- 58 Turner, B.J., Diffoot, N. and Rasch, E.M. (1992) The callichthyid catfish *Corydoras aeneus* is an unresolved diploid-tetraploid sibling species complex. Ichthyol. Explor. Freshwaters, 3: 17–23.

T- 59 Tsigenopoulos, C.S. and Berrebi, P. (2000) Molecular phylogeny of north Mediterranean freshwater barbs (genus *Barbus*: Cyprinidae) inferred from cytochrome *b* sequences: biogeographic and systematic implications. Mol. Phylogenet. Evol., 14: 165–179.

T- 60 Taki, Y., Urushido, T., Suzuki, A. and Serizawa, C. (1977) A comparative chromosome study of *Puntius* (Cyprinidae: Pisces). I. Southeast Asian species. Proc. Japan Acad., Ser. B, 53: 231–235.

T- 61 Taki, Y. and Suzuki, A. (1977) A comparative chromosome study of *Puntius* (Cyprinidae: Pisces). II. Indian and Ceylonese species. Proc. Japan Acad., Ser. B, 53: 282–286.

T- 62 Turner, B.J. (1974) Genetic divergence of Death Valley pupfish species: biochemical versus morphological evidence. Evolution, 28: 281–294.

T- 63 Thacker, C.E. (2009) Phylogeny of Gobioidei and placement within Acanthomorpha, with a new classification and investigation of diversification and character evolution. Copeia, 2009: 93–104.

T- 64 Terashima, H. and Ida, H. (1991) Karyotypes of three species of the family Cottidae (Scorpaeniformes). Japan. J. Ichthyol., 37: 358–362.

T- 65 Taylor, J.S., Braasch, I., Frickey, T., Meyer, A. and Peer, Y.V. de (2003) Genome duplication, a trait shared by 22,000 species of ray-finned fish. Genome Res., 13: 382–390.

T- 66 Tsukamoto, R.Y., Okazaki, I., Sakai, K., Strussmann, C.A., Takashima, F., Sato, Y., Shirahata, Y. and Sugiyama, H. (1993) Sex chromosomes of *Arctoscopus japonicus* (Trichodontidae, Perciformes). Advance Abstracts for the spring meeting of Japan Fisheries Society in 1993, p. 337.

T- 67 Takai, A. and Izutsu, H. (2008) Diversified chromosomal characteristics in *Centropyge* fishes (Pomacanthidae, Perciformes). Hydrobiologia, 603: 15–23.

T- 68 Tinni, S.R., Jessy, N.S., Hasan, M.M., Mustafa, Md.G. and Alam, S.S. (2007) Comparative karyotype analysis with differential staining in two forms of *Anabas testudineus* Bloch. Cytologia, 72: 71–75.

T- 69 Torres, R.A., Foresti, F. and Oliveira, C. (1998) Comparative cytogenetic studies in allopatric populations of *Trichomycterus paolence* Eigenmann, 1918 from southeastern Brazil (Siluriformes, Trichomycteridae). Cytologia, 63: 107–114.

T- 70 Takai, A. and Kosuga, S. (2008, (2007)) Karyotypes and banded chromosomal features in two anemonefishes (Pomacentridae, Perciformes). Chrom. Sci., 10: 71–74.

T- 71 Torres, R.A., Motta, T.S., Nardino, D., Adam, M.L. and Ribeiro, J. (2008) Chromosomes, RAPDs and evolutionary trends of the Neotropical fish *Mimagoniates microlepis* (Teleostei: Characidae: Glandulocaudinae) from coastal and continental regions of the Atlantic forest, southern Brazil. Acta Zool., 89: 253–259.

T- 72 Tiersch, T.R., Chandler, R.W., Kallman, K.D. and Wachtel, S.S. (1989) Estimation of nuclear DNA content by flow cytometry in fishes of the genus *Xiphophorus*. Comp. Biochem. Physiol., 94B: 465–468.

T- 73 Tiersch, T.R., Chandler, R.W., Wachtel, S.S. and Elias, S. (1989) Reference standards for flow cytometry and application in comparative studies of nuclear DNA content. Cytometry, 10: 706–710.

T- 74 Tiersch, T.R. and Chandler, R.W. (1989) Chicken erythrocytes as an internal reference for analysis of DNA content by flow cytometry in grass carp. Trans. Amer. Fish. Soc., 118: 713–717.

T- 75 Torres, R.A., Roper, J.J., Foresti, F. and Oliveira, C. (2005) Surprising genomic diversity in the Neotrop. fish *Synbranchus marmoratus* (Teleostei: Synbranchidae): how many species? Neotropical Ichthyol., 3: 277–284.

T- 76 Terencio, M.L., Schneider, C.H., Gross, M.C., Silva, A.M., Feldberg, E. and Porto, J.I.R. (2008) Comparative cytogenetics of *Carnegiella marthae* and *Carnegiella strigata* (Characiformes, Gasteropelecidae) and description of a ZZ/ZW sex chromosome system. Genet. Mol. Biol., 31 (suppl.): 231–234.

T- 77 Torres-Mariano, A.R. and Morelli, S. (2008) B chromosomes in a population of *Astyanax eigenmanniorum* (Characiformes, Characidae) from the Araguari river basin (Uberlândia, MG, Brazil). Genet. Mol. Biol., 31 (suppl.): 246–249.

T- 78 Treco, F.R., Malabarba, L.R., Giuliano-Caetano, L. and Dias, A.L. (2008) Cytogenetic study of two species of the family Pimelodidae (Siluriformes) collected in lago Guaiba, Rio Grande do Sul, Brazil. Neotrop. Ichthyol., 6: 87–92.

T- 79 Takai, A. (2007) Chromosomal distribution of C-banded heterochromatin and Ag-stained nucleolus organizer regions in two moray eels (Muraenidae, Anguilliformes). Chrom. Sci., 10: 15–19.

T- 80 Takahashi, E.K., Giuliano-Caetano, L. and Dias, A.L. (1994) Variabilidade cariotipica em *Astyanax bimaculatus* (Pisces, Tetragonopterinae) coletado no ribeirão três bocas—Londrina, PR. V. Simp. Citogenet. Evol. e Aplic. de Peixes Neotropicais, Botucatu, SP, p. 53 (abstract). (In Portuguese)

T- 81 Treco, F.R. and Dias, A.L. (2009) Karyotypes of two species of the genus *Pimelodus* (Siluriformes, Pimelodidae). Folia Biol. (Krakow), 57: 43–48.

T- 82 Teribele, R., Gravena, W., Carvalho, K., Giuliano-Caetano, L. and Dias, A.L. (2008) Karyotypic analysis in two species of fishes of the family Curimatidae: $AgNO_3$, CMA_3 and FISH with 18S probe. Caryologia, 61: 211–215.

T- 83 Tatewaki, R. and Kitada, J. (1994) Ag-NORs of the two species of *Lateolabrax* (Perciformes, Pisces). Chrom. Inform. Serv., 57: 18–19.

T- 84 Takai, A. (2010, personal communication)

T- 85 Takehana, Y., Hamaguchi, S. and Sakaizumi, M. (2008) Different origins of ZZ/ZW sex chromosomes in closely related medaka, *Oryzias javanicus* and *O. hubbsi*. Chrom. Res., 16: 801–811.

[U]
U- 1 Ueda, T. and Ojima, Y. (1978) Differential chromosomal characteristics in the funa subspecies (*Carassius*). Proc. Japan Acad., Ser. B, 54: 283–288.
U- 2 Ueda, T. and Ojima, Y. (1983) Karyotypes with C-banding patterns of two species in the genus *Salvelinus* of the family Salmonidae. Proc. Japan Acad., Ser. B, 59: 343–346.
U- 3 Ueda, T., Ojima, Y., Kato, T. and Fukuda, Y. (1983) Chromosomal polymorphisms in the rainbow trout (*Salmo gairdneri*). Proc. Japan Acad., Ser. B, 59: 168–171.
U- 4 Ueda, T. and Ojima, Y. (1983) Geographic and chromosomal polymorphisms in the Iwana (*Salvelinus leucomaenis*). Proc. Japan Acad., Ser. B, 59: 259–262.
U- 5 Ueda, T. and Ojima, Y. (1984) Sex chromosomes in the kokanee salmon *Oncorhynchus nerka*. Bull. Japan. Soc. Sci. Fish., 50: 1495–1498.
U- 6 Ueda, T. and Ojima, Y. (1984) Sex chromosomes in the rainbow trout *Salmo gaidneri*. Bull. Japan. Soc. Sci. Fish., 50: 1499–1504.
U- 7 Ueda, T. and Ojima, Y. (1984) Cytogenetical characteristics of the progeny from the heteroploidy in the rainbow trout. Proc. Japan Acad., Ser. B, 60: 183–186.
U- 8 Ueda, T., Ojima, Y. and Naka, K. (1985) Chromosomal polymorphisms in the Biwa trout (Salmonidae) with an increase or decrease in constitutive heterochromatin. Proc. Japan Acad., Ser. B, 61: 483–485.
U- 9 Ueda, T. and Kobayashi, J. (1988) The relation between the sex chromosome and the gonadal development in the triploid rainbow trout, *Salmo gairdneri*. Chrom. Inform. Serv., 44: 8–10.
U- 10 Ueda, T. and Kobayashi, J. (1988) Heteroploidy induced by production of triploids in the kokanee, *Oncorhynchus nerka*. Chrom. Inform. Serv., 44: 7–8.
U- 11 Ueda, T. and Kobayashi, J. (1990) Disappearance of Ag-NORs originated from maternal species in hybrids between female Atlantic salmon and male brown trout. Chrom. Inform. Serv., 49: 25–26.
U- 12 Ueda, T., Hayashi, M., Koide, N., Sofuni, T. and Kobayashi, J. (1991) Preliminary examination of the mutagenicity test using embryo cells of rose bitterling, *Rhodeus ocellatus ocellatus*. Chrom. Inform. Serv., 51: 12–14.
U- 13 Ueda, T. and Kobayashi, J. (1991) Chromosomal polymorphisms in oily bitterling *Acheilognathus limbatus*. Chrom. Inform. Serv., 50: 13–14.
U- 14 Ueda, T., Fukuda, H. and Kobayashi, J. (1991) Karyological study of the Dolly Varden *Salvelinus malma* from Alaska. Japan. J. Ichthyol., 37: 354–357.
U- 15 Ueda, T., Sato, R., Iwata, M., Komaru, A. and Kobayashi, J. (1991) The viable 3.5n trouts produced between diploid females and allotriploid males. Japan. J. Genet., 66: 71–75.
U- 16 Ueda, T., Onozaki, H. and Hayashi, F. (1994) Tetraploid hybrids between female *Rhodeus o. ocellatus* and male *Rhodeus atremius*. Bull. Dept. Educ. Utsunomiya Univ., 44: 65–72. (In Japanese with English abstract)
U- 17 Ueda, T. and Takahashi, Y. (1995) Chromosomal studies of three kinds of hybrids in acheilognathin fishes. Bull. Dept. Educ. Utsunomiya Univ., 45: 59–63. (In Japanese with English abstract)
U- 18 Ueda, T., Arai, R., Akai, Y., Ishinabe, T. and Wu, H.L. (1996) Karyological study of a bitterling *Rhodeus atremius* (Teleostei: Cyprinidae) collected in Zhejiang, China. Cytobios, 86: 265–268.
U- 19 Ueda, T. (1996) The possible induction of androgenetic diploid rainbow trout by an application of heat-shock. Chrom. Inform. Serv., 61: 12–13.
U- 20 Ueda, T. (1996) Chromosome aberrations in salmonid fish embryos using prolonged-stored eggs in coelomic fluid. Chrom. Inform. Serv., 61: 13–15.
U- 21 Ueda, T. (1996) Chromosomal aberrations induced by retention of cyprinid fish unfertilized eggs in freshwater. Cytologia, 61: 423–430.

U- 22 Ueda, T., Mashiko, N., Takizawa, H., Akai, Y., Ishinabe, T., Arai, R. and Wu, H. (1997) Ag-NOR variation in chromosomes of Chinese bitterlings, *Rhodeus lighti* and *Tanakia himantegus* (Cypriniformes, Cyprinidae). Ichthyol. Res., 44: 302–305.

U- 23 Ueda, T. and Naoi, H. (1999) BrdU-4Na-EDTA-Giemsa band karyotypes of 3 small freshwater fish, *Danio rerio, Oryzias latipes*, and *Rhodeus ocellatus*. Genome, 42: 531–535.

U- 24 Ueda, T., Naoi, H. and Arai, R. (2001) Flexibility on the karyotype evolution in bitterlings (Pisces, Cyprinidae). Genetica, 111: 423–432.

U- 25 Ueno, K. (1974) Chromosomal polymorphism and variant of isozymes in geographical populations of *Pseudobagrus aurantiacus*, Bagridae. Japan. J. Ichthyol., 21: 158–164.

U- 26 Ueno, K. (1976) Chromosomes of bitterlings. Tansuigyo, 2(1): 134–138. (In Japanese)

U- 27 Ueno, K. and Ojima, Y. (1976) Diploid-tetraploid complexes in the genus *Cobitis* (Cobitidae, Cyprinida). Proc. Japan Acad., Ser. B, 52: 446–449.

U- 28 Ueno, K. and Ojima, Y. (1977) Chromosome studies of two species of the genus *Coreoperca* (Pisces: Perciformes), with reference to the karyotypic differentiation and evolution. Proc. Japan Acad., Ser. B, 53: 221–225.

U- 29 Ueno, K., Iwai, S. and Ojima, Y. (1980) Karyotypes and geographical distribution in the genus *Cobitis* (Cobitidae). Bull. Japan. Soc. Sci. Fish., 46: 9–18. (In Japanese with English abstract)

U- 30 Ueno, K. (1981) Karyotypes of Cobitidae. Mar. Sci. Monthly, 13(1): 60–70. (In Japanese)

U- 31 Ueno, K. and Arimoto, B. (1982) Induction of triploids in *Rhodeus ocellatus ocellatus* by cold shock treatment of fertilized eggs. Experientia, 38: 544–546.

U- 32 Ueno, K. (1984) Induction of triploid carp and their haematological characteristics. Japan. J. Genet., 59: 585–591.

U- 33 Ueno, K. and Ojima, Y. (1984) A chromosome study of nine species of Korean cyprinid fish. Japan. J. Ichthyol., 31: 338–344.

U- 34 Ueno, K. (1985) Karyotypes of bagrid catfishes in Japan and Korea. Mar. Sci. Monthly, 17(2): 102–108. (In Japanese)

U- 35 Ueno, K., Senou, H. and Kim, I.S. (1985) A chromosome study of five species of Korean cobitid fish. Japan. J. Genet., 60: 539–544.

U- 36 Ueno, K., Ikenaga, Y. and Kariya, H. (1986) Potentiality of application of triploidy to the culture of Ayu, *Plecoglossus altivelis* Temminck et Schlegel. Japan. J. Genet., 61: 71–77.

U- 37 Ueno, K., Nagase, A. and Ye, Y.J. (1988) Tetraploid origin of the karyotype of the Asian sucker, *Myxocyprinus asiaticus*. Japan. J. Ichthyol., 34: 512–514.

U- 38 Ueno, K. and Ojima, Y. (1991) Notes on the chromosomes of *Girella melanichthys* and *Lutjanus russelli* (Pisces, Perciformes). Chrom. Inform. Serv., 52: 3–5.

U- 39 Ueno, K. and Jarayabhand, P. (1991) Karyotypes of a serranid, *Epinephelus malabaricus* and a centropomid, *Psammoperca waigiensis* (Pisces, Perciformes). Chrom. Inform. Serv., 51: 23–25.

U- 40 Ueno, K., Ye, Y.J. and Umeoka, T. (1992) A comparative study of chromosomes in the cyprinid fish genera *Gnathopogon* and *Squalidus* of Japan. Bull. Japan. Soc. Sci. Fish., 58: 1273–1277. (In Japanese with English abstract)

U- 41 Ueno, K. and Kang, J.H. (1992) Multiple sex chromosomes in the redfin velvetfish, *Hypodytes rubripinnis*. Japan. J. Ichthyol., 39: 170–173.

U- 42 Ueno, K. and Takai, A. (1997) Nucleolus organizer regions detected on three chromosome pairs in *Thalassoma cupido* (Pisces: Perciformes). Fish. Sci., 63: 476–477.

U- 43 Ueno, K. and Takai, A. (2000) Chromosome evolution involving Robertsonian rearrangements in *Xyrichthys* fish (Labridae, Perciformes). Cytobios, 103: 7–15.

U- 44 Ueno, K., Ota, K. and Kobayashi, T. (2001) Heteromorphic sex chromosomes of lizardfish (Synodontidae): focus on the ZZ-ZW_1W_2 system in *Trachinocephalus myops*. Genetica, 111: 133–142.

U- 45 Uribe-Alcocer, M., Arreguin-Espinosa, J., Torres-Padilla, A. and Castro-Pérez, A. (1983) Los cromosomas de *Dormitator latifrons* (Pisces: Gobiidae). An. Inst. Cienc. Mar Limnol. Univ. Nal. Autón. México, 10: 23–30. (In Spanish with English abstract)
U- 46 Uribe-Alcocer, M. and Ramirez-Escamilla, A. (1989) Comparación citogenética entre las especies del género *Dormitator* (Pisces: Gobiidae). An. Inst. Cienc. Mar Limnol. Univ. Nal. Autón. México, 16: 75–80. (In Spanish with English abstract)
U- 47 Uribe-Alcocer, M. and Arreguin-Espinosa, J. (1989) Los cromosomas de los peces *Oreochromis urolepis hornorum* y *Oreochromis mossambicus* (Pisces: Cichlidae). An. Inst. Cienc. Mar Limnol. Univ. Nal. Autón. México, 16: 189–198. (In spanish with English abstract)
U- 48 Uribe-Alcocer, M., Nader-Garcia, B. L. and Valdés-Morales, N. (1992) Karyotypes of two cichlid fishes from Mexico, *Cichlasoma ellioti* and *C. trimaculatum*. Japan. J. Ichthyol., 39: 174–177.
U- 49 Uribe-Alcocer, M., Tellez-Vargas, C. and Diaz-Jaimes, P. (1999) Chromosomes of *Cichlasoma istlanum* (Perciformes: Cichlidae) and karyotype comparison of two presumed subspecies. Rev. Biol. Trop., 47: 1051–1059.
U- 50 Uribe-Alcocer, M., Olvera-Garcia, H. and Diaz-Jaimes, P. (2002) Karyotype analysis in three species of the genus *Chirostoma* (Atheriniformes: Atherinidae). Ichthyol. Res., 49: 85–88.
U- 51 Urushido, T., Takahashi, E. and Taki, Y. (1975) Karyotypes of three species of fishes in the order Osteoglossiformes. Chrom. Inform. Serv., 18: 20–22.
U- 52 Urushido, T., Takahashi, E., Taki, Y. and Kondo, N. (1977) A karyotype study of polypterid fishes, with notes on their phyletic relationship. Proc. Japan Acad. Ser. B, 53: 95–98.
U- 53 Uwa, H., Iwamatsu, T. and Ojima, Y. (1981) Karyotype and banding analyses of *Oryzias celebensis* (Oryziatidae, Pisces) in cultured cells. Proc. Japan Acad., Ser. B, 57: 95–99.
U- 54 Uwa, H. and Ojima, Y. (1981) Detailed and banding karyotype analyses of the medaka, *Oryzias latipes* in cultured cells. Proc. Japan Acad., Ser. B, 57: 39–43.
U- 55 Uwa, H. and Iwata, A. (1981) Karyotype and cellular DNA content of *Oryzias javanicus* (Oryziatidae, Pisces). Chrom. Inform. Serv., 31: 24–26.
U- 56 Uwa, H., Iwata, A., Iwamatsu, T. and Ojima, Y. (1982) Chromosomal study of three species of the medaka, genus *Oryzias*. Medaka, 1: 11.
U- 57 Uwa, H., Tanaka, K. and Formacion, M.J. (1982) Karyotype and banding analyses of the Hainan medaka, *Oryzias curvinotus* (Pisces). Chrom. Inform. Serv., 33: 15–17.
U- 58 Uwa, H., Iwamatsu, T. and Saxena, O.P. (1983) Karyotype and cellular DNA content of the Indian ricefish, *Oryzias melastigma*. Proc. Japan Acad., Ser. B, 59: 43–47.
U- 59 Uwa, H. (1986) Karyotype evolution and geographical distribution in the ricefish, genus *Oryzias* (Oryziidae). In: Uyeno, T., Arai, R., Taniuchi, T. and Matsuura, K. eds. Indo-Pcific fish biology. Ichthyological Society of Japan, Tokyo, pp. 867–876.
U- 60 Uwa, H. and Magtoon, W. (1986) Description and karyotype of a new ricefish, *Oryzias mekongensis*, from Thailand. Copeia, 1986: 473–478.
U- 61 Uwa, H. and Jeon, S.R. (1987) Karyotypes in two divergent groups of a ricefish, *Oryzias latipes*, from Korea. Korean J. Limnol., 20: 139–147. (In Korean with English abstract)
U- 62 Uwa, H. and Parenti, L.R. (1988) Morphometric and meristic variation in ricefishes, genus *Oryzias*: a comparison with cytogenetic data. Japan. J. Ichthyol., 35: 159–166.
U- 63 Uwa, H., Wang, R.F. and Chen, Y.R. (1988) Karyotypes and geographical distribution of ricefishes from Yunnan, southwestern China. Japan. J. Ichthyol., 35: 332–340.
U- 64 Uyeno, T. and Miller, R.R. (1971) Multiple sex chromosomes in a Mexican cyprinodontid fish. Nature, 231: 452–453.
U- 65 Uyeno, T. and Miller, R.R. (1972) Second discovery of multiple sex chromosomes among fishes. Experientia, 28: 223–225.
U- 66 Uyeno, T. and Smith, G.R. (1972) Tetraploid origin of the karyotype of catostomid fishes. Science, 175: 644–646.

U- 67 Uyeno, T. (1972) Chromosomes of offspring resulting from crossing coho salmon and brook trout. Japan. J. Ichthyol., 19: 166–171.

U- 68 Uyeno, T. (1973) A comparative study of chromosomes in the teleostean fish order Osteoglossiformes. Japan. J. Ichthyol., 20: 211–217.

U- 69 Uyeno, T. and Miller, R.R. (1973) Chromosomes and the evolution of the plagopterin fishes (Cyprinidae) of the Colorado river system. Copeia, 1973: 776–782.

U- 70 Uyeno, T., Miller, R.R. and Fitzsimons, J.M. (1983) Karyology of the cyprinodontid fishes of the Mexican family Goodeidae. Copeia, 1983: 497–510.

U- 71 Ueda, T., Iijima, K., Naoi, H., Arai, R., Ishinabe, T. and Jeon, S.R. (2006) Karyotypes of three *Tanakia* bitterlings (Pisces, Cyprinidae) from East Asia. Cytologia, 71: 251–255.

U- 72 Ueda, T. (1985) Chromosomes of salmonid fishes. Mar. Sci. Monthly, 17(2): 95–101. (In Japanese)

U- 73 Ueda, T. and Kobayashi, J. (1990) Karyotype differentiation of Atlantic salmon, *Salmo salar*; especially the sequential karyotype change. La Kromosomo, II, 58: 1967–1972.

U- 74 Uwa, H. (1992) Live specimens of ricefishes and relatives suborder Adrianichthyoidei maintained in Shinshu University: a list with karyotypic data. Medaka, 4: 41–44.

U- 75 Uwa, H. (1992) Karyotype evolution of ricefishes in Sulawesi. In: Uwa, H. and Soeroto, B. eds. Phylogeny and species differentiation of Adrianichthyoidei in Indonesia. Interim report. Japan, pp. 145–148. (In Japanese with English abstract)

U- 76 Uwa, H. (1991) Cytosystematic study of the Hainan medaka, *Oryzias curvinotus*, from Hong Kong (Teleostei: Oryziidae). Ichthyol. Explor. Freshwaters, 1: 361–367.

U- 77 Ueda, T. and Ojima, Y. (1984) Karyological characteristics of the brown trout, the Japanese char and their hybrids. Proc. Japan Acad., Ser. B, 60: 249–252.

U- 78 Ueda, T., Sawada, M. and Kobayashi, J. (1987) Cytogenetical characteristics of the embryos between diploid female and triploid male in rainbow trout. Japan. J. Genet., 62: 461–465.

U- 79 Ueno, K. and Takai, A. (2008) Multiple sex chromosome system of $X_1X_1X_2X_2/X_1X_2Y$ type in lutjanid fish, *Lutjanus quinquelineatus* (Perciformes). Genetica, 132: 35–41.

U- 80 Ueda, T. (2007) Chromosomal differentiation in bitterlings (Pisces, Cyprinidae). In: Pisano, E., Ozouf-Costaz, C., Foresti, F. and Kapoor, B.G. eds. Fish cytogenetics. Science Publishers, Enfield, pp. 3–16.

U- 81 Ueda, T. (2010, personal communication)

U- 82 Uribe-Alcocer, M. and Diaz-Jaimes, P. (1996) Chromosome complements of *Gobionellus microdon* and *Eleotris picta* collected in Mexico. J. Fish Biol., 48: 796–798.

U- 83 Ueno, K. and Kang. J.H. (1992) Distribution of heterochromatin and nucleolar organizer regions in the chromosomes of two girellid and three kyphosid fishes. Advance abstracts for the 25th annual meeting, 1992, The Ichthyological Society of Japan, Tokyo, p. 46. (In Japanese)

[V]

V- 1 Van de Peer, Y., Taylor, J.S. and Meyer, A. (2003) Are all fishes ancient polyploids? J. Structural Functional Genomics, 3: 65–73.

V- 2 Vasiliev, V.P. (1975) Karyotypes of different forms of the Kamchatka trout, *Salmo mykiss* and the rainbow trout, *Salmo gairdneri*. J. Ichthyol., 15(6): 889–900.

V- 3 Vasiliev, V.P., Makeeva, A.P. and Ryabov, I.N. (1975) On the triploidy of remote hybrids of carp (*Cyprinus carpio* L.) with other representatives of Cyprinidae. Genetika, 11: 49–56. (In Russian with English summary)

V- 4 Vasiliev, V.P. (1978) Chromosome polymorphism in *Spicara smaris* (Pisces, Centracanthidae). Zool. Zhurnal, 57: 1276–1279. (In Russian with English summary)

V- 5 Vasiliev, V.P. (1978) Karyotypes of 5 species of fishes (Pisces) from the Black Sea. Tsitologiya, 20: 1092–1094. (In Russian with English summary)

V- 6 Vasiliev, V.P., Makeeva, A.P. and Ryabov, I.N. (1978) The study of chromosome complexes in cyprinid fish and their hybrids. Genetika, 14: 1453–1460. (In Russian with English summary)

V- 7 Vasiliev, V.P. (1979) On the role of karyotypes in the whole complex of isolating mechanisms in fishes. J. General Biol., 40: 623–628. (In Russian with English summary)

V- 8 Vasiliev, V.P. and Polykarpova, L.K. (1980) The karyotypes of the Black Sea species of the genera *Crenilabrus* and *Symphodus* (Perciformes, Labridae) and the evidence of natural hybridization between *C. ocellatus and C. quinquemaculatus.* Zool. Zhurnal, 59: 1334–1342. (In Russian with English summary)

V- 9 Vasiliev, V.P., Sokolov, L.I. and Serebryakova, E.V. (1980) Karyotype of the Lena River Siberian sturgeon *Acipenser baeri* Brandt and some aspects of the karyotype evolution in Acipenseriformes. Vopr. Ichthyol., 20: 814–822. (In Russian)

V- 10 Vasiliev, V.P. and Sokolov, L.I. (1980) A method of studying karyotypes in the Chondrostei. Tsitologiya, 22: 1106–1109. (In Russian with English summary)

V- 11 Vasiliev, V.P. (1980) The selective mortality of different chromosome morphs in *Spicara flexuosa* Raf. Genetica, 52/53: 327–331.

V- 12 Vasiliev, V.P., Ivanov, V.N. and Polykarpova, L.K. (1980) Frequencies of chromosomal morphs in various size groups of Black Sea bass (*Spicara flexuosa* Raf., Centracanthidae). In: Kirpichnikov, V.S. ed. Karyological variability, mutagenesis and gynogenesis in fishes. Inst. Cytol. Acad. Sci. USSR, Leningrad, pp. 43–49. (In Russian with English summary)

V- 13 Vasiliev, V.P. (1980) Chromosome numbers in fish-like vertebrates and fish. J. Ichthyol., 1980(3): 1–38.

V- 14 Vasiliev, V.P. and Vorobieva, E.I. (1981) Evolutionary aspect of natural hybrid fishes. God. Biol. Inst. Univ. Saraevu, 34: 171–194. (In Russian)

V- 15 Vasiliev, V.P. and Vasilieva, Y.D. (1982) New diploid-polyploid complex in fish. Dokl. Akad. Nauk SSSR, 266: 250–252. (In Russian)

V- 16 Vasiliev, V.P. (1982) Evolutionary aspects of the karyotypic diversity and the problem of stasipatric speciation in fishes. J. General Biol., 43: 455–469. (In Russian with English summary)

V- 17 Vasiliev, V.P., Vasilieva, Y.D. and Osinov, A.G. (1983) The first evidence favourable for the basic hypothesis of reticulated speciation in vertebrates. Dokl. Akad. Nauk SSSR, 271: 1009–1012. (In Russian)

V- 18 Vasiliev, V.P. and Grigoryan, K.A. (1990) Broad chromosome polymorphism in the Caucasus river goby. Dokl. Akad. Nauk SSSR, 311: 1509–1511. (In Russian)

V- 19 Vasiliev, V.P. (1990) Diploid-triploid complex in the genus *Cobitis* (Pisces, Cobitidae). Dokl. Akad. Nauk SSSR, 312: 249–252. (In Russian)

V- 20 Vasiliev, V.P., Vasilieva, Y.D. and Osinov, A.G. (1991) The problem of reticulate species formation in vertebrates of the diploid-triploid-tetraploid complex in the genus *Cobitis* (Cobitidae). IV. Tetraploid form. J. Ichthyol., 31(2): 21–35.

V- 21 Vasiliev, V.P. and Vasilieva, Y.D. (1992) Karyological evidence of the separate species status of *Neogobius kessleri* (Günther) and *Neogobius gorlap* Iljin (Pisces, Gobiidae). Dokl. Akad. Nauk SSSR, 324: 898–900. (In Russian)

V- 22 Vasiliev, V.P. and Grigoryan, K.A. (1993) Karyology of the Gobiidae. J. Ichthyol., 33(2): 1–16.

V- 23 Vasiliev, V.P. and Grigoryan, K.A. (1994) Chromosome polymorphism and karyological relations of gobies of the group *Neogobius cephalarges* Pallas—*Neogobius platyrostris* Pallas (Gobiidae). Russian J. Genet., 30: 1087–1094.

V- 24 Vasilieva, Y.D., Osinov, A.G. and Vasiliev, V.P. (1989) On the problem of reticular speciation in vertebrates: diploid-triploid-tetraploid complex in the genus *Cobitis* (Cobitidae). I. Diploid species. Vopr. Ichthyol., 29: 705–717. (In Russian)

V- 25 Vasilieva, Y.D., Collares-Pereira, M.J. and Madeira, J. (1992) Variability, divergence, and taxonomy of loach on the Iberian Peninsula. J. Ichthyol., 32(4): 69–84.

V- 26 Vasilieva, Y.D. and Ráb, P. (1992) Golden spiny loach, *Sabanejewia aurata* (Cobitidae), from the Laborec River. Vopr. Ichthyol., 32(2): 176–181. (In Russian)

V- 27 Vasilieva, Y.D. and Vasiliev, V.P. (1994) On the systematics of Caucasian gobies (Gobiidae): craniological and karyological analyses and distribution by biotope of some populations from the Black Sea and Caspian basins. Vopr. Ichthyol., 34(2): 187–194. (In Russian)

V- 28 Vasilieva, Y.D., Vasiliev, V.P. and Skomorokhov, M.O. (2003) Loaches (*Misgurnus*, Cobitidae) of Asiatic Russia: II. Morphology, synonymy, diagnoses, karyology, biology, and distribution. J. Ichthyol., 43: 501–510.

V- 29 Vega, L., Diaz, E., Cross, I. and Rebordinos, L. (2002) Cytogenetic and isozymatic characterization of the sole *Solea senegalensis* Kaup, 1858. Bol. Inst. Espanol Oceanogr., 18: 245–250. (In Spanish with English abstract)

V- 30 Venere, P.C. and Galetti Jr., P.M. (1985) Natural triploidy and chromosome B in the fish *Curimata modesta* (Curimatidae, Characiformes). Brazil. J. Genet., 8: 681–687.

V- 31 Venere, P.C. and Galetti Jr., P.M. (1995) Multiple longitudinal bands in fish chromosomes: comparison of structural G-banding and replication R bands among curimatids. Cytobios, 84: 71–78.

V- 32 Venere, P.C., Pfister, S.C., Moreira-Filho, O. and Galetti Jr., P.M. (1997) Chromosomal characterization in Characinae and Cynopotaminae (Pisces, Characidae). Cytobios, 92: 123–131.

V- 33 Verma, G.K. (1968) Studies on the structure and behaviour of chromosomes of certain fresh water and marine gobiid fishes. Ann. Nat. Acad. Sci. India, 1968: 178.

V- 34 Vervoort, A. (1979) Karyotype and DNA content of *Phractolaemus ansorgei* Blgr. (Teleostei: Gonorynchiformes). Experientia, 35: 479–480.

V- 35 Vervoort, A. (1980) The karyotypes of seven species of *Tilapia* (Teleostei: Cichlidae). Cytologia, 45: 651–656.

V- 36 Vervoort, A. (1980) Karyotype of *Caecobarbus geertsi* Boulenger, (Teleostei: Cyprinidae). Nucleus, 23: 76–77.

V- 37 Vervoort, A. (1980) Karyotypes and nuclear DNA contents of Polypteridae (Osteichthyes). Experientia, 36: 646–647.

V- 38 Vervoort, A. (1980) Tetraploidy in *Protopterus* (Dipnoi). Experientia, 36: 294–295.

V- 39 Vicente, V.E., Moreira-Filho, O. and Camacho, J.P.M. (1996) Sex-ratio distortion associated with the presence of a B chromosome in *Astyanax scabripinnis* (Teleostei, Characidae). Cytogenet. Cell Genet., 74: 70–75.

V- 40 Viktorovsky, R.M. (1978) The evolution of karyotypes in chars of the genus *Salvelinus*. Tsitologiya, 20: 833–838. (In Russian with English summary)

V- 41 Viktorovsky, R.M. and Maximova, T.G. (1978) Chromosomal complex of *Coregonus ussuriensis* and some aspects of the evolution of karyotypes of Coregonidae. Tsitologiya, 20: 967–970. (In Russian with English summary)

V- 42 Viktorovsky, R.M. and Ermolenko, L.N. (1982) The chromosomal complexes of *Coregonus nasus* and *C. lavaretus* and the problem of the *Coregonus* karyotype divergence. Tsitologiya, 24: 797–801. (In Russian with English summary)

V- 43 Viktorovsky, R.M., Ermolenko, L.N., Makoedov, A.N., Frolov, S.V. and Shevchishin, A.A. (1983) Karyotype divergence in the genus *Coregonus*. Tsitologiya, 25: 1309–1315. (In Russian with English summary)

V- 44 Viktorovsky, R.M., Makoedov, A.N. and Shevchishin, A.A. (1985) The chromosomal sets of *Brachymystax lenok* and *Hucho taimen* and the divergency of the salmonid genera. Tsitologiya, 27: 703–709. (In Russian with English summary)

V- 45 Vissotto, P.C., Foresti, F. and Oliveira, C. (1997) A ZZ/ZW sex chromosome system in *Imparfinis mirini* (Pisces, Siluriformes). Cytologia, 62: 61–66.

V- 46 Vissotto, P.C., Foresti, F. and Oliveira, C. (1999) Karyotype description of five species of Pimelodidae (Teleostei, Siluriformes). Chrom. Sci., 3: 1–7.

V- 47 Vissotto, P.C., Foresti, F. and Oliveira, C. (1999) Supernumerary chromosomes in two species of the family Pimelodidae (Teleostei, Siluriformes). Chrom. Sci., 3: 9–13.

V- 48 Vissotto, P.C., Foresti, F. and Oliveira, C. (2001) Karyotypic characterization of two species of the genus *Imparfinis* (Teleostei, Siluriformes, Heptapteridae). Chrom. Sci., 5: 97–103.

V- 49 Vitturi, R., Carbone, P., Catalano, E. and Macaluso, M. (1984) Chromosome polymorphism in *Gobius paganellus* Linneo 1758 (Pisces, Gobiidae). Biol. Bull., 167: 658–668.

V- 50 Vitturi, R., Carbone, P., Macaluso, M. and Catalano, E. (1986) Karyotypic variability in the gadid fish, *Gaidropsarus mediterraneus* (Linnaeus, 1758). Copeia, 1986: 826–829.

V- 51 Vitturi, R., Carbone, P., Catalano, E. and Macaluso, M. (1986) Karyotypes of five species of Blennioidea (Pisces, Perciformes). Caryologia, 39: 273–279.

V- 52 Vitturi, R., Mazzola, A., Macaluso, M. and Catalano, E. (1986) Chromosomal polymorphism associated with Robertsonian fusion in *Seriola dumerili* (Risso, 1810) (Pisces: Carangidae). J. Fish Biol., 26: 529–534.

V- 53 Vitturi, R., Catalano, E., Macaluso, M. and Zava, B. (1986) Karyotypes of nine species of the family Labridae (Pisces, Perciformes). Biol. Zentralbl., 105: 519–530.

V- 54 Vitturi, R. (1988) Karyotype analysis and evidence for Robertsonian fusion in *Trigla lucerna* L. and *Helicolenus dactylopterus* (Delar.) (Pisces, Scorpaeniformes). Biol. Zentralbl., 107: 553–558.

V- 55 Vitturi, R. and Catalano, E. (1988) Karyotype of *Carapus acus* (Brünnich, 1768) (Pisces, Carapidae). Caryologia, 41: 131–135.

V- 56 Vitturi, R. and Catalano, E. (1988) Karyotypes in two species of the genus *Hippocampus* (Pisces: Syngnathiformes). Mar. Biol., 99: 119–121.

V- 57 Vitturi, R., Zava, B., Catalano, E., Maiorca, A. and Macaluso, M. (1988) Karyotypes in five Mediterranean teleostean fishes. Biol. Zentralbl., 107: 339–344.

V- 58 Vitturi, R., Catalano, E., Maiorca, A. and Carollo, T. (1988) Karyological studies in *Coris julis* (Pisces, Labridae). Genetica, 76: 219–223.

V- 59 Vitturi, R. and Catalano, E. (1989) Multiple chromosome polymorphism in the gobiid fish *Gobius niger jozo* L. 1758 (Pisces, Gobiidae). Cytologia, 54: 231–235.

V- 60 Vitturi, R., Catalano, E., Lo Conte, M.R. and Spampinato, P. (1989) Ag-NORs and C-banding pattern of the labrid species *Xyrichthys novacula* (L.) (Pisces, Perciformes). Biol. Zentralbl., 108: 263–266.

V- 61 Vitturi, R., Mazzola, A. and Catalano, E. (1990) Karyotype analysis, Ag-NORs and C-bands location in *Pagellus bogaraveo* (Brünnich 1768) (Pisces, Sparidae). Biol. Zentralbl., 109: 223–226.

V- 62 Vitturi, R. and Rasotto, M.B. (1990) Karyotype analysis of *Cottus gobio* L. (Pisces, Cottidae). Cytobios, 62: 81–86.

V- 63 Vitturi, R., Mazzola, A., Catalano, E. and Lo Conte, M.R. (1990) Karyotype analysis, nucleolar organizer regions (NORs), and C-banding pattern of *Dicentrarchus labrax* (L.) and *Dicentrarchus punctatus* (Bloch, 1792) (Pisces, Perciformes) with evidence of chromosomal structural polymorphism. Cytologia, 55: 425–430.

V- 64 Vitturi, R., Catalano, E. and Lafargue, F. (1991) Evidence for heteromorphic sex chromosomes in *Zeus faber* (Pisces, Zeiformes): nucleolus organizer regions and C-banding pattern. Cytobios, 68: 37–43.

V- 65 Vitturi, R., Catalano, E. and Colombera, D. (1992) Karyotype analyses of the teleostean fish *Balistes carolinensis* and *Lophius piscatorius* suggest the occurrence of two different karyoevolutional pathways. Biol. Zentralbl., 111: 215–222.

V- 66 Vitturi, R., Catalano, E. and Barbieri, R. (1992) Karyological and molecular characterization of *Mullus surmuletus* and *Mullus barbatus* (Pisces, Mullidae). Cytologia, 57: 65–74.

V- 67 Vitturi, R. and Lafargue, F. (1992) Karyotype analyses reveal inter-individual polymorphism and association of nucleolus-organizer-carrying chromosomes in *Capros aper* (Pisces: Zeiformes). Mar. Biol., 112: 37–41.

V- 68 Vitturi, R., Colombera, D. and Catalano, E. (1993) Intra-populational chromosome polymorphisms in four teleost species. Cytobios, 75: 171–182.

V- 69 Vitturi, R., Catalano, E., Colomba, M.S., Montagnino, L. and Pellerito, L. (1995) Karyotype analysis of *Aphanius fasciatus* (Pisces, Cyprinodontiformes): Ag-NORs and C-band polymorphisms in four populations from Sicily. Biol. Zentralbl., 114: 392–402.

V- 70 Vitturi, R., Libertini, A., Campolmi, M., Calderazzo, F. and Mazzola, A. (1998) Conventional karyotype, nucleolar organizer regions and genome size in five Mediterranean species of Syngnathidae (Pisces, Syngnathiformes). J. Fish Biol., 52: 677–687.

V- 71 Völker, M., Ráb, P. and Kullmann, H. (2005) Karyotype differentiation in *Chromaphyosemion* killifishes (Cyprinodontiformes, Nothobranchiidae). I: Chromosome banding patterns of *C. alpha*, *C. kouamense* and *C. lugens*. Genetica, 125: 33–41.

V- 72 Vasiliev, V.P. (1985) Evolutionary karyology of fishes. Publishing House Nauka, Moscow. (In Russian)

V- 73 Vinogradov, A.E. (2004) Evolution of genome size: multilevel selection, mutation bias or dynamical chaos? Curr. Opinion Genet. Develop., 14: 620–626.

V- 74 Vasiliev, V.P. (1975) Karyotypes of some forms of arctic char *Salvelinus alpinus* (L.) from the Kamchatka water bodies. Vopr. Ichthyol., 15: 417–430. (In Russian)

V- 75 Vitturi, R. and Catalano, E. (1988) Karyotypic variability associated with Robertsonian fusion in two species of the family Ophidiidae (Pisces, Perciformes). Boll. Zool., 55: 55–58.

V- 76 Vila, I. (2006) A new species of killifish in the genus *Orestias* (Teleostei: Cyprinodontidae) from the southern high Andes, Chile. Copeia, 2006: 472–477.

V- 77 Valentim, F.C.S., Falcão, J.N., Porto, J.I.R. and Feldberg, E. (2006) Chromosomes of three freshwater stingrays (Rajiformes Potamotrygonidae) from the Rio Negro basin, Amazon, Brazil. Genetica, 128: 33–39.

V- 78 Valcarcel, A., Brunner, P. and Maggese, M.C. (1993) B-chromosome polymorphism in the South American catfish, *Rhamdia sapo*. Aquaculture, 110: 111–118.

V- 79 Völker, M., Ráb, P. and Kullmann, H. (2007) Banded karyotypes of *Chromaphyosemion poliaki* and *C. volcanum* (Cyprinodontiformes, Nothobranchiidae), with a discussion of the validity of *C. poliaki*. Ichthyol. Explor. Freshwaters, 18: 1–8.

V- 80 Völker, M., Sonnenberg, R., Ráb, P. and Kullmann, H. (2006) Karyotype differentiation in *Chromaphyosemion* killifishes (Cyprinodontiformes, Nothobranchiidae). II: Cytogenetic and mitochondrial DNA analyses demonstrate karyotype differentiation and its evolutionary direction in *C. riggenbachi*. Cytogenet. Genome Res., 115: 70–83.

V- 81 Völker, M., Sonnenberg, R., Ráb, P. and Kullmann, H. (2007) Karyotype differentiation in *Chromaphyosemion* killifishes (Cyprinodontiformes, Nothobranchiidae). III: Extensive karyotypic variability associated with low mitochondrial haplotype differentiation in *C. bivittatum*. Cytogenet. Genome Res., 116: 116–126.

V- 82 Vasilieva, Y.D., Vasiliev, V.P. and Boltachev, A.R. (2005) Taxonomic relationships of gudgeons (*Gobio*, Gobioninae, Cyprinidae) of Crimea. J. Ichthyol., 45: 730–743.

V- 83 Vasilieva, Y.D., Vasiliev, V.P. and Kuga, T.I. (2004) On taxonomy of gudgeons of the genus *Gobio* (Gobioninae, Cyprinidae) of Europe: a new gudgeon species *Gobio kubanicus* from the basin of the Kuban river. J. Ichthyol., 44: 716–731.

V- 84 Vasconcelos, C. and Martins-Santos, I.C. (2000) Chromosome polymorphism in species of the Pimelodidae family (Pisces, Siluriformes). Hereditas, 132: 103–109.

V- 85 Vieira Lacerda, M.C. and Maistro, E.L. (2007) Cytogenetic analysis of three sympatric *Gymnotus* species (Teleostei: Gymnotidae) from the Fundo stream, MG, Brazil. Cytologia, 72: 89–93.

V- 86 Vinogradov, A.E. (1998) Genome size and GC-percent in vertebrates as determined by flow cytometry: the triangular relationship. Cytometry, 31: 100–109.

V- 87 Vicari, M.R., Artoni, R.F., Moreira-Filho, O. and Bertollo, L.A.C. (2006) Basic and molecular cytogenetics in freshwater Cichlidae (Osteichthyes, Perciformes). Karyotypic conservationism and divergence. Caryologia, 59: 260–266.

V- 88 Vasiliev, V.P., Vasilieva, Y.D. and Osinov, A.G. (1990) Reticulate evolution in vertebrates: the diploid-triploid-tetraplod complex in the genus *Cobitis* (Cobitidae). III. Origin of the triploid form. J. Ichthyol., 30(6): 22–31.

V- 89 Vasiliev, V.P., Osinov, A.G. and Vasilieva, Y.D. (1991) Reticulate species formation in vertebrates: the diploid-triploid-tetraploid complex in the genus *Cobitis* (Cobitidae). V. Origin of even-polyploid species. J. Ichthyol., 31(4): 22–36.

V- 90 Venere, P.C., Souza, I.L., Martins, C. and Oliveira, C. (2008) Occurrence of ZZ/ZW sex chromosomes in *Thoracocharax stellatus* fish (Characiformes, Gasteropelecidae) from the Araguaia River, South America. Genetica, 133: 109–112.

V- 91 Vicari, M.R., Artoni, R.F., Moreira-Filho, O. and Bertollo, L.A.C. (2008) Diversification of a ZZ/ZW sex chromosome system in *Characidium* fish (Crenuchidae, Characiformes). Genetica, 134: 311–317.

V- 92 Venere, P.C., Souza, I.L., Silva, L.K.S., Dos Anjos, M.B., De Oliveira, R.R. and Galetti Jr., P.M. (2008) Recent chromosome diversification in the evolutionary radiation of the freshwater fish family Curimatidae (Characiformes). J. Fish Biol., 72: 1976–1989.

V- 93 Venere, P.C. and Galetti Jr., P.M. (1989) Chromosome evolution and phylogenetic relationships of some neotropical Characiformes of the family Curimatidae. Braz. J. Genet., 12: 17–25.

V- 94 Venere, P.C., Miyazawa, C.S. and Galetti Jr., P.M. (1999) New cases of supernumerary chromosomes in characiform fishes. Genet. Mol. Biol., 22: 345–349.

V- 95 Vidotto, A.P., Swarça, A.C., Fenocchio, A.S. and Dias, A.L. (2004) Cytogenetic studies in three *Pimelodella meeki* polulations (Pisces, Pimelodidae) from Tibagi River basin (Brazil). J. Heredity, 95: 517–520.

V- 96 Vitturi, R., Libertini, A., Mazzola, A., Colomba, M.S. and Sara, G. (1996) Characterization of mitotic chromosomes of four species of the genus *Diplodus*: karyotypes and chromosomal nucleolar organizer region phenotypes. J. Fish Biol., 49: 1128–1137.

V- 97 Völker, M., Ráb, P. and Kullman, H. (2008) Karyotype differentiation in *Chromaphyosemion* killifishes (Cyprinodontiformes, Nothobranchiidae): patterns, mechanisms, and evolutionary implications. Biol. J. Linn. Soc., 94: 143–153.

V- 98 Vasiliev, V.P. and Vasilieva, Y.D. (2008) Comparative karyology of species of the genera *Misgurnus* and *Cobitis* (Cobitidae) from the Amur river basin in connection with their taxonomic relations and the evolution of karyotypes. J. Ichthyol., 48: 1–13.

V- 99 Vasiliev, V.P. and Vasilieva, Y.D. (2008) Comparative karyological analysis of mud loach and spined loach species (genera *Misgurnus* and *Cobitis*) from the Far East region of Russia. Folia Zool., 57: 51–59.

V-100 Vitturi, R., Catalano, E. and Colombera, D. (1993) Chromosome analysis of *Bothus podas* (Pisces, Pleuronectiformes) from the Mediterranean Sea. J. Fish Biol., 43: 221–227.

V-101 Vinogradov, A.E. (2005) Genome size and chromatin condensation in vertebrates. Chromosoma, 113: 362–369.

V-102 Vasiliev, V.P., Vinogradov, A.E., Rozanov, Y.M. and Vasilieva, Y.D. (1999) Cellular DNA content in different forms of the bisexual-unisexual complex of spined loaches of the genus *Cobitis* and in Luther's spined loach *C. lutheri* (Cobitidae). J. Ichthyol., 39: 377–383.

V-103 Vasilieva, Y.D. and Vasiliev, V.P. (1995) Systematics of Caucasian freshwater gobies (Gobiidae) in the light of contemporary data, with a description of a new species, *Neogobius rhodioni*, sp. nov. J. Ichthyol., 35(2): 139–157.

V-104 Vitturi, R., Catalano, E., Lo Conte, M.R., Alessi, A.M., Amico, F.P. and Colombera, D. (1991) Intra-populational and intra-individual mosaicisms of *Uranoscopus scaber* L. (Perciformes, Uranoscopidae). Heredity, 67: 325–330.

V-105 Vitturi, R., Mazzola, A., Catalano, E. and Lo Conte, M.R. (1992) Karyotype characterization of four Mediterranean sparid fish (Pisces, Perciformes) using conventional and banding techniques. Cytobios, 72: 107–115.

V-106 Vasconcelos, A.J.M. and Molina, W.F. (2009) Cytogenetical studies in five Atlantic Anguilliformes fishes. Genet. Mol. Biol., 32: 83–90.

V-107 Vicari, M.R., Almeida, M.C., Bertollo, L.A.C., Moreira-Filho, O. and Artoni, R.F. (2006) Cytogenetic analysis and chromosomal characteristics of the polymorphic 18S rDNA in the fish *Prochilodus lineatus* (Characiformes, Prochilodontidae). Genet. Mol. Biol., 29: 621–625.

*V-108 Vasiliev, V.P. and Vasilieva, Y.D. (1994) The karyological diversity in spined loaches from genera *Cobitis* and *Sabanjewia*. Abstr. VIII Congr. Soc. Europ. Ichthyol., Oviedo, p. 67. (After K-114)

V-109 Venere, P.C., Ferreira, I.A., Martins, C. and Galetti Jr., P.M. (2004) A novel ZZ/ZW sex chromosome system for the genus *Leporinus* (Pisces, Anostomidae, Characiformes). Genetica, 121: 75–80.

V-110 Van Eenennaam, A.L., Murray, J.D. and Medrano, J.F. (1998) Synaptonemal complex analysis in spermatocytes of white sturgeon, *Acipenser transmontanus* Richardson (Pisces, Acipenseridae), a fish with a very high chromosome number. Genome, 41: 51–61.

V-111 Vila, I., Scott, S., Lam, N., Iturra, P. and Méndez, M.A. (2010) Karyological and morphological analysis of divergence among species of the killifish genus *Orestias* (Teleostei: Cyprinodontidae) from the southern Altiplano. In: Nelson, J.S., Schultze, H.P. and Wilson, M.V.H. eds. Origin and phylogenetic interrelationships of teleosts. Verlag Dr. Friedrich Pfeil, München, pp. 471–480.

[W]

W- 1 Wang, C.Y. and Li, Y.L. (1982) Studies on the karyotype of goldfish (*Carassius auratus*). I. A comparative study of the chromosomes in crucian and red dragon-eye goldfish. Acta Genet. Sinica, 9(3): 238–242. (In Chinese with English abstract)

W- 2 Wang, C.Y. and Li, Y.L. (1987) Studies on the karyotype of Xiji colour-crucian carp (*Carassius auratus*). Hereditas (Beijing), 9(2): 25–26. (In Chinese)

W- 3 Wang, D.X., Wang, J., Guo, F., Liang, J.R. and Qin, Y.X. (2002) Study on the karyotype in *Nibea miichtheoides*. Marine Sciences (Beijing), 26(11): 68–70. (In Chinese with English abstract)

W- 4 Wang, J. and Ojima, Y. (1988) Fish gene mapping using mouse-fish cell hybrids. Proc. Japan Acad., Ser. B, 64: 53–56.

W- 5 Wang, J. and Ojima, Y. (1989) Isolation and characterization of a BUdR-resistant fish cell line. Proc. Japan Acad., Ser. B, 65: 112–115.

W- 6 Wang, J.X. and Zhao, X.F. (1993) Chromosome study of three species of tetraodontiform fishes. Zool. Res., 14: 345–346. (In Chinese)

W- 7 Wang, J.X., Zhao, X.F., Wang, X.M. and Tian, M.C. (1994) Karyotype analysis for seven species of clupeiform and perciform fishes. Zool. Res., 15: 76–79. (In Chinese with English abstract)

W- 8 Wang, J.X., Zhao, X.F., Fan, C.L. and Sun, X.Y. (1994) Studies on karyotypes of two species of scorpaeniform fishes. Chinese J. Zool., 29(4): 14–17. (In Chinese)

W- 9 Wang, R.F., Shi, L.M. and He, W.S. (1985) Studies on nucleolus organizer regions in several species of carp (*Cyprinus*) by silver-staining. Zool. Res., 6: 391–398. (In Chinese with English abstract)

W- 10 Wang, R.F., Shi, L.M. and He, W.S. (1988) A comparative study of the Ag-NORs of *Carassius auratus* from different geographic districts. Zool. Res., 9: 165–169. (In Chinese with English abstract)

W- 11 Wang, R.F., Shi, L.M. and He, W.S. (1989) The karyotype of *Carassius auratus* from Er Hai Lake. Zool. Res., 10: 169–170. (In Chinese)

W- 12 Wang, R.F., He, W.S. and Wu, S.F. (1997) Study on the karyotype and banding of *Semilabeo prochilus* (Pisces). Zool. Res., 18: 411–414. (In Chinese with English abstract)

W- 13 Wang, S.F., Wang D.X., Su Y.Q. and Wang J. (2003) The karyotype of *Nibea diacanthus.* J. Xiamen Univ. (Nat. Sci.), 42: 682–684. (In Chinese with English abstract)

W- 14 Wang, X. and Shen, J.B. (1989) The study on the karyotypes of the fishes in the Heilongjiang River system I. The chromosome numbers and karyotypes of 11 species of Gobioninae. Hereditas (Beijing), 11(3): 23–25. (In Chinese)

W- 15 Wang, Z.X., Zhang, J.X. and Jin, G.Q. (1986) Study on the incompatibility of fish hybridization. Acta Hydrobiol. Sinica, 10: 171–179. (In Chinese with English abstract)

W- 16 Warkentine, B.E., Smith, C.L. and Rachlin, J.W. (1987) A reevaluation of the karyotype of the Atlantic silverside, *Menidia menidia.* Copeia, 1987: 222–224.

W- 17 Wasko, A.P. and Galetti Jr., P.M. (1998) Karyotype diversity in the neotropical fish *Bryconamericus* (Characidae, Tetragonopterinae). Cytobios, 94: 185–193.

W- 18 Wasko, A.P. and Galetti Jr., P.M. (1999) Extensive NOR variability in fishes of the genus *Bryconamericus* (Characidae). Cytologia, 64: 63–67.

W- 19 Wasko, A.P., Cesar, A.C.G., Martins, C. and Galetti Jr., P.M. (2001) A ZZ/ZW sex chromosome system in Cheirodontinae fish. Chrom. Sci., 5: 145–148.

W- 20 Webb, C.J. (1974) Fish chromosomes: a display by scanning electron microscopy. J. Fish Biol., 6: 99–100.

W- 21 Webb, C.J. (1980) Systematics of the *Pomatoschistus minutus* complex (Teleostei: Gobioidei). Phil. Trans. R. Soc., (B), 291: 201–241.

W- 22 Webb, C.J. (1986) Karyology of the Indo-Pacific *Parioglossus raoi* (Herre) (Teleostei: Gobioidei) from Fiji. Aust. J. Mar. Freshw. Res., 37: 347–351.

W- 23 Webb, C.J. (1986) Karyology of *Pomatoschistus microps* (Teleostei: Gobioidei). J. Mar. Biol. Ass. U.K., 66: 259–266.

W- 24 Wharton, J.H., Ellender, R.D., Middlebrooks, B.L., Stocks, P.K., Lawler, A.R. and Howse, H.D. (1977) Fish culture: characteristics of a cell line from the silver perch, *Bairdiella chrysura.* In Vitro, 13: 389–397.

W- 25 Wickbom, T. (1945) Cytological studies on Dipnoi, Urodela, Anura, and Emys. Hereditas, 31: 241–346.

W- 26 Wildrick, D.M., Kim, I.S. and Greenfield, D.W. (1985) A unique gambusiine karyotype and its relevance to the systematics of the *Gambusia punctata* species group. Copeia, 1985: 1053–1056.

W- 27 Wolters, W.R., Chrisman, C.L. and Libey, G.S. (1981) Lymphocyte culture for chromosomal analyses of channel catfish, *Ictalurus punctatus.* Copeia, 1981: 503–504.

W- 28 Woznicki, P., Jankun, M. and Luczynski, M. (1997) Chromosome studies in brown trout (*Salmo trutta* m. *fario*) from Poland: hypothetical evolution of the 11th, 12th and 14th chromosome pairs in the *Salmo* karyotype. Cytobios, 91: 207–214.

W- 29 Woznicki, P., Jankun, M., Kucharczyk, D., Boron, A. and Luczynski, M. (1999) Cytogenetic characterization of sea trout (*Salmo trutta*) from Poland. Copeia, 1999: 501–505.

W- 30 Wu, G.M., Xin, C.B. and Chen, Y.L. (1986) A comparative study on the karyotypes of four species of catfish. Acta Genet. Sinica, 13(3): 213–220. (In Chinese with English abstract)

W- 31 Wu, G.M., Zhu, X.P., Hu, G. and Luo, J.R. (1991) The karyotype of *Puntius gonionotus* (Cyprinidae). Chinese J. Zool., 26(1): 20–21. (In Chinese)

W- 32 Wu, Y.P., Ye, Y.Z. and Wu, Q.J. (2000) Studies on the induction of allotriploid using heat shock in zebrafish, *Brachydanio frankei* and *Brachydanio rerio*. Oceanol. Limnol. Sinica, 31: 465–470. (In Chinese with English abstract)

W- 33 Wu, Z.A. and Yan, H.Y. (1980) Cytogenetic studies on fishes. II. Karyotype analysis of *Cyprinus carpio* and *Carassius auratus* by peripheral blood lymphocyte culture. Acta Genet. Sinica, 7(4): 370–375. (In Chinese with English abstract)

W- 34 Wu, Z.A. and Yan, H.Y. (1981) Cytogenetical studies on fishes. I. The karyotype of *Rhodeus sinensis*. Sinozool., 1981(1): 167–172. (In Chinese with English abstract)

W- 35 Wildekamp, R.H., Shidlovskiy, K.M. and Watters, B.R. (2009) Systematics of the *Nothobranchius melanospilus* species group (Cyprinodontiformes: Nothobranchiidae) with description of two new species from Tanzania and Mozambique. Ichthyol. Explor. Freshwaters, 20: 237–254.

W- 36 Winkler, F.M., Garcia-Melys, D. and Palma-Rojas, C. (2004) Karyotypes of three South East Pacific flounder species of the family Paralichthyidae. Aquacult. Res., 35: 1295–1298.

W- 37 Woods, I.G., Wilson, C., Friedlander, B., Chang, P., Reyes, D.K., Nix, R., Kelly, P.D., Chu, F., Postlethwait, J.H. and Talbot, W.S. (2005) The zebrafish gene map defines ancestral vertebrate chromosomes. Genome Res., 15: 1307–1314.

[X]

X- 1 Xiao, H., Zhang, R.D., Feng, J.G., Ou, Y.M., Li, W.X., Chen, S.Y. and Zan, R.G. (2002) Nuclear DNA content and ploidy of seventeen species of fishes in *Sinocyclocheilus*. Zool. Res., 23: 195–199. (In Chinese with English abstract)

X- 2 Xu, K.S., Su, Z.J. and Bai, G.S. (1986) A karyotype study of *Acipenser sinensis* Gray. Zool. Res., 7: 262. (In Chinese)

X- 3 Xu, D.D., You, F., Wu, Z.H., Li, J., Ni, J., Xiao, Z.Z., Zhang, P.J. and Xu, Y.L. (2009) Genetic characterization of asymmetric reciprocal hybridization between the flatfishes *Paralichthys olivaceus* and *Paralichthys dentatus*. Genetica, 137: 151–158.

[Y]

Y- 1 Yabu, H. and Ishii, K. (1980) Chromosomes of flatfish *Glyptocephalus stelleri* (Schmidt). Chrom. Inform. Serv., 29: 16.

Y- 2 Yabu, H. and Ishii, K. (1981) Chromosomes in Pacific saury, *Cololabis saira*. Bull. Japan. Soc. Sci. Fish., 47: 559.

Y- 3 Yabu, H. and Ishii, K. (1982) Nuclear divisions in Pacific pomfret *Brama japonica* Hilgendorf. Bull. Japan. Soc. Sci. Fish., 48: 505–507.

Y- 4 Yabu, H. and Ishii, K. (1984) Chromosomes of the great blue shark *Prionace glauca* (Linnaeus). Bull. Japan. Soc. Sci. Fish., 50: 7–10.

Y- 5 Yamada, J. (1967) An observation of the chromosomes in the embryonic cells of a goby, *Chaenogobius urotaenia* (Hilgendorf). Bull. Fac. Fish. Hokkaido Univ., 18: 183–187.

Y- 6 Yamaha, E., Kimura, S., Tanaka, M., Sakao, S., Fujimoto, T. and Arai, K. (2002) Haploid-diploid mosaic embryos in cuttthroat trout, *Oncorhynchus clarki*. Fish Genet. Breed. Sci., 32: 121–126. (In Japanese with English abstract)

Y- 7 Yamamoto, K. and Ojima, Y. (1973) A PHA-culture method for cells from the renal tissue of teleosts. Japan. J. Genet., 48: 235–238.

Y- 8 Yamazaki, F. (1971) A chromosome study of the ayu, a salmonoid fish. Bull. Japan. Soc. Sci. Fish., 37: 707–710.

Y- 9 Yamazaki, F., Onozato, H. and Arai, K. (1981) The chopping method for obtaining permanent chromosome preparation from embryos of teleost fishes. Bull. Japan. Soc. Sci. Fish., 47: 963.

Y- 10 Yang, H.Y. (1982) Studies of the karyotype of *Siniperca chuatsi*. Acta Genet. Sinica, 9(2): 143–146. (In Chinese with English abstract)

Y- 11 Yang, H.Y. (1991) Chromosome G-band patterns of grass carp (*Ctenopharyngodon idellus*) and blunt-snout bream (*Megalobrama amblycephala*). Acta Zool. Sinica, 37: 431–437. (In Chinese with English abstract)

Y- 12 Yi, Y.L., Chen, H.X. and Chen, M.R. (1984) Distribution of constitutive heterochromatin in the genome of the symbranchoid eel, *Monopterus albus*. Acta Hydrobiol. Sinica, 8: 245–247. (In Chinese)

Y- 13 Yin, H.B., Sun, Z.W. and Pan, W.Z. (1996) The karyotype study on triploid catfish (*Silurus asotus*). J. Fish. China, 20: 178–182. (In Chinese)

Y- 14 Yu, X.J., Zhou, T., Li, K., Li, Y.C. and Zhou, M. (1987) On the karyosystematics of cyprinid fishes and a summary of fish chromosome studies in China. Genetica, 72: 225–235.

Y- 15 Yu, X.J., Zhou, T, Li, Y.C., Li, K. and Zhou, M. (1989) Chromosomes of Chinese freshwater fishes. Science Press, Beijing. (In Chinese)

Y- 16 Yu, X.Y., Li, Y.C. and Zhou, T. (1990) Karyotype studies of cyprinid fishes in China. Comparative study of the karyotypes of 8 species of schizothoracine fishes. J. Wuhan Univ. (Nat. Sci.), 1990(2): 97–104. (In Chinese with English abstract)

Y- 17 Yu, X.Y. and Yu, X.J. (1990) A schizothoracine fish species, *Diptychus dipogon*, with a very high number of chromosomes. Chrom. Inform. Serv., 48: 17–18.

Y- 18 Yu, Z.I., Kong, X.Y., Feng, D.Y. and Xie, Z.Y. (1992) Studies on the karyotypes of *Sebastes schlegeli* (Hilgendorf) and *Hexagrammos otakii* Jordan et Starks. J. Ocean Univ. Qingdao, 22(2): 118–124. (In Chinese with English abstract)

Y- 19 Yu, Z.I., Kong, X.Y., Xie, Z.Y. and Wang, Y.Y. (1993) Studies on the karyotypes and Ag-NOR banding of *Pagrosomus major* and *Sparus macrocephalus*. J. Ocean Univ. Qingdao, 23(3): 107–115. (In Chinese with English abstract)

Y- 20 Yu, Z.I., Kong, X.Y. and Xie, Z.Y. (1995) Studies on karyotypes of fishes of economic importance in coastal waters of Shandong Peninsula. J. Fish. Sci. China, 2(2): 1–6. (In Chinese with English abstract)

Y- 21 Yu, Z.I., Kong, X.Y., Xie, Z.Y. and Dai, J.X. (1996) Studies on the chromosomes of five species of marine fish. J. Ocean Univ. Qingdao, 26(1): 44–48. (In Chinese with English abstract)

Y- 22 Yokoyama, T., Ebitani, N. and Kubo, T. (1992) Karyotypes and banding patterns in eight species of the scopionfish (Scorpaenidae). Zool. Sci., 9: 1210.

Y- 23 Yüksel, E. and Gaffaroğlu, M. (2008) The analysis of nucleolar organizer regions in *Chalcalburnus mossulensis* (Pisces: Cyprinidae). J. FisheriesSciences.com, 2(3): 587–591.

Y- 24 Yüksel, E. and Gaffaroğlu, M. (2008) NOR phenotype of *Cyprinion macrostomus* (Osteichthyes, Cyprinidae). J. FisheriesSciences.com, 2(2): 114–117.

Y- 25 Yang, P.H., Wang, X.Y., Wu, W.X., Xie, C.H., Li, M.J. and Zhang, Q. (2007) Chromosome karyotype analysis of *Carassius auratus* var. Dongtingking and cultivar identification. Freshwater Fish., 37(3): 3–7. (In Chinese with English abstract)

Y- 26 Yin, S.W., Liao, J.Q., Deng, Q., Chen, G.H. and Liu, W.C. (2008) Karyotype of the hybrid F_1 of snapper (*Lutjanus erythropterus* male × *L. sebae* female) and their parents. Fish. Sci. (China), 27: 171–174. (In Chinese with English abstract)

[Z]

Z- 1 Zan, R.G. and Song, Z. (1979) Analysis and comparison between the karyotypes of *Ctenopharyngodon idellus* and *Megalobrama amblycephala*. Acta Genet. Sinica, 6(2): 205–210. (In Chinese with English abstract)

Z- 2 Zan, R.G. and Song, Z. (1980) Analysis and comparison between the karyotypes of *Cyprinus carpio* and *Carassius auratus* as well as *Aristichthys nobilis* and *Hypophthalmichthys molitrix*. Acta Genet. Sinica, 7(1): 72–77. (In Chinese with English abstract)

Z- 3 Zan, R.G. and Song, Z. (1980) Studies of the karyotypes of eight species of fishes in *Cyprinus* and *Anabarilius*. Zool. Res., 1: 141–150. (In Chinese with English abstract)

Z- 4 Zan, R.G. (1982) Studies of sex chromosomes and C-banding karyotypes of two forms of *Carassius auratus* in Kunming Lake. Acta Genet. Sinica, 9(1): 32–39. (In Chinese with English abstract)

Z- 5 Zan, R.G., Song, Z. and Liu, W.G. (1984) Studies of karyotypes of seven species of fishes in Barbinae, with a discussion on identification of fish polyploids. Zool. Res., 5(1) suppl.: 82–90. (In Chinese with English abstract)

Z- 6 Zan, R.G., Liu, W.G. and Song, Z. (1985) Tetraploid-hexaploid relationship in Schizothoracinae. Acta Genet. Sinica, 12(2): 137–142. (In Chinese with English abstract)

Z- 7 Zan, R.G. (1985) Fish polyploidy and its role in the evolution of fishes. J. Yunnan Univ., 7(2): 235–243 and 7(3): 331–338. (In Chinese with English abstract)

Z- 8 Zan, R.G., Song, Z. and Liu, W.G. (1986) Studies on karyotypes and nuclear DNA contents of some cyprinoid fishes, with notes on fish polyploids in China. In: Uyeno, T., Arai, R., Taniuchi, T. and Matsuura, K. eds. Indo-Pacific fish biology. Ichthyological Society of Japan, Tokyo, pp. 877–885.

Z- 9 Zelinsky, Yu.P., Svimonishvii, T.N. and Gorodilov, J.N. (1980) Karyotypes of Atlantic salmon (*Salmo salar* L.) populations from Keret' and Pechora Rivers. In: Kirpichnikov, V.S. ed. Karyological variability, mutagenesis and gynogenesis in fishes. Inst. Cytol. Acad. Sci. USSR, Leningrad., pp. 10–15. (In Russian with English summary)

Z- 10 Zelinsky, Yu.P. (1980) A chromosomal analysis of the Atlantic salmon population in the Ladoga Lake. Zool. Zhurnal, 59: 1260–1263. (In Russian with English summary)

Z- 11 Zelinsky, Yu.P. and Medvedeva, I.M. (1985) Analysis of chromosomal variability and polymorphism in Atlantic salmon, *Salmo salar*, of Lake Onega. J. Ichthyol., 25(2): 70–77.

Z- 12 Zelinsky, Yu.P., Dmitrenko, Y.Y. and Pervozvanskiy, V.Y. (1992) Evaluation of genome interaction from growth characteristics and karyological and morphological characters in triploid hybrids between wild goldfish and carp reared in Karelia. Vopr. Ichthyol., 32: 103–110. (In Russian)

Z- 13 Zelinsky, Yu.P. and Makhrov, A.A. (2001) Chromosomal variability, genome reorganization in phylogeny, and the systematics of *Salmo* and *Parasalmo* species (Salmonidae). J. Ichthyol., 41: 209–216.

Z- 14 Zenzes, M.T. and Voiculescu, I. (1975) C-banding patterns in *Salmo trutta*, a species of tetraploid origin. Genetica, 45: 531–536.

Z- 15 Zhang, H.Q. and Li, Q.W. (1991) First report of karyotypes of 5 marine fish species. J. Liaoning Normal Univ. (Nat. Sci.), 14(3): 263–264. (In Chinese)

Z- 16 Zhang, J.X., Liu, X.F., Wang, Z.X. and Jin, G.Q. (1984) A comparative study on the karyotypes among the hybrid fish (*Sinilabeo decorus tungting* male × *Cirrhinus molitorella* female) and its parental fishes. Acta Hydrobiol. Sinica, 8: 313–320. (In Chinese with English abstract)

Z- 17 Zhang, J.X. and Wang, Z.X. (1987) Studies on the karyotype of *Leuciscus waleckii* (Dybowski). Acta Hydrobiol. Sinica, 11: 367–370. (In Chinese)

Z- 18 Zhang, K.J., Gao, J., Zhang, J.L., He, Y.M. and Wang, W.S. (1995) Study on karyotypes of hybrid crucian carp (white crucian carp × scattered mirror carp) and its parents. J. Fish. China, 19: 305–309. (In Chinese with English abstract)

Z- 19 Zhang, Q.Q. and Arai, K. (1999) Distribution and reproductive capacity of natural triploid individuals and occurrence of unreduced eggs as a cause of polyploidization in the loach, *Misgurnus anguillicaudatus*. Ichthyol. Res., 46: 153–161.

Z- 20 Zhang, S.M. (1990) Karyotypic studies in two species of *Chanda* (Centropomidae: Pisces). Chrom. Inform. Serv., 49: 7–8.

Z- 21 Zhang, S.M., Zheng, Y., Yao, H. and Zhang, X.Z. (1992) Nucleolar organizer regions in four species of fishes. Chrom. Inform. Serv., 53: 6–8.

Z- 22 Zhang, S.M. and Zeng, Y. (1993) Ag-NORs of three cyprinid fishes. Chrom. Inform. Serv., 54: 19–20.

Z- 23 Zhao, J.L. (2000) A survey of karyotype study of marine and brackish water fish in China. J. Shanghai Fish. Univ., 9(4): 344–347. (In Chinese)

Z- 24 Zhou, J.S., Shen, J.B. and Liu, M.H. (1983) A cytological study on the gynogenesis of Fangzheng crucian carp of Heilongjiang Province. Acta Zool. Sinica, 29: 11–16. (In Chinese with English abstract)

Z- 25 Zhou, M., Kang, Y., Li, Y.C. and Zhou, T. (1988) Studies on the silver-stained karyotypes of 7 species in Cyprinidae (Pisces). Zool. Res., 9: 225–229. (In Chinese with English abstract)

Z- 26 Zhou, M., Li, Y.C., Zhou, T. and Yu, X.J. (1989) A BrdU-BSG method for G-banding in fish chromosomes and an idiogram of G-banded karyotype of silver carp. Acta Genet. Sinica, 16(3): 184–187. (In Chinese with English abstract)

Z- 27 Zhou, T., Ling, J.X. and Yang, Y.Q. (1980) The karyotype of black carp (*Mylopharyngodon piceus* (Rich)). J. Wuhan Univ. (Nat. Sci.), 1980(4): 112–116. (In Chinese with English abstract)

Z- 28 Zhou, T. (1984) Chromosome studies in Chinese fresh-water fishes. Zool. Res., 5 (suppl.): 38–51.

Z- 29 Zhu, X.P., Wu, G.M., Hu, G. and Luo, J.R. (1990) The karyotype of *Leptobarbus hoeveni* (Cyprinidae). Hereditas (Beijing), 12(3): 20–21. (In Chinese)

Z- 30 Zhu, X.P., Liu, Y.H., Chen, Y.L., Wei, Q.F. and Zhang, S.A. (2003) Morphological characters and karyotype of *Oxyeleotris marmoratus* Bleeker. J. Fish. Sci. China, 10(1): 85–86. (In Chinese with English abstract)

Z- 31 Zhuang, J.S. and Liu, L.Y. (1982) Karyotype analysis of the snake-head *Ophiocephalus argus*. J. Beijing Normal Univ. (Nat. Sci.), 1982(3): 81–83. (In Chinese with English abstract)

Z- 32 Zou, S.M., Li, S.F., Zhao J.L., Cai W.Q. and Liu J.F. (2003) Karyotype of *Pseudosciaena crocea* in Guanjingyang of Fujian. J. Shanghai Fish. Univ., 12: 179–181. (In Chinese)

Z- 33 Zelinsky, Yu.P. (1976) Chromosome complexes in the Atlantic salmon *Salmo salar* L. from the non-anadromous population of Lake Kujto (the White Sea basin). Vopr. Ichthyol., 16: 1119–1124. (In Russian)

Z- 34 Zhu, H.P., Ma, D.M. and Gui, J.F. (2006) Triploid origin of the gibel carp as revealed by 5S rDNA localization and chromosome painting. Chrom. Res., 14: 767–776.

Z- 35 Zhou, L.Q., Yang, A.G., Liu, X.Z., Du, W. and Zhuang, Z.M. (2005) The karyotype of the tonguefish *Cynoglossus semilaevis*. J. Fish. China, 29: 417–419. (In Chinese)

Z- 36 Zhou, L. and Gui, J.F. (2002) Karyotypic diversity in polyploid gibel carp, *Carassius auratus gibelio* Bloch. Genetica, 115: 223–232.

Z- 37 Zhao, X.F., Wang, J.X., Yang, Q. and Zhang, J.L. (1994) Karyotype analysis of five species of marine fishes. Zool. Res., 15 (suppl.): 103–106. (In Chinese with English abstract)

Z- 38 Zhang, D.H. (2007) The karyotypes, C-bands and Ag-NORs of *Ctenogobius giurinus*. Freshwater Fish., 37(2): 16–19. (In Chinese with English abstract)

Z- 39 Zhou, B.C., Shu, H., Liu, F., Cai, X.Y., Tan, J.H. and Zhang, H.F. (2009) Karyotypes in three marine important fish species. Fish. Sci., 28: 325–328. (In Chinese with English abstract)

*Z- 40 Zanandrea, G. and Capanna, E. (1964) Contributo alla cariologia del genere *Lampetra*. Boll. Zool., 31: 669–677. (After K-114)

Z- 41 Zoch, P.K., Hanks, B.G. and Gold, J.R. (1989) The standard and NOR-stained karyotype of *Rivulus agilae* (Rivulidae: Teleostei). Texas J. Sci., 41: 104–106.

Journal List

Acad. Nat. Lincei, ser. 8 = Atti della Accademia Nazionale dei Lincei, Serie Ottava (Roma)
Acta Amazonica
Acta Biol. Colombiana = Acta Biologica Colombiana (Bogotá)
Acta Biol. Debrecina = Acta Biologica Debrecina (Debrecen, Hungary)
Acta Biol. Exp. Sinica = Acta Biologiae Experimentalis Sinica
Acta Biol. Iugoslavica = Acta Biologica Iugoslavica
Acta Cient. Venezolana = Acta Cientifica Venezolana
Acta Genet. Sinica = Acta Genetica Sinica
Acta Hydrobiol. = Acta Hydrobiologica
Acta Hydrobiol. Sinica = Acta Hydrobiologica Sinica
Acta Sci. Nat. Brno = Acta Scientiarm Naturalium Academiae Scientiarum Bohemoslovacae Brno, Nova series
Acta Zool. = Acta Zoologica (Stockholm)
Acta Zool. Fennica = Acta Zoologica Fennica (Helsinki)
Acta Zool. Sinica = Acta Zoologica Sinica
Acta Zootax. Sinica = Acta Zootaxonomica Sinica
Advanced Aquarists Mag. = Advanced Aquarists Magazine
Afr. Zool. = African Zoology
AKA-KN = American Killifish Association-Killie Notes
Amer. Natur. = The American Naturalist
Amer. Zool. = American Zoologist
An. Acad. Brasil. Cienc. = Anais da Academia Brasileira de Ciencia (Rio de Janeiro)
An. Inst. Cienc. Mar Limnol. Univ. Nal. Autón. México = Anales del Instituto do Ciencias del Mar y Limnologia, Universidad Nacional Autónoma de México
Anim. Sci. = Animal Science (Sofia)
Animals and Nature = The Nature and Animals (Tokyo)
Ann. Acad. Reg. Sci. Upsalien. = Annales Academiæ Regiæ Scientiarum Upsaliensis
Ann. Mus. Royal Afr. Centrale = Annales. Musee Royal de l'Afrique Centrale
Ann. Nat. Acad. Sci. India = Annals of the National Academy of Sciences India
Ann. New York Acad. Sci. = Annals of the New York Academy of Sciences
Ann. Rep. Biol. Res., Jeonbug Natn. Univ. = Annual Report of Biological Research, Jeonbug National University (Korea)
Ann. Rep. Biwako Bunkakan = Annual Report of the Biwako Bunkakan (Japan)
Antarctic Sci. = Antarctic Science
Aquaculture
Aquacult. Res. = Aquaculture Research
Aquaria
Aquarien Terrarien = Aquarien und Terrarien

Aquarium
Aquarium-J. = The Aquarium Journal
Arch. FischWiss. = Archiv für Fischereiwissenschaft
Arch. Zootec. = Archivos de Zootecnia
Arquivos Mus. Bocage = Arquivos do Museu Bocage
Asian Fisheries Science (Manila)
Atti Soc. Ital. Sci. nat. Museo civ. Stor. nat. Milano = Atti della Società Italiano di Scienze Naturali e del Museo Civico di Storia Naturale di Milano
Aust. J. Mar. Freshw. Res. = Australian Journal of Marine and Freshwater Research

Basic Sci. Rev., Jeonbug Natn. Univ. = Basic Science Review, Jeonbug National University (Korea)
Bilješke notes, Inst. Oceanograf. Ribarstyo, Yugoslavija = Bilješke Notes. Instituto za Oceanografiju i Ribarstyo, Jugoslavija
Biochem. Syst. Ecol. = Biochemical Systematics and Ecology
Biol. Bratislava = Biologia, Bratislava
Biol. Bull. = Biological Bulletin
Biol. J. Linn. Soc. = Biological Journal of the Linnean Society
Biol. Morya = Biologiya Morya
Biol. Rev. = Biological Review of the Cambridge Philosophical Society
Biol. Zentralbl. = Biologisches Zentralblatt
BKA-Killi News = British Killifish Association-Killi News
BMC Evol. Biol. = BMC Evolutionary Biology
BMC Genetics
Bol. Inst. Espanol Oceanogr. = Boletin del Instituto Español de Oceanografia
Bol. Inst. Oceanogr. Univ. Oriente = Boletin del Instituto Oceanografico, Universidad de Oriente, Venezuela
Bol. Inst. Pesca = Boletim do Instituto de Pesca
Boll. Zool. = Bollettino di Zoologia
Bolm Inst. Oceanogr. S Paulo = Boletim do Instituto Oceanografica. São Paulo
Brazil. Archiv. Biol. Technol. = Brazilian Archives of Biology and Technology
Brazil. J. Biol. = Brazilian Journal of Biology
Brazil. J. Genet. = Brazilian Journal of Genetics
Bull. Aichi Univ. Educ. (Nat. Sci.) = Bulletin of Aichi University of Education (Natural History) (Japan)
Bull. Biogeogr. Soc. Japan = Bulletin of the Biogeographical Society of Japan
Bull. Dept. Educ. Utsunomiya Univ. = Bulletin of the Department of Education, Utsunomiya University (Japan)
Bull. Fac. Educ. Yamaguchi Univ. = Bulletin of the Faculty of Education, Yamaguchi University (Japan)
Bull. Fac. Fish. Hokkaido Univ. = Bulletin of the Faculty of Fisheries, Hokkaido University (Japan)
Bull. Fr. Pêche Piscic. = Bulletin Français de la Pêche et de la Pisciculture
Bull. Hiroshima Women's Univ. = Bulletin of the Hiroshima Women's University (Japan)
Bull. Inst. Basic. Sci., Inha Univ. = Bulletin of the Institute for Basic Science, Inha University (Korea)
Bull. Inst. Zool. Academia Sinica, Monograph = Bulletin of the Institute of Zoology, Academia Sinica, Monograph
Bull. Japan. Soc. Sci. Fish. = Bulletin of the Japanese Society of Scientific Fisheries
Bull. Korean Fish. Soc. = Bulletin of the Korean Fisheries Society
Bull. Mar. Sci. = Bulletin of Marine Science
Bull. Natl. Res. Inst. Aquaculture = Bulletin of National Research Institute of Aquaculture (Japan)
Bull. Natn. Sci. Mus. Tokyo = Bulletin of the National Science Museum, Tokyo

Bull. Natn. Sci. Mus. Tokyo, (A) = Bulletin of the National Science Museum, Tokyo. Series A (Zoology)
Bull. Sci. Yougoslavie, Sec. A = Bulletin Scientifique, Yougoslavie. Section A
Bull. Soc. Hist. Nat. Toulouse = Bulletin de la Société d'Histoire Naturelle de Toulouse

Calif. Fish Game = California Fish and Game
Can. J. Fish. Aquat. Sci. = Canadian Journal of Fisheries and Aquatic Sciences
Can. J. Genet. Cytol. = Canadian Journal of Genetics and Cytology
Can. J. Zool. = Canadian Journal of Zoology
Caribbean J. Sci. = Caribbean Journal of Science
Caryologia
Chinese J. Zool. = Chinese Journal of Zoology
Chrom. Inform. Serv. = Chromosome Information Service
Chrom. Res. = Chromosome Research
Chrom. Sci. = Chromosome Science
Chromatin
Chromosoma
Cien. Cultura = Ciência e Cultura
College Rev. College Liberal Arts Sci., Seoul Natn. Univ. = College Review of College of Liberal Arts and Sciences, Seoul National University (Korea)
Comp. Biochem. Physiol. = Comparative Biochemistry and Physiology
Contri. Sci. Nat. Hist. Mus. Los Angeles County = Contribution in Science. Los Angeles County Museum of Natural History
Copeia
Curr. Biol. = Current Biology
Curr. Opinion Genet. Develop. = Current Opinion in Genetics and Development
Curr. Sci. = Current Sciences (India)
Cybium
Cybium, 3rd ser.
Cytobios
Cytogenet. Cell Genet. = Cytogenetics and Cell Genetics
Cytogenet. Genome Res. = Cytogenetic and Genome Research
Cytogenetics
Cytologia
Cytometry

Dokl. Akad. Sci. USSR = Doklady Akademii Nauk SSSR
Doklady Acad. Sci. Ukrain. SSR, Ser. B = Doklady Academii Nauk Ukrainskoy SSR, Ser. B
Dokl. Biol. Sci. = Doklady Biological Sciences
Doñana, Acta Vertebrata = Doñana - Acta Vertebrata

Environ. Biol. Fishes = Environmental Biology of Fishes
Environ. Ecol. = Environment and Ecology
Evolution
Exp. Cell Res. = Experimental Cell Research
Experientia

Fish Genet. Breed. Sci. = Fish Genetics and Breeding Science
Fish. Bull. = Fishery Bulletin
Fish. Sci. = Fisheries Science (Tokyo)
Fish. Sci. (China) = Fisheries Science (China)

Folia Biol. (Krakow) = Folia Biologica (Kraków)
Folia Zool. = Folia Zoologica (Brno)
Freshwater Fish. = Freshwater Fisheries (China)
Fujian Fish. = Fujian Fisheries (China)

Gayana
Gene
Genen Phaenen
Genet. Mol. Biol. = Genetics and Molecular Biology (Brazil)
Genet. Mol. Res. = Genetics and Molecular Research (Brazil)
Genet. Res. Camb. = Genetical Research, Cambridge
Genet. Sel. Evol. = Genetics Selection Evolution
Genetica
Genetics
Genetika = Russian Journal of Genetics
Genetika, Acta Biol. Iugoslavica = Acta Biologica Iugoslavica, serija F, Genetika (Beograd)
Genome
Genome Res. = Genome Research
Geobios
God. Biol. Inst. Univ. Saraevu = Godišnjaka Biološkog Instituta Univerziteta u Sarajevu

Hereditas
Hereditas (Beijing)
Heredity
Hydrobiol. J. = Hydrobiological Journal (Gidrobiologichesky Zhurnal)
Hydrobiologia

Ichthyol. Explor. Freshwaters = Ichthyological Exploration of Freshwaters
Ichthyol. Res. = Ichthyological Research (Japan)
Ichthyologia = Acta Biologica Iugoslavica, Serija E, Ichthyologia
Iden = Iden (Tokyo)
In Vitro
Ind. Biologist = Indian Biologist
Ind. J. Anim. Sci. = Indian Journal of Animal Sciences
Ind. J. Exp. Biol. = Indian Journal of Experimental Biology
Ind. J. Zool. = Indian Journal of Zoology
Ind. Vet. J. = Indian Veterinary Journal (Madras)
Interciencia (Caracas)
Intl. J. Acad. Ichthyol. = International Journal of Academy of Ichthyology (U.P., India)
Issled. Fauny Morei = Issledovaniia Fauny Morei
Ital. J. Zool. = The Italian Journal of Zoology

J. Annamalai Univ. Sci. = Journal of the Annamalai University. Part B, Science (India)
J. Aquaculture = Journal of Aquaculture (Korea)
J. Beijing Normal Univ. (Nat. Sci.) = Journal of Beijing Normal University (Natural Science)
J. Cytol. Genet. = The Journal of Cytology and Genetics (India)
J. Dalian Fish. Univ. = Journal of Dalian Fisheries University (China)
J. Dalian Fish. College = Journal of Dalian Fisheries College (China)
J. Exp. Mar. Biol. Ecol = Journal of Experimental Marine Biology and Ecology.
J. Exp. Zool. = The Journal of Experimental Zoology

J. Fac. Sci. Hokkaido Univ., Ser. Zool. = Journal of the Faculty of Science, Hokkaido University, Ser. 6, Zoology (Japan)
J. Fish Biol. = Journal of Fish Biology
J. Fish. China = Journal of Fisheries of China
J. Fish. Res. Board Can. = Journal of the Fisheries Research Board of Canada
J. Fish. Sci. China = Journal of Fishery Sciences of China
J. Fish. Sci. Technol. = Journal of Fisheries Science and Technology (Korea)
J. FischeriesSciences.com = Journal of FischeriesSciences.com
J. General Biol. = Journal of General Biology
J. General Physiol. = The Journal of General Physiology
J. Genet. = Journal of Genetics
J. Heredity = The Journal of Heredity
J. Ichthyol. = Journal of Ichthyology
J. Inland Fish. Soc. India = Journal of the Inland Fisheries Society of India
J. Liaoning Normal Univ. (Nat. Sci.) = Journal of Liaoning Normal University (Natural Science) (China)
J. Mar. Biol. Ass. India = Journal of the Marine Biological Association of India
J. Mar. Biol. Ass. U.K. = Journal of Marine Biological Association of the United Kingdom
J. Morphol. = Journal of Morphology
J. Ocean Univ. Qingdao = Journal of Ocean University of Qingdao (China)
J. Shanghai Fish. Univ. = Journal of Shanghai Fisheries University (China)
J. Shimonoseki Univ. Fish. = Journal of Shimonoseki University of Fisheries (Japan)
J. Structural Functional Genomics = Journal of Structural and Functional Genomics
J. Wuhan Univ. (Nat. Sci.) = Journal of Wuhan University (Natural Science Edition) (China)
J. Xiamen Univ. (Nat. Sci.) = Journal of Xiamen University (Natural Science) (China)
J. Yunnan Univ. = Journal of Yunnan University (China)
J. Zhanjiang Fish. Coll. = Journal of Zhanjiang Fisheries College (China)
J. Zhejiang College Fish. = Journal of Zhejiang College of Fisheries (China)
J. Zool. Lond. = Journal of Zoology, London
J. Zool. Syst. Evol. Res. = Journal of Zoological Systematics and Evolutionary Research
Japan Women's Univ. J. = Journal of Japan Women's University
Japan. J. Genetics = Japanese Journal of Genetics
Japan. J. Ichthyol. = Japanese Journal of Ichthyology

Korean J. Genet. = Korean Journal of Genetics
Korean J. Ichthyol. = The Korean Journal of Ichthyology
Korean J. Limnol. = Korean Journal of Limnology
Korean J. Zool. = Korean Journal of Zoology

La Kromosomo
La Kromosomo, II = La Kromosomo, Series II
Lat. Am. J. Aquat. Res. = Latin American Journal of Aquatic Research
Life Sci. Adv. = Life Science Advances (India)

Mar. Biol. = Marine Biology
Mar. Freshwater Res. = Marine and Freshwater Research
Mar. Sci. Monthly = Marine Sciences, Monthly (Tokyo)
Mar. Sci. Bull. = Marine Science Bulletin (China)
Marine Sciences (Beijing)
Matsya (India)
Medaka = The Fish Biology Journal Medaka (Japan)

Mem. Hyogo Univ. Agricul. = Memoirs of the Hyogo University of Agriculture (Japan)
Mitt. Hamburg. Zool. Mus. Inst. = Mitteilungen aus dem Hamburgischen Zoologischen Museum und Institut
Mol. Phylogenet. Evol. = Molecular Phylogenetics and Evolution

Nat. Acad. Sci. India, Ann. Num. = National Academy of Sciences India, Ann. Num.
Nat. Acad. Sci. Letters = National Academy Science Letters (Allahabad, India)
Nat. Hist. Mus. Stadt Bern, Jahrbuch = Naturhistorisches Museum der Stadt Bern, Jahrbuch
Naturalia = Naturalia, São Paulo
Naturaliste can. = Le Naturaliste Canadien
Nature
Neotrop. Ichthyol. = Neotropical Ichthyology
Notulae Naturae = Notulae Naturae of the Academy of Natural Sciences of Philadelphia
Nucleus = The Nucleus

Occ. Pap. Calif. Acad. Sci. = Occasional Papers of the California Academy of Sciences
Occ. Pap. Mus. Zool. Univ. Michigan = Occasional Papers of the Museum of Zoology, University of Michigan
Oceanol. Limnol. Sinica = Oceanologia et Limnologia Sinica
Oikos

Pakistan J. Zool. = Pakistan Journal of Zoology
Perspect. Cytol. Genet. = Perspectives in Cytology and Genetics (New Delhi)
Phil. Trans. R. Soc. London, (B) = Philosophical Transactions of the Royal Society of London, Ser. B
Physiol. Ecol. Japan = Physiology and Ecology Japan
Polar Biol. = Polar Biology
Postilla
Proc. Ind. Acad. Sci., Sec. B = Proceedings of the Indian Academy of Sciences, Section B
Proc. Japan Acad. = Proceedings of the Japan Academy
Proc. Japan Acad., Ser. B = Proceedings of the Japan Academy. Ser. B, Physical and Biological Sciences
Proc. Japan. Soc. Syst. Zool. = Proceedings of the Japanese Society of Systematic Zoology
Proc. Louisiana Acad. Sci. = Proceedings of the Louisiana Academy of Sciences
Proc. Nat. Acad. Sci. India = Proceedings of the National Academy of Sciences, India
Proc. R. Soc. Lond., B = Proceedings of the Royal Society of London. Series B, Biological Sciences
Proc. Soc. Exper. Biol. Med. = Proceedings of the Society for Experimental Biology and Medicine
Proc. Zool. Inst. Leningrad = Proceedings of the Zoological Institute, Leningrad
Proc. Zool. Soc. Calcutta = Proceedings of the Zoological Society, Calcutta
Progress Modern Biol. = Progress of Modern Biology (Uspekhi Sovremennoi Biologii)
Progressive Fish-Culturist = The Progressive Fish-Culturist (Washington, D.C.)

Radovi Anubih
Rapp. Comm. Int. Mer. Médit. = Rapport du Commission International de la Mer Méditerranee
Rep. Mishima Res. Inst. Sci. Liv., Nihon Univ. = Report of the Mishima Research Institute of Sciences for Living, Nihon University (Japan)
Res. Bull. (N.S.) Panjab Univ. = Research Bulletin (N.S.) of the Panjab University (India)
Res. Rev. BioSciences (India) = Research and Reviews in BioSciences (India)
Rev. Biol. Trop. = Revista de Biologia Tropical (San José)

Rev. Biol. Uruguay = Revista de Biologia del Uruguay
Rev. Fish Biol. Fish. = Reviews in Fish Biology and Fisheries
Rev. fr. Aquariol. = Revue Française D'aquariologie
Rev. Hydrobiol. Trop. = Revista Hydrobiologia Tropical
Russian J. Genet. = Russian Journal of Genetics
Russian J. Mar. Biol. = Russian Journal of Marine Biology

Saber, Univ. Oriente, Venezuela = Saber, Universidad de Oriente, Venezuela
Sci. Culture = Science and Culture (Calcutta)
Sci. Rep. Res. Inst. Evol. Biol. = Science Report of the Research Institute of Evolutionary Biology (Tokyo)
Sci. Mar. = Scientia Marina
Science
Sinozool. = Sinozoologia
Southwest. Nat. = The Southwestern Naturalist
Stain Technol. = Stain Technology
Swedish. J. Agric. Res. = Swedish Journal of Agricultural Research

Tansuigyo (Osaka)
Texas J. Sci. = The Texas Journal of Science
Texas Rep. Biol. Med. = Texas Reports on Biology and Medicine
TFH = Tropical Fish Hobbyist
Theor. Appl. Genet. = Theoretical and Applied Genetics
Trans. Amer. Fish. Soc. = Transactions of the American Fisheries Society
Trans. Chinese Ichthyol. Soc. = Transactions of the Chinese Ichthyological Society
Travaux Mus. Natl. Hist. Nat. "Grigore Antipa" = Travaux du Museum National d'Histoire Naturelle "Grigore Antipa"
Trends Genet. = Trends in Genetics
Tropic Oceanology (Beijing)
Trudy Zool. Inst. = Trudy Zoologicheskogo Instituta
Tsitol. Genet. = Tsitologiya i Genetika
Tsitologiya
Turk. J. Biol. = Turkish Journal of Biology
Turk. J. Zool. = Turkish Journal of Zoology

Veterinar. Arhiv = Veterinarski Arhiv
Vidensk. Medd. Dansk naturh. Foren. = Videnskabelige Meddelelser fra Dansk Naturhistorisk Forening
Vie Milieu = Vie et Milieu
Vopr. Ichthyol. = Voprosy Ikhtiologii

Z. Binnenfisch. = Zeitschrift für die Binnenfischerei
Z. Fisch. = Zeitschrift für Fischerei
Z. Fischkunde = Zeitschrift für Fischkunde
Z. Zool. Syst. Evol.-Forsch. = Zeitschrift für Zoologische Systematik und Evolutionsforschung
Zool. Abhand. Staat. Mus. Tier. Dresden = Zoologische Abhandlungen Staatliches Museum für Tierkunde in Dresden
Zool. Anz. = Zoologischer Anzeiger
Zool. Mag. Japan = Zoological Magazine, Japan
Zool. Orientalis = Zoologica Orientalis
Zool. Res. = Zoological Research (Kunming, China)
Zool. Sci. = Zoological Science (Japan)
Z l. Zh l = Zoologicheskii Zhurnal

Index